PRINCIPLES OF SOLAR CELLS

Connecting Perspectives on Device, System, Reliability, and Data Science

PRINCIPLES OF SOLAR CELLS

Connecting Perspectives on Device, System, Reliability, and Data Science

Muhammad A. Alam
Purdue University, USA

M. Ryyan Khan
Purdue University, USA

World Scientific

NEW JERSEY • LONDON • SINGAPORE • BEIJING • SHANGHAI • HONG KONG • TAIPEI • CHENNAI • TOKYO

Published by

World Scientific Publishing Co. Pte. Ltd.

5 Toh Tuck Link, Singapore 596224

USA office: 27 Warren Street, Suite 401-402, Hackensack, NJ 07601

UK office: 57 Shelton Street, Covent Garden, London WC2H 9HE

Library of Congress Cataloging-in-Publication Data

Names: Alam, Muhammad Aftab, author. | Khan, M. Ryyan, author.

Title: Principles of solar cells : connecting perspectives on device, system, reliability, and data science / Muhammad A. Alam, Purdue University, USA, M Ryyan Khan, Purdue University, USA.

Description: Hackensack, NJ : World Scientific, [2022] | Includes bibliographical references and index.

Identifiers: LCCN 2022008888 | ISBN 9789811231537 (hardcover) | ISBN 9789811233029 (paperback) | ISBN 9789811231544 (ebook for institutions) | ISBN 9789811231551 (ebook for individuals)

Subjects: LCSH: Solar cells.

Classification: LCC TK2960 .A39 2022 | DDC 621.31/244--dc23/eng/20220502

LC record available at https://lccn.loc.gov/2022008888

British Library Cataloguing-in-Publication Data

A catalogue record for this book is available from the British Library.

For any available supplementary material, please visit
https://www.worldscientific.com/worldscibooks/10.1142/12139#t=suppl

The book is dedicated to our children
Nira, Raha, and Rida and *Safwan*

Contents

Preface

This is a book about the physics of solar cells, an electronic device that converts sunlight directly to electricity. In the 1950s, three scientists at Bell Laboratories (David Chaplin, Calvin Fuller, and Gerald Pearson) invented solar cells as a "long-lived battery" to power telephone systems in remote locations in America. Everyone soon realized that only a fraction (e.g., 0.0001%) of the 147 petawatts of solar power incident on earth would satisfy the energy needs of the global population. At the time, solar cells were too expensive and could only be used in niche applications. Since then, successive national initiatives have supported the development of solar cell technology to power satellites, reduce dependence on foreign oil, combat global warming, and satisfy the basic energy needs of grid-disconnected poor across the world.

As a result, the price of solar cells has reduced to a point that we find solar cells everywhere: installed in large arrays in vast solar farms, floating on lakes and rivers, covering rooftops, and so on. How does a solar cell work? How efficient can it be? Why are there intricate patterns of metal lines decorating the surface of a solar module? How do they arrange the modules in a solar farm to maximize energy harvesting? How do you store sunlight during the day so that you can use it at night? And how long can a solar farm operate profitably? The book will answer all these questions, and many more.

Many books have been written about solar cell technology. This book is different in several ways:

1. **An end-to-end perspective.** The physics of solar cells spans a fantastic range in space and time. There exist 17 orders of magnitude difference in length scale between the sub-nanometer scale photon–atom interaction and the panel design of the Mars Rover or Juno spacecraft to Jupiter, working hundreds of millions of miles away from the sun. Similarly, the sub-nanosecond radiative recombination events differ from decades-long degradation processes (due to corrosion, for example) by 15–16 orders of magnitude in time. Most researchers focus on a small block of the space-time range and communicate by jargon inaccessible to outsiders. The book unifies these diverse specialized topics within an end-to-end modeling framework, because significant advancement is possible only with an appreciation of the broader context.

2. **An new conceptual infrastructure.** For the unified treatment of the field, we needed to develop a simple analytical infrastructure that does away with the traditional approaches to various topics. We describe the thermodynamics of solar cell operation, the diffusion of photogenerated carriers across a junction, the drift of current across an optimized grid, the layout of a solar farm, all without using a single differential equation. Each concept is presented in a simple, crystal-clear form, so that the implications of the ideas are immediately obvious.

3. **Fundamental limits as a beacon and a guardrail.** The second law of thermodynamics has helped discredit numerous proposals for "perpetual motion machines." Today, the Nernst limit for electrochemical systems, Boltzmann limit for transistors, Landauer limit for energy dissipation, diffusion limit for biosensors, and Shockley–Queisser (S-Q) limit for solar cells, etc., play the same role in scrutinizing new ideas. Since the S-Q limit is a composite limit — built on the corresponding limits of short-circuit current, open-circuit voltage, and fill factor — it can be used to identify conceptual errors and/or measurement artifacts. Indeed, the S-Q limits for solar farms, energy storage, and reliability can guard against overenthusiastic predictions or unsubstantiated claims regarding new PV technologies. In this book, you will find repeated discussion regarding these limits so that you can read the fast-moving literature more thoughtfully.

4. **Analogies for inspiration.** The ability to harvest light for energy is a pervasive capability in biology. It is not therefore surprising that photosynthesis in plants and cyanobacteria has many similarities to photon harvesting by solar cells: biomimetic nanophotonic structures improve light collection; the veins in a leaf have similar fractal dimensions as the grids on a solar module; and the undergrowth harvesting the light diffusing through a forest canopy is not so different from solar farms harvesting diffused light in smog-filled cities with a low clearness index. These analogies allow us to borrow conceptual techniques already developed in other fields and inspire innovative design in solar farms.

The book will not teach you how to make a solar cell, but it will teach you how to make a solar cell better, to trace and reclaim the photons that would have been lost otherwise. The book will show how collaboration across multiple disciplines makes photovoltaics real. The book will also tell you regarding the importance of reliability in reducing the overall cost of solar energy. In short, we have a lot to talk about in this book. Please get a pencil and paper, and let us begin.

M. A. Alam and M. R. Khan
West Lafayette, IN
2022

Acknowledgments

This book is a culmination of decade-long research and extensive discussion with many collaborators. First, we wish to acknowledge the students in our group whose work appears in this book: Reza Asadpour, Tahir Patel, Xingshu Sun, Raghu V. Chavali, Biswajit Ray, Sourabh Dongaonkar, and James Moore.

Among our collaborators from academia, we learned much from discussion with Prof. Mark Lundstorm (Purdue), Peter Bermel (Purdue), Rakesh Agrawal (Purdue), Jeff Gray (Purdue), Richard Schwartz (Purdue), B.J. Stanbery (Colorado), Ned Ekins-Daukes (UNSW), Aditya Mohite (Rice), Wanyi Nie (Sandia), Sean E. Shaheen (Colorado), Tobin Marks (Northwestern), Mark Ratner (Northwester), Rebecca Saive (U. of Twente), Mike Scarpulla (Utah) Tonio Bonassisi (MIT), Vikram Dalal (Iowa), Stefaan De Wolf (Kaust), Juzer Vasi (IIT Bombay), Pradeep Nair (IIT Bombay), Souvik Mahapatra (IIT Bombay), Anisul Haque (East West), M. Rezwan Khan (UIU), and Nauman Butt (LUMS). Our colleagues from industry and national laboratories were particularly helpful in answering questions. They include Chris Deline (NREL), Sarah Kurtz (NREL), Larry Kazmerski (NREL), Bill Tumas (NREL), Joshua Stein (Sandia), Cliff Hansen (Sandia), Paul Stradins (NREL), Pietro Altermatt (Trina Solar), Jim Joseph John (DEWA), M. Frie (AMAT), and Mark Pinto (AMAT).

We are also grateful to Vicky Johnson and Lori Carte who proofread the initial drafts of the book. Lakshmi Narayanan and Zvi Ruder at World Scientific were unfailingly supportive as they guided us through this publication process. Finally, we are grateful to our family, in particular to Salmina Sadeq and Saimah Tahsin, for their patience in the long process of writing this book.

CHAPTER 1

Overview: Sun, Earth, and Solar Cell

Chapter Summary

- Sun is a nuclear reactor that showers energy throughout the solar system.
- The incident energy on earth depends on longitude, latitude, and the seasons. Due to atmospheric scattering, sunlight arrives on ground as direct and diffusion light.
- A solar cell can convert only a fraction of the incident energy. Thus, a 1 MW solar farm would require approximately 4 acres of land.
- There are web-based calculators to determine the location-specific energy output and the corresponding cost of solar energy.
- Solar cell technologies depend on the applications: modules deployed in terrestrial solar farm are different from those deployed in space.

Sunlight may be free, but the solar energy is not! Depending on your location on earth, you may expect 1–2 kW-hr/m^2/year from the sun. With a 20–22% efficient commercial solar panels, this translates to $\sim$200–400 kW-hr/m^2/year of energy from a solar farm, i.e., a 1 MW solar farm would require approximately 2–4 acres of land. In addition to the land cost, the panel price depends on material and manufacturing cost as well as its lifetime. To understand the operating principles and economic viability of solar energy, therefore, we need to understand how the solar illumination depends on the location and the season, how efficiently can solar cell/module/farm convert this incident energy, how long does the solar module survive, and how much does the conversion cost. This overview chapter will qualitatively connect the four ideas. Subsequently, the four parts of the book will discuss the ideas quantitatively in a greater detail.

1.1 Introduction: A self-driven nuclear reactor fueling the solar system

Any story of solar cells must begin with the sun and the intensity and the wavelength of photons it showers across the solar system. The sun is a gas sphere of radius $r_s \approx 7 \times 10^8$ m, composed mostly of hydrogen (~75%) and helium (~25%). The enormous gravity at the center of the sun creates a plasma 100 times denser than water (~10^5 kg/m^3). The pressure ignites a thermonuclear reaction that converts hydrogen into helium, with temperature approaching 20 million degrees. Gamma rays generated in the core begin a long random walk involving repeated absorption and isotropic emission by the molecules in the outer layer until they reach the surface of the sun a million years later. By that time, the temperature (T_S) has reduced to 5777 K. Given the temperature, the radiation intensity (I_s) from the sun can be calculated by the Stefan–Boltzmann law:

$$I_s = \sigma T_S^4, \tag{1.1}$$

where $\sigma = 5.67 \times 10^{-8}$ W m^{-2} K^{-4}. Since $T_S = 5777$ K, $I_s = 63.15 \times 10^6$ W/m^2.

1.1.1 Extraterrestrial solar intensity is easily calculated

Once the photons are emitted from the sun, they move in the free space at the velocity of light (c). They will reach Earth in less than 10 minutes, and Neptune in a few hours. At the distance d from the sun, the intensity of sunlight decays as $1/d^2$, so that

$$I_0(d) = \sigma T_S^4 \times (r_s/d)^2. \tag{1.2}$$

Given the distance of a planet from the sun (d), Fig. 1.1 and Table 1.1 summarize the intensity of the sunlight incident on the planet. For example, the formula correctly predicts that the irradiance above Earth's atmosphere is ~1380 W/m^2. By the time the photons reach Jupiter, the intensity reduces to ~50 W/m^2. With

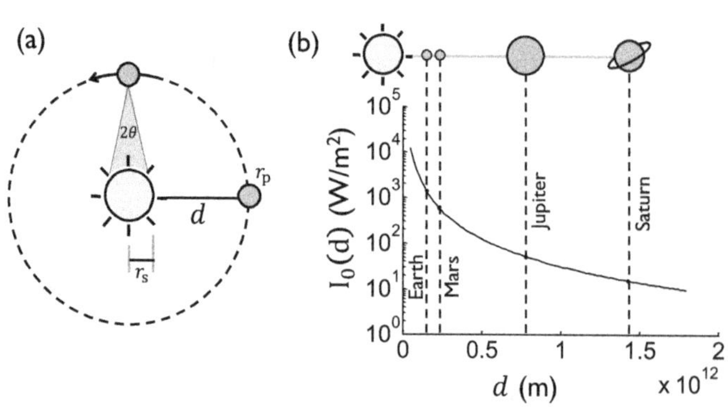

Figure 1.1: (a) Average orbit of a planet around the sun. (b) Solar intensity as a function of distance away from the sun's surface.

Table 1.1: Extraterrestrial intensities on planets calculated by Eq. (1.2).

Planet	$d(\times 10^9)$ m	$I_0(d)$ W/m^2
Venus	108.2	2643.22
Earth	149.6	1382.69
Mars	227.9	595.8
Jupiter	778.3	51.1
Saturn	1427	15.2

this 25-fold reduction in intensity, no wonder the Juno spacecraft needed a complex triple junction solar cell to produce just a few watts of power. In general, the extraterrestrial intensities are useful to calculate the energy output of solar cells mounted on satellites and spacecrafts. To calculate the energy output of ground-based solar cells (e.g., installed on our rooftop, or on a Mars rover), we need to calculate the fraction of light that transmits through the planetary atmosphere before reaching our panels, a topic we will discuss in the next section.

Homework 1.1: The solid angle of the sun seen from a planet

The sun appears as a small disk in the sky. Show that the solid angle of a sphere of radius r_s at a distance d is given by $\theta_S \sim \pi(r_s/d)^2$ steradians.

Solution. By definition, the solid angle of a sphere of radius r_s, seen at distance d, is given by $\theta_S = 2\pi[1 - \cos\theta]$, where $\theta = r_s/d$ radians (see Fig. 1.1). Since $d \gg r_s$, therefore

$$\theta_S = 2\pi[1 - \cos\theta] \approx \pi\theta^2 = \pi\left(\frac{r_s}{d}\right)^2.$$

From the earth, the solid angle is $\theta_S = \pi\left(7 \times 10^8/149.6 \times 10^9\right)^2 \approx 6.5 \times 10^{-5}$ steradians.

1.1.2 A planet's temperature is defined by the light transmitted through its atmosphere

Interestingly, given the average temperature of a planet t_p, we can determine the fraction of direct extraterrestrial light (τ) that travels through the atmosphere and reaches the planet surface. To understand this remarkable relationship, let us first calculate the energy emitted by the sun, namely,

$$P_{s,\,emit} = A_s \times \sigma T_S^4 = 4\pi r_s^2 \times \sigma T_S^4.$$

Homework 1.2: Measuring the sun's surface temperature remotely

You do not need to travel to the sun with a thermometer to measure its surface temperature, (T_S). Show that you can determine T_S by measuring the intensity of the sunlight just outside Earth's atmosphere (i.e., $I_0 = 1382.69\ \mathrm{W/m^2}$ from Table 1.1), and the angular diameter of the sun measured from the earth (i.e., $2\theta = 9.3 \times 10^{-3}$ radians), and the earth-to-sun distance ($d = 149.6 \times 10^9$ m).

Solution. First, let us calculate the radius of the sun, r_s. For a disk of radius r_s placed at a distance d, $\theta = r_s/d$ by definition. Given θ and d, we find that $r_s = 6.97 \times 10^8$ m. The flux emitted by the sun, $I_s \times 4\pi r_s^2$, must equal the flux that crosses the imaginary sphere containing the earth on its surface, i.e., $I_0 \times 4\pi d^2$. Moreover, $I_s = \sigma T_S^4$ by Eq. (1.1). Putting these relationships together, we find that $T_S = 5777$ K.

Here, $A_s \equiv 4\pi r_s^2$ is the surface area of the sun. The fraction of solar radiation reaching the planet is

$$P_{\mathrm{p,inc}} = A_{\mathrm{p}} \left(\frac{P_{\mathrm{s,emit}}}{4\pi d^2} \right).$$

Here, $A_p \equiv \pi r_p^2$ is the cross-sectional area of the planet (with radius r_p) that intercepts the light from the sun. Unlike a black hole, a planet reflects (e.g., by ice or cloud) a fraction of the energy incident on it, defined by its albedo R_A. Therefore, the total energy absorbed by the planet is

$$P_{\mathrm{p,abs}} = (1 - R_{\mathrm{A}})P_{\mathrm{p,inc}} = (1 - R_{\mathrm{A}}) \times A_p \left(\frac{P_{\mathrm{s,emit}}}{4\pi d^2} \right).$$

To calculate the energy flux emitted by the planet, we will consider it as an imperfect blackbody, characterized by emissivity $\varepsilon < 1$. For a planet with constant temperature (T_p), the power emitted is given by

$$P_{\mathrm{p,emi}} = \varepsilon \times \sigma T_{\mathrm{p}}^4 \times 4\pi r_{\mathrm{p}}^2.$$

A planet maintains its constant temperature by balancing the absorption and emission fluxes. Equating $P_{\mathrm{p,abs}} = P_{\mathrm{p,emi}}$, we get

$$T_{\mathrm{p}} = T_{\mathrm{S}} \sqrt{\frac{r_{\mathrm{s}}}{2\,d} \sqrt{\frac{1 - R_{\mathrm{A}}}{\varepsilon}}}. \tag{1.3}$$

Interestingly, the absorption and emission fluxes are both proportional to r_p^2. Therefore, the radius of the planet does not play any role in determining its temperature.

Assuming that the light absorption by the atmosphere is relatively small, we can rewrite Eq. (1.3) to calculate the fractional light transmission through the atmosphere:

$$\tau \simeq 1 - R_{\mathrm{A}} = \varepsilon \left(\frac{2d}{r_{\mathrm{s}}}\right)^2 \left(\frac{T_{\mathrm{P}}}{T_{\mathrm{S}}}\right)^4. \tag{1.4}$$

The average intensity of direct sunlight normally incident on the ground (I_{DNI}) is given by

$$I_{\mathrm{DNI}} = \tau\, I_0(d) = 4\,\sigma\, \varepsilon T_{\mathrm{p}}^4. \tag{1.5}$$

Curiously, the formula appears to be independent of the properties of the sun! Actually, the properties of the sun are implicitly hidden in the expression for T_{p}, given by Eq. (1.3).

1.2 Intensity of sunlight depends on the geographical location, the season, and the hour of the day

The ground sunlight intensity, given by Eq. (1.5), is based on the integrated (spatially averaged) absorption and emission fluxes over the entire planet. Clearly, the local irradiance at the equator differs significantly from that at the North Pole. In general, the output of a solar cell will depend on its geographical location, the month of the year, the hour of the day, and so on. In this section, we will derive a set of formulas to calculate local irradiance.

Homework 1.3: A planet's temperature as a measure of its albedo

Use Eq. (1.4) to calculate the average amount of light transmitted through the atmosphere around the earth.

Solution. The average temperature of the earth is $T_{\mathrm{p}} = 288\,\mathrm{K}$ and its emissivity is $\varepsilon \approx 0.612$. We also know that $T_{\mathrm{S}} = 5777\,\mathrm{K}$, $r_{\mathrm{s}} = 6.96 \times 10^8$ m, and $d = 149.6 \times 10^9$ m. Therefore, Eq. (1.4) predicts

$$\tau = 1 - R_{\mathrm{A}} = 0.70.$$

The average intensity of direct, normally incident light reaching the earth's surface is given by Eq. (1.5), namely, $I_{\mathrm{DNI}} = \tau\, I_0(d) = 0.7 \times 1382.62 \approx 967\ \mathrm{W/m^2}$. Detailed calculations show that the albedo ($R_{\mathrm{A}} = 1 - \tau = 30\%$) consists of reflection from the atmosphere (~6%), clouds (~20%), and the ground (~4%).

Homework 1.4: Accumulation of CO_2 and global warming

Based on Eq. (1.4), derive the sensitivity of T_p to the emissivity ε. If the accumulation of CO_2 decreases the emissivity by 0.02, then calculate the increase in the earth's temperature. (Ref.: S. Arrhenius, "On the Influence of Carbonic Acid in the Air upon the Temperature of the Ground," *Phil. Mag.*, April 1896.)

1.2.1 Seasonal variation depends on the distance from the sun

The orbit of the earth around the sun is elliptical (Fig. 1.1 is an approximation): the earth is closest to the sun in January ($d_{\min}$), and furthest in July ($d_{\max}$) as shown in Fig. 1.2(a). The extraterrestrial intensity (i.e., solar power/area) follows the same pattern (see Fig. 1.2(b)). The extraterrestrial normally incident radiation on the D_n-th day of the year is approximately given by:

$$I_0(D_n) = I_0(d)\left(1 + \Delta \cdot \cos\frac{2\pi D_n}{D}\right). \tag{1.6}$$

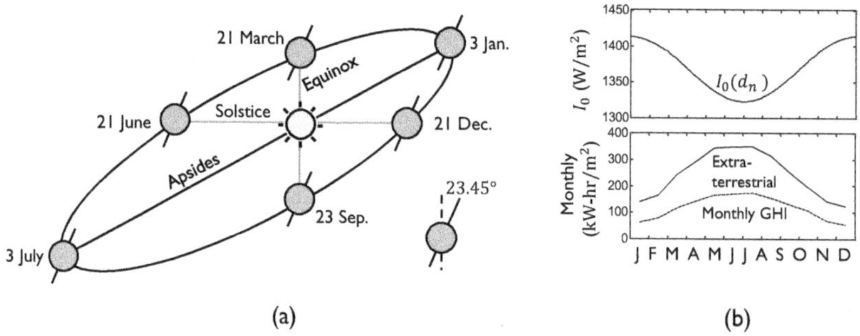

Figure 1.2: (a) The earth's orbit around the sun is elliptical and tilted at 23.45° with respect to the solar plane. The maximum distance is $d_{\max} = 152 \times 10^9$ m, while the minimum distance is $d_{\min} = 147 \times 10^9$ m. (b) Extraterrestrial solar intensity (top) and monthly integrated solar energy (bottom).

Here, $I_0(d)$ is the extraterrestrial irradiance for the planet located at a distance d from the sun, D is the number of days in a year for that planet, and $\Delta \equiv 2(d_{\max} - d_{\min})/d$ defines the asymmetry of the elliptical orbit. For earth, $\Delta = 2(1.52 \times 10^{11} - 1.46 \times 10^{11})/1.49 \times 10^{11} \approx 0.04$, and $I_0 = 1382.69$ W/m^2. Interestingly, the top graph of Fig. 1.2(b), calculated based on Eq. (1.6), shows that $I_0(D_n)$ is the lowest during the peak of the summer (in the Northern Hemisphere)! To resolve this puzzle, in the next section, we will discuss the daily and seasonal sun path across the sky and the altitude-dependent attenuation through the atmosphere.

1.2.2 Extraterrestrial intensity is reduced as sunlight travels through the atmosphere

We have already seen in Sec. 1.1.2 that approximately 70% of the *normally incident* light transmits through the atmosphere (i.e., $\tau \approx 0.7$). We know from the colors of the rainbow that the sunlight consists of photons of different wavelengths (or energy) and the earth is surrounded by a thin layer (~15 km) of atmosphere composed of nitrogen (~75%), oxygen (~20%) and trace amounts of argon, carbon dioxide, and moisture. Among these elements, oxygen, nitrogen, and argon absorb very little sunlight, but they do strongly scatter high-energy photons. This energy-dependent Rayleigh scattering makes the sky blue and the sunset orange. In addition, carbon dioxide and moisture absorb low-energy photons. As a result, the global (i.e., total) illumination on the earth's surface is reduced by 30% (or more) compared to the extraterrestrial irradiance.

Unfortunately, we cannot use the formula for normal transmission through the atmosphere (i.e., Eq. (1.5)) to calculate the location-specific energy yield of a solar farm. Instead, we need to focus on longitude- and time-dependent trajectory of the sun (i.e., sun path) across the sky. Figure 1.3(a) shows that the instantaneous position of the sun within the sun path is given by the azimuth ($\gamma_s(t)$) and zenith ($\theta_Z(t)$) angles. As the day progresses from sunrise till sunset, the time-dependent angle of incidence $\theta_Z(t)$ first reduces from 90° to its minimum value at the solar noon, then increases back to 90°. The sun path depends on the location (i.e., longitude; Fig. 1.3(b)) and time of the year (i.e., month; Fig. 1.3(c)).

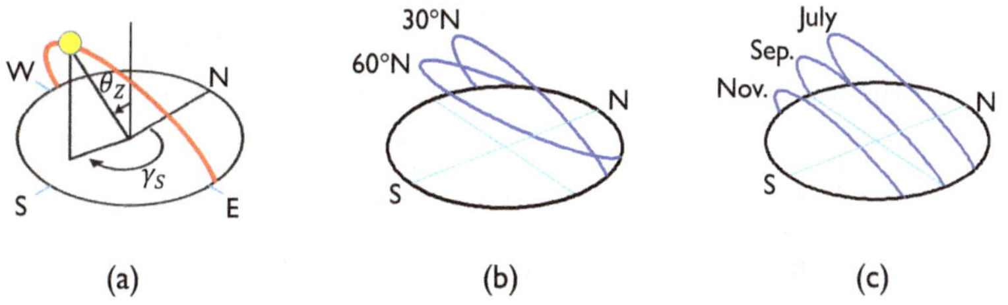

Figure 1.3: (a) The instantaneous position of the sun is defined by its azimuth (γ_s) and zenith (θ_Z) angles. (b) The sun path depends on the latitude (e.g., 30°N and 60°N in July). (c) The sun path depends on the month of the year (e.g., July, September, and November at 30°N).

As shown in Fig. 1.4(a), the angle-dependent transmission of light $T(\theta_Z)$ through the atmosphere determines the amount of extraterrestrial sunlight reaching the earth's surface. The direct normal illumination (DNI) on a plane on the ground perpendicular to the sunrays is approximated by

$$I_{\mathrm{DNI}}(D_{\mathrm{n}}, \theta_Z(t)) = I_0(D_{\mathrm{n}})T(\theta_Z) \sim I_0(D_{\mathrm{n}})\, c_1 \tau^{\mathrm{AM}(t)^{c_2}}. \tag{1.7}$$

We have already seen that $\tau = 1 - R_{\mathrm{A}} \approx 0.7$ is the average transmittance of the atmosphere with normal incidence, the exponent $c_2 = 0.678$ is an empirical

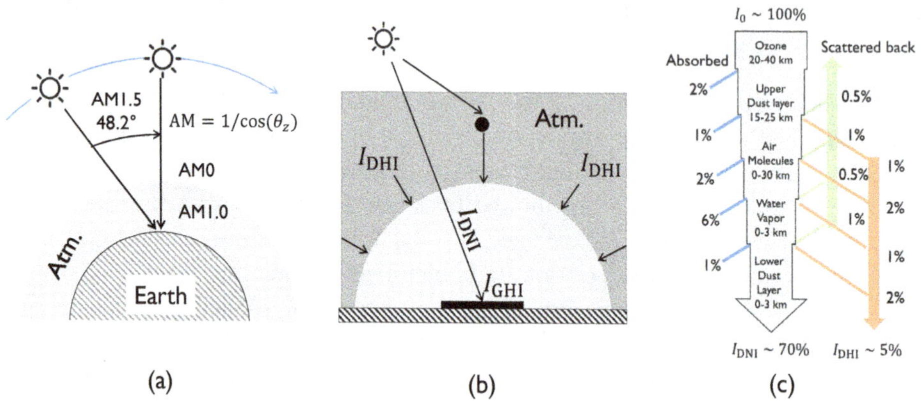

Figure 1.4: (a) Based on the position of the sun and the location on the earth, AM measures the distance traveled by the direct component of the sunlight. (b) A horizontal surface illuminated by direct and diffuse light. (c) The atmospheric attenuation is related to the absorption and scattering of the direct and diffuse light.

parameter that depends on the composition of the atmosphere, and the prefactor $c_1 = 1.0$ for clear, cloudless days. The term

$$\mathrm{AM}(t) = 1/\cos(\theta_Z(t))$$

accounts for the relative increase in the atmospheric distance (air mass) a photon needs to travel when the sun is not directly overhead. The summer is hotter because the smaller θ_Z (e.g., July sun path in Fig. 1.3(c)) enhances $T(\theta_Z)$ and compensates for the lower extraterrestrial insolation $I_0(D_n)$, as shown in Fig. 1.2(b).

Homework 1.5: Sun path from NOAA database

Find the longitude and latitude of the city you were born in. Find the sun path on your birthday by using either the Solarpath calculator from PV-LightHouse (`https://goo.gl/MxwEc7`) or the NOAA (National Oceanic and Atmospheric Administration) website. In both cases, download the Excel spreadsheet to calculate θ_Z at solar noon. The NOAA datasheet can be downloaded from `http://www.esrl.noaa.gov/gmd/grad/solcalc/calcdetails.html` under (NOAA_Solar_Calculations_day.xls).

Continuously orienting a solar module toward the sun to calculate I_{DNI} is difficult. A simpler measurement involves determining the intensity of sunlight on a fixed, ground-mounted horizontal plane, as shown in Fig. 1.4(b). The instantaneous intensity is given by

$$I_b(D_n, \theta_Z) \sim I_{\mathrm{DNI}}(D_n, \theta_Z(t)) \cos\theta_Z(t). \tag{1.8}$$

As expected, the intensity vanishes at sunrise/sunset when $\theta_Z = 90°$, and it peaks during the solar noon when θ_Z is the smallest.

> **Homework 1.6: Irradiance depends on the zenith angle of the sun**
>
> When the sun rises in the morning or sets in the evening, $\theta_Z = 90°$. As the sun reaches the peak, it reaches the minimum θ_Z. Use Eq. (1.7) to calculate the terrestrial solar power when the sun is located at the following zenith angles: $\theta_Z = 0°, 30°, 48.2°, 60°, 75°$, and $90°$. Make an intensity vs. time sketch to explain how the intensity varies throughout the day.

1.2.3 Nonzero insolation even when you cannot see the sun: Contributions from diffused light

If you measure the insolation on a horizontal plane on any given day as shown in Fig. 1.4(b), you may be surprised to find that the instantaneous power (also known as Global Horizontal Irradiance (GHI), I_{GHI}) is larger than the direct normal irradiance (DNI), calculated from Eq. (1.8). We should not be surprised. Even on an overcast day, when the sun cannot be seen, we can still tell apart day from night by the uniform (isotropic) illumination of the sky. The extraterrestrial photons are absorbed and re-emitted by the molecules of the atmosphere. After multiple scattering and a random walk through the atmosphere, they arrive on the horizontal surface isotropically. This diffused horizontal irradiance (DHI) contributes to I_{GHI} as follows:

$$I_{\text{GHI}} = I_b + I_{\text{DHI}} = I_{\text{DNI}} \cos\theta_Z(t) + I_{\text{DHI}}. \tag{1.9}$$

To determine I_{DHI}, therefore, we simply need to measure the location- and time-dependent I_{GHI} and use Eq. (1.7) to calculate I_{DNI}. Figure 1.5(a) shows the three components of the sunlight at Washington DC (latitude ~38°N) on a clear,

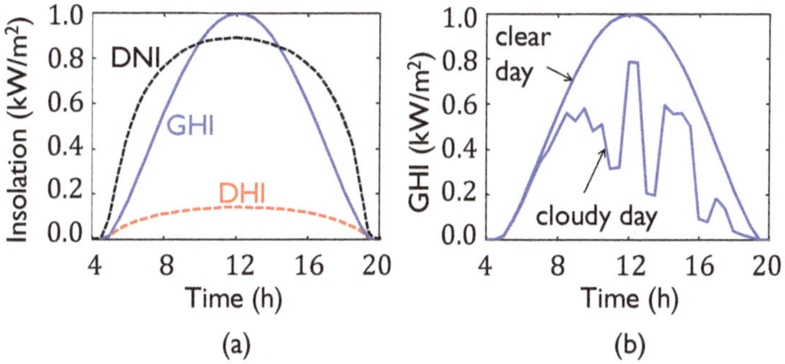

Figure 1.5: (a) Insolation components on a clear day (June 15, 2014) in Washington DC. (b) GHI for a clear (June 15, 2014) and a cloudy (June 19, 2014) day in Washington DC are compared.

cloudless day. Figure 1.5(b) shows that the total irradiance is very different on a cloudy day, as expected.

Homework 1.7: A proper definition of air mass and why it does not matter

Did you notice that $\mathrm{AM}(t) \to \infty$ as $\theta_Z \to 90$ degrees during sunrise and sunset? This cannot be right! For a planet of radius r_p and an atmosphere of thickness $h \ll r_\mathrm{p}$, show that $\mathrm{AM}(\theta_Z = 90) \approx \sqrt{2r_\mathrm{p}/h}$ is large, but finite.

Solution. Given two spheres with radii r_p and $r_\mathrm{p} + h$, a tangent at any point P of the smaller sphere intersects the larger sphere at point Q, where PQ is the maximum distance traveled by a photon to reach point P. Normalized to h, $\mathrm{AM}(\theta_Z = 90) = \sqrt{(r_\mathrm{p} + h)^2 - r_\mathrm{p}^2}/h = \sqrt{2(r_\mathrm{p}/h) + 1} \sim \sqrt{2r_\mathrm{p}/h}$. The general expression

$$\mathrm{AM}(\theta_Z) = \sqrt{(r_\mathrm{p}/h)^2 \cos^2(\theta_Z) + 2(r_\mathrm{p}/h) + 1} - (r_\mathrm{p}/h)\cos(\theta_Z)$$

reproduces the limits $\mathrm{AM} \sim 1/\cos(\theta_Z)$ and $\sqrt{2r_\mathrm{p}/h}$ respectively for $\cos(\theta_Z) \gg \sqrt{2(r_\mathrm{p}/h) + 1}/(r_\mathrm{p}/h)$ and $\cos(\theta_Z) \ll \sqrt{2(r_\mathrm{p}/h) + 1}/(r_\mathrm{p}/h)$. The first approximate expression is widely used because the intensity of sunlight during sunrise and sunset (i.e., $\theta_Z = 90$) is essentially negligible for the overall energy output of a solar cell.

Fortunately, given its importance in agriculture, global warming, and many other fields, I_GHI has been measured and tabulated in terms of the clearness index (k_t) throughout the world.

$$k_t \equiv \frac{I_\mathrm{GHI}}{I_0(D_\mathrm{n})\cos(\theta_Z)} \tag{1.10}$$

where $I_0(D_\mathrm{n})$ is the extraterrestrial intensity, given by Eq. (1.6). Inserting Eq. (1.10) into Eq. (1.9), we can calculate I_DNI (see Homework 1.8).

Equation (1.11) in Homework 1.8 describes the clearness index empirically. To calculate the clearness index theoretically, we can decompose the GHI into direct and diffused components by assuming that the diffused light is isotropic. In practice, the diffused light is somewhat anisotropic, because it includes a circumsolar component aligned with the direct component. In addition, when $\theta_Z \approx 90°$ in early morning and late afternoon, refraction and multiple scattering of light brighten the horizon (see C.G. Bohren, "Colors of Sky," *The Physics Teacher*, 1985, p. 267). A sophisticated theoretical model of diffuse light considers all three components, namely, circumsolar, horizontal, and isotropic.

Homework 1.8: Clearness index is a measure of diffused light

A particularly simple form of clearness index is described by the empirical relationship

$$k_t = \tau^{\mathrm{AM}} + \beta(1 - \tau^{\mathrm{AM}}), \tag{1.11}$$

where the first term denotes the direct light transmission through the atmosphere. The second term explains that the scattered light $(1 - \tau^{AM})$ is not lost completely: a fraction (β ~0.3) that depends on the geographical location, average cloud-cover, atmospheric contamination, etc., returns to the ground as diffused light (see Fig. 1.4(c)). Given Eq. (1.11) (and $c_1 = 1$, and $c_2 = 1$), show that the diffused light component is given by

$$I_{\mathrm{DHI}} = \beta(1 - \tau^{\mathrm{AM}}) I_0(D_{\mathrm{n}}) \cos(\theta_Z). \tag{1.12}$$

For normal incidence (AM = 1), show that an additional 9% of the light is scattered back to surface. Calculate the fraction of the diffused light for various times of the day with $\theta_Z = 0°, 30°, 48.2°, 60°, 75°$, and $90°$.

A planar surface placed flat on the ground can only see the direct and sky-diffused light, as discussed above. However, a third contribution, called ground albedo, involving direct and sky-diffused light isotropically reflected from the ground (e.g., sand, grass), can illuminate a surface tilted with respect to the horizontal plane. Every photon matters; therefore we will need to calculate the direct sun, sky-diffused, and ground albedo contributions carefully to predict the energy yield of a solar cell.

1.2.4 Irradiance and insolation are slightly different concepts

In the discussion above, we have used the terms irradiance and insolation interchangeably. We could have been more careful. Irradiance is an instantaneous measure of solar power per unit area ($\mathrm{W/m^2}$), while insolation is a measure of the cumulative energy on a surface for a defined period of time (e.g., daily, monthly, or annually), expressed in kWh per unit area. To calculate the total energy output from a solar farm, for example, we should focus on integrated insolation (composed of DHI, DNI, and GHI), rather than instantaneous irradiance.

1.2.5 There are many different ways to harvest solar energy, but we will focus on solar cells

Given the average $I_{\mathrm{GHI}} = 1000\ \mathrm{W/m^2}$, the radius of the earth ($r_{\mathrm{p}} = 6.37 \times 10^6$ m), the land area fraction (~29%), it is easy to show that more than 25 petawatts of solar power is available for conversion into other forms of energy. For example,

biofuel technology relies on photosynthesis by crops. Chemically extracted biofuels from crops have high enough energy density for many applications. Unfortunately, natural photosynthesis is inefficient, and biofuel conversion may not be cost-effective. Solar illumination can also be used to initiate and accelerate certain chemical reactions for generating end products (e.g., aluminum), or intermediate chemicals for energy storage. This process is called **artificial photosynthesis**. As another example, **solar thermal** technology uses concentric mirrors to concentrate solar irradiance onto a small area at the focal point at the top of a tower to heat water, salt, or other phase change materials. The steam produced can run a turbine or power an electrical grid during the daytime. The molten salt can store energy for the nighttime. Finally, **photovoltaics** or solar cells are widely used for terrestrial as well as extraterrestrial applications to convert sunlight directly to electricity. In the following discussion, let us focus on photovoltaic energy conversion by solar cells.

1.3 A solar cell can use only part of the solar spectrum

One needs a semiconductor solar cell to convert the incident sunlight into useful electrical energy. To produce electricity, a photon must be able to excite an electron from the valence band to the conduction band across the bandgap, E_g. Photons with $E < E_g$ transmit through the solar cell unabsorbed. Therefore, the efficiency of a solar cell depends not only on the number of photons received, but also on the energy distribution of photons with $E \geq E_g$.

1.3.1 The extraterrestrial radiation carries the imprint of the solar emission

The (quasi) detailed balance between emission and absorption during this long random walk within the sun makes the solar spectrum appear as if it came out of a blackbody radiator, with a peak temperature of ~5777 K (see Fig. 1.6(a)). These free ballistic photons will now travel to the planets carrying the imprint of the $T_S = 5777\,\text{K}$ blackbody spectrum:

$$U_{BB}(E)dE = \frac{2\theta E^3 dE}{h^3 c^3} \frac{1}{e^{E/k_B T_S} - 1} \tag{1.13}$$

where $U_{BB}(E)dE$ is amount of energy flux carried by photons with energy between E and $E + dE$ and solid angle θ. Here, E is the photon energy, and c is the velocity of light.

Homework 1.9: Meteorological data can be obtained from NASA database

The website `https://power.larc.nasa.gov/data-access-viewer/` contains a NASA database for a variety of meteorological information. Enter the longitude and latitude for the city you are currently in. Click the **Submit** button. Study the list of parameters to see the types of information available.

Let us choose a few for illustrative purposes.

1. *Parameter for Sizing and Pointing Solar Panels:* Midday insolation (kW/m^2).
2. *Solar Geometry:* Hourly solar angle relative to the horizon.
3. Parameters for Tilted Solar Panels: Radiation on equator-tilted surfaces.
4. Diurnal Cloud Information: Cloud amount 3 hourly intervals.
5. Metrology (Temperature): Air temperature at 10 m.
6. Metrology (Wind): Wind speed at 50 m.
7. Metrology (Wind): Wind speed for several vegetation and surface types.
8. Metrology (other): Relative humidity.

Download the information and plot the data.

Based on Eq. (1.13), the 5777 K solar blackbody spectrum is plotted as the black dashed line in Fig. 1.6(a). One expects to measure this spectrum just outside the earth's atmosphere. Since the sun is not an ideal blackbody, the actual extraterrestrial solar spectrum is slightly different with a large number of spikes, as shown by the red solid line marked AM0 in Fig. 1.6(a). These Fraunhofer lines (approximately 25,000) arise from the a variety of interactions of photons with various forms of He and H, and other molecules within the sun. The AM0 spectrum is measured *before* the photons have interacted with the earth's atmosphere or the "air mass" — hence the name Air Mass-0 or AM0. The solar spectrum remains essentially unchanged as it travels through the empty space between the sun and the earth. Therefore, the solar spectrum is the scaled version of AM0 everywhere in space, with the peak intensity defined by the distance from the sun. For example, the solar intensity on Mars is $\sim$43% compared to earth (see Figs. 1.6(b, c).

Homework 1.10: Photon number, flux, and energy

Using Eq. (1.13), show that the number density associated with the photons with energy E and $E + dE$ is given by

$$N_{\mathrm{BB}}(E)dE = \frac{U_{\mathrm{BB}}(E)}{E}dE = \frac{2\theta E^2 dE}{h^3 c^3} \frac{1}{e^{E/k_{\mathrm{B}}T_{\mathrm{S}}} - 1}. \tag{1.14}$$

Also show that the peak and average photon energy associated with Eq. (1.13) are given by

$$E_{\mathrm{peak}} \approx 2.82\, k_{\mathrm{B}} T_{\mathrm{S}} \tag{1.15}$$

$$E_{\mathrm{avg}} \approx 2.71\, k_{\mathrm{B}} T_{\mathrm{S}}. \tag{1.16}$$

Next show that the average energy of the photons in the solar spectrum is 1.34 eV. Naturally, the most efficient solar cells must have a bandgap to match this energy, i.e., $E_{\mathrm{g}} \sim E_{\mathrm{avg}}$. Finally, derive Eq. (1.1) by calculating the power radiated by the blackbody

$$I_{\mathrm{s}} = I_{\mathrm{BB}} = \int_0^\infty E\, cN_{\mathrm{BB}}(E)dE \equiv \sigma T_{\mathrm{S}}^4.$$

If all the photons had $E = E_{\mathrm{avg}}$, then show that the average number of photons per area integrated over solid angle 2π is

$$N_{\mathrm{avg}} = \frac{\pi^2 (k_{\mathrm{B}} T_{\mathrm{S}})^3}{2.71 \hbar^3 c^2}.$$

Homework 1.11: Puzzle of wavelength vs. energy-resolved blackbody distribution

The solar spectrum is generally plotted *not* as a function of photon energy E, but as a function of wavelength λ (see Fig. 1.6). Using the fact that $E = hc/\lambda$, show that the energy-resolved blackbody distribution given by Eq. (1.13) can be converted into a wavelength-resolved blackbody distribution given by

$$U_{\mathrm{BB}}(\lambda)d\lambda = \frac{2\theta hc^2}{\lambda^5} \frac{d\lambda}{e^{hc/\lambda k_{\mathrm{B}} T_{\mathrm{S}}} - 1}.$$

Plot both functions to show that the peaks of the spectrum do not occur at the same energy or wavelength. Explain why this result invalidates the argument that vision is optimized for solar spectrum.

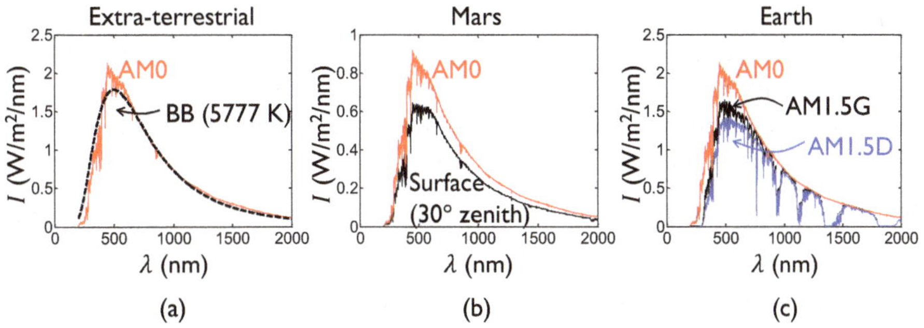

Figure 1.6: Comparison of solar spectrum. (a) Blackbody (at 5777 K) radiation compared to the extraterrestrial AM0 spectrum measured outside the earth's atmosphere. The blackbody temperature is chosen so that it emits the same integrated energy flux of AM0, i.e., the area under the red and black curves are the same. (b) Extraterrestrial ground-level solar spectra on Mars. (c) Extraterrestrial (AM0), global terrestrial (AM1.5G), and direct terrestrial (AM1.5D) solar spectra at the sea level on earth. The AM1.5G spectrum includes both direct and diffuse light, while the AM1.5D spectrum only accounts for direct light.

1.3.2 The spectrum changes significantly as it passes through the atmosphere

We used Eq. (1.7) to explain why the intensity of sunlight is attenuated as the photons travel through the atmosphere. The actual reduction depends on the wavelength (λ); therefore the spectrum of the sunlight at the ground level (e.g., AM1.5G) is very different compared to that of extraterrestrial spectrum (AM0) (see Fig. 1.6(c)). Sunlight is partially absorbed and scattered by the atmosphere, which reduces the amplitude of the spectrum received on the earth's surface. The large dips in the spectrum are correlated to the strong absorption by atmospheric water (H_2O), oxygen (O_2), and carbon dioxide (CO_2). In contrast, Mars has a thin atmosphere mostly composed of dust particles. The light scattering is weak and large wavelength absorption nonexistent; therefore the surface spectrum of Mars is similar to the AM0 spectrum (see Fig. 1.6(b)).

1.3.3 Standardized spectrum is used to compare solar cell technologies

The radiation intensity of the AM0 spectrum shown in Fig. 1.6(a) can be obtained by numerically integrating over the wavelength, i.e., $I_0 = 1366.8\,\text{W/m}^2$ (or, $1367\,\text{W/m}^2$ as per World Meteorological Organization, WMO). This flux is also called the **solar constant**. The solar constant is slightly lower than that calculated using the Stefan–Boltzmann law for an ideal blackbody at 5777 K ($I_0 = 1382.69\ \text{W/m}^2$, Table 1.1), because the actual spectrum differs slightly compared to the blackbody radiation, as shown in Fig. 1.6(a). We have also seen that the photons are scattered and attenuated by the earth's atmosphere before reaching the ground. Therefore, the intensity of sunlight depends on the location, hour of the day, and the season, making comparisons of different PV technologies difficult. Therefore, the PV industry, along with American Society for Testing and Materials (ASTM) and various government research and development laboratories, has defined two specific standard terrestrial spectra shown in Fig. 1.6(c). They are:

AM1.5G is the global, hemispherical irradiance (2π steradian view) on the specified surface. The net intensity of the AM1.5G spectrum is 1000 W/m^2.

AM1.5D is the direct (i.e., beam) irradiance normal to the specified surface. The exclusion of the diffuse light makes the integrated intensity of the AM1.5D spectrum (900 W/m^2) slightly lower than AM1.5G (1000 W/m^2). The direct insolation is used to calculate the efficiency of concentrator PV systems.

To develop these standards, one must consider factors involving the atmospheric condition as well as the receiving surface. Among the atmospheric conditions, it is assumed that the sun's zenith is $\theta_Z = 48.19°$ — which yields in the effective light path of $1/\cos\theta_Z = 1.5$ times the atmospheric depth. Hence the name AM1.5 or air mass 1.5. The receiving surface is assumed to be $37°$ tilted toward the equator to face the sun. This tilt is chosen to represent the average latitude of the 48 contiguous states of the United States of America.

Homework 1.12: Approximate intensities of the AM1.5D and the AM1.5G spectrum

Use Eqs. (1.7) and (1.12) to show that we can predict the approximate intensities specified in the AM1.5G and AM1.5D standards reasonably well.

1.4 Technology, energy yield, and cost of solar cells

Given the solar spectrum, we are now ready to discuss the photovoltaic technology that converts sunlight to electricity. The energy yield and the cost of the photovoltaic technology will determine if it has the potential to be a viable renewable energy source for the world.

1.4.1 A technology spanning over magnitudes of length scales

We will see in the rest of the book that photovoltaic energy conversion can be maximized only if we understand the physical processes that span from nanometers (e.g., optical transition at the atomic length scale) to kilometers (e.g., optimal topology of a solar farm) length scales (see Fig. 1.7). In addition, some processes (such as repeated generation of electron–hole pairs by a single photon, i.e., photon recycling) occur at a nanosecond timescale, while others erode system efficiency involving decades-long processes (e.g., corrosion or potential-induced degradation).

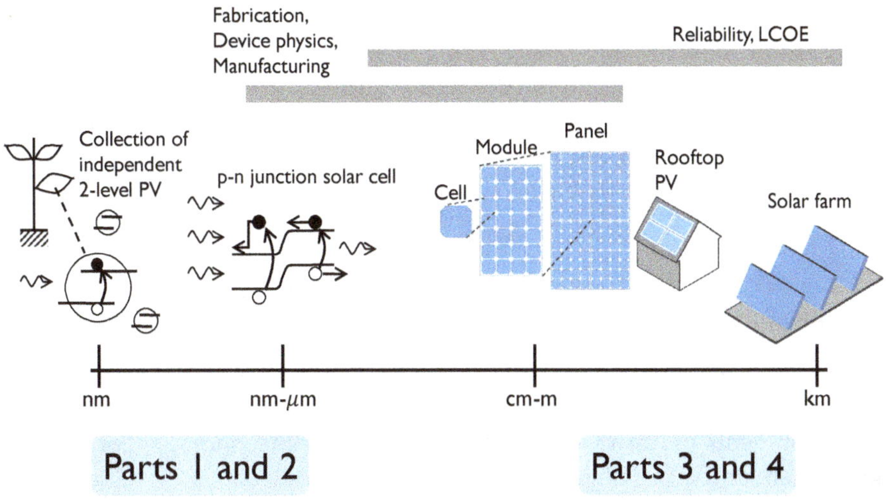

Figure 1.7: From understanding the photon–atom interaction of a two-level atom (∼nm) to the design of a solar farm (∼km), the development of a photovoltaic technology demands deep appreciation of various physical processes that span length scales of 12–14 orders of magnitudes. Parts I and II of the book will focus on the physics and technology of solar cells, while Parts III and IV will discuss the physics of solar farms and the degradation of the fielded modules.

At each length- and timescale, one wishes to suppress the losses, reduce the cost of manufacturing and operation, and maximize the end-to-end system efficiency. A more efficient system would meet the energy demand with smaller footprint and lower cost of electricity. Before getting into the details, however, it is helpful to answer two simple questions: How much energy output do we expect, and how much would the energy cost?

1.4.2 How much energy output should we expect?

Of the 1000 W/m^2 (or, 1 kW/m^2) AM1.5G power incident on earth, only a fraction can be converted to electricity by a photovoltaic system shown in Fig. 1.7. Currently, commercially available solar panels are approximately 20–22% efficient. If the average illumination is 0.5 kW/m^2 and the day is 8 hours long (typical for 40–60° latitudes), the daily energy output from a 1 m^2, 20% efficient solar cell is 0.5 kW/m^2 × 8 h × 20% = 0.8 kWh/m^2. In a solar farm, panels are tilted toward the sun and arranged in well-separated rows (more on this in Part III of the book). If ∼10% of the sunlight is lost on the ground between the rows, the annual electricity production of this farm would be 0.5 kW/m^2 × 8 h × 20% × 365 × 9/10 = 262.8 kWh per unit of farm area. This is not the net energy added to the grid: there are additional losses due to elevated operating temperature (∼10%), soiling (∼5–20%, depending on the location), inverter (∼4–6%), etc. In several PV systems, the solar panel array is connected to a storage for uniform power delivery throughout the day and night. The coupling between the panels and the storage also adds ~5–10% loss even in an optimized system. With everything included, one expects approximately 200 kWh/m^2 of energy generated by a solar farm annually. To generate 1 MW of power, therefore, one needs $(10^3/200/(8 \times 365) = 14,600\ \text{m}^2$ or approximately 4 acres of land area.

Homework 1.13: Purdue solar installation: Seeing a solar cell in action

Purdue University is located in West Lafayette, IN, USA (longitude: 40.4259°N; latitude: 86.9081°W). The university runs an experimental PV installation that allows students to learn various aspects of solar cell technology. For a background about the installation, see `https://polytechnic.purdue.edu/newsroom/knoy-hall-first-campus-use-solar-power`.

1. Monitor the system-level performance as well as irradiance and temperature (both ambient and module) as a function of time using the below link.

   `https://easyview.auroravision.net/easyview/index.html?entityId=1986490&lang=en`

2. How does the results compare between actual measurement from the weather station at Purdue vs. the NASA web calculator discussed in Homework 1.9.

1.4.3 How much does a unit of solar energy cost?

At the end, the viability of photovoltaic technology is driven by economics, more specifically by the levelized cost of energy (LCOE). The LCOE is the average cost of unit energy (kWh) over the operating life of the solar panels. Therefore, we want the solar farms to be more efficient, less costly, and to have a longer lifetime. Optimized device design improves energy conversion efficiency and effective manufacturing reduces cost. The reliability is equally important. Currently, the best panels have a lifetime of 20–25 years. The cell efficiency may degrade due to increase in the defect density, interconnect corrosion, encapsulant discoloration, and so on. The daily and seasonal variation in temperature also exert mechanical stress and cell fracture. Understanding and analyzing reliability physics will enable the design of panels with a longer lifetime. If the lifetime could be increased from 25 to 50 years, the LCOE would reduce by a factor of 2!

Homework 1.14: There are a variety of online calculators to estimate the solar energy yields

A number of online calculators, such as **PVWatts, PV-GIS, PV*Sol, PVSyst, Energy3D, PVSim,** etc., can estimate the potential energy yield of a simple installation at a specific geographical location. For this homework, let us use **PVWatts** `https://pvwatts.nrel.gov/pvwatts.php` to calculate the energy yield for three locations (you can draw the system while specifying `system information`):

1. A rooftop installation at your home with `"standard"` modules and `"fixed roof-mount"` array type.

2. A rooftop installation at your office building with `"premium"` modules and `"open-rack fixed"` arrays.

3. Purdue University Knoy Hall installation, discussed in Homework 1.13. Compare the results by plotting the predicted and actual monthly energy yields. What is the approximate area of the Knoy system?

4. How much more energy could the Knoy Hall system get with `"one-axis tracking"` arrays? Using the `System Loss` tab, make the DC-to-AC inverter lossless. How does the output increase as a result?

Homework 1.15: Driving a solar car

If the surface area of a typical car ($\sim$8 m^2) is covered with 30% efficient solar cells, how far would one be able to drive the car exclusively by solar energy?

Solution. Based on the parameters in the previous example, the total power output of the solar panel is $\sim 1000 \times 0.3 \times\ 0.3 \times 8 = 720$ W. Here we have assumed that the average insolation is approximately 30% of the AM1.5G illumination (i.e. 1000 W/m^2). With 5 hours of sunlight a day, the total energy output is 3.6 kWh. A gallon of gasoline has approximately 4 kWh of stored energy; therefore the solar energy collected has equivalent energy content of a gallon of gasoline. A car drives approximately 20–25 miles per gallon of gasoline. Since most people live within a commuting distance of less than 10 miles, a solar-driven car (with gasoline and battery backup) may not be impossible. For these applications, the solar cells must be highly efficient, lightweight, and flexible to conformally cover the surface.

Homework 1.16: Area needed to supply US energy demand exclusively by solar cells

The energy demand of the US population is approximately 3 TW (i.e., 300 million people, each consuming 10 kW on average). What is the size of a solar farm needed to produce 3 TW of power to supply the energy demand of the USA population? The area of USA is approximately 9.86 million km^2.

Solution. By the end of 2017, USA has installed 50 GW solar power. This is the nameplate capacity based on 1 kW/m^2 peak insolation. The average insolation is only 20–30% of the peak insolation. Therefore, these solar farms produce 50 GW $\times$ 0.3 = 15 GW of energy. This is approximately 0.5% of the total 3 TW power needed.

With 20% efficient solar cells, the area needed to supply US energy demand is approximately equal to

$$\frac{3 \times 10^{12}\ \text{W}}{(1000 \times 0.3 \times 0.2\ \text{W/m}^2)(10^3\ (\text{m/km})^2)} = 50,000\ \text{km}^2.$$

Since the modules must be spaced by a factor of 2, we need 100,000 km^2 of solar farm, roughly the size of Indiana. This is approximately $10^5/9.86 \times 10^6 \approx$ 1% of the US land area.

For an interesting comparison, we know that silicon is the material of choice for solar cells and computer integrated circuits. The total silicon wafer produced in 2017 for integrated circuits is approximately 1.5 km^2. The two areas (i.e., 100,000 km^2 vs. 1.5 km^2) differ by a factor of 66,000! Fortunately, making a solar cell is easier and less expensive than making an integrated circuit, but the ratio does illustrate the challenge of creating an economy based exclusively on renewable energy.

Homework 1.17: LCOE can be calculated by a web calculator, such as `https://www.nrel.gov/pv/lcoe-calculator/`

In the Baseline column, click **Preset** in the first row to open a dialog box to define the system parameters: mono-Si, glass polymer backsheet, fixed tilt utility-scale system for Indianapolis, USA. Using predefined parameters, calculate the LCOE of the city. In the **Proposed** section, reduce the cell cost to zero. How much does the LCOE reduce by? What does it mean for the price of cell technology regarding the overall viability of solar cell technology?

Homework 1.18: Weight and specific power of a solar module

How much does a solar module weigh (kg/m^2) and what is its specific power (W/kg)?

Solution. The paper by Haegel et al. published in Nature Energy, vol. 3, pp. 1002–1012, Nov. 2018, summarizes the information needed to answer this question. A typical solar module consists of a semiconductor (e.g. Si, 700–900 g/m^2; GaAs,CdTe, CIGS, 40–50 g/m^2; Perovskite, 20–30 g/m^2), coated by encapsulants (typically 1000 μm thick with 1000 g/m^2), protected by glass front cover (3.2 mm, 8000 g/m^2) and PET/PVF backsheet (350 μm, 450 g/m^2), and encased by Aluminum frame (1800 g/m^2).

A typical Si-module (with Si, encapsulant, glass, Al frame, and backsheet) weights approximately 12 kg/m^2, while a thin-film solar cells (with front and back glass-support) weights 17–18 kg/m^2. Clearly, the frame and the glass-cover are important contributors to the total weight of the modules. In principle, a flexiglass-based perovskite solar cell (with 40 g/m^2 front and back protective cover), without Aluminum framing, could weigh less than 1 kg/m^2.

Given the typical efficiency of 20%, the specific power $\mathrm{kg/W}$ for AM1.5G illumination is $1000 \times 0.2/12$ ~17 W/kg for Silicon modules and $1000 \times 0.2/17 = 15$ W/kg for thin-film modules. In principle, the flexiglass-protected 15% efficient ultrathin solar cell may approach specific power of $1000 \times 0.15/0.62 = 250$ W/kg. Specific weight is an important figure of merit for solar cells for cars, satellites, movable systems, and so on. For satellite applications, specific power exceeding 100 W/kg is typical, based on AM0 illumination, light concentration, and higher efficiency multijunction solar cells.

1.5 Overview of the book

In the next chapter, we will begin by explaining how a photon generates an electron–hole pair in an idealized two-level solar cell. We will discuss the thermodynamic limits of energy conversion, as well as the energy conversion efficiency of practical solar cells. Then we will show how the physics evolves as the cells are combined into modules and panels, and panels are installed in a solar farm. We will conclude the book by exploring the reliability of the solar cells. The book is organized as follows:

Part I: Thermodynamic Limits of Energy Conversion. We can qualitatively explain the performance of a solar cell by viewing it as an idealized two-level photon engine. This approach provides an intuitive understanding of photovoltaic operation and its fundamental limits. Moreover, a solar cell cannot absorb all the sunlight incident on it: we explain the finite absorption limit in a finite sized solar cell. Given the intuitive understanding of a two-level system, it is easy to derive the well-known Shockley–Queisser (S-Q) thermodynamic limit of practical 3D solar cells. The derivation of the S-Q formula assumes that the cell operates at 300 K ambient temperature. In practice, the cells operate much hotter (~40°C above the ambient temperature). We will explain how this self-heating reduces the energy output of solar cells and the strategies needed to reduce the cell temperature.

Part II: Physics of Practical Solar Cells. In the next part of the book, we focus on practical solar cells. We discuss the physics of ideal solar cells and derive the corresponding I-V characteristics for p-i-n, p-n, heterojunction, perovskite, and organic photovoltaics. The idealized solar cell performance is compromised by the formation of *shunts*. We explain the characteristics, physics, and universality of shunts across various PV technologies. Cells are connected in series and encapsulated to form a module. We analyze the physics of the grid pattern on a cell and architecture for a cell-to-module connection to minimize the *series resistance* loss.

Part III: Physics and Optimization of Solar Farms. Rows of panel arrays are used to structure a solar farm. We discuss the basic physics and present a global perspective of various solar farm configurations. The energy generated during the day needs to be stored for nighttime demands and to even out the output in the presence of sunlight fluctuations. We explain the basic principle of coupling energy storage to PV systems.

Part IV: Reliability Physics of Solar Cells. Finally, long-lived, reliable modules make PV technology an economically viable source of renewable energy. In the final part of the book, we discuss different degradation modes that determine the lifetime of the modules. These degradation modes include corrosion of the metal grids, potential-induced degradation, yellowing of the encapsulant, and so on. We explain how the modules are tested for reliability and its degradation monitored in the field.

To summarize, the book aims to provide a deep and intuitive understanding of a PV system, starting from the ideal limits, practical devices, to panels. It provides insights to explain as well as characterize the behavior starting from the device all the way to the system level.

References

[1] John A. Duffie and William A. Beckman. *Solar Engineering of Thermal Processes: Duffie/Solar Engineering 4e*. John Wiley & Sons, Inc., Hoboken, NJ, USA, April 2013.

[2] Antonio Luque and Steven Hegedus, editors. *Handbook of Photovoltaic Science and Engineering, Second Edition*. March 2011.

[3] J.E. Hay and J.A. Davies. Calculation of the solar radiation incident on an inclined surface. pages 59–72, Downsview, Canada, 1980.

[4] R. R. Perez, Pierre Ineichen, R. D. Seal, E. L. Maxwell, and A. Zalenka. Dynamic global-to-direct irradiance conversion models. *ASHRAE Transactions*, 98(1):354–369, 1992.

[5] Kenneth M. Edmondson, David E. Joslin, Chris M. Fetzer, Richard R. King, Nasser H. Karam, Nick Mardesich, Paul M. Stella, Donald Rapp, and Robert Mueller. Simulation of the mars surface solar spectra for optimized performance of triple junction solar cells. September 2005.

[6] Eugene Parker. Mysteries of the sun. *Physics World*, 10(10):35–40, October 1997. Publisher: IOP Publishing.

[7] Bernard H. Soffer and David K. Lynch. Some paradoxes, errors, and resolutions concerning the spectral optimization of human vision. *American Journal of Physics*, 67(11):946–953, October 1999. Publisher: American Association of Physics Teachers.

[8] B. H. Suits. The color of the sun — perception and spectral density. *The Physics Teacher*, 56(9):600–602, November 2018. Publisher: American Association of Physics Teachers.

Part I

Thermodynamics of Solar Cells

CHAPTER 2

A Two-level Solar Cell

Chapter Summary

- A simple 2-level atom can be used to understand some of the fundamental limits of a solar cell, including how PV efficiency relates to the Carnot efficiency.
- Above band-gap thermalization, below-bandgap transmission, and angle-entropy losses define the ultimate efficiency of a solar cell.
- Thermalization and below bandgap losses can be reduced by series-connecting solar cells with different bandgaps.
- Angle-entropy loss, associated with the solid angle difference of the absorbed and emitted light, can be reduced by concentrating the sunlight or by restricting the emission angle.
- Advanced solar cell concepts, such as intermediate-bandgap solar cells, can also be understood by a simple generalization of the 2-level model.

2.1 Introduction: How efficient can a solar cell be?

In Chapter 1, we saw that a solar spectrum is described by the intensity and the energy distribution of its photons (spectrum). When sunlight falls onto a semiconductor of bandgap, E_g, the below-bandgap photons ($E < E_g$) are refracted, but not absorbed. The above bandgap photons ($E > E_g$), however, excite electron–hole pairs at various energy levels. Eventually, the electron–hole pairs relax to the band-edge before being collected by the electrical contacts. Let us consider an idealized solar cell where all the above-bandgap photons are absorbed and none of the photogenerated electron–hole pairs recombine through defects. What is the maximum light-to-electricity conversion efficiency (i.e., thermodynamic limit) of such an ideal solar cell? After all, any effort to improve the practical cell efficiency will be guided by this thermodynamic limit.

Even for an idealized system, the efficiency calculation is nontrivial, but the final answer is shocking: an ideal solar cell is only 30–33% efficient! In the fog of nontrivial integration, differentiation, and numerical analysis of a "proper" calculation, we often lose track of the key insights that lead to the surprising limit, the root causes of inefficiency, and if we can do better. Therefore, let us begin with the basics and calculate the efficiency of the simplest system imaginable where single energy ($E_{\mathrm{ph}} \equiv \hbar\omega$) photons from a monochromatic "sun" isotropically excite a collection of atoms. Each atom has just two energy levels exactly matched to the incident light, i.e., $\hbar\omega = E_{\mathrm{ph}} = E_1 - E_2$ (see Fig. 2.1(a)).

In this chapter, we will see that these two-level atoms can be viewed as a "photon engine" that accepts energy from the sun and rejects the waste heat to the environment surrounding the photoexcited molecules. We will find that the model anticipates — transparently and intuitively — the fundamental issues of efficiency loss of a solar cell. Except for a few correction factors, the functional relationships derived for the two-level model correctly anticipate the corresponding results for more complicated two- and three-dimensional solar cells, shown in Figs. 2.1(b)–(d) and discussed in detail in Chapters 3 and 4.

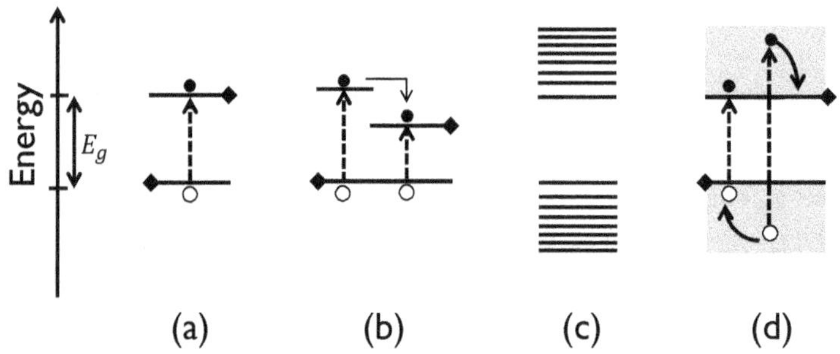

Figure 2.1: Photons absorbed by (a) a two-level system, (b) a three-level system, (c, d) a multi-level system. The diamonds refer to carrier extraction points. This chapter focuses on few-level systems (a, b). The physics of solar cells involving a multi-level system is discussed in Chapters 3 and 4.

2.2 Thermodynamic performance limit of an isolated two-level system

2.2.1 A two-level system

Figure 2.2(a) shows a set of two-level "atoms" immersed in an *isotropic*, three-dimensional field of photons (i.e., the atoms are illuminated from all directions). For example, the photosynthesis in marine diatoms involves pigment molecules immersed in a fluid, illuminated by multiply reflected, diffuse (isotropic) light. The energy levels need not be discrete, just much narrower than the bandgap. In this section, our goal is to show that if we could connect these atoms to pairs of imaginary electrical contacts to extract the photogenerated electrons, as in

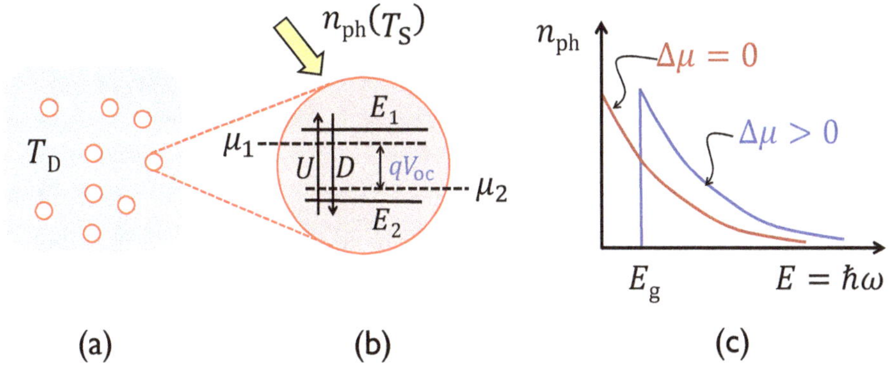

Figure 2.2: (a) A collection of two-level atoms immersed in a field of isotropic monochromatic photons. (b) A photoexcited two-level atom (energy levels E_1 and E_2) is illuminated by photons from the "sun." The difference in the electrochemical potentials $\Delta\mu \equiv \mu_1 - \mu_2$ at the open-circuit condition is indicated by qV_{oc}. One can extract power from this photo-illuminated atom by attaching two probes to μ_1 and μ_2, allowing the photoexcited electrons to leave the atom and flow in the external circuit. (c) The Bose–Einstein distributions with two different (i.e., zero and finite) chemical potentials.

Figs. 2.2(b)–(c), we would have achieved (or even exceed!) the Carnot efficiency — the ultimate limit of energy conversion in any thermodynamic engine.

2.2.2 A two-level system illuminated by a monochromatic sun

Typically, if the atoms remain in equilibrium with their surrounding phonons and photons, the relative populations of the atoms in the ground state E_2 versus those in the excited state E_1 are governed by the Fermi–Dirac (FD) distribution

$$f_i = \frac{1}{e^{(E_i - \mu_i)/k_B T_D} + 1}, \tag{2.1}$$

where T_D is the absolute temperature of the two-level system, k_B is the Boltzmann constant, and μ_i is the electrochemical potential (or quasi-Fermi level) associated with the energy level i (1 or 2). Note that $(\mu_1 - \mu_2)_D$ is not necessarily zero. Here, the subscript D represents the "device" (i.e., the two-level system).

The FD distribution defines the probability of occupation of a state in a bulk semiconductor. What does FD distribution mean for a two-level system, where an atom can either be in the excited state or in the ground state? Here the probability of occupation should be understood as a collective property of the atoms. The fraction of atoms in the up (or down) state is characterized by the Fermi–Dirac occupation probability of those states, appropriately normalized so that the sum of the atoms equals N.

The external isotropic, monochromatic illumination (see Fig. 2.2(b)) changes the relative populations of excited vs. ground-state atoms by rebalancing the absorption and emission rates. The absorption or "up" transition is given by

$$U(E_2 \to E_1) = Af_2(1 - f_1)n_{\text{ph}}, \tag{2.2}$$

while the emission or "down-transition" is given by

$$D(E_1 \to E_2) = Af_1(1 - f_2)(n_{\text{ph}} + 1). \tag{2.3}$$

Here, A is a constant, the extra $+1$ on the right-hand side of the down-transition describes the spontaneous emission (see *Feynman Lectures*, vol. 3, Chap. 4 for a more detailed discussion), and n_{ph} is the Bose–Einstein (BE) distribution for isotropic photons, i.e.,

$$n_{\text{ph}}(T_{\text{S}}) = \frac{1}{e^{\left[(E_1 - E_2) - (\mu_1 - \mu_2)_S\right]/k_B T_{\text{S}}} - 1}, \tag{2.4}$$

see Fig. 2.2(c). Here, the subscript S is a reminder that we are talking about photons coming from a "monochromatic" sun. Equation (2.4) may be unfamiliar but is easily derived. Assume that the sun is an isolated ball of atoms and photons in equilibrium with each other at the absolute temperature T_{S}, then equate Eqs. (2.2) and (2.3) using Eq. (2.1), and finally, solve for $n_{\text{ph}}(T_{\text{S}})$. Although the sun is powered by internal nuclear reactions, measurement of the solar spectrum shows that $(\mu_1 - \mu_2)_{\text{S}} \equiv \Delta\mu_{\text{S}} \approx 0$. We will use this assumption for the following discussion.

At the "open-circuit" condition shown in Fig. 2.2(b), we do not extract the photogenerated carriers. Therefore, the absorption (U) must be balanced by emission (D), so that

$$f_1(1 - f_2)(n_{\text{ph}} + 1) = f_2(1 - f_1)n_{\text{ph}}. \tag{2.5}$$

Inserting Eqs. (2.1) and (2.4) into Eq. (2.5), we find that

$$\frac{E_2 - \mu_2}{T_{\text{D}}} + \frac{E_1 - E_2}{T_{\text{S}}} = \frac{E_1 - \mu_1}{T_{\text{D}}}, \tag{2.6}$$

or, equivalently,

$$qV_{\text{oc}} \equiv (\mu_1 - \mu_2)_D = (E_1 - E_2)\left[1 - \frac{T_{\text{D}}}{T_{\text{S}}}\right]. \tag{2.7}$$

Here, V_{oc} is the open-circuit voltage of the two-level system and q is the electron charge. We note that the "Carnot factor," involving the ratio of the temperatures of the device and the sun, shows up in the expression.

If we now short-circuit the electrical contacts attached to each atom, one exchanging electrons exclusively with E_1 and the other with E_2 (see Fig. 2.2(c)), and if the photon flux R is small, then the energy input to the collection of atoms is $(E_1 - E_2) \times R \times N$, while the maximum energy output is $\sim V_{\text{oc}} \times qR \times N = (\mu_1 - \mu_2)_D \times R \times N$, so that the efficiency η is given by

$$\eta_{\text{c}} = \frac{(\mu_1 - \mu_2)_D R \times N}{(E_1 - E_2)\, R \times N} = \left[1 - \frac{T_{\text{D}}}{T_{\text{S}}}\right]. \tag{2.8}$$

This derivation suggests that the energy is dissipated irreversibly when electrons transfer from the atom to the contacts. The input and output powers used to derive Eq. (2.8) can be represented as a Carnot photon engine. The schematic in Fig. 2.3(b) makes the analogy between a solar cell and a photon engine explicit. In this limit, therefore, a photon engine is just another form of heat engine connected between two reservoirs of temperature T_S and T_D, described by the Carnot formula in Eq. (2.8). It is possible to exceed the Carnot limit by attaching the photon engine between two non-equilibrium reservoirs (see Homework 2.2).

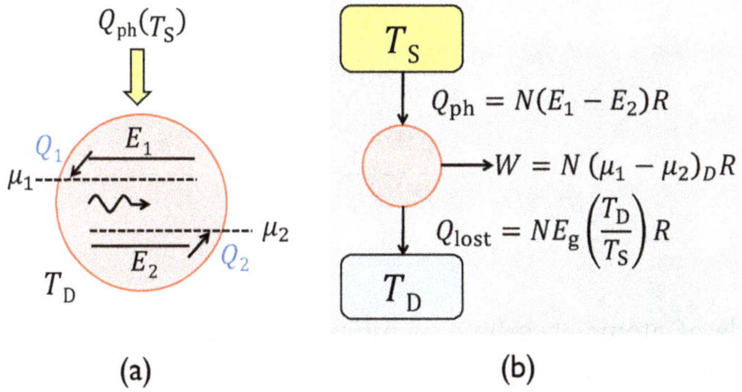

Figure 2.3: (a) The energy band for the two-level system; (b) the energy flux balance of a "photon engine."

2.2.3 Two-level atoms with different energy gaps

Let us return to our original discussion of two-level atoms illuminated by isotropic sunlight. In the previous discussion, the atoms had identical energy gap and absorbed single-energy photons to achieve the Carnot efficiency. Let us generalize the problem so that the ensemble includes N_1 atoms with $E_{G,1} \equiv (E_1^{(1)} - E_2^{(1)})$, N_2

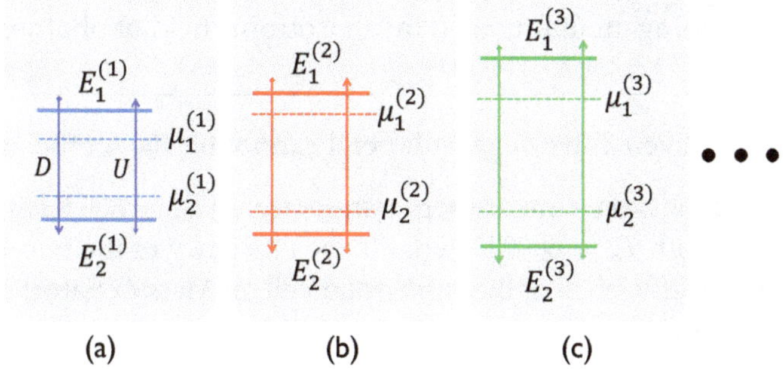

Figure 2.4: An ensemble of non-interacting two-level systems having different energy gaps.

atoms with $E_{G,2} \equiv (E_1^{(2)} - E_2^{(2)})$, etc. (see Fig. 2.4). Can the ensemble, as a whole, still achieve the Carnot efficiency?

The total power input to the system is $P_{in} = \sum_{i=1}^{M}(E_1^{(i)} - E_2^{(i)})N_i R$, while the total power output is $P_{out} = \sum_{i=1}^{M}(\mu_1^{(i)} - \mu_2^{(i)})N_i R$. As mentioned earlier, R is the photon flux from the sun. The principle of detailed balance requires that each group of atoms be in equilibrium with the corresponding set of incident photons, i.e., $(\mu_1^{(i)} - \mu_2^{(i)}) = \eta_i(E_1^{(i)} - E_2^{(i)})$. Each two-level atom operates at the Carnot efficiency ($\eta_i = \eta_1$). Thus, the ratio of P_{in}/P_{out} is given by

$$\begin{aligned}\eta_S &= \frac{N_1(\mu_1^{(1)} - \mu_2^{(1)}) + N_2(\mu_1^{(2)} - \mu_2^{(2)}) + \cdots}{N_1(E_1^{(1)} - E_2^{(1)}) + N_2(E_1^{(2)} - E_2^{(2)}) + \cdots} \\ &= \frac{\eta_1 N_1(E_1^{(1)} - E_2^{(1)}) + \eta_1 N_2(E_1^{(2)} - E_2^{(2)}) + \cdots}{N_1(E_1^{(1)} - E_2^{(1)}) + N_2(E_1^{(2)} - E_2^{(2)}) + \cdots} \\ &= \eta_C. \end{aligned} \tag{2.9}$$

The ensemble of atoms absorbing at different frequencies can still achieve the Carnot efficiency, provided the atoms are isolated and the power from each atom is independently collected by weakly coupled probes attached to these "atoms." In the PV literature, solar cells based on such a "spectral splitting technique" have been discussed in the context of very high efficiency cells, see Chapter 4.

2.3 Thermalization energy loss in a two-level "molecular" solar cell

In the beginning of the chapter, we said that the efficiency limit of a solar cell is shockingly low, i.e., just 30–33%. And yet, in the last section, we saw that a collection of atoms illuminated by isotropic and monochromatic sunlight achieves the Carnot efficiency, the maximum any engine can achieve. Actually, we over-idealized the solar cell. To see how the efficiency drops from 95% to 33%, we need to consider a bulk semiconductor with a collection of atoms. Let us begin with a two-atom molecule, again illuminated by an isotropic field of photons.

Homework 2.1: Even a two-level solar cell cannot be 100% efficient

Assuming that the atoms are at room temperature ($T_D = 300$ K) and the sun is a blackbody with $T_S = 6000$ K, what is the efficiency of the two-level solar cell? What is the efficiency of the same solar cell on Mars or Saturn? Does the solar cell get more efficient away from the sun?

(continued on the next page)

Homework 2.1 (*continued from the previous page*)

Solution. The cell efficiency on the earth is

$$\eta_{\rm c} = \left[1 - \frac{300}{6000}\right] = 0.95.$$

In general, the cell efficiency in any planet is obtained by assuming $T_{\rm p} \approx T_D$ and insert Eq. (1.3) in Eq. (2.8):

$$\eta_{\rm c} = \left[1 - \sqrt{\frac{R_{\rm s}}{2d}}\sqrt{\frac{1-R_A}{\varepsilon}}\right]. \tag{2.10}$$

Since d increases faster than other parameters, $T_{\rm p}$ decreases for planets further away from the sun. As a result, a solar cell will be more efficient on Saturn compared to Earth.

Finally, a nighttime "solar" cell that operates between the temperature of the earth (~300 K) and that of the outer space (~3 K) would have an efficiency of

$$\eta_{\rm c} = \left[1 - \frac{3}{300}\right] = 0.99,$$

which is higher than the efficiency of a daytime solar cell! Obviously, higher efficiency does not imply higher power output!

Homework 2.2: A solar cell illuminated by light-emitting diodes

Let us assume that a collection of light-emitting diodes (LEDs) surround and isotropically illuminate a box of two-level atoms. Unlike a blackbody source, the LED photons have chemical potential equal to the LED bandgap, i.e., $(\mu_1 - \mu_2)_{\rm S} = (\mu_1 - \mu_2)_{\rm LED}$.

1. Reevaluate Eq. (2.5) by assuming that the photon distribution given by Eq. (2.4) has non-zero $\Delta\mu_{\rm S}$ to obtain an expression for the efficiency $\eta_{\rm c}$ as a function of $T_{\rm D}, T_{\rm LED}, (E_1 - E_2),$ and$(\mu_1 - \mu_2)_{\rm LED}$.

2. Rearrange your equation in the form of the Carnot formula, $\eta = (1 - T_{\rm D}/T_{\rm S}^*)$, where $T_{\rm S}^*$ is the effective temperature of the LED source. Does the source effectively appear hotter or cooler due to the non-zero chemical potential?

2.3.1 Thermalization loss in a "bilayer" solar cell

Consider a collection of two-atom molecules, with the donor and acceptor atoms in each molecule linked together as a common unit, as shown in Fig. 2.5. In process 1, a photon is absorbed in the donor atom, generating an electron–hole pair. In process 2, the electron transfers very quickly from the energy $E_1^{(1)}$ to the energy level $E_1^{(2)}$ of the acceptor atom. Finally, in process 3, the free electron at $E_1^{(2)}$ recombines with the empty state at $E_2^{(1)}$ at the cross gap, giving away a photon of energy $(E_1 - E_2)_D$. The cross-gap recombination reflects the fact that inter-level relaxation of electrons to the lowest energy level (process 2) is much faster than the cross-gap radiative recombination (process 3).

The up- and down-transitions are given by

$$U(E_2^{(1)} \to E_1^{(1)}) = Af_2^{(1)}(1 - f_1^{(1)})n_{\rm ph}^U \tag{2.11}$$

and

$$D(E_1^{(2)} \to E_2^{(1)}) = Af_1^{(2)}(1 - f_2^{(1)})(n_{\rm ph}^D + 1). \tag{2.12}$$

Here, $n_{\rm ph}^U$ and $n_{\rm ph}^D$ are the Bose–Einstein distributions corresponding to photons having energies $(E_1^{(1)} - E_2^{(1)})$ and $(E_1^{(2)} - E_2^{(1)})$, respectively. An ultrafast transition from $E_1^{(1)}$ to $E_1^{(2)}$ implies $f_1^{(1)} \approx f_1^{(2)}$. Using this relationship and equating the up and the down-transitions given by Eqs. (2.11) and (2.12), we find

$$(\mu_1^{(2)} - \mu_2^{(1)}) = (E_1^{(2)} - E_2^{(1)}) - \left(\frac{T_{\rm D}}{T_{\rm S}}\right)(E_1^{(1)} - E_2^{(1)}). \tag{2.13}$$

The corresponding cell efficiency is

$$\eta_{\rm b} = \frac{\mu_1^{(2)} - \mu_2^{(1)}}{E_1^{(1)} - E_2^{(1)}} = \frac{E_1^{(2)} - E_2^{(1)}}{E_1^{(1)} - E_2^{(1)}} - \frac{T_{\rm D}}{T_{\rm S}} < \left(1 - \frac{T_{\rm D}}{T_{\rm S}}\right) \equiv \eta_{\rm c}. \tag{2.14}$$

In other words, bilayer efficiency is reduced ($\eta_{\rm b} < \eta_{\rm c}$) due to thermalization energy loss related to the fact that the emission energy is lower than the absorption energy,

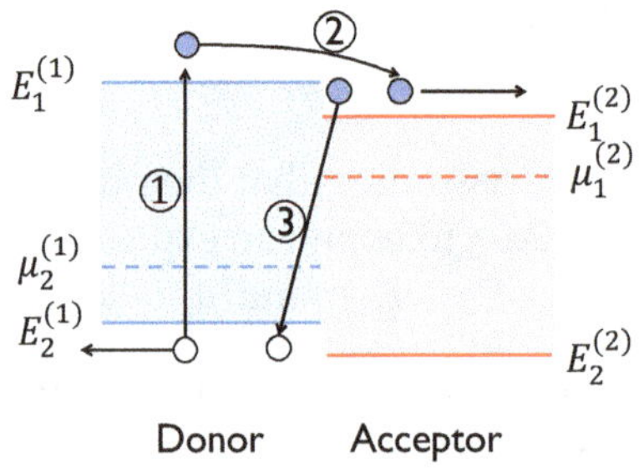

Figure 2.5: Schematic diagram showing photon absorption (process 1), inter-level energy relaxation (process 2), and cross-gap recombination (process 3) in a two-atom "bilayer" solar cell. The relaxation step leads to irreversible energy loss and reduced cell efficiency.

i.e., $(E_1^{(2)} - E_2^{(1)}) < (E_1^{(1)} - E_2^{(1)})$. We will discuss the topic in more detail in Sec. 2.3.3.

2.3.2 Thermalization loss in series-connected atoms

Consider a molecule with two atoms operating in series as shown in Fig. 2.6. For example, the photosynthesis process involves a sequential absorption of photons in photosynthesis centers, PS1 and PS2. Another example would be a solar module consisting of several solar cells connected in series. The atoms have the same bandgap; therefore, atoms (1) and (2) share the incident photon flux equally — i.e., each absorbs half the photon flux indicated by the upward arrow 1. The up-transitions are

$$U^{(1)}(E_2^{(1)} \to E_1^{(1)}) = \frac{1}{2} A f_2^{(1)} (1 - f_1^{(1)}) n_{\text{ph}}^U, \tag{2.15}$$

$$U^{(2)}(E_2^{(2)} \to E_1^{(2)}) = \frac{1}{2} A f_2^{(2)} (1 - f_1^{(2)}) n_{\text{ph}}^U. \tag{2.16}$$

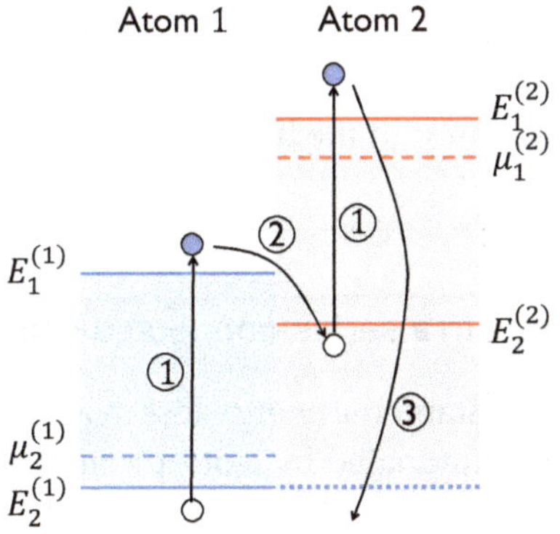

Figure 2.6: Schematic diagram showing the probabilistic absorption of photons in two series-coupled atoms.

The cells are identical; therefore, $(E_1^{(1)} - E_2^{(1)}) = (E_1^{(2)} - E_2^{(2)}) = E_{\text{ph}}$. Therefore, the total up-transition is

$$U = U^{(1)} + U^{(2)} = A f_2^{(1)} (1 - f_1^{(1)}) n_{\text{ph}}^U. \tag{2.17}$$

The coupling between the cells occurs through electron relaxation from $E_1^{(1)}$ to $E_2^{(2)}$. Similarly, the down-transition in the coupled system (arrow 3) is given by

$$D(E_1^{(2)} \to E_2^{(1)}) = A f_1^{(2)} (1 - f_2^{(1)}) (n_{\text{ph}}^D + 1). \tag{2.18}$$

Finally, by equating the up- and down-transitions, we obtain

$$\left[\frac{E_2^{(1)} - \mu_2^{(1)}}{k_B T_D} + \frac{E_{ph}}{k_B T_D}\right] = \frac{E_1^{(2)} - \mu_1^{(2)}}{k_B T_D},$$

$$\therefore (\mu_1^{(2)} - \mu_2^{(1)}) = E_{ph} - \Delta E - \left(\frac{T_D}{T_S}\right) E_{ph}. \tag{2.19}$$

The efficiency of the series coupled system is

$$\eta_s = \frac{E_{ph} - \Delta E}{E_{ph}} - \frac{T_D}{T_S} < \left(1 - \frac{T_D}{T_S}\right) = \eta_c. \tag{2.20}$$

Here, the energy loss during the electron transfer, i.e., $\Delta E \equiv E_1^{(1)} - E_2^{(2)}$, explains the efficiency loss below the Carnot limit.

Homework 2.3: Difference between a bilayer and a series-connected system

What is the basic difference between a bilayer and a series-connected system? Which type of solar cell would have a higher output voltage?

Hint. Note that the bilayer system extracts electrons through the lower E_1, and the efficiency is limited by the cross-gap. However, the series-coupled system collects electrons through the higher E_1. The series-connected system will have higher V_{oc}.

2.3.3 Thermalization loss in a series-connected cell with many atoms

Finally, as a prelude to understand the energy loss in an actual solid, let us consider an ensemble of molecules illuminated by isotropic sunlight. The two-level atoms within a molecule are not independent, but are coupled in a way that the electrons can transfer from one atom to the next (see Fig. 2.7). The transfer of electrons from an atom with a larger bandgap to an atom with a smaller bandgap is accompanied by the emission of phonons to the environment. Let us assume that all the atoms can absorb photons, but inter-atom energy relaxation is so fast that photon emission is only possible for atoms with the smallest energy gap $(E_1^{(1)} - E_2^{(1)})$. In other words, electrodes are attached and energy is extracted from the atom with the smallest gap. Balancing the upward and downward fluxes, we can write

$$\begin{aligned}\eta_s &= \frac{(N_1 + N_2 + N_3 + \cdots)(\mu_1^{(1)} - \mu_2^{(1)})}{N_1(E_1^{(1)} - E_2^{(1)}) + N_2(E_1^{(2)} - E_2^{(2)}) + \cdots} \\ &= \eta_1 \frac{(N_1 + N_2 + N_3 + \cdots)(E_1^{(1)} - E_2^{(1)})}{N_1(E_1^{(1)} - E_2^{(1)}) + N_2(E_1^{(2)} - E_2^{(2)}) + \cdots} < \eta_c.\end{aligned} \tag{2.21}$$

The efficiency is reduced because the energy absorbed by atoms with larger bandgaps has been lost to thermalization (i.e., phonon emission). The process generates entropy and the loss is irreversible. In Homework 2.4, we will see that thermalization loss reduces η_b to 65–70%.

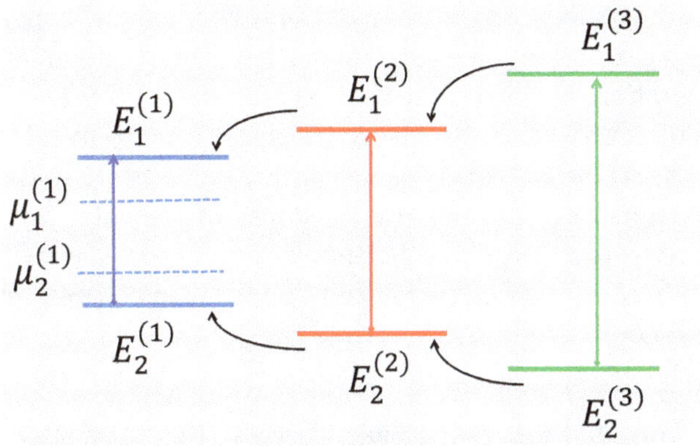

Figure 2.7: An ensemble of interacting two-level systems having different energy gaps. The number of atoms with bandgap E_i is n_i. Every atom absorbs photons equal to its bandgap, but the electrons are extracted only from the lowest-bandgap atoms.

Homework 2.4: Approximate efficiency loss due to thermalization

Show that the efficiency of a system with $N_1 = N_2 = N_3 = \cdots$ is given by

$$\eta_s = \eta_c \times \frac{1}{1 + \Delta E_{avg}/(E_1^{(1)} - E_2^{(2)})}$$

where ΔE_{avg} is the average excess bandgap of atoms. The relevant energy levels must be able to absorb the blackbody radiation, i.e., $\Delta E_{avg} \approx k_B T_S \approx 0.5$ eV. Assuming that a typical solar cell has a bandgap ($E_g \equiv E_1^{(1)} - E_2^{(2)} = 1.33$ eV), show that the thermalization loss reduces the efficiency by 25–30%. In Chapter 3, we will see that the exact thermodynamic calculation gives similar results.

2.3.4 Reducing the thermalization loss

Thermalization involves energy loss to the environment as electrons relax from one atom to the next. There are several ways to reduce the thermalization loss. For now, let us discuss the approaches qualitatively, we will consider these techniques quantitatively in Chapter 4.

1. *Multiple exciton generation (MEG).* In this scheme, the excess energy released when an electron relaxes from the higher to the lower energy level is not lost to phonons, but instead is transferred to the acceptor atom itself. The energy liberated generates a new electron–hole pair. Thus, the excess energy is not lost to thermalization, but it is recycled to increase the photocurrent.

2. *Hot-electron PV.* Another method to decrease the thermalization loss involves extraction of carriers in the excited states before they thermalize. An important consideration for the design of these hot-carrier PV is that the thermalization time (10–100 ps) and the carrier transport time to the electrical contacts must be comparable so that the hot electrons do not relax to the lower-energy state. This constraint dictates the thickness of the solar cell and defines a trade-off between absorption efficiency and the efficiency of hot-carrier collection. The trade-off could be relaxed by increasing the thermalization time by phononic confinement.

3. *Hybrid photovoltaic/thermal (PV/T).* Another approach to reduce thermalization loss is based on a hybrid photovoltaic/thermal (PV/T) system. In this scheme, a circulating fluid collects the waste heat generated by the PV module and uses the heated fluid to run an engine. Such integrated systems return the efficiency toward the Carnot limit for systems containing multiple atoms with different bandgaps.

Homework 2.5: Thermalization in the bilayer solar cell

To confirm that Eq. (2.14) is consistent with Eq. (2.21), recall that for a pair of donor and acceptor atoms we have

$$\eta_{\mathrm{b}} = \frac{(N_1 + N_2)\left(\mu_1^{(2)} - \mu_2^{(1)}\right)}{N_1(E_1^{(1)} - E_2^{(1)}) + N_2(E_1^{(2)} - E_2^{(2)})}. \tag{2.22}$$

In this particular case, $N_2 = 0$ as there is no absorption in material (2), i.e., there is no photon absorption involving $(E_1^{(2)} - E_2^{(2)})$. Thus, from Eq. (2.22), we find that

$$\eta_{\mathrm{b}} = \frac{\mu_1^{(2)} - \mu_2^{(1)}}{E_1^{(1)} - E_2^{(1)}}, \tag{2.23}$$

which is precisely the term following the first equal sign in Eq. (2.14).

2.4 Efficiency loss due to angular anisotropy of sunlight

Taken together, Carnot and thermalization losses reduce the efficiency to 65–70%, which is still much higher than the ultimate efficiency of 30–33%. To understand

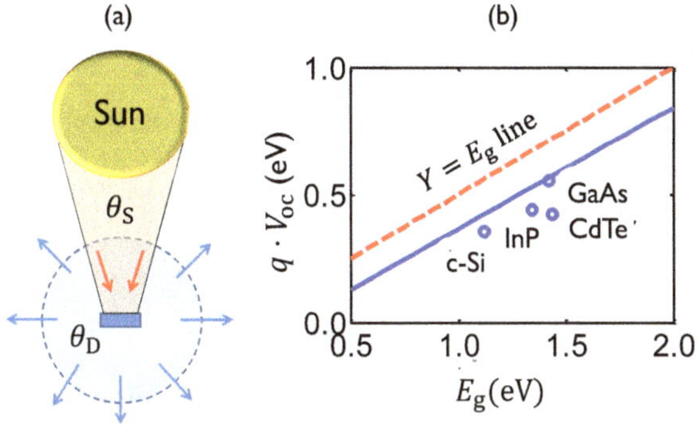

Figure 2.8: (a) Angle mismatch between the sun and the solar cell. (b) The blue line, based on Eq. (2.26), defines the thermodynamic limit of V_{oc} as a function of bandgap $E_g = E_1 - E_2$. The red dashed line is included as a reference to indicate that $V_{oc} < E_g$. The best solar cells (symbol: circles) are beginning to approach the thermodynamic limit of V_{oc}.

the other sources of losses, recall that we have so far assumed that the solar cell is illuminated by multiply scattered, diffused (i.e., isotropic) sunlight. In practice, a solar cell illuminated directly by the sun has a lower efficiency due to "angle entropy loss." Let us explain.

The sun is approximately 150×10^6 km away; therefore, it appears as a small disk in the sky. We recall from Homework 1.1 that the solid angle subtended by the sun for an observer on the earth is only $\theta_S = 6.5 \times 10^{-5}$ steradians, see Fig. 2.8(a). The angle is so small that the sunrays are essentially parallel (and hence the shadow behind an object). On the other hand, when the photons absorbed by the atoms are re-emitted, they are radiated isotropically (radiation angle $\theta_D \approx 4\pi$ steradians). Therefore, Eq. (2.5) must be rewritten as

$$\theta_D\, f_1(1 - f_2)(n_{ph} + 1) = \theta_S\, f_2(1 - f_1)n_{ph} \tag{2.24}$$

or

$$-\ln\left(\frac{\theta_D}{\theta_S}\right) + \left(\frac{E_2 - \mu_2}{k_B T_D}\right) + \left(\frac{E_1 - E_2}{k_B T_S}\right) = \left(\frac{E_1 - \mu_1}{k_B T_D}\right),$$

which leads to

$$\eta_a = \frac{\mu_1 - \mu_2}{E_1 - E_2} = \left(1 - \frac{T_D}{T_S}\right) - \frac{k_B T_D}{E_1 - E_2}\ln\left(\frac{\theta_D}{\theta_S}\right). \tag{2.25}$$

This remarkable formula says a photon engine working with direct sunlight will always have lower power conversion efficiency compared to a photon engine operating with isotropic light. With $E_g \equiv E_1 - E_2$, the efficiency degradation is directly related to the loss in open-circuit voltage:

$$qV_{oc} \equiv \mu_1 - \mu_2 = \left(1 - \frac{T_D}{T_S}\right) E_g - k_B T_D \ln\left(\frac{\theta_D}{\theta_S}\right). \tag{2.26}$$

Homework 2.6: Open-circuit voltage vs. the bandgap

Show that the open-circuit voltage, V_{oc}, cannot exceed the bandgap, E_g.

Solution. To estimate the difference between V_{oc} and E_g, we can use Eq. (2.26). If $T_S \approx 6000$ K and $T_D = 300$ K, then

$$qV_{oc} = 0.95 \times E_g - 0.296 \text{ eV}.$$

In other words, for a typical solar cell with a bandgap between 1.2 and 1.5 eV, almost 15–17% of the open-circuit voltage is lost due to the mismatch between θ_S and θ_D. We will see in the next chapter that the exact calculation of V_{oc} for a 3D solar cell leads to a formula with a slightly different constant (see Fig. 2.8(b)):

$$qV_{oc} = 0.95 \times E_g - 0.201 \text{ eV}, \tag{2.27}$$

because the loss due to angular anisotropy is partially compensated by the contribution from 3D photonic density of states. This compensation reduces the angle entropy loss to ~9–11%.

2.4.1 Discussion: Isotropic vs. direct sunlight

What is the difference between isotropic and direct sunlight that reduces the PV efficiency so significantly? An atom has certain directivity in the radiation pattern in vacuum. In the derivation above, however, we have assumed that the photons arrive and are absorbed in a narrow (solid) angle θ_S, while they re-radiate in a broader angle of $\theta_D \approx 4\pi$. This can only happen if the phases of the atoms excited to level 1 are subsequently randomized by the collision among the atoms so that the atoms eventually re-emit at random angles. The entropy gain of a system is $k_B T_D \ln(N_{\text{final}}/N_{\text{initial}})$. Here, the number of radiating angular states is given by $(N_{\text{final}}) \approx 4\pi$ and the number of angular states occupied by the incoming sunlight is given by $(N_{\text{initial}}) \approx \theta_S$. We see that the extra loss term can be viewed as an irreversible entropy gain due to angular mismatch between incident and re-radiated photons. A complex derivation of this entropy loss exists, but the use of two-level PV makes the physical interpretation intuitive and transparent.

What does it mean to "lose energy" due to angle anisotropy? Let us say that a number of photons enter the solar cell at normal incidence; such photons occupy a specific point in k-space, as shown by the black circle in Fig. 2.9(a). These photons are absorbed and the atoms excited. The atoms then go through a momentum scattering process, indicated by the arrows in Fig. 2.9(b). Since the endpoint of each random walk is different, each atom will emit a photon at a random angle. Individually, the photons emitted have the same energy as the photons absorbed, so that there is no loss of real energy. However, we should recognize that it takes energy to create collimated photons (similar to incident sunlight) from random photons emitted by the cell. When the collimated photons are scattered, the corresponding entropy increase indicates that some part of the incident energy cannot be converted to work. This energy has been irreversibly lost in the process of momentum scattering or angle randomization.

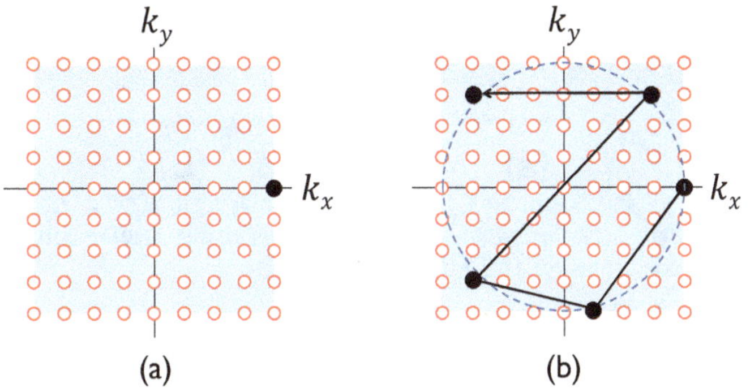

Figure 2.9: (a) A single state occupied by a photon (approximately) normally incident from the sun. (b) The momentum randomization of the photon inside the solar cell is indicated by the black arrows. Although we have used a two-dimensional momentum space for simplicity, the photon incidence and re-emission are intrinsically three-dimensional.

Homework 2.7: Direct vs. diffuse light

In Chapter 1, we explained that atmospheric scattering converts a fraction of the direct light to diffuse light. Once absorbed by the solar cell, the direct light has angle entropy loss, but isotropic diffuse light does not. Does it imply that light scattering will improve the efficiency of the solar cell?

Hint. Argue why light scattering by the atmosphere will **not** improve the efficiency of the solar cell.

2.4.2 Recovery of angle entropy loss

The efficiency of the solar cell will improve if we can reduce the entropy loss due to angle mismatch. We can either make the absorption angle larger or the emission angle narrower, and both approaches are in practical use today.

1. *Mirrors.* Solar cells use mirrors at the back surface to reflect light and to reduce the emission angle from 4π to 2π. Inserting this new angle in Eq. (2.26), we find that the open-circuit voltage increases by $(k_B T_D/q) \times \ln(2)$ or 17 mV at room temperature. This leads to a slight improvement in efficiency.

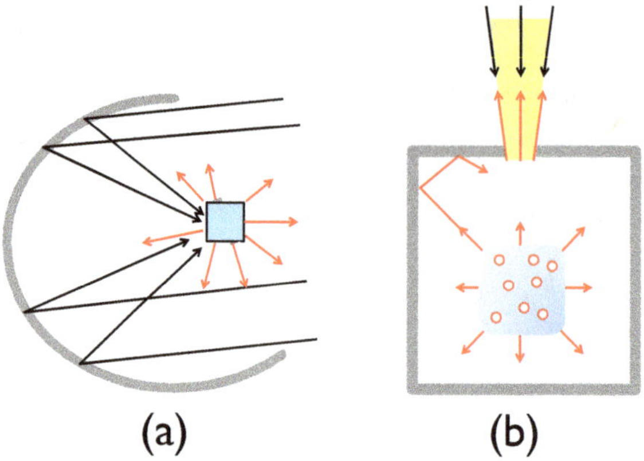

Figure 2.10: (a) Angle broadening of incident photons using a solar concentrator. (b) A scheme of limiting the emission angle of a solar cell.

2. *Solar concentrator.* If the atoms are placed in a small sphere at the focus of a concentric hemisphere, the atoms will be illuminated from all sides with $(2\pi/\theta_S) \approx 10^5$ suns (see Fig. 2.10(a)). The incident angle is now 2π, matching exactly the angle of the radiated photons. In this case, the angular anisotropy term disappears and V_{oc} once again reaches the values corresponding to the Carnot limit. Therefore, the essence of the concentrator solar cells lies in suppressing the angle entropy generated in a typical solar cell illuminated by direct sunlight.

3. *Narrow emission angle.* It might be possible to create a set of optical structures so that illumination and emission are possible only with a narrow solid angle (see Fig. 2.10(b)). Depending on the narrowness of the angle, the angle entropy term will be reduced, and the efficiency will approach the Carnot efficiency for a collection of two-level atoms.

2.5 Energy loss due to below-bandgap light transmission

We are now ready to remove the final assumption that led to the Carnot limit for two-level solar cells, i.e., all the photons incident on a solar cell are actually absorbed. Recall that the efficiency of a solar cell is traditionally *defined* by the ratio of energy converted to electricity to the total incident energy from the sun. If a photon with energy below (or above) the bandgap passes right through the atoms (without interacting with the atoms themselves), the solar cell will still be held responsible for not being able to convert those photons to electrical energy. This below-bandgap loss is really not a loss at all, because the photons still carry the memory of the sun and have the ability to do work. In any case, the classical definition presumes that the transmitted energy will be irretrievably lost and, therefore, should be rightfully chalked up as a loss mechanism. The failure to convert the below-bandgap photons accounts for 25–30% reduction in efficiency.

2.5.1 Recovery of below-bandgap loss

1. *Tandem cells.* Consider, for example, that a quasi-transparent PV has been integrated with the structure of a greenhouse. The below-bandgap photons that escape through the solar cell (e.g., large-bandgap organic solar cells or OPVs) can still be used to drive the photosynthesis of the plants placed behind the quasi-transparent solar cells. Indeed, we can replace the plants by a smaller-bandgap solar cell (e.g. c-Si) to convert the below-bandgap energy transmitted through the larger bandgap solar cell. In fact, the concept can be generalized to include a series of solar cells with decreasing bandgaps stacked on top of each other. We will discuss the physics of these highly efficient tandem cells in Chapter 4.
2. *Thermal photovoltaics.* Another interesting scheme to utilize the below-bandgap loss involves thermal PVs (TPVs). Here, the first layer absorbs the full-spectrum of sunlight and then re-emits it at a lower energy. The second selective emitter layer transmits only those photons that are easily absorbed by the PV layer at the bottom. This selective emission ensures the recycling of the below- and above-bandgap photons that had previously been lost to below-bandgap transmission and above-bandgap thermalization. The recycling keeps the top absorber layer hot and improves the conversion efficiency of the PV layer at the bottom.

2.6 Conclusions: Insights from a two-level model of a solar cell

In this chapter, we discussed the fundamental limits of energy conversion efficiency of a solar cell. An idealized two-level solar cell, illuminated by isotropic light, is shown to achieve the thermodynamic Carnot efficiency of ~95%. In practice, however, three additional loss mechanisms reduce the efficiency far below the Carnot limit (see Fig. 2.11). These include (i) the angle mismatch loss between absorbed and emitted photons, ~10–12%; (ii) the below-bandgap loss associated with unabsorbed photons, ~20–25%; and (iii) the thermalization loss of above-bandgap high-energy photons, ~20–25%. We find that it is fundamentally impossible to convert more than 30–33% sunlight into useful energy.

In Chapter 3, we will derive the exact expressions for the thermodynamic limit and loss mechanisms for a 3D solar cell. Remarkably, the results will be essentially identical to the results derived in this chapter based on an idealized two-level model! A broad range of strategies have been developed to reduce the losses of a single-junction solar cell. We will discuss these strategies in Chapter 4.

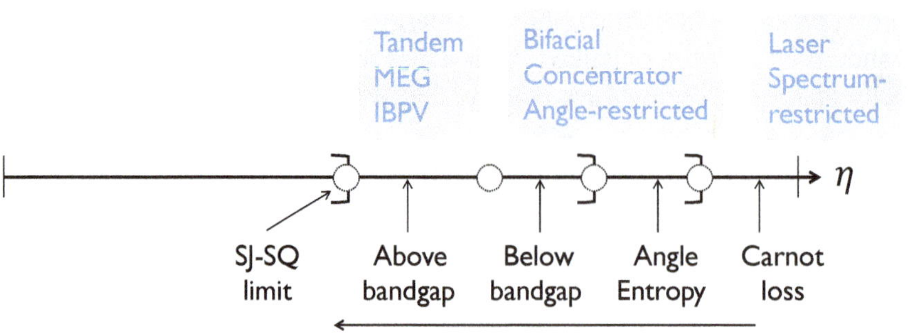

Figure 2.11: A summary of various loss mechanisms that reduce the efficiency of a solar cell is listed below the efficiency line. The techniques to improve the cell performance by suppressing specific loss mechanisms are listed above the efficiency line. These include tandem solar cells, multiple-exciton generation (MEG) solar cells, intermediate-bandgap solar (IBPV) cells, bifacial solar cells (which accept light from both sides), concentrator and angle-restricted solar cells, etc. Finally, one can exceed the Carnot limit by replacing the sunlight by spectrum-restricted sources, such as lasers.

References

[1] Muhammad A. Alam and M. Ryyan Khan. Fundamentals of PV efficiency interpreted by a two-level model. *American Journal of Physics*, 81(9):655–662, September 2013.

[2] Joseph Noyes, Manfred Sumper, and Pete Vukusic. Light manipulation in a marine diatom. *Journal of Materials Research*, 23(12):3229–3235, January 2011.

[3] Richard P. Feynman, Robert B. Leighton, and Matthew Sands. *The Feynman Lectures on Physics, Vol. 3*. Addison Wesley, January 1971.

[4] Robert F. Pierret. *Advanced semiconductor fundamentals*. Prentice Hall, Upper Saddle River, N.J., 2003.

[5] F. Herrmann and P. Wurfel. Light with nonzero chemical potential. *American Journal of Physics*, 73(8):717, 2005.

[6] A. Barnett, C. Honsberg, D. Kirkpatrick, S. Kurtz, D. Moore, D. Salzman, R. Schwartz, J. Gray, S. Bowden, K. Goossen, M. Haney, D. Aiken, M. Wanlass, and K. Emery. 50% Efficient Solar Cell Architectures and Designs. In *Photovoltaic Energy Conversion, Conference Record of the 2006 IEEE 4th World Conference on*, volume 2, pages 2560–2564, May 2006.

[7] Stuart K. Stubbs, Samantha J. O. Hardman, Darren M. Graham, Ben F. Spencer, Wendy R. Flavell, Paul Glarvey, Ombretta Masala, Nigel L. Pickett, and David J. Binks. Efficient carrier multiplication in InP nanoparticles. *Physical Review B*, 81(8):081303, February 2010.

[8] Sung Jin Kim, Won Jin Kim, Yudhisthira Sahoo, Alexander N. Cartwright, and Paras N. Prasad. Multiple exciton generation and electrical extraction from a PbSe quantum dot photoconductor. *Applied Physics Letters*, 92(3):031107–031107–3, January 2008.

[9] D. König, K. Casalenuovo, Y. Takeda, G. Conibeer, J.F. Guillemoles, R. Patterson, L.M. Huang, and M.A. Green. Hot carrier solar cells: Principles, materials and design. *Physica E: Low-dimensional Systems and Nanostructures*, 42(10):2862–2866, September 2010.

[10] T.T. Chow. A review on photovoltaic/thermal hybrid solar technology. *Applied Energy*, 87(2):365–379, February 2010.

[11] W. Spirkl and H. Ries. Solar thermophotovoltaics: An assessment. *Journal of Applied Physics*, 57(9):4409–4414, May 1985.

[12] Louise C. Hirst and Nicholas J. Ekins-Daukes. Fundamental losses in solar cells. *Progress in Photovoltaics: Research and Applications*, 19(3):286–293, 2011.

[13] William H. Press. Theoretical maximum for energy from direct and diffuse sunlight. *Published online: 23 December 1976; | doi:10.1038/264734a0*, 264(5588):734–735, December 1976.

[14] Jonathan P. Dowling, Marlan O. Scully, and Francesco DeMartini. Radiation pattern of a classical dipole in a cavity. *Optics Communications*, 82(5–6):415–419, May 1991.

[15] Nathaniel M. Gabor, Zhaohui Zhong, Ken Bosnick, Jiwoong Park, and Paul L. McEuen. Extremely Efficient Multiple Electron-Hole Pair Generation in Carbon Nanotube Photodiodes. *Science*, 325(5946):1367–1371, September 2009.

[16] Biswajit Ray, Pradeep R. Nair, R. Edwin Garcia, and Muhammad A. Alam. Modeling and optimization of polymer based bulk heterojunction (BH) solar cell. pages 1–4. IEEE, December 2009.

CHAPTER 3

Thermodynamic Limits of 3D Solar Cells

Chapter Summary

- Practical solar cells have two bands of energies (i.e. conduction and valence bands). The efficiency calculation involves a simple generalization of the concepts already developed for 2-level systems.
- The short-circuit current reduces and open-circuit voltage increases with increasing bandgap. The efficiency is maximized when the bandgap equals the average photon energy from the sun.
- An inventory of the various sources of energy losses provides a deep insight regarding the operation of the solar cells and opportunities for efficiency improvement.
- Exact efficiency calculation involves complex integrals, yet with a few simple approximations, the final results can be expressed in simple analytical forms.

3.1 Introduction: Real solar cells have more than two levels

In Chapter 2, we explained that the thermalization, angle entropy, and sub-bandgap losses erode the efficiency of an idealized (i.e., two-level) solar cell. The explanation was physically intuitive but somewhat contrived and qualitative. In this chapter, we quantitatively explain why the thermodynamic efficiency limit of a **single-junction** solar cell is approximately 30–33%. Chapter 4 focuses on the techniques to improve the conversion efficiency.

We first calculate the photon fluxes absorbed and emitted by a perfectly absorbing (i.e., blackbody) 3D solar cell. The difference between the voltage-independent absorption and voltage-dependent emission fluxes defines the current–voltage characteristics (J–V) extracted from a solar cell. We explain how these theoretical J–V characteristics relate to the actual J–V characteristics measured in the laboratory. The *J–V* characteristics determine the power output ($P_{out} = J \times V$) and the efficiency ($\eta = P_{out}/P_{in}$) of the solar cell. A careful analysis of the power loss components explains how η reduces from 100% to 30–33%.

3.2 The *J–V* characteristics of a solar cell

The photocurrent of a solar cell is given by

$$J(V) = J_{abs} - J_{emi}(V) \tag{3.1}$$

i.e., the difference between absorption and emission fluxes (see Fig. 3.1(a)). A semiconductor can absorb and emit photons only with $E \geq E_g$; therefore, Eq. (3.1) can be rewritten as

$$J(V) = q \int_{E_g}^{\infty} (n_S(E) + n_{amb}(E) - n_D(E,V))\, dE \tag{3.2}$$

where $n_S(E)$ and $n_{amb}(E)$ are the fluxes of solar and ambient photons with energy E absorbed by the solar cell, and $n_D(E,V)$ is the flux of photons with energy E emitted by the solar cell.

To calculate $n_S(E)$ and $n_D(E)$, we use the generalized Planck's law (see Eq. (2.4)) for the flux of the photons emitted by a blackbody within solid angle θ:

$$n_{ph}(E,T,\mu,\theta) = \tilde{n}^2 \frac{2\theta}{c^2h^3} \frac{E^2}{\exp((E-\Delta\mu)/k_BT) - 1}. \tag{3.3}$$

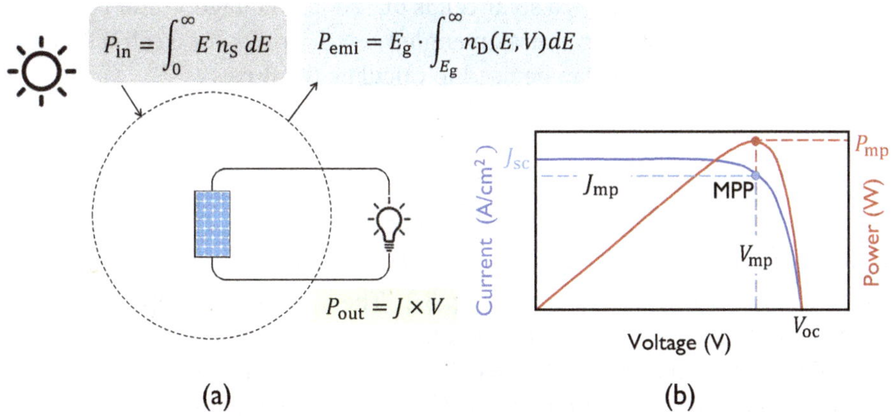

Figure 3.1: (a) Power flux balance of a solar cell. The power fluxes are obtained by integrating over the photon spectrum shown in Fig. 3.2 and Table 3.1. (b) The current–voltage and power–voltage characteristics of a solar cell operating at the thermodynamic limit. The efficiency is maximized when the solar cell operates at the maximum power point voltage (V_{mp}) and maximum power point current (I_{mp}).

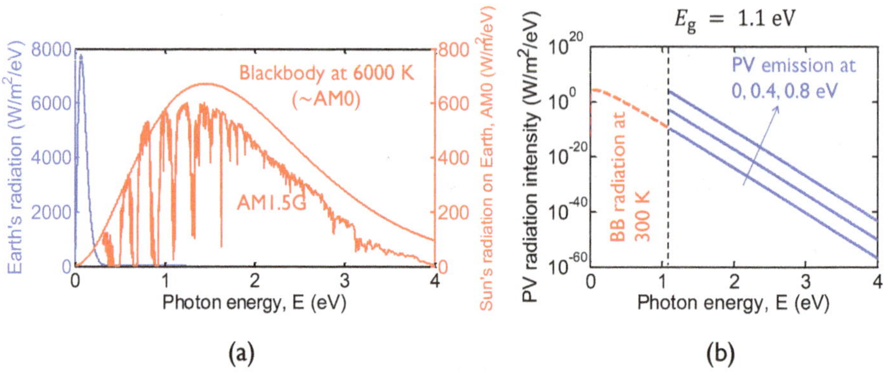

Figure 3.2: (a) The red line indicates the emission spectra ($E \cdot n_S$) of an idealized blackbody at 6000 K, radiating within a solid angle of $\theta_S \approx 6.5 \times 10^{-5}$. For reference, the spectrum that transmits through the atmosphere (AM1.5) is also shown. In contrast, the blue line on the far left indicates the earth's radiation ($E \cdot n_D(E, V = 0)$), i.e., radiation from an imperfect blackbody (BB) at 300 K emitting at a solid angle $\theta_D = 2\pi$ sr. (b) The earth's radiation is replotted in a log-linear scale. Any material with a bandgap (E_g) radiates only above its band edge, e.g., $E > E_g$ eV; therefore, the device emission spectra (blue line) coincides with the blackbody spectra (dashed red line) only above the bandgap. The device radiation $E \times n_D(V)$ rises exponentially with the applied voltage, e.g., $V = 0.4$ V and $V = 0.8$ V.

Here, k_B is the Boltzmann constant, T is the blackbody temperature, $\tilde{n}$ is the refractive index of the material, and μ is the chemical potential. The incoming and outgoing fluxes are calculated across a plane in free space; therefore $\tilde{n} \approx 1$.

We recall from Fig. 2.2(c) that the sun is a blackbody at $T_S \approx 6000$ K, and $\mu \approx 0$. Sunlight is incident on Earth with a very small solid angle $\theta_S \approx 6.5 \times 10^{-5}$ steradians. On the other hand, a solar cell is biased at a voltage V and operates at $T_D \approx 300$ K, with an isotropic emission angle $\theta_D \approx 4\pi$ (or 2π for a cell with a back mirror). Therefore, Eq. (3.3) can be used to calculate the fluxes:

$$
\begin{aligned}
n_S(E) &\equiv n_{ph}(E, T_S, \Delta\mu = 0, \theta_S) \\
n_{amb}(E) &\equiv n_{ph}(E, T_D, \Delta\mu = 0, \theta_D) \\
n_D(E, V) &\equiv n_{ph}(E, T_D, \Delta\mu = qV, \theta_D).
\end{aligned}
\tag{3.4}
$$

The incident solar intensity spectrum ($I_0 = E \times n_S$) and solar cell radiation spectrum ($I_{emi} = E \times n_D$) are shown in Fig. 3.2. The spectrum I_0 approximates the extraterrestrial spectrum AM0 (air mass 0), as was discussed in Sec. 1.3.1.

Inserting Eqs. (3.4) in Eq. (3.2), we calculate the *J*–*V* characteristics of the solar cell. For a semiconductor with bandgap E_g, the integral is evaluated numerically: Assume V, calculate $n_S(E)$ and $n_D(E, V)$ for each energy $E > E_g$, and then integrate over the energies to calculate $J(V)$. A typical result is shown in Fig. 3.1(b).

The technique we just used to calculate the *J*–*V* characteristics is sometimes called a "detailed balance" approach. The term is misleading: the principle of detailed balance, namely, that every process and its inverse must balance in every point in space and for every pair of energies, holds only in equilibrium. A solar cell under illumination is obviously not in equilibrium. The term loosely describes the fact that at the open-circuit condition ($J(V_{oc}) = 0$), the fluxes generated by photons and lost due to radiative recombination cancel each other. In this sense, one could suggest that the notion of detailed balance holds, even though fluxes do not balance between every pair of energy states. For $V \neq V_{oc}$, an external current flows in the circuit and photon fluxes are no longer balanced and the system cannot be described by the principle of detailed balance. Therefore, the general use of the term "detailed balance model" to describe the operation of a solar cell under all bias conditions is a misnomer and should be treated as such.

3.3 Power output and conversion efficiency of a solar cell

The numerical *J–V* characteristics from the previous section allow us to calculate the bandgap- and voltage-dependent efficiency of the solar cell:

$$\eta(E_g, V) = \frac{J(E_g, V) \times V}{P_{in}}, \tag{3.5}$$

where the input power, P_{in}, is obtained by summing over the entire solar spectrum, namely,

$$P_{in} = \int_0^\infty E \cdot n_S(E)\, dE. \tag{3.6}$$

The conversion efficiency $\eta_{max}(E_g, V_{mp})$ is maximized at maximum power point voltage (V_{mp}) and current (J_{mp}) (see Fig. 3.1(b)).

We can calculate the *J–V* characteristics and plot $P_{out}^{max}(E_g, V_{mp})$ for various semiconductors with bandgap E_g; see the green-shaded curve marked P_{out} in Fig. 3.3(a). We see that $\eta_{max}(E_g)$ reaches its maximum of $\eta_{SQ} \approx 30\%$ at $E_g \approx 1.3$ eV. This limiting performance of a single-material solar cell is known as the Shockley–Quiesser (SQ) limit. The result was first published in 1961, just a few years after solar cells were invented at Bell Laboratories. Since the *J–V* characteristics and P_{in} both depend on the solar spectrum (e.g., AM0, AM1.5G, etc.), so do the efficiency (~30–33%).

3.4 Power budget: How did we lose 70% of the incident sunlight?

If the maximum efficiency of a solar cell is 33%, then the most efficient solar cell still fails to convert two-thirds of the solar energy. In Chapter 2, we identified the key loss mechanisms: thermalization, angle entropy, and sub-bandgap losses (although it is difficult to identify these losses from Eq. (3.2)). Actually, the losses can be attributed to the seven factors summarized in Table 3.1 and shown in Fig. 3.3. These include Carnot loss (P_{Carnot}) due to temperature mismatch, sub-bandgap

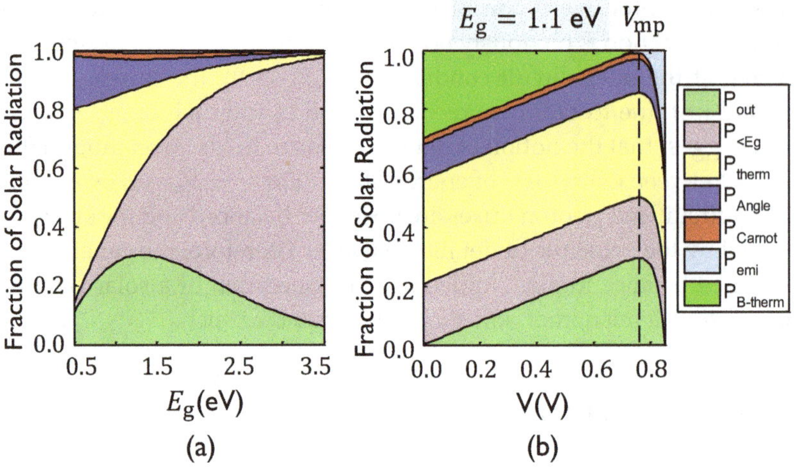

Figure 3.3: (a) Solar cell power components at the maximum power point plotted as a function of bandgap E_g. (b) The power budget for $E_g = 1.1\,\text{eV}$ for various bias conditions. The components at the maximum power point are indicated by the dashed line. Table 3.1 summarizes the formula used to calculate the power components.

loss ($P_{<E_g}$) due to unabsorbed photons, angle entropy loss (P_{angle}) due to angle randomization of the emitted photons, first thermalization loss (P_{therm}) associated with carrier relaxation to the band edge, and the subsequent thermalization loss $P_{\text{B-therm}}$ due to carrier relaxation from the band edge to the contacts. If we sum up the output power and the losses for a given material (i.e., given E_g) at any operating voltage, we can account for every photon in the solar spectrum:

$$P_{\text{in}} = P_{\text{out}} + P_{<E_g} + P_{\text{therm}} + P_{\text{emi}} + P_{\text{angle}} + P_{\text{Carnot}} + P_{\text{B-therm}}. \tag{3.7}$$

These losses are unavoidable for a single-junction solar cell even with ideal (perfectly absorbing, infinite mobility) materials.

At the maximum power point defined by V_{mp} and J_{mp}, the below-bandgap loss $P_{<Eg}$ and emission loss P_{emi} restrict the number of carriers collected (i.e., limit the current), and P_{therm}, P_{angle}, and P_{Carnot} restrict the average energy of the collected carriers (i.e., reduce the voltage). The power budget at the optimum power point is shown in Fig. 3.3(a).

Figure 3.3(b) and Eq. (3.15) show that $P_{\text{B-therm}} \neq 0$ for $V < V_{\text{mp}}$. Here, $P_{\text{B-therm}}$ indicates the extra carrier relaxation due to band bending. For $V > V_{\text{mp}}$, the current reduces quickly, and the increased recombination (i.e., emission) is reflected in P_{emi}. Note that the band bending is in the opposite direction for $V > V_{\text{mp}}$ — thus carriers gain energy as they "climb up" the band before extraction, which in turn gives a (small) negative value for $P_{\text{B-therm}}$. The reduced net thermalization ($P_{\text{therm}} + P_{\text{B-therm}}$) at $V > V_{\text{mp}}$ is positive.

Table 3.1: Power budget for solar cell operation.

Incident power	$$P_{\text{in}} = \int_0^{\infty} n_S E\, dE \quad (3.8)$$ The total incident power ($T_S = 6000$ K) is $\sim 1500\ \text{W/m}^2$. However, in practice, AM1.5G integrates to $1000\ \text{W/m}^2$.
Power out	$$P_{\text{out}} = J \times V \quad (3.9)$$ The output power varies with V and maximizes at V_{mp}.
Below bandgap	$$P_{<E_g} = \int_0^{E_g} n_S E\, dE \quad (3.10)$$ The photons with $E < E_g$ cannot be absorbed by the solar cell.
Thermalization loss	$$P_{\text{therm}} = \int_{Eg}^{\infty} ((E - E_g) n_S) dE \quad (3.11)$$ Carriers generated by high energy ($E > E_g$) photons thermalize to the band edge giving out energy to phonons.
Emission loss	$$P_{\text{emi}} = E_g \times \int_{E_g}^{\infty} n_D(V) dE \quad (3.12)$$ This counts for the power lost to the unavoidable radiative recombination.
Angle anisotropy	$$P_{\text{angle}} = k_B T_S \ln\left(\frac{\theta_D}{\theta_S}\right) \times \left(\frac{J}{q}\right) \quad (3.13)$$ The mismatch in solid angle of the incident and emitted light (θ_S and θ_D) increases entropy.
Carnot loss	$$P_{\text{Carnot}} = E_g \left(\frac{T_D}{T_S}\right)\left(\frac{J}{q}\right) \quad (3.14)$$ This is the fraction of power that cannot be converted to work due to finite difference in temperatures of T_S and T_D.
Thermalization at band edge	$$P_{\text{B-therm}} = J \times (V_{\text{mp}} - V) \quad (3.15)$$ This is the extra thermalization of carriers due to band bending away from V_{mp}.

Homework 3.1: PVLimits: A thermodynamic limit calculator

Use **PVLimits** to re-create Fig. 3.3.

1. Log in to `https://nanohub.org/`. Go to the link `https://nanohub.org/resources/pvlimits`.
2. Download the user manual posted at https://nanohub.org/resources/24377/download/Documentation_pvlimits-v2.pdf.
3. Read Secs. 1 and 2 to become familiar with the operation of the simulator and the meaning of various symbols.
4. Work out the examples in Sec. 3.
5. Redefine the input parameters to re-create Fig. 3.3.
6. How would the numbers change if you were trying to operate a Mars rover using a solar cell?
   Hint: You will need to change the spectrum from AM1.5G to blackbody spectrum and change the "Distance from the sun" to 227.9 million km (see Chapter 1; Table 1.1 and Fig. 1.6). On Mars, $T_D \approx 293\,\text{K}$ at midday near the equator (but it plummets to 200 K during the night).

Homework 3.2: Solar cell in another solar system

What would be the ideal bandgap of a solar cell if we traveled to an exo-planet orbiting a red dwarf? Assume that the "solar" spectrum is given by a black-body temperature $T_S = 3000\,\text{K}$ and the solar cell is kept at $T_D = 300\,\text{K}$ and illuminated by the "sunlight" intensity of 1 kW/m^2.

Solution. We could answer the question by recalculating and replotting Fig. 3.3 or using **PVLimits** program, discussed in Homework 3.1. Interestingly, we can obtain an approximate answer as follows: The efficiency of a solar cell is maximized when its bandgap equals the average energy of the blackbody spectrum; see Eq. (1.16). In other words, $E_g^{\text{opt}} = E_{\text{avg}} = 2.71 k_B T_S$ applies over a fairly broad range of source temperature T_S. The optimum bandgap of a solar cell illuminated by the sun is $E_g^{\text{opt}} = 2.71 \times 8.617 \times 10^{-5} \times 5777 = 1.35$ eV, approximately. The result compares very well with the numerically computed optimum bandgap shown in Fig. 3.3(a). Similarly, for the red dwarf, $E_g^{\text{opt}} = 2.7 \times 8.61 \times 10^{-5} \times 3000 = 0.7$ eV. When the source temperature is reduced from 6000 K to 3000 K, the numerical simulator **PVLimits** also finds that $E_g^{\text{opt}} \approx 0.7$ eV. The *exact* match is accidental — after all, Eq. (1.16) does not account for smaller corrections associated with the device temperature T_D and the intensity of the red dwarf.

Homework 3.3: It is important to carefully define the cell efficiency

A filter is placed in front of a solar cell illuminated by blackbody radiation, but the short-circuit current of the solar cell remains unchanged. Can you explain why? Since the power incident on the solar cell is lower, would you say that the cell is more efficient?

Solution. The filter must have cut off the below-bandgap photons that contributed to P_{in}. Any modification of the above-bandgap absorption or radiative emission would have changed the short-circuit current, or equivalently, P_{out}. The cell appears "more efficient" because it produces the same P_{out} with reduced P_{in}. Clearly, this efficiency increase is misleading.

Similarly, the SQ limit of a solar cell illuminated by the AM1.5G spectrum increases to 33% simply because the atmosphere acts as a high-pass filter, i.e., it scatters and absorbs below-bandgap (larger wavelength) photons more effectively than above-bandgap photons. Here, both P_{in} and P_{out} are reduced, but P_{out} reduces slightly less than P_{in} does. This also explains the puzzle why the solar cell is "more efficient" in the early morning and late afternoon — because sunlight takes a longer path through the atmosphere at an angle and the below-bandgap photons are disproportionately suppressed.

3.5 Key features of the *J*–*V* characteristics can be calculated analytically

The *J*–*V* curve based on Eq. (3.2) can be calculated analytically if we make the "Boltzmann approximation" and neglect the -1 term in the denominator of Eq. (3.3). The three terms now involve simple integrals of the form $E^2\, e^{-E/k_{\text{B}}T}$. Therefore, Eqs. (3.4) are easily integrated to analytically calculate the *J*–*V* characteristics as well as other quantities, such as voltage- and bandgap-dependent efficiency $\eta(E_{\text{g}}, V) = JV/P_{\text{in}}$ (see Homework 3.4). Once the integration is complete, you will find that the definition

$$\gamma_i(E_{\text{g}}) \equiv \frac{2k_{\text{B}}T_{\text{i}}}{c^2h^3}\left(E_{\text{g}}^2 + 2k_{\text{B}}T_{\text{i}}E_{\text{g}} + 2k_{\text{B}}^2T_{\text{i}}^2\right),$$

related to 3D blackbody spectrum and evaluated either for the sun (γ_{S} with $T_i = T_{\text{S}}$) or for the solar cell (γ_{D} with $T_i = T_{\text{D}}$), allows us to write the analytical expression for *J*–*V* and η compactly. Based on the analytical expression (which is left as an exercise for you), you will be able to calculate the following characteristic points of the *J*–*V* curve:

- Since $\eta(E_{\text{g}}, V)$ is maximized at the maximum power point. Therefore, we can calculate the maximum power point voltage (V_{mp}) for any E_{g} by setting

$(\partial\eta/\partial E_g)_V = 0$ or, $(\partial\eta/\partial V)_{E_g} = 0$:

$$V_{mp} = \frac{E_g}{q}\left(1 - \frac{T_D}{T_S}\right) - \frac{k_B T_D}{q}\ln\left(\frac{\theta_D}{\theta_S}\right). \tag{3.16}$$

The fraction $(E_g/q)(T_D/T_S)$ in the first term is the Carnot loss, and the second term, involving the ratio of the solid angles, is the angle entropy loss. This interpretation should sound familiar, because we qualitatively derived exactly the same expression for the two-level system (see Eq. (2.26) in Chapter 2).

- The corresponding maximum power point current J_{mp} is obtained by setting $J_{mp} = J(V = V_{mp})$, and we find that

$$J_{mp}(E_g) = q\theta_S\left[\gamma_S(E_g) - \gamma_D(E_g)\right] e^{-E_g/k_B T_S}. \tag{3.17}$$

- The short-circuit current (J_{sc}) is obtained by setting $V = 0$ in Eq. (3.2), i.e.,

$$J_{sc} = q\theta_S\gamma_S(E_g)e^{-Eg/k_B T_S}. \tag{3.18}$$

- Finally, the open-circuit voltage V_{oc} is obtained by setting $J = 0$ in Eq. (3.2). After some careful algebra, you will find that

$$V_{oc} = \frac{E_g}{q}\left(1 - \frac{T_D}{T_S}\right) - \frac{k_B T_D}{q}\ln\left(\frac{\theta_D}{\theta_S}\right) + \frac{k_B T_D}{q}\ln\left(\frac{\gamma_S}{\gamma_D}\right). \tag{3.19}$$

Comparing these results, we see that $J_{sc} \simeq J_{mp}$: they differ by the (small) device-specific emission term $\gamma_D(E_g)$ at V_{mp}. Moreover, V_{oc} and V_{mp} are described by very simple and nearly identical expressions. The difference between them is given by the third term in Eq. (3.19). The third term describes an increase in the free energy per carrier, resulting from the mismatch between the temperatures of the absorbed and emitted photon distributions.

Given the critical points of the *J–V* characteristics, the maximum cell efficiency is obtained by evaluating Eq. (3.5) at the maximum power point:

$$\eta_{max}(E_g) = \frac{J_{mp}(E_g)V_{mp}(E_g)}{P_{in}}. \tag{3.20}$$

Similarly, the fill factor (FF) can be calculated by

$$\text{FF} \equiv \frac{J_{mp}V_{mp}}{J_{sc}V_{oc}} \approx \left[1 - \frac{\gamma_D(E_g)}{\gamma_S(E_g)}\right]\left[\frac{V_{mp}}{V_{mp} + (k_B T_D/q)\ln(\gamma_S/\gamma_D)}\right]. \tag{3.21}$$

The analytical expressions are correct within 1–2% of the exact result for the practical bandgaps, i.e., $E_g > 0.5$ eV.

Homework 3.4: An analytical expression for the thermodynamic efficiency

Calculate the expressions for P_{in}, P_{out}, and η.

Solution. From the discussion above, we know that

$$\begin{aligned}P_{\text{in}} &= \int_0^\infty \theta_{\text{S}}\, n_{\text{ph}}(E, T_{\text{S}}, \mu = 0, \theta_{\text{S}}) dE \\ &= \int_0^\infty \frac{2\theta_{\text{S}}}{c^2 h^3} \frac{E^2}{\exp(E/k_{\text{B}}T) - 1} dE \\ &= \frac{\pi^4}{15} \frac{2\theta_{\text{S}}}{c^2 h^3} (k_{\text{B}} T_{\text{S}})^4. \because \int_0^\infty x^3 dx/(e^x - 1) = \pi^4/15. \end{aligned} \tag{3.22}$$

You may recognize the result as the Stefan–Boltzmann equation (i.e., $P_{\text{in}} = \sigma T_{\text{S}}^4$) discussed in Chapter 1. Similarly,

$$\begin{aligned}P_{\text{out}} &= \frac{2qV}{c^2 h^3} \left[\int_{E_{\text{g}}}^\infty \frac{\theta_{\text{S}} E^2 dE}{\exp(E/k_{\text{B}}T_{\text{S}}) - 1} - \int_{E_{\text{g}}}^\infty \frac{\theta_{\text{D}} E^2 dE}{\exp((E - qV)/k_{\text{B}}T_{\text{D}})} \right] \\ &\sim \frac{2qV}{c^2 h^3} \left[\int_{E_{\text{g}}}^\infty \frac{\theta_{\text{S}} E^2 dE}{\exp(E/k_{\text{B}}T_{\text{S}})} dE - \left(e^{qV/k_{\text{B}}T_{\text{D}}} \right) \int_{E_{\text{g}}}^\infty \frac{\theta_{\text{D}} E^2 dE}{\exp(E/k_{\text{B}}T_{\text{D}})} \right] \end{aligned} \tag{3.23}$$

$$= \frac{2qV}{c^2 h^3} (k_{\text{B}} T_{\text{S}})^3 \theta_{\text{S}} \gamma_{\text{S}} e^{-E_{\text{g}}/k_{\text{B}}T_{\text{S}}} \left[1 - \frac{\theta_{\text{D}}}{\theta_{\text{S}}} \frac{\gamma_{\text{D}}}{\gamma_{\text{S}}} \left(\frac{T_{\text{D}}}{T_{\text{S}}} \right)^3 F \right] \tag{3.24}$$

where $F \equiv e^{qV/k_{\text{B}}T_{\text{D}}} \times e^{-(E_{\text{g}}/k_{\text{B}}T_{\text{D}})(1 - T_{\text{D}}/T_{\text{S}})}$. The efficiency is defined by the ratio of the input and the output power.

3.6 Conclusions

In this chapter, we derived the *J–V* relationship of a solar cell by considering the balance of incoming and outgoing power fluxes. An analysis of the power budget (Eq. (3.7), Table 3.1, and Fig. 3.3) explains how various loss components (thermalization loss, angle entropy loss, below-bandgap loss, etc.) restrict the power conversion efficiency to ~30% under 6000 K blackbody radiation (or, ~33% under AM1.5G). The efficiency limit is variously known as Shockley–Quiesser limit (because they first derived it), the radiative limit (because the idealized cell had no other recombination or loss mechanism), and single-junction thermodynamic limit (because the analysis is based on a single semiconductor with bandgap, E_{g}).

In practice, a practical cell operates considerably below the SQ limit due to the increased cell temperature due to thermalization loss (to be discussed in Chapter 5), incomplete photon absorption (to be discussed in Chapter 6), and so on.

In Chapter 2, we qualitatively explained that we can improve the power conversion efficiency by using a number of strategies, such as using solar cells with multiple bandgaps (tandem PV), concentrating the incident sunlight (concentrator PV), or restricting the emission angle (see Fig. 2.11). In the next chapter, we will use the analytical formula derived in Sec. 3.5 for $J_{\rm mp}$ and $V_{\rm mp}$ to calculate the efficiency limits of these advanced solar cell concepts (e.g., multi-junction tandem, concentrator PV, bifacial solar cells, etc.) that are far more efficient in converting sunlight to electricity. We will also introduce a new graphical technique called **S-Q triangle** to calculate the thermodynamic efficiency limit of these highly efficient multi-junction solar cells.

Homework 3.5: $V_{\rm oc}$, $V_{\rm mp}$, and FF can be expressed in simple forms

Find simple analytical expressions for $V_{\rm mp}$, $V_{\rm oc}$, and FF for a terrestrial solar cell. Determine the maximum FF for the bandgap range of 0.5 eV $< E_{\rm g} <$ 2.0 eV. How would the expressions change for a solar cell placed on a Mars rover? Recalculate the expressions when the back-mirror is absent and emission is isotropic, i.e., $\theta_{\rm D} = 4\pi$. How do the results compare to those from a two-level system discussed in Chapter 2, Eq. (2.27)?

Solution. For $T_{\rm S} \approx 6000\,\text{K}$, $T_{\rm D} = 300\,\text{K}$, and $1 < E_{\rm g} < 2$ eV, we find that $\gamma_{\rm S}/\gamma_{\rm D} \approx 40 \pm 10$. Therefore, $(k_{\rm B}T_{\rm D}/q)\ln(\gamma_{\rm S}/\gamma_{\rm D}) \approx 0.095$ V. If a back-mirror is present, $\theta_{\rm D}(=2\pi)$. Inserting these values in Eqs. (3.16), (3.19), and (3.21), we find that

$$V_{\rm mp} = 0.95\frac{E_{\rm g}}{q} - 0.296$$

$$V_{\rm oc} = 0.95\frac{E_{\rm g}}{q} - 0.201$$

$$\text{FF} \approx 0.975 \times \frac{V_{\rm mp}}{V_{\rm mp} + 0.095}$$

FF increases with the bandgap and approaches 90% for $E_{\rm g} \approx 2$ eV.

Homework 3.6: J_{sc} and J_{mp} depend on the incident spectrum

Find approximate expressions for J_{sc} photo-excited by (a) a blackbody (AM0) spectrum vs. (b) the AM1.5G spectrum.

Solution. We can use Eq. (3.18) to calculate and plot J_{sc} under AM0 illumination (see the figure below). If $x \equiv E_g/k_BT_S \sim 2E_g$, then J_{sc} is approximated by

$$J_{sc} \simeq \begin{cases} 120(1-0.25x), & \text{if } x \leq 3 \\ 44.33(x^2+2x+2)e^{-x}, & \text{if } x \geq 3 \end{cases} \tag{3.25}$$

Similarly, there is no analytical expression for J_{sc} with AM1.5G spectrum. However, Web-based calculators (e.g., `https://www.pveducation.org/pvcdrom/solar-cell-operation/short-circuit-current`) can be used to plot J_{sc} vs. E_g (see the figure below). The maximum current saturates to ~70 mA/cm^2, and the following expression holds for $0.5 < E_g < 2$ eV:

$$J_{sc} \simeq J_{sun}(1-\beta_{sun}E_g). \tag{3.26}$$

Here, J_{sun} is the projected current at $E_g \to 0$, and $\beta_{sun}^{-1} \sim 4.7k_BT_S = 2.35\,\text{eV}^{-1}$ accounts for the reduction of J_{sc} with increasing bandgap.

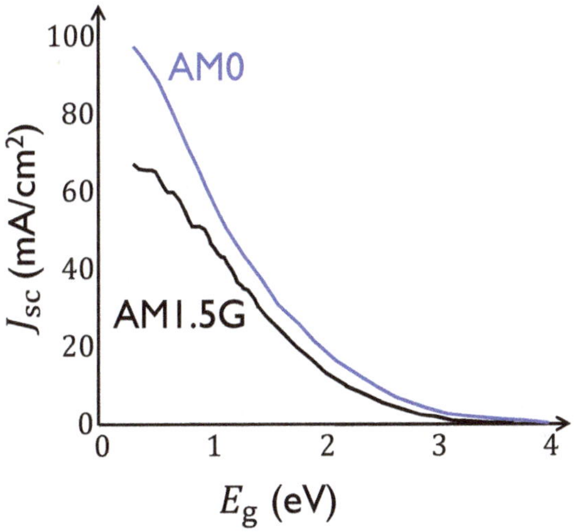

The short-circuit current plotted as a function of the bandgap.

Homework 3.7: Solar cells are more efficient on earth than in outer space!

A cell is tested with AM0 (extraterrestrial) and AM1.5G (terrestrial) spectra. How would the efficiencies compare?

Solution. AM1.5G-illuminated solar cells will be more efficient. Figure 3.2(a) shows that the scattering and absorption by the atmosphere reduce the overall intensity of AM1.5G spectrum. Still, the efficiency is higher because P_{in} (in the denominator) reduces faster than P_{out} (current × voltage). This is because lower-energy sub-bandgap photons are disproportionately affected by absorption, see the spikes in Fig. 3.2(a) spectrum.

Homework 3.8: Thermodynamic vs. classical *J–V* characteristics

Show that Eq. (3.2) can be expressed in a more familiar form in terms of photocurrent (J_{ph}) and dark current (J_{dark}):

$$J(V) = J_{\text{ph}} - J_{\text{dark}}(V)$$

where $J_{\text{dark}} \equiv J_0 \left(e^{qV/k_B T_D} - 1\right)$.

Solution. Equation (3.2) includes three terms: absorption from the sun $n_S(E)$, absorption from the environment $n_{\text{amb}}(E)$, and the emission by the solar cell $n_D(E, V)$. We first rearrange the terms:

$$J(V) = q \int_{E_g}^{\infty} n_S(E)dE - \int_{E_g}^{\infty} (n_D(E,V) - n_{\text{amb}}(E))dE.$$

Then we define the photocurrent as

$$J_{\text{ph}} \equiv q \int_{E_g}^{\infty} n_S(E)dE$$

and the dark current as

$$\begin{aligned} J_{\text{dark}}(V) &= q \int_{E_g}^{\infty} \left[n_D(E,V) - n_{\text{amb}}(E)\right] dE \\ &\approx q \int_{E_g}^{\infty} \left[n_D(E, V=0)e^{qV/k_B T_D} - n_{\text{amb}}(E)\right] dE \\ &= \left[q \int_{E_g}^{\infty} n_{\text{amb}}(E)dE\right] \left(e^{qV/k_B T_D} - 1\right) \\ &\equiv J_0 \left(e^{qV/k_B T_D} - 1\right). \end{aligned} \quad (3.27)$$

(continued on the next page)

Homework 3.8 *(continued from the previous page)*

Here, J_0 is known as the reverse saturation current density. In the derivation above, we have made two assumptions. First, $n_\text{D}(V) = n_\text{D}(V = 0)e^{qV/k_\text{B}T_\text{D}}$ is obtained by using Eq. (3.4) and neglecting the -1 term in the denominator of Eq. (3.3) (Boltzmann approximation). Second, we have assumed that $T_\text{D} = T_\text{amb}$ so that $n_\text{D}(E, V = 0) = n_\text{amb}(E)$. After all, in equilibrium (without sunlight and $V = 0$), the absorption from the environment must balance the emission from the solar cell, so that $J_\text{dark} = 0$.

The thermodynamic and classical $J(V)$ characteristics have the same form, but J_0 is numerically different. The difference reflects a deep and subtle issue involving how photons are absorbed and recycled in a solar cell. The lower thermodynamic J_0 suggests a more optimistic V_oc compared to a classical model. We will return to this topic in Part II of the book.

References

[1] William Shockley and Hans J. Queisser. Detailed Balance Limit of Efficiency of p-n Junction Solar Cells. *Journal of Applied Physics*, 32(3):510, 1961.

[2] Louise C. Hirst and Nicholas J. Ekins-Daukes. Fundamental losses in solar cells. *Progress in Photovoltaics: Research and Applications*, 19(3):286–293, 2011.

[3] Muhammad Ashraful Alam and M. Ryyan Khan. Thermodynamic efficiency limits of classical and bifacial multi-junction tandem solar cells: An analytical approach. *Applied Physics Letters*, 109(17):173504, October 2016.

[4] Robert F. Pierret. *Advanced semiconductor fundamentals.* Prentice Hall, Upper Saddle River, N.J., 2003.

[5] G. Yu, J. Gao, J. C. Hummelen, F. Wudl, and A. J. Heeger. Polymer Photovoltaic Cells: Enhanced Efficiencies via a Network of Internal Donor-Acceptor Heterojunctions. *Science*, 270(5243):1789–1791, December 1995.

[6] L. J. A. Koster, E. C. P. Smits, V. D. Mihailetchi, and P. W. M. Blom. Device model for the operation of polymer/fullerene bulk heterojunction solar cells. *Physical Review B*, 72(8):085205, August 2005.

[7] Thomas Kirchartz, Kurt Taretto, and Uwe Rau. Efficiency Limits of Organic Bulk Heterojunction Solar Cells. *The Journal of Physical Chemistry C*, 113(41):17958–17966, October 2009.

[8] M. C. Scharber, D. Mühlbacher, M. Koppe, P. Denk, C. Waldauf, A. J. Heeger, and C. J. Brabec. Design rules for donors in bulk-heterojunction solar cells — towards 10% energy-conversion efficiency. *Advanced Materials*, 18(6):789–794, 2006.

[9] Barry P. Rand, Diana P. Burk, and Stephen R. Forrest. Offset energies at organic semiconductor heterojunctions and their influence on the open-circuit voltage of thin-film solar cells. *Physical Review B*, 75(11):115327, March 2007.

Appendix 1: Derivation of the 3D emission spectrum of a blackbody radiator

To derive Eq. (3.3), we need to calculate the number of energy levels that emit photons within the energy band E and $E + \Delta E$. The term blackbody may be scary, but the calculation of the number of atoms emitting photons between E and $E + \Delta E$ is relatively simple. Consider a blackbody box with sides (a_x, a_y, a_z), so that $V = a_x a_y a_z$. We would like to count the number of photonic modes or states inside the box, within a wave-vector range of $[0, k]$, as follows.

$$N_{\mathrm{ph},V} = 2 \times (\text{volume in } k\text{-space}) \times (\text{no. of modes per k-space vol.}). \tag{3.28}$$

The factor of 2 accounts for the two types of light polarizations, namely, transverse electric (TE) and transverse magnetic (TM) modes. Figure 3.4 shows that the spherical volume in the k-space is given by $(4/3)\pi k^3$. The number of relevant modes per k along each axis are $a_x/2\pi$, $a_y/2\pi$, and $a_z/2\pi$ (see Chapter 4 in [4]). The number of modes $N_{\mathrm{ph},V}$ is essentially the mode count inside the sphere in the

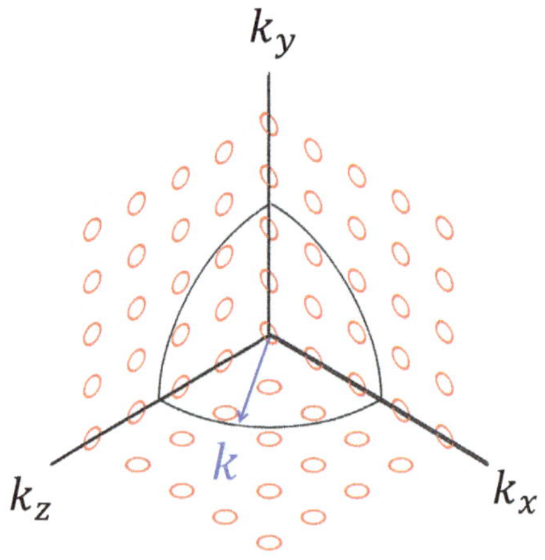

Figure 3.4: The k-space diagram is shown here. The relevant modes are inside the sphere of radius k in the first quadrant.

first quadrant in k-space. Finally, we need to normalize the mode count by the volume of the blackbody.

$$n_{\text{ph}} = \frac{N_{\text{ph},V}}{V} = \frac{2 \times \frac{4}{3}\pi k^3 \times \frac{a_x}{2\pi}\frac{a_z}{2\pi}\frac{a_z}{2\pi}}{V} = \frac{1}{3\pi^2}k^3 \tag{3.29}$$

is the number of photonic modes per unit volume of the blackbody.

The dispersion relationship of light is $\omega = (c/\widetilde{n})k$, or $E = (c/\widetilde{n})\hbar k$. Here, E (or ω) is the photon energy (or angular frequency), and $\widetilde{n}$ is the refractive index of the medium where we measure the radiation (not necessarily inside the blackbody itself). Furthermore, c is the speed of light in free space, and $h = 2\pi\hbar$ is Planck's constant. Therefore, we can find n_{ph} as a function of energy (E):

$$n_{\text{ph}}(E) = \frac{1}{3\pi^2}((\widetilde{n}/c)E/\hbar)^3 = \frac{8\pi}{3}\frac{\widetilde{n}^3}{(hc)^3}E^3. \tag{3.30}$$

All these modes create a combined isotropic (i.e., 4π sr.) radiation. Finally, the photonic density of states, defined as the number of photonic modes per energy and per solid angle can be calculated as follows:

$$D_{\text{ph}}(E) = \frac{1}{4\pi}\frac{dn_{\text{ph}}}{dE} = \frac{2\widetilde{n}^3}{(hc)^3}E^2. \tag{3.31}$$

We have determined the number of states at various energies; however, we still need to find the occupancy probability of photons at each state. Photons are bosons, and their occupancy is determined by the Bose–Einstein (BE) distribution. The generalized BE distribution is given by

$$f_{\text{BE}}(E,T) = \frac{1}{\exp((E-\mu)/k_{\text{B}}T) - 1}. \tag{3.32}$$

Here, k_{B} is the Boltzmann constant, μ is the chemical potential of photons, and T is the temperature of the blackbody. Now, the total photon emission flux in the solid angle θ, per energy interval is given by

$$n_{\text{ph}}(E,T,\mu,\theta) = \theta \times D_{\text{ph}}(E) f_{\text{BE}}(E,T) \times (c/\widetilde{n}) \tag{3.33}$$

$$= \widetilde{n}^2 \frac{2\theta}{c^2h^3}\frac{E^2}{\exp((E-\mu)/k_{\text{B}}T) - 1}. \tag{3.34}$$

This is a generalized form of Planck's law. The photon fluxes in most cases are calculated and compared in free space. Otherwise, unless explicitly mentioned, we will set $\widetilde{n} = 1$.

Homework 3.9: Solar cell is an n-D world

In the previous homework, we calculated the photon density of states for isotropic emission into x, y, z directions (3D). Imagine the solar cell in an n-dimensional world ($n = 1, 2, 3$). We have already explained the solar cell for $n =$ 3-D case. Assume that the refractive index of the surrounding medium is 1.

1. Show that $D_{\mathrm{ph}} = (2/(hc)^n)E^{n-1}$ in n-D. For $n = 3$, the expression reproduces Eq. (3.31).
2. What is the emission and acceptance angle in n-D? [Hint: Use Homework 1.1 to generalize the definition of θ_{S}, and use in a mirrorless case with $\theta_{\mathrm{D}} = 1, \pi, 2\pi$ in 1D, 2D, and 3D, respectively.]
3. Now find the $J(V)$ in the n-D case.
4. Numerically find efficiency vs. E_{g} for n-D.

Appendix 2: Single-bandgap vs. donor–acceptor heterojunction solar cell

In Sec. 2.3.1, we derived the efficiency limit (see Eq. (2.14)) of a diatomic molecular solar cell, i.e.,

$$\eta_{\mathrm{b}} = \left(1 - \frac{T_{\mathrm{D}}}{T_{\mathrm{S}}}\right) - \frac{\Delta E}{E_{\mathrm{g}}}$$

where the direct bandgap E_{g} dictates the photon absorption, while the cross-gap E_{c} defines the radiative recombination so that $\Delta E(\equiv E_{\mathrm{g}} - E_{\mathrm{c}})$ is the energy lost by electrons as they transfer from the donor to the acceptor atom. It turns out that the diatomic molecules serve as an excellent model for bulk-heterojunction organic solar cells. In this section, we will use the detailed balance approach to derive the SQ limit or radiative limit of bulk-heterojunction solar cells. The exercise will highlight the importance of carefully writing the limits of the incoming and outgoing fluxes and how the efficiency loss due to cross-gap radiation is ultimately reflected in the maximum power point voltage, V_{mp}.

The role of heterojunctions in OPVs

A bulk-heterojunction organic solar cell (BHJ-OPV) consists of two demixed, phase-segregated, bi-continuous, donor–acceptor organic semiconductors capped by a transparent anode and a metallic cathode. The operation of a BHJ-OPV is often explained as follows (see Chapter 8 for additional details): excitons generated by sunlight in such organic semiconductor composites are localized in a single

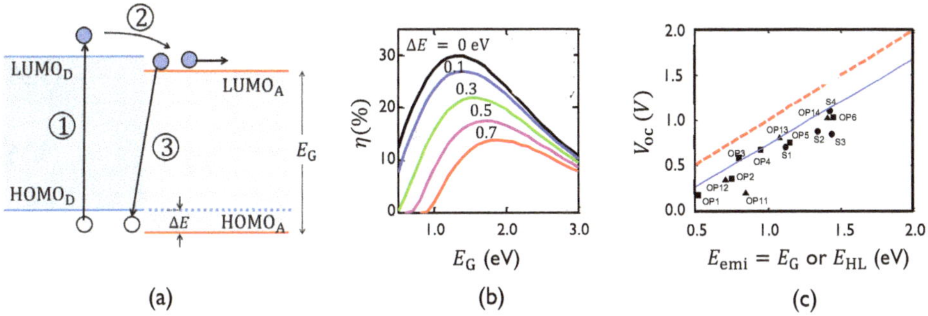

Figure 3.5: (a) A schematic band diagram of HJ-OPV. Absorption starts beyond E_g (step 1), the carrier separates at D/A interface (step 2), and emission happens at the cross-gap (step 3). The SQ efficiency limit for an OPV in (b, c). (b) When the band discontinuity is small, classical SQ analysis suggests that peak efficiency is obtained close to the bandgap of ~1.45 eV. The band discontinuity between the organic semiconductors reduces efficiency (primarily through a loss in V_{oc}) and shifts the peak efficiency curve to higher bandgaps. For $\Delta E = 0.5$ eV, maximum efficiency occurs at ~1.7 eV. (c) The relationship between V_{oc} vs. E_{emi}. Here, $E_{emi} = E_g$ for the single-bandgap semiconductor or $E_{emi} = E_c$, the cross-gap of OPV. The blue solid line represents the calculated (theoretical) approximation for V_{oc} for different E_g. The symbols are experimental values obtained from literature [8, 9].

conjugated unit of a few nanometers and held together by undiluted Coulomb attraction between the charges, typical of low-dielectric-constant materials. Therefore, if the excitons are not dissociated into free electrons and holes by the atomically sharp quasi-electric field of the donor (D)/acceptor (A) interface, they would be lost to self-recombination. Once dissociated, electrons and holes are spatially separated in their respective polymer channels and remain isolated from each other by the D-A heterojunction (which is typically a type II heterobarrier) with staggered bandgaps (see Fig. 3.5(a)). The lack of minority carriers in the electron-transporting and hole-transporting regions suppresses bulk recombination, and the high internal field created by the electrode work function difference sweeps the free carriers out before they can recombine at the interface. This efficient exciton dissociation, spatial free carrier isolation, and drift-dominated transport of photogenerated carriers explains the exceptionally higher efficiency of a BHJ-OPV compared to a single or bi-layer OPV.

Derivation of the SQ limit for an OPV

For a material with bandgap E_g, Eq. (3.2) uses the flux balance between absorption (J_{abs}) and emission (J_{emi}) to calculate the J–V characteristics of a solar cell. The fluxes are given by

$$J_{\text{abs}} = q\int_{E_g}^{\infty} (n_S(E) + n_{\text{amb}}(E))dE \tag{3.35}$$

$$J_{\text{emi}} = q\int_{E_g}^{\infty} n_D(E, V)dE. \tag{3.36}$$

Then, the radiative efficiency was determined by the ratio of the maximum power output (P_{out}) to the spectrum-integrated power input (P_{in}), i.e.,

$$\eta_{SQ} = \frac{P_{\text{out}}}{P_{\text{in}}} = \frac{V \times (J_{\text{abs}} - J_{\text{emi}}(V))}{P_{\text{in}}}$$

where $P_{\text{in}} = \int_{E_g}^{\infty} n_S E dE$. For this condition, V_{mp} and V_{oc} are, respectively, given by Eq. (3.16) and Eq. (3.19), i.e.,

$$V_{\text{mp}} = \frac{E_g}{q}\left(1 - \frac{T_D}{T_S}\right) - \frac{k_B T_D}{q}\ln\left(\frac{\theta_D}{\theta_S}\right) \tag{3.37}$$

$$V_{\text{oc}} = \frac{E_g}{q}\left(1 - \frac{T_D}{T_S}\right) - \frac{k_B T_D}{q}\ln\left(\frac{\theta_D}{\theta_S}\right) + \frac{k_B T_D}{q}\ln\left(\frac{\gamma_S}{\gamma_D}\right). \tag{3.38}$$

We have already discussed that a BHJ-OPV is different from a classical single-bandgap solar cell because its operation requires an energy discontinuity at the heterojunction given by the difference between the lowest unoccupied molecular orbitals between donor and acceptor molecules, i.e., $\Delta E(\equiv \text{LUMO}_D - \text{LUMO}_A)$. The photons are still absorbed at $E_g(\equiv \text{LUMO}_D - \text{HOMO}_D)$, but they recombine primarily at the D/A interface with cross-gap $E_c \equiv \text{LUMO}_A - \text{HOMO}_D \equiv E_g - \Delta E$ (see Fig. 3.5(a)). This reduction in emission bandgap increases the dark current and reduces efficiency. In other words, everything should remain the same except the emission flux in Eq. (3.36), which must be rewritten as

$$J_{\text{emi}}^{\text{BHJ}} = q\int_{E_c}^{\infty} n_D(V)dE. \tag{3.39}$$

Figure 3.5(b) shows the OPV-specific thermodynamic efficiency limit for the donor/acceptor OPV obtained by solving for the J–V relationship, and the blue solid line of Fig. 3.5(c) shows the corresponding thermodynamic limit of V_{oc}, with E_g for the standard solar cell replaced by the cross-gap $E_c \equiv E_g - \Delta E$ for the organic solar cell.

The V_{oc} for the OPVs reported in the literature (filled square symbols in Fig. 3.5) are close to their thermodynamic limit. The gap between the red and the blue lines is attributed to fundamental angle entropy of the system (the middle term in Eq. (3.38)), which cannot be reduced by exciton engineering. In this regard, V_{oc} in OPVs, despite its morphological complexity, is no different from that of inorganic semiconductors, such as Si, GaAs, CIGS, etc.

The key to improving V_{mp} or V_{oc} (as well as the efficiency) of an OPV is to increase the cross-gap $E_{\mathrm{c}} \equiv E_{\mathrm{g}} - \Delta E$. However, to prevent back-injection and to ensure efficient exciton dissociation, we need $\Delta E \gg k_{\mathrm{B}}T_{\mathrm{D}}$. For a typical discontinuity of $\Delta E(\equiv \mathrm{LUMO_D} - \mathrm{LUMO_A}) \approx 0.7$ eV, the efficiency at 2 eV (bandgap of light-absorbing P3HT layers) is reduced from ~25% to ~15% (see Fig. 3.5(a)). This result should be viewed as an upper limit, because we have assumed that the thermalization loss and the free energy gain from the band discontinuity (ΔE) are sufficient to dissociate excitons into free electron–hole pairs.

CHAPTER 4

Thermodynamic Limits of Tandem, Bifacial, and Concentrator Solar Cells

Chapter Summary

- In this chapter, we will show that the analytical expressions calculated in Chapter 3 have simple geometrical interpretations.
- Once interpreted geometrically as a Shockley–Quiesser (S-Q) triangle, the graphical technique predicts the efficiency of single junction, multi-junction, concentrator, and bifacial solar cells.
- Practical solar cells have additional losses, such as, non-radiative recombination. Simple modification of the S-Q triangle predicts the performance of these practical solar cells.
- S-Q triangle can also be used to predict the efficiency of the third-generation solar cells, obtaining the results far more economically and transparently compared to traditional techniques.

4.1 Introduction

We saw in the last chapter that a single-junction (SJ) solar cell fails to convert 65–70% of the incident sunlight into useful electrical energy. In fact, as we will see in the next chapter, these unconverted sub-bandgap (sub-BG) and above-bandgap (above-BG) photons further degrade the performance and reliability through self-heating. When everything about solar cells are said and done, we will show in Part III of the book that a fixed-tilt single-junction (SJ) solar module converts only a small fraction of photon energy incident on a solar farm to electricity. Nonetheless, economics still works in our favor because sunlight is an abundant, free, and enduring energy source!

4.1.1 PV efficiency is improved by reducing energy and entropy losses

Over the years, a number of innovative concepts have been proposed to improve the conversion efficiency of solar cells (see Fig. 4.1). For example, as shown in Fig. 4.2(b), multi-junction tandem cells (MJT-PV) use a sequence of absorbers with decreasing bandgaps to harvest the energies associated with sub-BG and above-BG photons, ultimately doubling the efficiency over its SJ counterpart. Figure 4.2(c) shows a more recent variant called bifacial multi-junction tandem cell (B-MJT), which provides further gain because it can accept light from both the front and the back, thereby converting some of the ground-reflected photons (albedo) that would have been otherwise wasted in between the rows of a solar farm. Newer

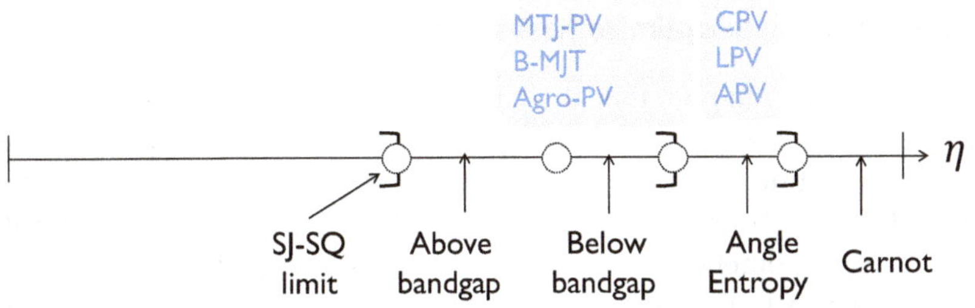

Figure 4.1: The relatively low efficiency of a single-junction solar cell reflects various loss mechanisms shown in the bottom row. The losses can be reduced by various innovative solar cell concepts shown in the top row.

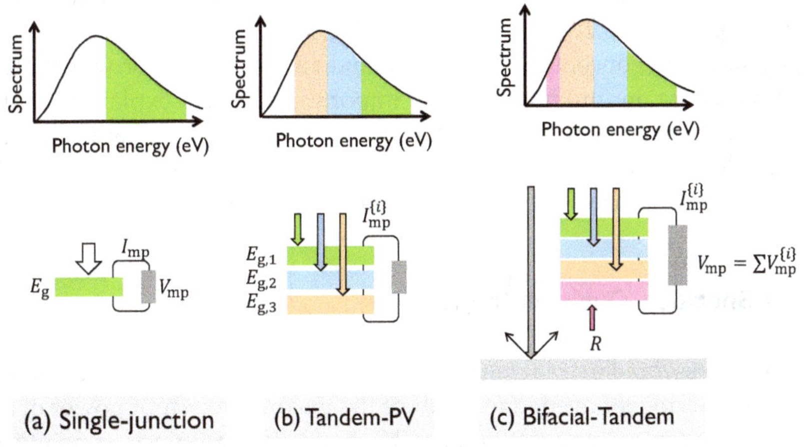

Figure 4.2: A variety of high-efficiency solar cell concepts have been proposed: (a) A single-junction solar cell absorbs photons with $E > E_g$. (b) A tandem solar cell uses a set of absorbers with different bandgaps to utilize a broader range of photons. In the series-connected configuration, the bandgaps are chosen to produce the same output current, $I_{mp}^{\{i\}}$. (c) A bifacial tandem solar cell can accept light directly from the top surface and the ground-scattered albedo from the bottom side.

system concepts such as agro-PV or aqua-PV, where higher-energy photons are used to generate electricity and the lower-energy photons are used to support photosynthesis, could also be viewed in the framework of tandem solar cells. Finally, the energy of the above-BG photons can also be harvested by a hot-carrier PV and multiple-exciton solar cell, while the energy of the sub-BG photons can be recouped by an impurity-band solar cell.

Instead of the energy loss associated with sub-BG and above-BG photons, concentrator PV (CPV), luminescent-coupled solar cells (LPVs), and emission angle-restricted PV (APV) focus on reducing the angle entropy loss of solar cells. One can also mix and match these energy loss and entropy reduction strategies to generate a variety of concepts to improve the conversion efficiency of solar cells.

4.1.2 It is not easy to optimize next-generation solar cells

We saw in the last chapter that the calculation of the efficiency limit of an SJ solar cell involves complicated integrals and difficult book-keeping of multiple fluxes. Now imagine the challenge of optimizing multi-junction bifacial tandem cells (e.g., Fig. 4.2(c)) under concentrated sunlight using a similar formulation. Even if we could somehow formulate the problem properly and determine the efficiency numerically, the fog of mathematics is likely to obscure the insights necessary for a practical design.

In this chapter, we develop an intuitive but powerful graphical approach called Shockley–Queisser triangle (S-Q triangle). The approach will unify the thermodynamic efficiency results of various types of solar cells scattered in the literature through simple scaling relationships. It will also predict the efficiency limits of emerging solar cell concepts (e.g., bifacial tandem solar cells) for which the thermodynamic limits are unknown. More importantly, it will explain the intrinsic trends of nonlinear efficiency gain with increasing cell number, how a two-junction bifacial tandem cell may outperform a four-junction monofacial tandem cell, the effect of series resistance on the choice of cell configuration, and so on.

4.2 The Shockley–Queisser triangle

The analysis presented in this chapter relies on two key observations related to the voltage and the current needed to produce the maximum output power of a solar cell, i.e., maximum power point voltage (V_{mp}) and maximum power point current (I_{mp}). We recall from Eq. (3.16) that at the radiative limit, V_{mp} of a solar cell with bandgap E_{g} is given by the relationship involving the Carnot factor η_{c} and the angle entropy factor:

$$V_{\mathrm{mp}} = \frac{E_{\mathrm{g}}}{q}\left(1 - \frac{T_{\mathrm{D}}}{T_{\mathrm{S}}}\right) - \frac{k_{\mathrm{B}}T_{\mathrm{D}}}{q}\ln\left(\frac{1}{c}\frac{\theta_{\mathrm{D}}}{\theta_{\mathrm{S}}}\right). \tag{4.1}$$

Here, T_D and T_S are the temperatures of the solar cell and the sun, respectively. The Carnot factor is $\eta_c \equiv (1 - T_D/T_S) = 1 - 300/6000 = 0.95$. The angle entropy factor, $\Delta \equiv (k_B T_D/q)\ln(\theta_D/c\theta_S)$, depends on the angular radiation from the solar cell (θ_D) and the size of the solar disk (θ_S). The concentration factor c effectively increases the size of the solar disk seen by the solar cell, see Fig. 2.10. Ideally, $\theta_D = 2\pi$ or 4π depending on the back reflector. Thus, $\Delta \simeq 0.31$ at one-sun concentration (i.e., $c = 1$).

Furthermore, recall from Eq. (3.17) and Eq. (3.18) that the maximum power point current I_{mp} is well approximated by the corresponding short-circuit current I_{sc}. With AM1.5G illumination, I_{sc} is approximated by Eq. (3.26):

$$I_{mp} \simeq I_{sc} = cI_{sun}(1 - \beta_{sun}E_g) \tag{4.2}$$

where c is the solar concentration, and $I_{sun}(= 83.75\ \text{mA/cm}^2)$ is the projected current at $E_g \to 0$, and $\beta_{sun} \sim (4.7k_BT_S)^{-1} = 0.425\ \text{eV}^{-1}$ is the effective loss coefficient of photocurrent with increasing bandgap. The linear approximation holds for $0.5\,\text{eV} < E_g < 2\,\text{eV}$. The nonlinearity of I_{mp} under arbitrary blackbody illumination is easily analyzed by a simple one-to-one mapping between E_g and its linear approximation.

Homework 4.1: It is easy to check the thermodynamic limit of J_{sc}

An experimental group reports that an organic solar cell with 1.8 eV bandgap and illuminated by AM1.5G solar spectrum, produces $J_{sc} = 35\ \text{mA/cm}^2$. Is this current reasonable?

Solution. For a 1.8 eV bandgap solar cell, Eq. (4.2) can be used to calculate the upper limit of the short-circuit current, namely, $J_{sc} \sim 83.75(1 - 0.425E_g) = 83.75(1 - 0.425 \times 1.8) = 19.23\ \text{mA/cm}^2$. Clearly, it is impossible to get a $35\ \text{mA/cm}^2$ current from a 1.8 eV bandgap solar cell under AM1.5G illumination. The paper must have made a mistake in reporting the results.

Inserting Eq. (4.1) into Eq. (4.2) to eliminate E_g, and defining $i_{mp} \equiv I_{mp}/I_0$ and $v_{mp} \equiv V_{mp}/V_0$, we obtain the equation for the S-Q triangle, namely,

$$i_{mp} = 1 - v_{mp}. \tag{4.3}$$

Here, $I_0 \equiv cI_{sun}(1 - \beta\Delta)$ and $V_0 \equiv (1 - \beta\Delta)/\beta$, with $\beta = \beta_{sun}/\eta_c$.

In Figs. 4.3(b, c), Eq. (4.3) defines the S-Q triangle. Each point on the diagonal represents a material with bandgap E_g. Specifically, given a bandgap E_g, Eqs. (4.1) and (4.2) can be used to calculate V_{mp} and I_{mp}. Once these quantities are normalized by V_0 and I_0, respectively, v_{mp} and i_{mp} define a point along the diagonal. The power output is obtained by drawing the green rectangle associated with this point (see Fig. 4.3(b)). Similarly, one can solve the optimum bandgaps of an N-junction

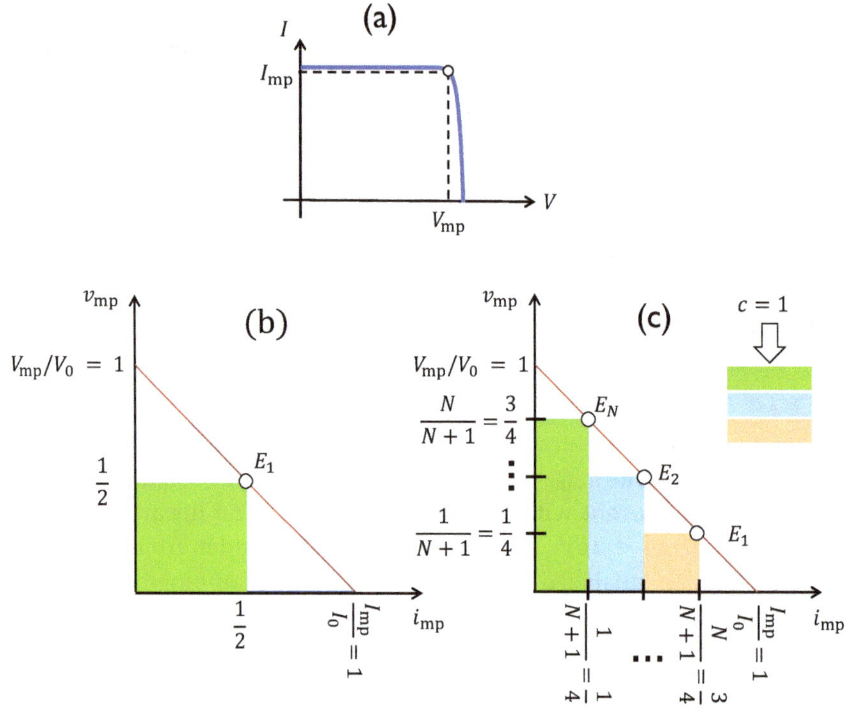

Figure 4.3: (a) The I–V characteristic of a solar cell with a known bandgap E_g. The maximum power point $(V_{mp}(E_g), I_{mp}(E_g))$ is identified. (b, c) Each point on the S-Q line represents a unique material with bandgap E_g. The axes correspond to the normalized v_{mp} and i_{mp}. The S-Q triangles for (b) single-junction ($N = 1$) and (c) triple-junction ($N = 3$) solar cells.

tandem solar cell, shown in Figs. 4.2(b) and 4.3(c). Each of the series-connected solar cells absorbs different parts of the solar spectrum, and the bandgaps must be optimized to ensure that the photocurrents generated are equal. The optimum bandgaps and the thermodynamic efficiency of an N-junction solar cell are obtained by tiling the S-Q triangle by rectangular boxes (equal base implies equal current) that maximize the triangle coverage. It follows that

$$V_{mp}^{\{i\}} = \frac{iV_0}{N+1}, \tag{4.4a}$$

$$I_{mp}^{\{i\}} = \frac{I_0}{N+1}. \tag{4.4b}$$

Equation (4.4b) expresses the fact that the tandem subcell currents ($x \equiv I_{mp}^{\{i\}}/I_0$) must be equal and should not depend on i. Moreover, once $V_{mp}^{\{i\}}$ is known, Eq. (4.1) identifies the material of interest with specific bandgap, $E_{g,i} = (V_{mp}^{\{i\}} + \Delta)/\eta_c$.

To calculate the overall efficiency, η_N, we first sum over the areas of the N boxes within the normalized S-Q triangle ($s_N = N/[2(N+1)]$). We then multiply the result by the corresponding scale factors I_0 and V_0 to calculate the total output power, $P_{\text{out}} = I_0 V_0 s_N$. Finally, the ratio of P_{out} to the incident $(P_{\text{in}} \times c)$ determines the efficiency of the N-junction tandem cell with concentrated sunlight c:

$$\eta_N(c) = \frac{I_0 V_0}{P_{\text{in}}} \times \frac{2N}{(N+1)c}. \tag{4.5}$$

Here, specifically, P_{in} (kW/m^2) is the power input under AM1.5G illumination.

Homework 4.2: Current in a two-terminal series-connected multi-junction PV

What is the I_{mp} for an optimized triple-junction tandem cell under AM1.5G illumination.

Solution. With $c = 1, N = 3$, and $I_{\text{sun}} = 83.75$ mA/cm^2, we find that $I_0 = 71.92$ mA/cm^2 and $V_0 = 1.904$ V. By Eq. (4.4), an N-junction tandem divides this current into $(N+1)$ parts, so that the currents in the subcells are equal. Therefore, for N-junction tandem, $I_{\text{mp}} = I_0/(N+1) = 71.92/4 \sim 18\,\text{mA/cm}^2$.

4.3 Application of a model to a variety of PV concepts

4.3.1 Efficiency of a single-junction PV with $c = 1$

The essential correctness of Eqs. (4.4) and (4.5) can be established by calculating the optimum efficiency of an SJ cell under AM1.5G illumination. With $c = 1$, $N = 1$, and $I_{\text{sun}} = 83.75$ mA/cm^2, we find $I_0 = 71.92$ mA/cm^2 and $V_0 = 1.904$ V. Therefore, $\eta_1 = 34.2\%$ occurs at $I_{\text{mp}} = I_0/2 = 35.92$ and $V_{\text{mp}} = 1.92/2 = 0.96$ eV (or $E_{\text{g}} = 1.34$ eV), shown by the green box in Fig. 4.3(b). The result is physically justified because $E_{\text{g}} \sim 2.7k_{\text{B}}T$ is the average photon energy of the solar spectrum. In addition, the results compare very well with the efficiency derived in Chapter 3, i.e., $\eta_1 = 33.7\%$ occurs at $E_{\text{g}} = 1.34$ eV. The S-Q triangle also explains the flatness of the efficiency between 1.1 eV $< E_{\text{g}} <$ 1.6 eV, shown in Fig. 3.3(a). After all, the normalized output power obtained from $p_{\text{mp}} = v_{\text{mp}} i_{\text{mp}} = v_{\text{mp}}(1 - v_{\text{mp}})$ is relatively insensitive to V_{mp} (or equivalently E_{g}) for a wide variety of the bandgaps close to 1.34 eV. This is very good news, because a variety of semiconductors (e.g., Si (1.1 eV), InP (1.35 eV), GaAs (1.45 eV), CdTe (1.5 eV), etc.) can in principle achieve high power conversion efficiency.

4.3.2 Thermodynamic efficiency of an N-junction tandem cell

A second approach to improve the conversion efficiency of solar cells involves choosing a series of absorbers with different bandgaps so that they all produce equal amounts of current. The absorbers are then connected optically and electrically in series to improve photoconversion efficiency. Traditional optimization involves an iterative search to find the bandgap combination for maximum efficiency. In contrast, Eq. (4.5) predicts that

$$\eta_N(c) = \eta_1(c) \times \left[\frac{2N}{N+1}\right]. \tag{4.6}$$

Figure 4.4 compares Eq. (4.6) (solid line) with the exact numerical results (symbols) available in the literature: the results agree within a few percent. Interestingly, Eq. (4.6) identifies the efficiency gain scaling factor for tandem cells (i.e. $2N/(N+1)$) that had been hidden in plain sight in all the numerical results. The scale factor anticipates a well-known result that $\eta_{N\to\infty} = 2\eta_1$. Graphically, the triangle in Fig. 4.3(c) is fully tiled with boxes for $N \to \infty$. The factor of 2 increase in efficiency of an infinite-junction tandem solar cell simply reflects the fact that the triangle has double the area of the square embedded in it, as shown in Fig. 4.3(b).

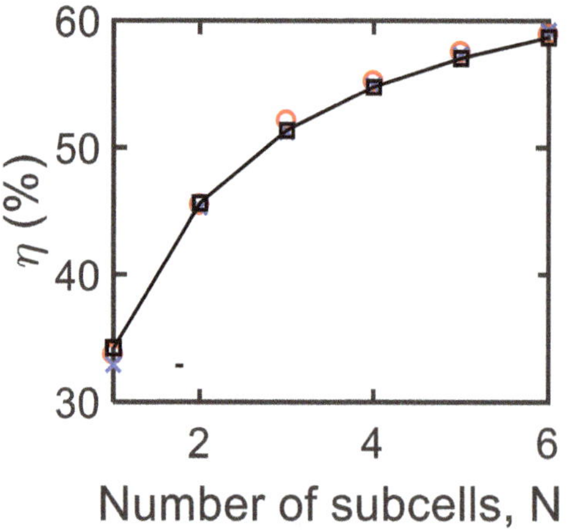

Figure 4.4: The simple N-dependence predicted by Eq. (4.5) (black square) reproduces the numerical results from Ref. [5] (red circles) and Ref. [13] (blue crosses).

Finally, although the scaling factor is specifically derived for AM1.5G, the result is general and this model can capture the essential trends in other spectrum (including blackbody radiation) as well. The key point to remember is that, the S-Q triangle method can be applied as long as there is linearity between V_{mp} and I_{mp} for any spectrum. The scaling factor set (V_0, I_0) needs to be calculated separately for each spectrum.

The power of the S-Q approach is now obvious: Eq. (4.6) anticipates the nonlinear dependence of η_N vs. N and predicts the bandgaps for any arbitrary N-junction tandem cell.

Homework 4.3: A modified N-junction formula to calculate the infinite junction limit

Show that Eq. (4.6) can be rewritten as $\eta_N = \eta_{N=\infty} - 2\eta_1/(N+1)$. Validate the formula for AM0 (Blackbody) and AM1.5G spectrum by using the web-enabled thermodynamic calculator **PVLimits**.

Solution. By Eq. (4.6), $\eta_{N=\infty} = 2\eta_1$. Therefore, $\eta_{N=\infty} - \eta_N = 2\eta_1/(N+1)$, that is, $\eta_N = \eta_{N=\infty} - 2\eta_1/(N+1) \sim \eta_{N=\infty} - 2\eta_1/N$ for $N \gg 1$. Although Eq. (4.6) (and its reformulation here) is derived for the idealized AM1.5G spectrum, in practice, it can be shown that the

$$\eta_N \sim \eta_{N=\infty} - C/N^{\alpha}$$

with $\alpha \sim 1$ holds for a broad range of solar spectrum (even those that do not obey time-reversal symmetry). In other words, simply by plotting as a straight line the numerically calculated η_N (for any arbitrary solar spectrum and concentration) as a function of $N^{-\alpha}$, we can determine $\eta_{N=\infty}$ from the y-intercept, with α being the constant that makes the line straight.

4.3.3 Non-optimum E_g in tandem PV

So far, we have assumed that the subcell bandgaps have been chosen to maintain current matching among the cells. In practice, one may not be able to integrate optimum bandgap materials into a single stack. What would be the output power if the subcell currents are mismatched?

In general, for an N-junction conventional tandem, the current is limited by the subcell with the lowest current contribution (i.e., the one that has the lowest absorption):

$$s_N = \left[\sum_{i=1}^{N} v_i\right] \times \min\{x_i\} \tag{4.7}$$

where, the normalized voltage of the i-th subcell is v_i and the corresponding normalized current is x_i. Once one has chosen a sequence of bandgaps, the set of $\{v_i\}$ is defined. We can then find the corresponding current $I_{\rm mp}$ (from Eq. (4.5)) and x_i:

$$x_i = x_{i+1} + v_{i+1} - v_i \text{ and,}$$
$$x_N = 1 - v_N.$$

The corresponding efficiency can be written as

$$\eta_N = \frac{V_0 I_0}{c\,P_{\rm in}} s_N \approx \frac{1}{c\,P_{\rm in}} \left[\sum_{i=1}^{N} V_{\rm mp}^i\right] \times \min\{I_{\rm mp}^i\}. \tag{4.8}$$

Here, we have assumed that each subcell will operate close to $V_{\text{mp}}^{\{i\}}$, although the currents are lower than $I_{\text{mp}}^{\{i\}}$. This approximation can be made due to the logarithmic relation between $V_{\text{mp}}^{\{i\}}$ and $I_{\text{mp}}^{\{i\}}$.

The tandem efficiency with non-optimum bandgaps can be increased by improved current matching, or making current matching irrelevant! The techniques include:

- **Photon Sharing.** A top cell with smaller-than-ideal bandgap intercepts photons meant for the bottom cell. The photon-starved bottom cell limits the overall efficiency. If the top cell is thinned, the photons transmitted can be absorbed by the bottom subcell. The overall efficiency improves because the subcell currents are better matched. The thermodynamic efficiency cannot be reached because the transmitted photons have a higher thermalization loss in the lower-bandgap bottom subcell.

- **Multi-terminal Configuration.** The current mismatch among the subcells can be reduced by utilizing 3- or 4-terminal tandem cells to electrically decouple the mismatched subcells. The excess current produced by a non-ideal cell is siphoned off to a separate load connected to the cell (instead of being dissipated within the cell itself). Ideally, such multi-terminal tandems can approach the thermodynamic limit. However, such structures are difficult to fabricate and the contacts increase electrical loss.

- **Segmented Cells.** If a subcell is segmented into M thinner sections and reconnected in series, I_{mp} reduces by and V_{mp} increases by a factor of M. If these segmented subcells are stacked in tandem, the current mismatch is suppressed or eliminated. For example, if two subcells produce 9 mA/cm^2 and 15 mA/cm^2 currents, the traditional tandem current is limited to 9 mA/cm^2. The additional 6 mA/cm^2 current produced by the second cell is wasted. With $M_1 = 3$ and $M_2 = 5$, the current mismatch is eliminated, because all cells now carry 3 mA/cm^2 current. In general, the segments can be connected laterally (as in a module) or vertically (as in a traditional tandem cell). Even without perfect current matching, efficiency can be improved significantly. A number of research groups across the world have already implemented the approach and demonstrated improved performance.

The efficiency gain associated with these strategies is intuitively interpreted by the S-Q triangle. While strategies are sound, their implementation could be nontrivial and expensive.

Homework 4.4: Performance of tandem cells with non-optimum bandgaps

In 2019, Fraunhofer Laboratory fabricated a wafer-bonded triple-junction solar cell composed of Si, $Al_{0.06}Ga_{0.94}As$, and $Ga_{0.51}In_{0.49}$ P, with approximate bandgaps of 1.1 eV, 1.46 eV, and 1.77 eV, respectively. Under AM1.5G illumination and 298 K ambient temperature, the cell produces $I_{sc} = 12.4\,\text{mA/cm}^2$, $V_{oc} = 3.18$ V, and FF $= 86.4\%$, with an overall efficiency of 34.1%. Show that the results are consistent with Eq. (4.8).

Solution. For an ideal (current-matched) triple-junction solar cell, V_{mp} calculated by Eq. (4.4) are 0.48 V, 0.96 V, and 1.44 V. The corresponding bandgaps are given by Eq. (4.1): 0.83 eV, 1.33 eV, and 1.84 eV. For this ideal cell, it is found in Homework 4.2 that $I_{mp} = I_0/(N+1) = 18\ \text{mA/cm}^2$, $V_{mp} = \sum_{i=1}^{N} V_{mp}^{\{i\}} = NV_0/2 \sim 2.82\,\text{V}$, and $\eta_N = \eta_1 \times 2N/(N+1) = 51.4\%$.

For the non-ideal "III–V on Si" tandem cell reported by Fraunhofer, the maximum current produced by each subcell is $I_{mp} = 20.74$, 31.78, and $44.6\,\text{mA/cm}^2$. The middle cell produces the rate-limiting current of $\min\{I_{mp}^i\} = 31.78 - 20.74 \approx 11\ \text{mA/cm}^2$ (cf. $I_{sc} = 12.4\ \text{mA/cm}^2$ from the experiment). Similarly, from Eq. (4.1), $\sum_{i=1}^{3} V_{mp}^i = 0.95(1.1+1.46+1.77) - 0.31\times 3 = 3.18\,\text{V}$. The efficiency predicted by Eq. (4.8), namely $\eta_N = 3.18\times 11/1000 = 33.95\%$, is comparable to $\eta \approx 34.1\%$ observed in the experiment. The analysis suggests that the efficiency can also be improved by reducing the middle cell bandgap closer to 1.33 eV. The efficiency can be improved by reducing current mismatch either by photon sharing among the cells or by segmentation of cells, so that each segment produces approximately $6\ \text{mA/cm}^2$.

Homework 4.5: Agro-photovoltaics can be viewed as an non-optimum N-junction tandem cell

Let us assume that we wish to share a tract of land to grow crops under solar panels. Many crops mostly require high-energy photons $E > E_{cut}$ (i.e., $\lambda < \lambda_{cut}$ ~750 nm). See part (a) of the figure below. If we take away this part of the solar spectrum for plants and crops, how should we redesign the N-junction tandem cells?

Solution. Considering the E_g–I_{sc} plot below, we can find I_{sc} corresponding to E_{cut}:

$$I_{sc}^{cut} = I_{sun}(1 - \beta_{sun} E_{cut}).$$

(*continued on the next page*)

Homework 4.5 (*continued from the previous page*)

These photons are needed by the plants and therefore not available to the solar cells. Therefore, the new E_g–I_{sc} for the solar cell should be

$$I_{sc}^{new}(E_g) = I_{sun}(1 - \beta_{sun}E_g) - I_{sc}^{cut}$$
$$= I_{sun}^{new}(1 - \beta_{new}E_g). \tag{4.9}$$

The corresponding S-Q triangle is shown in part (c) of the figure below. We can now follow the same method for designing tandems discussed in this chapter using this new relation.

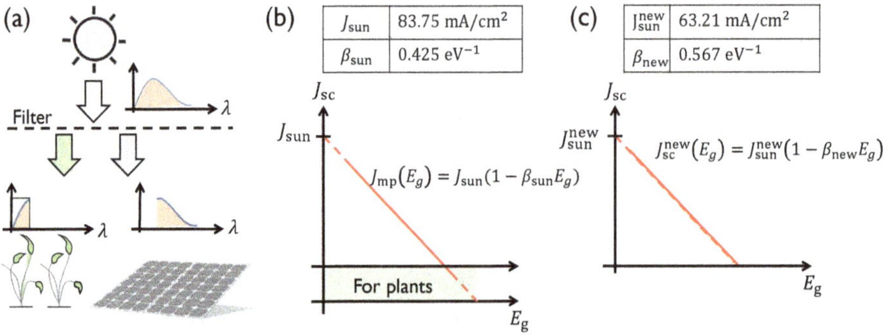

(a) Partitioning of the solar spectrum between plants and solar cells. (b) The original S-Q triangle of the full solar spectrum. (c) The new S-Q triangle associated with the scaled solar spectrum.

Homework 4.6: Agro-PV is a special case of the spectral-splitting approach

It is often difficult to fabricate the solar cells with different bandgaps on top of each other because their lattice constants may not match. In this case, it may be possible to sub-divide the cells into multiple two-terminal tandem stacks. For example, stack 1 will contain the N_1 subcells with the lowest bandgaps ($0 < E_g < E_{cut,1}$); stack 2, the next N_2 bandgaps ($E_{cut,1} < E_g < E_{cut,2}$); stack 3, the last N_3 bandgaps ($E_{cut,3} < E_g < E_{max}$), etc. One would then use a lens to spectrally split the solar spectrum so that each stack is illuminated by the part of the spectrum it can most efficiently convert to electricity. How would you determine the optimum bandgaps for this multiple-stack tandem cell?

(*continued on the next page*)

Homework 4.6 (*continued from the previous page*)

Solution. The S-Q triangle provides an intuitive answer to this complicated optimization problem. We should first split the S-Q triangle into three parts (corresponding to three stacks) by inserting lines at $E_{\text{cut},1}$ and $E_{\text{cut},2}$ and determine the corresponding currents $I_{\text{cut},1}$ and $I_{\text{cut},2}$. Current continuity within the two-terminal tandem stack requires that $I_{\text{sc},1} = (I_{\text{max}} - I_{\text{cut},1})/(N_1 + 1)$, $I_{\text{sc},2} = (I_{\text{cut},1} - I_{\text{cut},2})/N_2$, and $I_{\text{sc},3} = (I_{\text{cut},2})/N_3$. The corresponding bandgaps can be read off the S-Q triangle by inspection.

Homework 4.7: Multi-junction tandem cells for a submerged structure

The intensity of transmitted sunlight reduces as we go deeper into water (e.g., in oceans). For example, only 50% of the sunlight penetrates $d = 4$ m. And water also absorbs most photons with energies $E < E_{\text{w}}$ (typically, $E_{\text{w}} \approx 1.83$ eV for a depth of several meters). How would you redesign a tandem cell so that it works under water?

Solution. First, if the spectrum intensity is uniformly scaled for all photon energies (same absorption coefficient at wavelengths), the tandem bandgaps will be determined by calculating $c(< 1)$. The renormalized S-Q triangle will determine the new bandgaps.

Second, if only the very high-energy photons are absorbed (at relatively shallow depth) and $E_0 > E_{\text{w}}$, then the tandem design will remain the same.

Finally, if $E_0 < E_{\text{w}}$ with increasing depth, the tandem will be redesigned assuming $E_0^{new} = E_{\text{w}}$. Note that agro-photovoltaics uses high-energy photons to generate energy and lower-energy photons for photosynthesis. For water-submerged structures, however, the higher-energy photons are absorbed by water, while the lower-bandgap photons are used for photovoltaic energy conversion. In other words, the two approaches use complementary parts of the S-Q triangle.

4.4 Concentrator solar cells reduce entropy loss of a solar cell

4.4.1 Efficiency of single-junction concentrated solar cells

Since an SJ solar cell operates far below the Carnot limit (~95%) and only converts one-third (~33%) of the incident energy into useful power output, many solar cell innovations since the 1960s have focused on improving the efficiency of a photovoltaic converter. One of these approaches involves using a parabolic mirror to concentrate sunlight onto a solar cell.

The calculation of efficiency limits of concentrator solar cells is difficult; the numerical results are available only for specific concentrations. Fortunately, the bandgaps and efficiency predicted by Eqs. (4.4) and (4.5) hold for any arbitrary concentration, and therefore the model can be used to calculate the efficiency at arbitrary concentration as well. For example, for $c = 300$, $\Delta \equiv (k_B T_D/q)\ln(\theta_D/c\theta_S) = 0.16$. Therefore, $V_0 = 2.06$ V and $I_0/c = 77.8$ mA/cm^{-2}. The corresponding efficiency $\eta_1(c = 300) = 40.2\%$ compares well with $\eta = 41.1$ reported in the literature. In addition, the increase in V_{mp} to $2.06/2 = 1.03$ eV and the reduction in the bandgap to $E_g = 1.25$ eV to maximize efficiency are consistent with the values predicted by the web-enabled thermodynamic calculator **PVLimits**.

Homework 4.8: Concentrator Solar Cell

The increase in efficiency of a concentrator solar cell arises from the suppression of which loss component?

Solution. The suppression of angle entropy loss has been discussed in Chapter 2. As explained in Fig. 2.10(a), the cell "thinks" that sunlight is coming from everywhere ($\theta_S \to 4\pi$), not just one little disk in the sky. The effective absorption angle ($c\,\theta_S$) is better matched to emission angle (θ_D). This angle matching improves V_{mp} (see Eq. (4.1) or (3.16)) and the efficiency.

Homework 4.9: Efficiency of concentrated solar cells

Compare two different solar cells A and B. Cell A is 28% efficient under 1 sun. Cell B is also 28% efficient, but under 10 suns (i.e., $c = 10$). Under 1 sun, which cell is more efficient?

Solution. Cell A. Efficiency increases with increased solar concentration due to reduced angle entropy. Therefore, the efficiency of cell B will reduce at 1-sun concentration.

Homework 4.10: Angle-restricted solar cell

Show that restricting the emission angle, as in Fig. 2.10(b), offers an alternative to concentrator PV.

(continued on the next page)

Homework 4.10 (*continued from the previous page*)

Solution. Since the angle entropy loss of V_{mp} depends on the angle ratio (i.e., $f_c^{-1} \equiv \theta_{\text{D}}/\theta_{\text{S}}$ in Eq. (4.1), we can improve cell efficiency either by increasing θ_{S} (concentrator PV, CPV) or by restricting the emission angle θ_{D} (Angle-restricted PV, APV). In the radiative limit, APV and CPV (or a combination thereof) offer comparable performance. As an additional benefit, APV operates at 1-sun concentration ($c = 1$) with significantly reduced self-heating, a very important topic to be discussed in Chapter 5. The estimated boost in efficiency with increased f_c is shown in the figure below. In the inset, we see that efficiency scales logarithmically with f_c, as expected from Eq. (4.1). We will see in Chapter 6 that APV requires a nontrivial optical design to restrict the emission angle. The increase in the carrier concentration enhances non-radiative Auger and trap-assisted recombination. The practical benefit of APV is yet to be demonstrated.

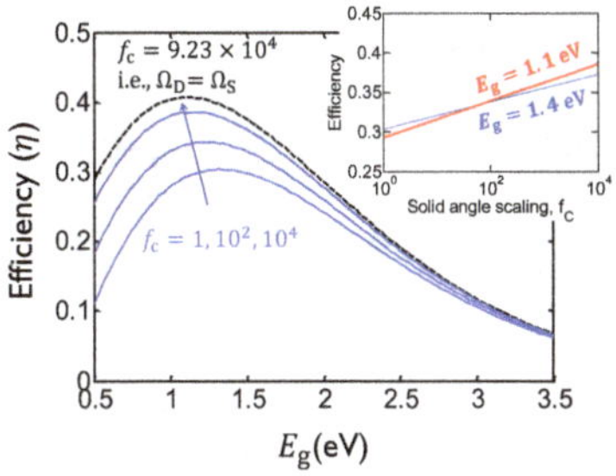

Improvement in efficiency by reducing the incidence and emission angle mismatch by a solid angle scaling factor of f_c. The calculation was done for AM0.

4.4.2 Multi-junction concentrator tandem cells address both energy and entropy losses

In the preceding section, we focused on recovering the energy loss in an SJ solar cell due to sub-BG and above-BG photons. The conversion efficiency is further improved in a tandem cell placed under concentrated sunlight. Based on the discussion in the preceding three chapters, we know that concentrated sunlight reduces angle entropy loss. However, the combined optimization of energy loss (with tandem cells) and entropy loss (by concentrated sunlight) is so difficult that it has been reported for only a few specialized cases. The optimum bandgaps to maximize efficiency is obtained by an iterative search over the bandgap space. Since Eqs. (4.4) and (4.5) apply to any N-junction tandem cell under arbitrary illumination, we can confirm its validity by comparing it to a few specific results for 4-junction and 5-junction cells. Table 4.1 and Fig. 4.5 show that both the bandgap sequence and the thermodynamic efficiencies compare well with the values reported in the literature.

Table 4.1: Effect of concentration on optimized bandgaps.

$N = 4, c = 300$							
	1	2	3	4		I_{mp}	η
V_{mp}	0.41	0.82	1.23	1.65		16.3	63.5
E_g	0.60	1.04	1.47	1.90			63.5
E_g [14]	0.52	0.97	1.38	1.89			63.7
$N = 5, c = 300$							
	1	2	3	4	5	I_{mp}	η
V_{mp}	0.34	0.69	1.03	1.37	1.71	12.9	66.2
E_g	0.53	0.89	1.25	1.61	1.98		66.2
E_g [14]	0.52	0.92	1.21	1.57	2.03		66.2

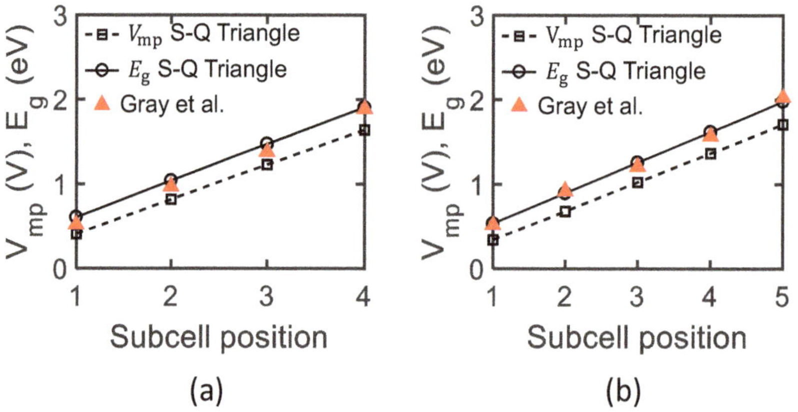

Figure 4.5: The analytical results predicted by Eqs. (4.4)–(4.6) compared to the numerical results published in the literature. (a) Four-junction tandem cell. (b) Five-junction tandem cell.

4.5 Bifacial tandem solar cells: An emerging solar cell technology

Although the concept of a bifacial solar cell, shown in Fig. 4.2(c), originated in the 1980s, recent technological innovations have made it competitive compared to traditional monofacial solar cells. Despite significant commercial interest (some say that the technology may capture 30% market share by 2030), the general thermodynamic limit of bifacial tandem solar cells is not known. In this section, we will use the S-Q triangle to calculate the thermodynamic limits of various types of bifacial solar cells.

Homework 4.11: Ultimate efficiency limit of a solar cell

The single-junction solar cell efficiency limit is $\eta_1 \approx 33\%$ under AM1.5G. This can be improved by $\sim$10% if we eliminate the angle anisotropy using solar concentrators or angle restriction, as discussed in Homeworks 4.9 and 4.10. So, the limit for single-junction cells without angle mismatch is $\eta_1^* \approx 43\%$. Show that for an infinite-junction tandem (i.e., $N \to \infty$) without any angle mismatch, Eq. (4.6) predicts that $\eta = 2 \times \eta_1^* = 86\%$. A more detailed numerical analysis shows that this "ultimate multicolor efficiency" of solar cell under AM0 illumination is 86.8%.

Figure 4.6 shows the generalization needed to calculate the efficiency of a bifacial tandem cell. The generalized S-Q triangle accommodates the cells illuminated both from the top (right triangle) and the bottom (left triangle). With an albedo of R, the maximum normalized current for a rear subcell should be $I_{\mathrm{mp}}/I_0 = R$. The extended S-Q triangle equations can therefore be rewritten from Eq. (4.5) as follows:

$$v_{\mathrm{mp}} = \begin{cases} 1 - i_{\mathrm{mp}} & \text{for top subcells} \\ 1 - \dfrac{i_{\mathrm{mp}}}{R} & \text{for rear subcells} \end{cases}.$$

For optimal output, all the subcells in the bifacial tandem must have equal current (current-matched). Assuming a subcell current of x, we can now find the corresponding v_{mp} of each subcell using the extended S-Q triangle shown in Fig. 4.6.

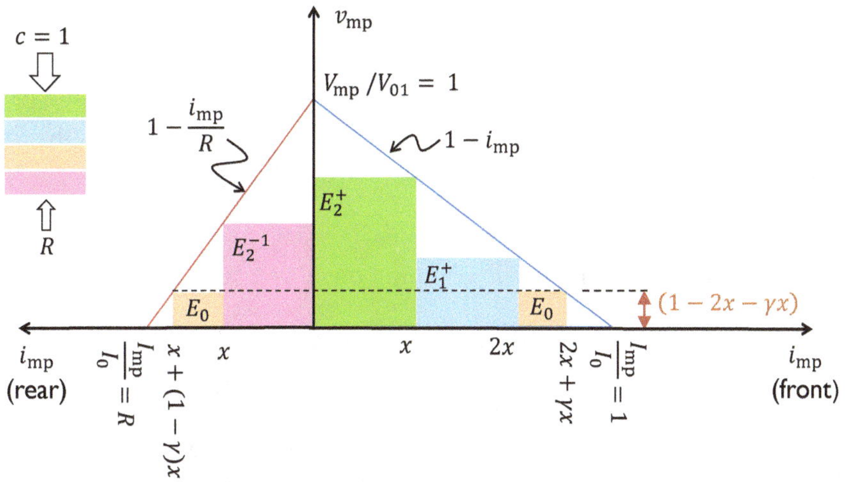

Figure 4.6: Bifacial solar cells can be optimized with an extended S-Q triangle, where the bottom cell uses the albedo light.

An interesting aspect of bifacial cells is that depending on the albedo, the cell with the smallest bandgap E_0 may have to be placed in the middle of the stack, allowing it to collect sunlight from the front as well as the rear. E_0 should be chosen such that the total normalized photocurrent is x — let us assume that the current contributions in this subcell from the front and rear sunlight are γx and $(1-\gamma)x$, respectively. By using these conditions in the extended S-Q triangle in Fig. 4.6, we can find γ.

In the bifacial tandem, there are U cells above and D cells below the E_0 cell, so that $N = U + D + 1$. Recalling the derivation of Eq. (4.5), the sum of the boxes gives the normalized power output: $s_N(U, D, R)$.

$$\begin{aligned} s_N &= \sum_{i=1}^{U} x(1-ix) + \sum_{j=1}^{D} x\left(1-\frac{jx}{R}\right) + x(1-Ux-\gamma x) \\ &= Nx - x^2 A. \end{aligned} \tag{4.10}$$

Here,

$$\gamma \equiv (1+D-Ux)/(1+R) \tag{4.11}$$

and

$$A \equiv \frac{U(U+1)}{2} + \frac{D(D+1)}{2R} + \frac{N}{1+R}. \tag{4.12}$$

The power is maximized for the current

$$\frac{I_{\text{mp}}}{I_0} \equiv x_0 = \frac{N}{2A}. \tag{4.13}$$

Note that we can relate the normalized power s_N to the efficiency as: $\eta_N = (V_0 I_0 / cP_{\text{in}}) \times s_N$. Thus, using x_0 from Eq. (4.13) in Eq. (4.10), we find that

$$\frac{\eta_N}{\eta_1} = \frac{s_N}{s_1} = 2N\, x_0. \tag{4.14}$$

Also, $ds_N/dU = 0$, for a fixed N and R defines the number of cells in the upper stack, U:

$$U = \frac{2N-R-1}{2(1+R)}. \tag{4.15}$$

Equations (4.11)–(4.15) define the maximum power from a stack of N cells illuminated by albedo R.

In addition, Eq. (4.10) reduces to limiting expressions: $\eta_1(U=0, D=0, R) = (1+R)\eta_1(R=0)$ and $\eta_N(U, D=0, R=0)/\eta_1(R=0) = 2N/(N+1)$ (see Eq. (4.6)). The gain gradually diminishes at higher N as larger-bandgap boxes tile the original triangle, consistent with Fig. 4.4. The triangles anticipate that the bottom cell can have the smallest bandgap (i.e., $E_2 \geq E_1$) provided $N \leq N_{crit} \equiv 1 + R^{-1}$. This sudden change in the optimum tandem topology (and the corresponding discontinuous jump in the efficiency) has no equivalence in traditional solar cells.

Using Eqs. (4.10)–(4.13) and (4.15) in (4.14), we find that for $N > N_{\text{crit}}$ and $D > 0, R \neq 0$):

$$\frac{\eta_N}{\eta_1(R=0)} = \frac{8R(1+R)N^2}{2R(2N^2+4N-1)-R^2-1}, \tag{4.16}$$

and for $N < N_{crit}, D = 0$, the corresponding equation is

$$\frac{\eta_N}{\eta_1(R=0)} = \frac{2(1+R)N}{2+(N-1)(1+R)}. \tag{4.17}$$

The bifacial tandem efficiencies calculated using these equations assuming $\eta_1(R = 0) = 33.7\%$ are shown in Fig. 4.7. The results obtained by Eqs. (4.16) and (4.17) compare very well with the numerical results published in the literature. The expression reduces to the limiting case of a traditional tandem cell for $R = 0$.

Homework 4.12: A simple formula describes the optimum bandgaps of a tandem solar cell

Based on the S-Q triangle, one can show that for ($N < N_{crit}$), the optimum bandgaps for a bifacial cell are given by

$$E_i = \frac{i}{\beta_{\text{sun}}N} + \frac{(N-i)\left[(1+R)\beta_{\text{sun}}E_0 - R\right]}{\beta_{\text{sun}}N},$$

where

$$E_0 = E_{\text{SJ}} - \frac{(N-1)(1-R)}{2\beta_{\text{sun}}N} \times \frac{2N}{N(1+R)+(1-R)},$$

with $E_{\text{SJ}} = 1.34$ eV and the total voltage across the series-connected subcells is given by

$$\frac{qV_{\text{mp}}}{N} = E_{g,a}\left(1 - \frac{T_{\text{D}}}{T_{\text{S}}}\frac{E_{\text{g,p}}}{E_{g,a}}\right) - k_{\text{B}}T_{\text{D}}\ln\left(\frac{\theta_{\text{D}}}{\theta_{\text{S}}}\right)$$

where $E_{g,a}$ and $E_{\text{g,p}}$ are the average and the largest bandgaps, respectively.

Use these formulas to calculate the optimum bandgaps and efficiency of a triple-junction ($N = 3$) tandem cell for (i) $R = 0$ and (ii) $R = 0.2$.

Figure 4.7 shows the relative efficiency gain of monofacial vs. bifacial tandem solar cells. For example, considering a typical albedo of $R \approx 0.3$, we see that a simple (and relatively easy to fabricate) 2-junction bifacial tandem achieves performance comparable to that of a far more complicated 4-junction monofacial ($R = 0$) tandem cell.

4.6 Thermodynamic limits of non-ideal solar cells

In the previous sections, we used the S-Q triangle to calculate the thermodynamic (radiative) limit of ideal single-junction, tandem, bifacial, and concentrator solar

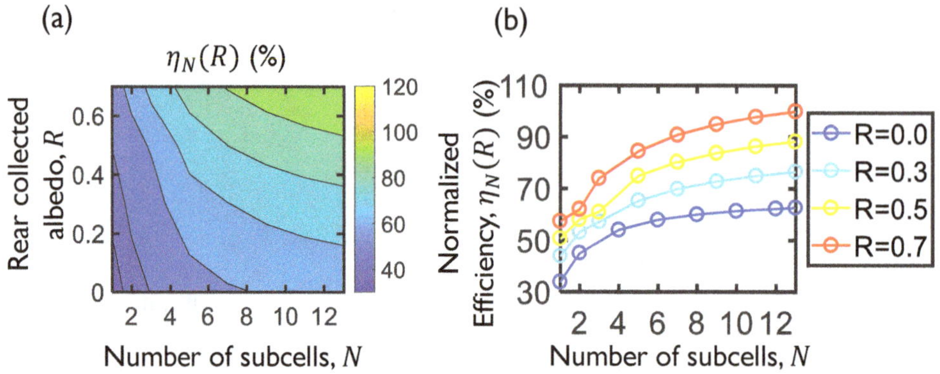

Figure 4.7: (a) Efficiency gain of a bifacial tandem solar cell depends on the number of subcells in the stack (N) and the albedo parameter (R). (b) A replot of (a) shows how the efficiency increases with N and R.

cells. In practice, it will be helpful to modify the S-Q analysis to calculate the corresponding "practical" thermodynamic limit that accounts for the inability to fabricate semiconductors with ideal bandgaps, material-specific losses, such as self-heating (Chapter 5), incomplete absorption in materials with finite thickness (Chapter 6), non-radiative losses (Chapter 7), etc. Although we will discuss the physics and the magnitude of these loss mechanisms later, in this section, we show that the S-Q triangle can easily account for these loss mechanisms and predict the corresponding performance.

4.6.1 Imperfect EQE and ERE in a tandem solar cell

A solar cell cannot convert all the incident above-bandgap photons to a useful photocurrent. For a single-junction device, we can write

$$I^*_{\mathrm{mp}} = \eta_Q \, I_{\mathrm{mp}} \tag{4.18}$$

where η_Q is the average external quantum efficiency (EQE) that accounts for the combined effects of imperfect absorption in finite-thickness cell and the loss due to the failure of photogenerated carriers to reach the contact due to electron–hole recombination. In addition, all solar cells suffer from non-radiative recombination. This reduces the steady-state carrier concentration and photon density inside the device. The lower photon density translates to a lowered external radiative efficiency (ERE). Therefore, V_{mp} is simultaneously affected by imperfect EQE and ERE, namely

$$V^*_{\mathrm{mp}} = V_{\mathrm{mp}} - \frac{k_{\mathrm{B}}T_{\mathrm{D}}}{q}\ln\frac{1}{\eta_R} - \frac{k_{\mathrm{B}}T_{\mathrm{D}}}{q}\ln\frac{1}{\eta_Q}. \tag{4.19}$$

For high-efficiency solar cells, $\eta_Q \sim 1$, we therefore need not consider for EQE explicitly in calculating the practical thermodynamic limit of a solar cell. We also know that $\eta_R < 1\%$ for indirect-bandgap semiconductors (e.g., Si) and $\eta_R \sim$10–20% for direct-bandgap semiconductors (e.g., GaAs). The following analysis accounts for the imperfect ERE.

Imperfect ERE due to non-radiative recombination reduces the operating voltage by

$$\Delta V_R = \frac{k_B T_D}{q} \ln \frac{1}{\eta_R}. \tag{4.20}$$

For example, $\Delta V_R \approx 130$ mV for Si cells, and $\Delta V_R \approx 40$ mV for high-efficiency GaAs cells. Since each material has a different ΔV_R, v_{mp} of the i-th subcell of a tandem cell will be reduced by $\Delta_i = \Delta V_R / V_0$. We can now rewrite the normalized output of the bifacial tandem cell including the effect of non-radiative recombination as follows:

$$s_N^* = s_N - x \sum_{i=-D}^{U} \Delta_i - x\Delta_0 \tag{4.21}$$

where s_N is the output when η_R is 100% in all subcells. Maximizing s_N^* with respect to x defines the optimum bifacial tandem configuration.

For example, consider a conventional tandem solar cell where Δ is the ERE loss of the subcells. By setting $ds_N^*/dx = 0$, we find that

$$\frac{I_{mp}}{I_0} = x_0 = \frac{1-\Delta}{(N+1)} \tag{4.22}$$

and

$$\max\{s_N^*\} = s_N^*(x = x_0) = \frac{N}{2(N+1)}(1-\Delta), \tag{4.23a}$$

$$\eta_N = \frac{V_0 I_0}{c P_{in}} s_N = \frac{V_0 I_0}{c P_{in}} \frac{N}{2(N+1)}(1-\Delta). \tag{4.23b}$$

Therefore, in the limit when $\eta_R = 1$ and $\Delta = 0$,

$$\eta_N = \frac{V_0 I_0}{c P_{in}} \times \max\{s_N^*\} = \frac{V_0 I_0}{c P_{in}} \times \frac{N}{2(N+1)}, \tag{4.24}$$

we recover the ideal limit, Eq. (4.5), as expected.

4.7 Third-generation solar cells

In the preceding sections, we discussed first- and second-generation solar cells, shown in Fig. 4.2. The **first-generation** solar cells include single-junction devices illuminated by unconcentrated sunlight with a maximum conversion efficiency of ~33%. These solar cells have been commercially successful and are used worldwide, but they are reaching the energy conversion limit. The **second-generation** solar cells focus on improving energy conversion by tandem solar cells, concentrator solar cells, bifacial cells, and so on. They offer much higher efficiency, but the complexity and cost have limited them to niche applications. The **third-generation** solar cells focus on some of the limitations of the second-generation solar cells in achieving higher efficiency. For example, an intermediate-bandgap solar cell can achieve performance comparable to a triple-junction cell, but using a single material. A luminescent planar collector can achieve performance comparable to that of a concentrator tandem cell, but does not need a large parabolic concentrator. In this section, we will use the S-Q triangle to explore the physics and fundamental limits of some of the third-generation solar cells. Once you learn how to use the S-Q triangles for these systems, you will be able to analyze any new concept.

Homework 4.13: Thermodynamic efficiency with AM0 illumination

In the discussion above, we calculated the thermodynamic limits with AM1.5G illumination, where J_{sc} reduces linearly with increasing E_g. How would one determine of efficiency and bandgaps of various types of solar cells under AM0 illumination, where the J_{sc} is approximately linear for $E_g \leq 1.5$ eV, but deviates significantly from linearity for $E_g > 1.5$ eV?

Solution. We recall from Homework 3.6 that if $x \equiv E_g/k_B T_S \sim 2E_g$, then the photocurrent generated by the AM0 spectrum is given by

$$J_{sc} = \begin{cases} 120(1-0.25x), & \text{if } x \leq 3 \\ 44.33(x^2+2x+2)e^{-x}, & \text{if } x \geq 3. \end{cases}$$

We can use the linear approximation, i.e. $J_{sc} = 120(1-0.25x) = 120(1-0.5E_g)$ to approximately determine the bandgaps ($E_{g,n} \equiv x_{0,n}\, k_B T_S$). For $E_g > 1.5$ eV, we determine the correct bandgaps ($E_{g,n} = x\, k_B T_S$) by requiring that the current be the same for the approximate and exact bandgaps, i.e.,

$$120(1-0.25x_0) = 44.33(x^2+2x+2)\,e^{-x}.$$

4.7.1 Intermediate-band solar cell as a split-spectrum tandem cell

As shown in Figs. 4.8(a, b), an intermediate-band solar cell has two parts. The high-energy photons ($\hbar\omega > E_{12}$) are absorbed by a material with bandgap $E_g = E_2 - E_1 \equiv E_{12}$. The sub-BG photons are then absorbed by a second material with the same bandgap E_g, but with added impurities at energy level E_3 that capture two sub-BG photons with energies $E_{13} + E_{32} = E_g$. The impurity band can be viewed as a "tunnel junction" that allows the electrons excited during E_{13} excitation to "recombine" with the hole left behind by the E_{32} excitation. Let us use

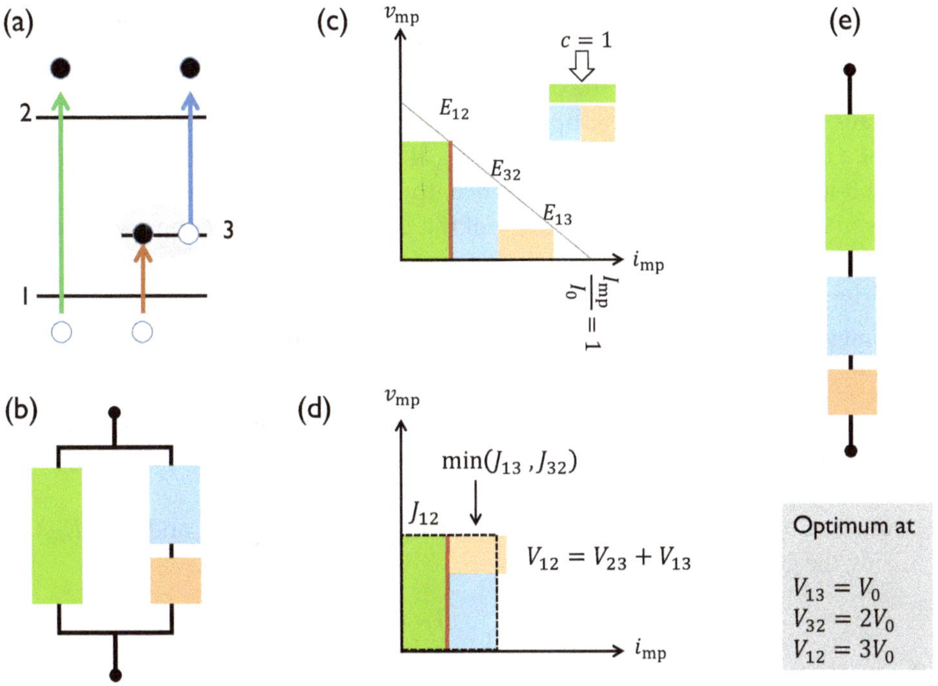

Figure 4.8: (a) Three transitions of an intermediate-band solar cell. (b) The color-coded electrical equivalent circuit highlights the series–parallel combination of the three transitions. (c) The red lines indicate that E_{12} transition is independent of E_{13} and E_{32} transitions. (d) The E_{13} and E_{32} transitions occur in series, therefore the corresponding rectangles from part (c) are stacked on top of each other. (e) The optimum transitions satisfy the same bandgap sequence of a traditional triple-junction tandem cell.

the S-Q triangle in Fig. 4.8 to show that the approach cannot exceed the efficiency of a standard 3-junction tandem cell.

The points of the triangle in Fig. 4.8(c) indicate the maximum power point voltages and currents, if the transitions acted independently. In practice, E_{12} transition is independent of the other two transitions, a fact indicated by the red line. Using the split-spectrum approach, we use the $E_g = E_{cut}$ line to divide the S-Q triangle into two parts. The power output of the first cell is defined by the point $(V_{mp}(E_{cut}), J_{sc}(E_{cut}))$ on the S-Q line. The two sub-BG photons correspond to the points $(V_{mp}(E_{13}), J_{sc}(E_{13}))$ and $(V_{mp}(E_{32}), J_{sc}(E_{32}))$. The E_{13} and E_{32} transitions occur in series, and therefore their corresponding rectangles must be stacked on top of each other (see Fig. 4.8(d)). The current in these series-connected transitions is constrained by the smaller of the two currents, indicated by the dashed line. We can now sum the rectangles to calculate the power output of the cell

$$P = (J_{sc}(E_{cut}) + \min(\Delta J_{sc}(E_{13}), \Delta J_{sc}(E_{32}))) \times V_{mp}(E_{cut}) \tag{4.25}$$

where $\Delta J_{sc}(E_{13}) \equiv J_{sc}(E_{13}) - \Delta J_{sc}(E_{23})$ and $\Delta J_{sc}(E_{32}) \equiv J_{sc}(E_{32}) - \Delta J_{sc}(E_g)$.

The energy conversion would be maximized if E_g, E_{13} and E_{23} are chosen such that (i) the currents associated with the transitions are equal (i.e., $J_{sc}(E_g) = J_{sc}(E_{13}) = J_{sc}(E_{32})$) and (ii) the open-circuit voltages V_{13}, V_{32}, V_{12} associated with the transitions are integer multiples of each other. This condition is identical to that of a triple-junction tandem cell (see Fig. 4.8(e)). This is the limiting efficiency of the 3-junction tandem cell. In practice, the finiteness of the impurity energy bands makes the conversion efficiency of the impurity-band PV somewhat lower than that of the corresponding 3-junction tandem cell.

Homework 4.14: An intermediate-bandgap cell needs an energy ratchet

Show that the 3-level system shown in Fig. 4.8 can reach its thermodynamic efficiency limit only if the angle entropy loss is eliminated at an extremely high concentration.

Solution. Let us write the expression for $V_{mp}(E_{12})$, $V_{mp}(E_{13})$, and $V_{mp}(E_{23})$ using the general formula:

$$V_{mp}(E_{ij}) = \left(1 - \frac{T_D}{T_S}\right) E_{ij} - \Delta,$$

where $\Delta \equiv (k_B T_D/q)\ln(\theta_D/c\theta_S)$ is the angle entropy factor.

It is easy to show that, for $E_{12} \equiv E_{13}+E_{32}$, it is impossible to match the optimum voltage, i.e., $V_{12} \neq V_{13}+V_{32}$, unless the angle entropy loss is neglected. The system can be designed only if a new level E_4 is introduced slightly below E_3 to account for the entropy mismatch, and photo-excited carriers are transferred from E_3 to E_4 in a *single* direction (which may be viewed as an "energy ratchet"). Thereafter, a second photon will be absorbed for a transition from E_4 to E_2 without allowing it to return to E_3.

4.7.2 Multiple-exciton generation solar cells (MEG-PV) compared to a double-junction tandem cell

One can prevent the thermalization loss of a high-energy photon by using the excess energy to excite multiple electron–hole pairs (excitons) in the valance and conduction bands. Let us use the triangle diagram in Fig. 4.9 to explain the performance limit of a MEG-PV solar cell. Note that in Fig. 4.8, we contacted energy levels 1 and 2 to extract carriers from intermediate-bandgap solar cells. The internal defect level 3 was not contacted. A MEG involves contacting levels 1 and 3, but keeping level 2 uncontacted.

Two types of excitations are expected. The first involves the direct excitation from level 1 to level 3 by the solar photons of the corresponding energy. The second process involves photoexcitation from level 1 to level 2, and then relaxation from

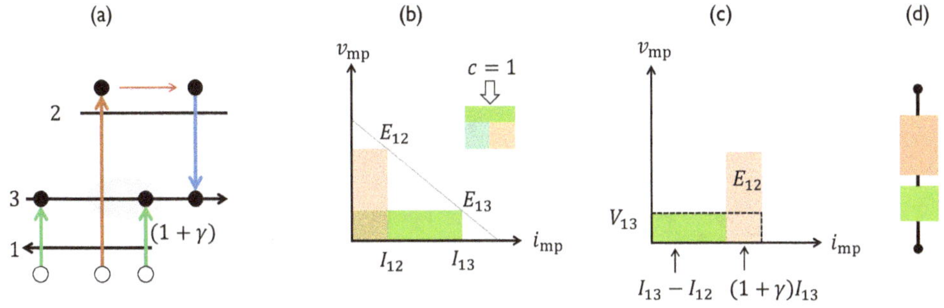

Figure 4.9: (a) The transition diagram for an excitonic solar cell. (b) The transitions are mapped on the S-Q line. (c) The current and voltage of the integrated system. (d) Performance comparison with a double-junction tandem cell.

2 to 3. The energy released during the 2–3 transition creates a new 1–3 transition. The first process involves a single electron–hole pair generation, while the second process produces two electron–hole pairs. In Fig. 4.9(b), the points along the S-Q line define the maximum power point current and voltages for E_{12} and E_{13} transitions. Figure 4.9(c) shows that the output voltage of the cell is defined by E_{13} transition, and the total current is given by the sum of the currents associated with the two transitions. In essence, the performance of a three-level excitonic cell falls between a single-junction and a double-junction tandem cell. For $\gamma = 0$ (i.e., exciton yield is zero), the MEG-PV must be optimized as a single-junction solar cell. For $\gamma = 1$ (i.e., every hot electron excites an additional electron), an optimized MEG-PV has the same current and voltage matching condition as that of a two-junction solar cell.

4.7.3 A hot-electron solar cell converts the thermalization energy of a solar cell

In Fig. 4.10, energy levels 1 and 3 are conduction and valence bands of a semiconductor, so that $E_g = E_3 - E_1$. A hot-electron solar cell distributes the excess energy available to above-bandgap photons ($\Delta E_{23} = E_2 - E_3$) among all the electrons excited to the conduction band, E_3. The hot electrons are collected from E_4, where $E_{34} \propto E_{23}$. The efficiency increases because while the short-circuit current is determined by the smaller bandgap, E_{31}, its open-circuit voltage is determined by a larger bandgap, E_{14}. The approach works only if the transition from E_3 to E_4 is highly efficient. Otherwise, the slow build-up of carriers at E_3 will suppress absorption and eventually the short-circuit current will be determined by E_{14}, not E_{13}. The corresponding S-Q triangle explains the improved efficiency of the idealized hot-carrier solar cell.

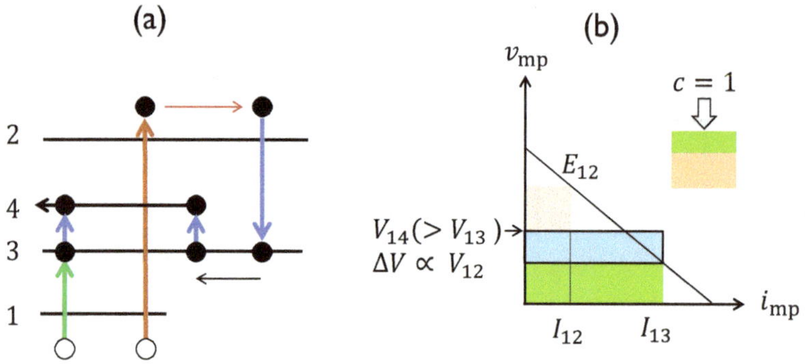

Figure 4.10: (a) A hot-electron solar cell involves transition between four states. (b) The corresponding S-Q triangle defines the excess energy available to heat the electrons.

4.7.4 Flat-plate luminescent concentrator solar cell

A concentrator solar cell uses a parabolic reflector to concentrate sunlight. The reflector focuses normally incident sunlight on the foci of the parabola. This implies that the concentrator solar cell must track the sun across the sky, and it will not work if the fraction of diffuse light is high. A luminescent collector concentor solar cell offers an opportunity to circumvent this limit. We can use the S-Q triangle to explain the operation of a luminescent concentrator solar cell.

For the luminescent concentrator cell shown in Fig. 4.11, the absorber region consists of atoms (or molecules, or quantum dots, or nanocrystals) that can absorb sunlight at one wavelength, but re-emit at a different wavelength. The re-emitted photons are guided toward the solar cell on the right. The intensity of the photons incident on the solar cell is defined by the area ratio of the absorber and the cell, i.e. $c \propto A_r/A_c$, as shown in Fig. 4.11(a). The effect is analogous to a concentrator solar cell (except for the wave-guiding and other losses (e.g., reabsorption, non-unity QE, etc.) and the narrowness of the luminescent re-emission), where the intensity increase is proportional to the parabolic reflector (see Fig. 4.11(b)) and can be designed by using the S-Q triangle for a concentrator solar cell, as shown in Fig. 4.11(c).

The luminescent concentrator cell can also be used in a tandem configuration. As shown in Fig. 4.11(d), the longer-wavelength photons transmit through the upper layer and are absorbed by the typical flat-plate solar cell at the bottom. The shorter-wavelength photons are absorbed by the atoms of the top absorber, and re-emitted and guided to the larger-bandgap solar cell on the right. Figure 4.11(e) shows an analogous concentrator cell where longer-wavelength photons are absorbed by the parabolic plate, while shorter-wavelength photons are concentrated onto the top cell. The small-area, short-wavelength cell can be designed as a concentrator cell (Fig. 4.11(c)), while the longer-wavelength cell can be designed

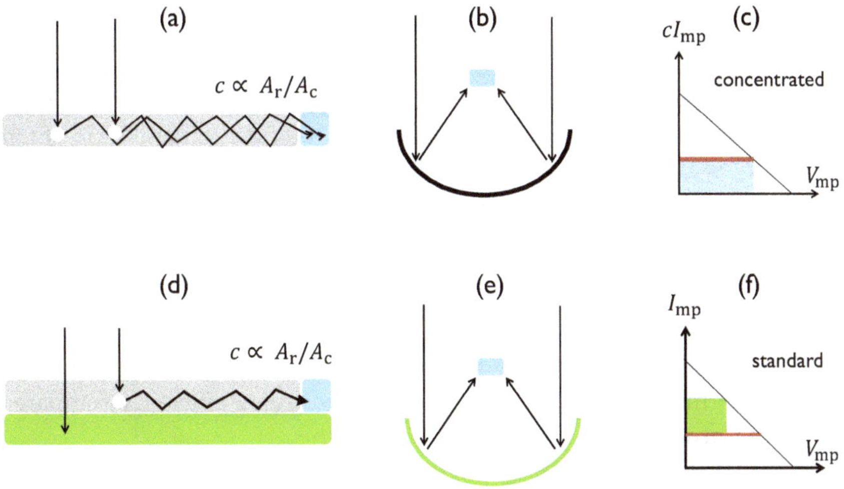

Figure 4.11: (a) A typical luminescent concentrator solar cell. The arrows indicate absorption and wave-guiding of the photons. (b) A typical concentrator solar cell. (c) Design of a concentrator solar cell using the S-Q triangle. (d) A tandem luminescent concentrator solar cell. The longer-wavelength (smaller-bandgap) cell operates at one-sun concentration, while the shorter-wavelength (larger-bandgap) cell operates at multiple-sun concentration. (d) A tandem cell where the top cell operates as a luminescent concentrator solar cell. (e) A hypothetical concentrator solar cells, with the blue cell in the center operating in the concentrator mode, while the other cell embedded in the reflector operates at one-sun concentration. (f) Design of the one-sun bottom cell using the S-Q triangle.

as a traditional solar cell (Fig. 4.11(f)). The red lines define the wavelength that separate the photoexcitation of the two cells.

4.8 Conclusions

The S-Q triangle offers an efficient and powerful technique to derive the thermodynamic efficiency limits a variety of classical (e.g., single-junction, tandem, and concentrator cells) and emerging (e.g., bifacial tandem cells) technologies. Third-generation solar cell concepts, such as intermediate-bandgap cells, multiple-exciton PV, luminescent concentrator, etc., can be interpreted by using this approach. The sequence of optimum bandgaps and the thermodynamic limits of currents and voltages are easily derived and can serve as an intuitive check for the experimental data. The approach provides, as a function of subcell number, a scaling justification for the improvement in the tandem cell efficiency and abrupt increase in bifacial tandem cell efficiency. We also explained how to include various non-ideal effects related to finite absorption, radiative and non-radiative recombinations, AM0 illumination, etc.

Homework 4.15: Carnot vs. Landsberg limits

In Chapter 2, we found that the ultimate conversion efficiency of a two-level atom kept at temperature $T_{\rm D}$ and excited by a "monochromatic" sun at temperature $T_{\rm S}$ is given by the Carnot limit, i.e., $\eta = (1 - T_{\rm D}/T_{\rm S}) \approx 95\%$. Landsberg *et al.* emphasized that an engine driven by radiative photon fluxes is different from a Carnot engine and its ultimate efficiency is given by

$$\eta = 1 - \frac{4}{3}\frac{T_{\rm D}}{T_{\rm S}} + \frac{1}{3}\left(\frac{T_{\rm D}}{T_{\rm S}}\right)^4. \tag{4.26}$$

Provide a thermodynamic derivation of the efficiency limit and physically interpret the terms of Eq. (4.26). How does the Landsberg efficiency compare to that of the Carnot limit?

Solution. Gordon *et al.* published an elegantly simple derivation of the Landsberg limit in the *American Journal of Physics*, 61(9), 1993. We recall from Homework 1.10 that the power radiated by a blackbody is given by $I_s = \sigma T^4$. Similarly, the energy density of photons is obtained by integrating Eq. (1.13) over energy and solid angle 4π, i.e., $Q_{ph} = \sigma^* T^4$, where $\sigma^* \equiv \frac{\pi^2 k_{\rm B}^4}{15\,\hbar^3 c^3}$ and $\sigma^*/\sigma = 4/c$. The key observation is that a radiation source has a large, but finite, temperature-dependent heat capacity, namely, $C_{\rm v}(T) \equiv dQ_{\rm ph}/dT = 4\sigma^* T^3$. For maximum energy output W, one replaces a single Carnot engine (e.g., a two-level atom in Fig. 2.3) with an infinite number of Carnot engines, each operating between reservoirs with infinitesimal temperature differential T and $T + \Delta T$ and then sums up the energy output of each engine. In other words,

$$W = \int_{T_{\rm D}}^{T_{\rm S}} \left(1 - \frac{T_{\rm D}}{T}\right) \cdot C_{\rm v}(T)dT = \sigma^* T_{\rm S}^4 \left(1 - \frac{4}{3}\frac{T_{\rm D}}{T_{\rm S}} + \frac{1}{3}\left(\frac{T_{\rm D}}{T_{\rm S}}\right)^4\right).$$

The efficiency $\eta \equiv W/Q_{\rm ph}(T_{\rm S})$ then gives us Eq. (4.26). With $T_{\rm S} = 6000$ K and $T_{\rm D} = 300$ K, $\eta = 93.3\%$. This Landsberg efficiency is slightly lower than the Carnot limit, but is higher than the ultimate energy conversion limit (~86.6%) under AM0 illumination of an infinite-junction tandem cell. (The S-Q triangle predicts a slightly higher efficiency for AM1.5G illumination.) The linear temperature-dependent term of Eq. (4.26) can also be interpreted by equating the entropy of a radiation source (i.e., $S_{\rm S} \equiv \int_0^{T_{\rm S}} d{\rm v}/T = \int_0^{T_{\rm S}} c_{\rm v}(T)\,dT/T = (4/3)\sigma^* T_{\rm S}^3$) to the entropy of the device (i.e., $S_D = Q_{lost}/T_{\rm D}$). Since energy conversion requires $Q_{\rm ph} = Q_{\rm lost} + W$, therefore $\eta = W/Q_{ph}(T_{\rm S}) = 1 - (4/3)(T_{\rm D}/T_{\rm S})$. The second temperature-dependent term of Eq. (4.26) is interpreted as radiation from the heat engine returning to the reservoir.

(*continued on the next page*)

Homework 4.15 (*continued from the previous page*)

The Landsberg limit of 93.3% is higher than the ultimate multicolor efficiency limit 86% discussed in Homework 4.11. A paper by Martin Green (*Nano Letters*, 11, 5985–5988, 2012) suggests a strategy to close this gap by using magneto-optical meta-materials that break the symmetry between absorption and emission angles and recycle/reabsorb the emitted photons due to radiative recombination.

In deriving the thermodynamic limit, we assumed that the cell temperature equals the ambient temperature, implying an infinitely fast heat dissipation. In practice, the heat dissipation rates due to conduction, convection, and/or radiation are finite. Therefore, the module temperature will always be higher than the cell temperature. In the next chapter, we will discuss the implications of self-heating on the fundamental limits of cell efficiency achievable in an otherwise ideal solar cell.

Homework 4.16: A different model of maximum power

Read the paper by A. Sergeev and K. Sablon to show that

$$\frac{qV_{\rm mp}}{k_{\rm B}T} = \mathrm{LW}(\gamma Ae) - 1$$

$$\eta_m = \gamma(k_{\rm B}T/e^*)[\mathrm{LW}(\gamma Ae) - 2 + (1/\mathrm{LW}(\gamma Ae))]$$

where LW is Lambert's W function, e^* is the average energy per phonon, $e = 2.7183$ is Euler's number, $A \equiv k_{\rm l}R_{\rm ab}/R_{\rm e}m(T)$, $k_{\rm l}$ is the luminescence quantum yield, and γ is the inverse Auger process, which can be set to 1.

References

[1] William Shockley and Hans J. Queisser. Detailed Balance Limit of Efficiency of p-n Junction Solar Cells. *Journal of Applied Physics*, 32(3):510, 1961.

[2] Louise C. Hirst and Nicholas J. Ekins-Daukes. Fundamental losses in solar cells. *Progress in Photovoltaics: Research and Applications*, 19(3):286–293, 2011.

[3] S. Dongaonkar, C. Deline, and M. A. Alam. Performance and Reliability Implications of Two-Dimensional Shading in Monolithic Thin-Film Photovoltaic Modules. *IEEE Journal of Photovoltaics*, 3(4):1367–1375, October 2013. Conference Name: IEEE Journal of Photovoltaics.

[4] T.J. Silverman, M.G. Deceglie, Xingshu Sun, R.L. Garris, M.A. Alam, C. Deline, and S. Kurtz. Thermal and electrical effects of partial shade in monolithic thin-film photovoltaic modules. *IEEE Journal of Photovoltaics*, 5(6):1742–1747, November 2015.

[5] C. H. Henry. Limiting efficiencies of ideal single and multiple energy gap terrestrial solar cells. *Journal of Applied Physics*, 51(8):4494–4500, August 1980.

[6] U.A. Yusufoglu, T.M. Pletzer, L.J. Koduvelikulathu, C. Comparotto, R. Kopecek, and H. Kurz. Analysis of the Annual Performance of Bifacial Modules and Optimization Methods. *IEEE Journal of Photovoltaics*, 5(1):320–328, January 2015.

[7] A. Luque, E. Lorenzo, G. Sala, and S. López-Romero. Diffusing reflectors for bifacial photovoltaic panels. *Solar Cells*, 13(3):277–292, January 1985.

[8] Reza Asadpour, Raghu V. K. Chavali, M. Ryyan Khan, and Muhammad A. Alam. Bifacial Si heterojunction-perovskite organic-inorganic tandem to produce highly efficient (33%) solar cell. *Applied Physics Letters*, 106(24):243902, June 2015.

[9] M. Ryyan Khan and Muhammad A. Alam. Thermodynamic limit of bifacial double-junction tandem solar cells. *Applied Physics Letters*, 107(22):223502, November 2015.

[10] Emre Gençer, Caleb Miskin, Xingshu Sun, M. Ryyan Khan, Peter Bermel, M. Ashraf Alam, and Rakesh Agrawal. Directing solar photons to sustainably meet food, energy, and water needs. *Scientific Reports*, 7(1):3133, June 2017.

[11] M. T. Patel, M. R. Khan, and M. A. Alam. Thermodynamic Limit of Solar to Fuel Conversion for Generalized Photovoltaic-Electrochemical Systems. *IEEE Journal of Photovoltaics*, 8(4):1082–1089, July 2018.

[12] Martin A. Green. *Third Generation Photovoltaics*. Springer-Verlag Berlin Heidelberg, [New York], 2006.

[13] Andrew S. Brown and Martin A. Green. Detailed balance limit for the series constrained two terminal tandem solar cell. *Physica E: Low-dimensional Systems and Nanostructures*, 14(1–2):96–100, April 2002.

[14] J. L. Gray and J. R. Wilcox. The design of multijunction photovoltaic systems for realistic operating conditions. In *2013 IEEE 56th International Midwest Symposium on Circuits and Systems (MWSCAS)*, pages 697–700, August 2013.

[15] Muhammad Ashraful Alam and M. Ryyan Khan. Thermodynamic efficiency limits of classical and bifacial multi-junction tandem solar cells: An analytical approach. *Applied Physics Letters*, 109(17):173504, October 2016.

CHAPTER 5

Self-heating of Solar Cells

Chapter Summary

- Thermalization losses make solar modules operate 20–40 degrees hotter than the ambient.
- The degree of heating depends on the balance of energy absorbed by the module and the cooling provided by convection and radiation.
- The temperature-coefficient of power loss is an important parameter (defined primarily by the temperature-dependent loss of open-circuit voltage) and it can be easily calculated numerically and analytically.
- The temperature-sensitivity of the open-circuit voltage can be used to determine the operating temperature of the solar cells.
- The solar cells can be cooled by various ways, including innovative approaches based on radiative cooling or series-connecting an energy harvester.

5.1 Introduction

In Chapters 2 through 4, we calculated the efficiency of an idealized solar cell by assuming that the device temperature (T_D) is the same as the ambient temperature (T_A). This is fundamentally impossible! A single-junction solar cell converts only a fraction (e.g., ~30–33%) of the incident energy into electrical output. Even if a fraction of the remaining energy is dissipated within the cell as heat, the cell temperature will rise considerably. Indeed, experiments show that the cell is always significantly hotter than the environment, i.e., $T_D - T_A \approx 20 - 40$ K.

The self-heating of PV modules reduces both short-term and long-term power outputs. In the short term, the energy output is reduced because the cell efficiency decreases with temperature, i.e.,

$$\eta(T_D) = \eta(T_A)\left[1 + \beta(T_D - T_A)\right]. \quad (5.1)$$

Here, $\eta(T_A)$ is the output power at the standard testing conditions (e.g., at $T_A = 298\,\mathrm{K}$) and β is the experimentally measured temperature coefficient: β is negative because $\eta(T_D)$ decreases with increasing temperature.

In the long term, a hotter module degrades faster, because corrosion, polymer degradation, etc., are thermally activated. We will discuss in Part IV of the book (Chapters 21–25) that if the degradation rate, R, at T_D, is related to that at T_A by the Arrhenius factor, then

$$R(T_D) = R(T_A)e^{+(E_A/k_B)(1/T_A - 1/T_D)}. \tag{5.2}$$

For typical $E_A \approx 0.8\,\mathrm{eV}$, even a $10\,\mathrm{K}$ heating increases $R(T_D)$ (and thereby decreases the lifetime) by more than a factor of 2! Field surveys show that solar modules in very hot climates degrade at ~1.5%/year, eight times faster than the ones installed in cold climates (~0.2 %/year).

In this chapter, we will discuss the physics of self-heating of a PV module through Eq. (5.1) by first calculating T_D (Sec. 5.2) and then $\eta(T_D)$ (Sec. 5.3). Since T_D itself depends on $\eta(T_D)$, the calculation must be self-consistent. We will conclude the chapter by explaining the various techniques used to cool solar cells and increase their efficiency.

5.2 Self-heating is defined by a complex balance of multiple fluxes

A fraction of the solar energy absorbed by the solar cell is converted to output electrical energy, and the rest is dissipated as heat. Figure 5.1 and Eq. (5.3) show that the incoming (absorbed light) and outgoing energy fluxes must be balanced

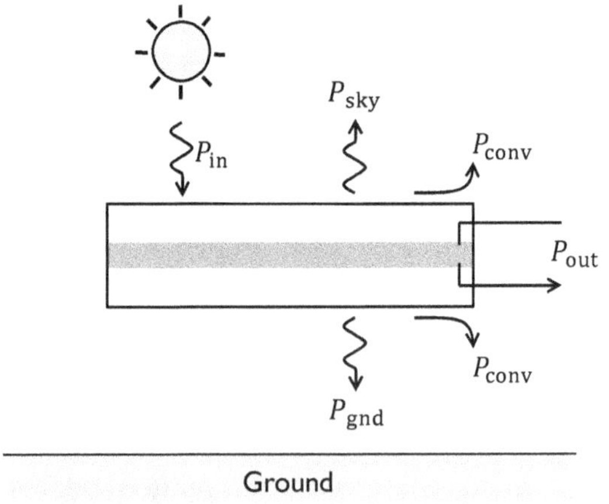

Figure 5.1: Schematic of a terrestrial PV module, where we identify the incoming and outgoing energy fluxes. Table 5.1 summarizes the terms of the energy balance equation for a solar module.

for a module to reach its steady-state temperature.

$$P_{\text{abs}} = \eta P_{\text{in}}(T_{\text{D}}) + 2P_{\text{conv}} + P_{\text{sky}} + P_{\text{gnd}}. \tag{5.3}$$

Here, the absorbed solar irradiance, P_{abs}, is determined by the solar spectrum (e.g., AM1.5G) and the absorptance of the solar cell (more on this in Chapter 6). The output power delivered by PV modules to the external load is $P_{\text{out}} \equiv \eta P_{\text{in}}$. The convective cooling by air at the top and bottom surfaces and conductive heat transfer through the aluminum frames are given by $2P_{\text{conv}}$. The sky cooling, P_{sky}, accounts for the radiative energy exchange with the atmosphere from the side facing the sky. Similarly, P_{gnd} accounts for the energy transfer to the ground from the back side. We will discuss these terms in more detail and summarize them in Table 5.1.

Table 5.1: Energy balance to calculate PV self-heating.

Energy balance	$$P_{\text{abs}} = P_{\text{out}} + 2P_{\text{conv}} + P_{\text{sky}} + P_{\text{gnd}}. \tag{5.4}$$ The output power is $P_{\text{out}} = \eta(T_{\text{D}})P_{\text{in}}$, and P_{in} is the solar irradiance.
Convection	$$P_{\text{conv}} = h(T_{\text{D}} - T_{\text{A}}). \tag{5.5}$$ The convection coefficient h is a function of wind velocity v_{w}.
Radiation/emission from module	$$P_{\text{emi}}(T) = \int d\Omega \cos\theta \int_0^\infty dE P_{\text{BB}}(T,E) \times \epsilon(E,\Omega). \tag{5.6}$$ Here $\epsilon(E,\Omega)$ is the angular emissivity of the module; $P_{\text{BB}}(T,E)$ the blackbody radiation spectrum per solid angle.
Net radiation to sky	$$P_{\text{sky}}(T,T_{\text{A}}) = \text{VF} \times (P_{\text{emi}}(T_{\text{D}}) - P_{\text{atm}}(T_{\text{A}})), \tag{5.7}$$ where thermal radiation from the atmosphere to PV modules is $$P_{\text{atm}}(T_{\text{A}}) = \int d\Omega \cos\theta \int_0^\infty dE\, P_{\text{BB}}(T_{\text{A}},E)\,\epsilon(E,\Omega)\,\epsilon_{\text{atm}}(E,\Omega). \tag{5.8}$$ Using Kirchhoff's law and the Beer–Lambert law, the angular emissivity of the atmosphere $\epsilon_{\text{atm}}(\lambda,\Omega)$ can be written as $\epsilon_{\text{atm}}(\lambda,\Omega) = 1 - t_{\text{atm}}(\lambda)^{1/\cos\theta}$, where $t_{\text{atm}}(\lambda)$ is the atmospheric transmittance.
Net radiation to ground	$$P_{\text{gnd}} = \text{VF} \times \epsilon\sigma(T_{\text{D}}^4 - T_{\text{amb}}^4). \tag{5.9}$$ Remember that ϵ is the emissivity averaged over E and integrated over the hemisphere. Here, VF is the view factor.

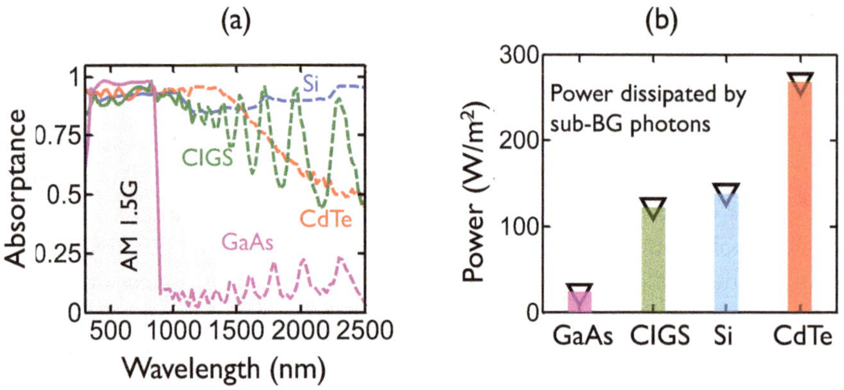

Figure 5.2: (a) Measured absorptance for different solar absorber materials vs. photon wavelength (solid lines, above-bandgap photons; dashed lines, below-bandgap photons). The light pink region in the background is the AM1.5G spectrum. (b) Power dissipated within the module due to sub-BG photons for different technologies.

5.2.1 Self-heating: The absorbed photon flux heats the solar cell

We recall from Table 3.1 that if the solar photon flux is given by $n_S(E)$, the corresponding input power is

$$P_{\text{in}} = \int_0^\infty E\, n_S(E) dE. \tag{5.10}$$

A fraction of these photons are absorbed, defined by the energy-dependent absorptance, $\epsilon(E)$. Figure 5.2 shows a typical wavelength-dependent absorptance profile of various solar cell technologies. Most of the above-bandgap photons are absorbed, i.e., $\epsilon(E > E_g) \sim 1$. Some of the below-bandgap photons, not absorbed by the semiconductor, are partially absorbed by the front and back electrodes, protective backsheets, so that $\epsilon(E < E_g) < 1$. Therefore,

$$P_{\text{abs}} = \int_0^\infty \epsilon(E) E\, n_S(E) dE.$$

The power absorbed is dissipated by various loss mechanisms, as discussed below.

5.2.2 Self-cooling: A module can be cooled by convection

A simple formula describes the energy loss by air convection from the top and the bottom surfaces of a solar module, namely,

$$2P_{\text{conv}} = 2h(T_D - T_A). \tag{5.11}$$

The convection coefficient h depends nonlinearly on the wind velocity (v_w) close to the module surface. Empirically,

$$2h = 10.45 - v_w + 10\sqrt{v_w}\ \text{W/K}\cdot\text{m}^2 \tag{5.12}$$

is valid up to $v_\mathrm{w} \approx 20$ m/s. Physically, the constant (i.e., 10.45) at $v_\mathrm{w} \to 0$ can be derived from the physics of heat transfer from a heated surface, i.e., $h \approx 0.5\left((T_\mathrm{D} - T_\mathrm{A})/\delta\right)^{0.25}$, where δ is the thin interfacial layer at the module/air interface that dictates the heat transfer. The second and third terms at higher wind speeds approximate

$$h \approx c_2 \frac{k_\mathrm{air}}{L_\mathrm{m}} \left(\frac{v_\mathrm{w} L_\mathrm{m}}{\nu_\mathrm{air}}\right)^m P_\mathrm{r}^n \propto v_\mathrm{w}^m,$$

where $c_2 = 133$, $m = 1/5$, and $n = 1/3$ are fitting constants, and $k_\mathrm{air} = 0.026$ W/m K is thermal conductivity, $\nu_\mathrm{air} = 1.57 \times 10^{-5}$ m^2 s^{-1} is the kinematic viscosity, $P_\mathrm{r} = 0.707$ is the Prandtl number of dry air, and $L_\mathrm{m} = 1.5$ m is the size of a typical module. The wind speed v_w is measured at the module height.

Homework 5.1: Heat dissipation within a module

Consider a 20% efficient solar module that reflects or transmits 30% of the incident power without absorption. The electrical contacts and the protective backsheet absorb the remaining sub-bandgap photons. If illuminated by the AM1.5G spectrum, how much heat is dissipated within the module?

Solution. The amount of sunlight absorbed is given by $P_\mathrm{abs} = (1 - R - T) \times P_\mathrm{in} = 0.7P_\mathrm{in}$. The electrical power output is $P_\mathrm{out} = \eta P_\mathrm{in} = 0.2P_\mathrm{in}$. Therefore,

$$P_\mathrm{heat} = P_\mathrm{abs} - P_\mathrm{out} = 0.5I_0 = 500 \text{ W/m}^2.$$

In other words, 50% of the 1000 W/m^2 AM1.5G solar illumination is dissipated as heat within the solar module.

Homework 5.2: Cooling due to air convection can be significant

From Homework 5.1, we see that the heat dissipation within the module is $P_\mathrm{heat} = 500$ W/m^2. Assume that the ambient temperature is $T_\mathrm{A} = 300$ K, the temperature coefficient $\beta = -0.41$ K^{-1}, and the convective heat transfer coefficient is given by Eq. (5.12). Determine the module temperature and temperature-dependent efficiency loss for $v_\mathrm{w} = 0$ and $v_\mathrm{w} = 10.45$ m/s.

Solution. For $v_\mathrm{w} = 0$, $2h = 10.45$ by Eq. (5.12). Using Eq. (5.11), we find that $(T_\mathrm{D} - T_\mathrm{A}) = P_\mathrm{heat}/2h = 500/10.45 \approx 50$ K. Therefore, $T_\mathrm{D} = 298 + 50 = 348$ K. If $\beta = -0.41$ (see Eq. (5.1)), the self-heating will reduce the normalized efficiency by $\Delta\eta(T_\mathrm{D})/\eta(T_\mathrm{A}) = -0.41 \times 50 \approx 20.5\,\%$. Similarly, for $v_\mathrm{w} = 10.45$ m/s, $(T_\mathrm{D} - T_\mathrm{A}) = 15$ K. In this case, the relative efficiency loss is $\Delta\eta(T_\mathrm{D})/\eta(T_\mathrm{A}) = -0.41 \times 15 \approx 6.15\,\%$. Convective cooling by wind reduces T_D and improves efficiency $\eta(T_\mathrm{D})$.

5.2.3 Self-cooling: A solar cell is also cooled by radiation

Radiative heat exchange between the module and the sky/ground can further reduce the device temperature. The radiation can be subdivided into two parts: *net* radiation toward the sky, P_{sky}, and *net* radiation toward the ground, P_{gnd}. The radiation from the module should be $\epsilon\sigma T_{\text{D}}^4$ following the Stefan–Boltzmann law, where ϵ accounts for the imperfect emission from the heated body. The fraction of this radiation from the tilted module that goes toward the sky and the ground is quantified by the view factor, VF, as explained in Homework 12.3 and Fig. 12.5. The module also receives radiation from the sky (σT_{sky}^4) and the ground (σT_{gnd}^4). Therefore, we can write

$$P_{\text{sky}} = \text{VF} \times \epsilon\sigma \left(T_{\text{D}}^4 - T_{\text{sky}}^4\right), \tag{5.13}$$

$$P_{\text{gnd}} = \text{VF} \times \epsilon\sigma \left(T_{\text{D}}^4 - T_{\text{gnd}}^4\right). \tag{5.14}$$

Typically, $T_{\text{gnd}} \approx T_{\text{A}}$, and $T_{\text{sky}} \approx 250$ K. For a module tilt angle of β, the view factor from the sky to the front surface of the module is given by $\text{VF} = (1 + \cos\beta)/2$ (see Fig. 12.4). For example, the front face of a horizontal module ($\beta = 0$) has the full "view" of the sky; therefore, $\text{VF} = 1$.

Homework 5.3: Multiple channels for heat dissipation

The temperature of a PV module is $T_{\text{D}} = 320$ K. Let us assume that a windless day is characterized by the ambient temperature $T_{\text{A}} = 300$ K, the sky temperature $T_{\text{sky}} \sim 250$ K, and $2h \approx 10.45$ W/m^{2}K. Compare the heat dissipation fluxes by calculating P_{conv}, P_{gnd}, and P_{sky}.

5.2.4 Device temperature must be computed self-consistently

By inserting Eqs. (5.10), (5.11), (5.13), and (5.14) into Eq. (5.3), we can self-consistently determine the module temperature. The various self-heating and self-loss components are summarized in Table 5.1. Note that the model can be applied to single-junction, multi-junction, concentrator, and bifacial cells with appropriate change in P_{in} and η, as discussed in Chapter 4. In practice, Eq. (5.3) is sometimes simplified by setting $P_{\text{abs}} = P_{\text{in}}$, $P_{\text{sky}} \sim 0$, and $P_{\text{gnd}} \sim 0$ to empirically relate P_{in} directly to T_{D}:

$$T_{\text{D}} - T_{\text{A}} = P_{\text{in}}(1 - \eta) \times f(v_{\text{w}}), \tag{5.15}$$

where $f(v_{\text{w}}) \sim (2h)^{-1}$ depends on the convective heat transfer proportional to the wind velocity (v_{w}). Instead of Eq. (5.12), two widely used empirical formula, evaluated at open-circuit condition ($\eta = 0$), are given by:

$$f(v_{\text{w}}) = \begin{cases} e^{a+bv_{\text{w}}}, & \text{Sandia model} \\ 1/(a + bv_{\text{w}}) & \text{Faiman model} \end{cases}$$

where a and b are empirical coefficients.

5.2.5 Importance of wavelength-dependent radiation and absorption

The simplified flux balance allows us to approximately calculate T_D and explain why $T_D > T_A$. For a more precise calculation, we need to use a wavelength-dependent generalization of radiative heat transfer Eqs. (5.13) and (5.14). First, the sky is not really a blackbody at $T_{sky} \sim 250$ K. The peak emission from a 300 K blackbody occurs at 10 μm wavelength. However, the atmosphere is transparent (i.e., less emissive) between 8 and 13 μm (see Fig. 5.3(a)). This reduction in emission from the atmosphere makes the sky seem "cooler," approximated by a 250 K blackbody radiator. Moreover, we assumed emission from the PV module characterized by ϵ. For any unencapsulated solar absorber (e.g., Si, GaAs, CIGS, CdTe, etc.), the radiation peak at 300 K lies within the absorber bandgap; therefore, $\epsilon \approx 0$. In other words, we expect no radiative heat dissipation for a bare absorber. A glass-encapsulated module, however, does have high emissivity ($\epsilon \approx 0.84$) within the atmospheric window. Therefore, the glass-encapsulated module runs at a lower temperature (10–20 K) than a bare absorber, as shown in Fig. 5.3(b). In order to accurately model the radiative heat transfer, we need to integrate the relevant emission spectrum over the wavelengths and emission angles shown in Fig. 5.3(a), based on the equations in Table 5.1. Such an analysis shows that reduced sub-bandgap absorption (Fig. 5.2(a)) allows GaAs modules to operate at much lower temperatures (~310°K) compared to other technologies.

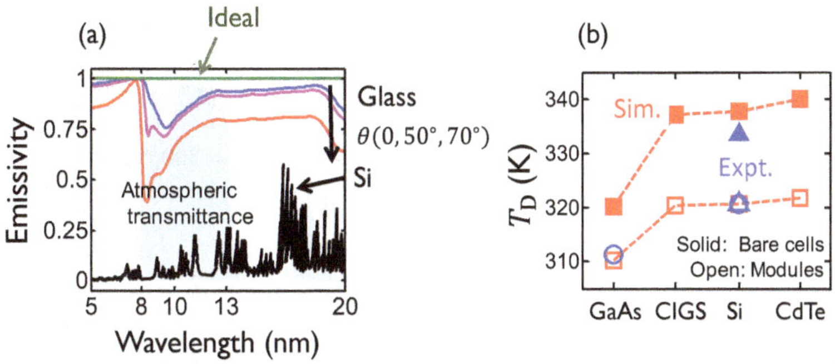

Figure 5.3: (a) Simulated emissivity (ϵ) profile of glass at different incident angles θ. The ideal emissivity for radiative cooling is shown by the green line. The light blue window defines the atmospheric transmittance. The bare silicon emissivity (black line) is significantly lower than that of glass. (b) The outdoor operating temperature of bare cells and encapsulated cells of GaAs, CIGS, Si, and CdTe. The square symbols with dashed lines are results simulated by the modeling framework summarized in Table 5.1. The experimental results (blue symbols) are taken from the literature.

5.3 Temperature-dependent efficiency of solar cells

Now that we know T_D, let us calculate $\eta(T_D)$ either numerically or analytically.

5.3.1 Numerical calculation of $\eta(T_\mathrm{D})$

We can use the numerical model described in Chapter 3 to directly calculate the cell efficiency as a function of T_D, except that T_D is now self-consistently obtained from Eq. (5.3). Figure 5.4(a) shows that the peak efficiency shifts at a lower bandgap, and the efficiency reduces with temperature. The efficiency loss arises from the redistribution of power budget (for $E_\mathrm{g} = 1.1$ eV, at an optimum power point) as a function of the solar cell temperature T_D is shown in Fig. 5.4(b). It is clear that P_Carnot and P_angle losses increase almost linearly with T_D (also obvious from the formula in Table 3.1). P_emi increases as well, but its contribution is negligible. As a result, the efficiency decreases almost linearly with increased temperature, which justifies Eq. (5.1). In the next section, we will analytically calculate the linear temperature coefficient β.

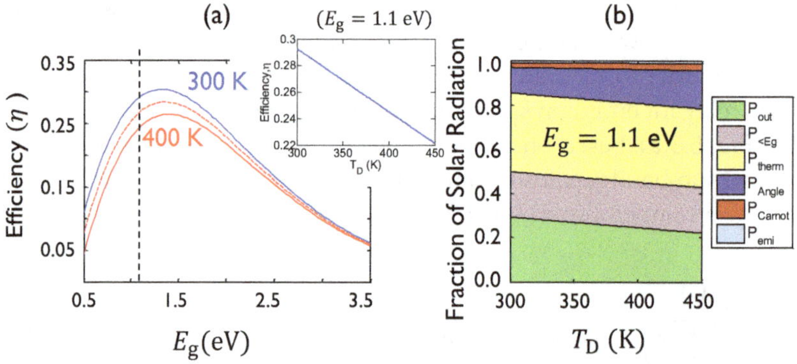

Figure 5.4: (a) A plot of $\eta(T_\mathrm{D})$ for various bandgaps for $T_\mathrm{D} = 300, 350, 400$ K. (b) Power budget of an $E_\mathrm{g} = 1.1$ eV solar cell as a function of T_D, showing the increasing contributions from P_Carnot and P_angle losses.

5.3.2 Analytical calculation of $\eta(T_\mathrm{D})$: Thermodynamic limit of temperature coefficient, β

Since $P_\mathrm{out} = V_\mathrm{mp} I_\mathrm{mp}$, therefore

$$\beta \equiv \frac{1}{\eta_\mathrm{STC}} \frac{d\eta}{dT_\mathrm{D}} = \frac{1}{P_\mathrm{out}} \frac{dP_\mathrm{out}}{dT_\mathrm{D}} = \frac{1}{V_\mathrm{mp}} \frac{dV_\mathrm{mp}}{dT_\mathrm{D}} + \frac{1}{I_\mathrm{mp}} \frac{dI_\mathrm{mp}}{dT_\mathrm{D}}. \tag{5.16}$$

We know that at the thermodynamic limit given by Eq. (4.1),

$$qV_\mathrm{mp} = \left(1 - \frac{T_\mathrm{D}}{T_\mathrm{S}}\right) E_\mathrm{g} - k_\mathrm{B} T_\mathrm{D} \left(\ln \frac{\theta_\mathrm{D}}{c\,\theta_\mathrm{S}}\right). \tag{5.17}$$

Taking a derivative with respect to T_D and evaluating the expression for a solar cell on Earth (i.e., $T_D \approx 300$ K and angle entropy of 0.314 at c = 1), we find that

$$\frac{1}{V_{mp}}\frac{dV_{mp}}{dT_D} = k_B\left(-\frac{E_g}{k_B T_S} - \frac{0.314}{k_B T_D}\right)/(0.95E_g - 0.314). \tag{5.18}$$

Similarly, we know from Eq. (4.2) that $I_{mp} = cI_{sun}(1-\beta_{sun}E_g)$, where $\beta_{sun} = 0.425$ for AM1.5G illumination. Taking the derivative with temperature, we find that

$$\frac{1}{I_{mp}}\frac{dI_{mp}}{dT_D} = \frac{-\beta_{sun}}{1-\beta_{sun}E_g}\frac{dE_g}{dT_D}. \tag{5.19}$$

We can add the contributions from Eqs. (5.18) and (5.19) in Eq. (5.16) to find β numerically. Interestingly, Eq. (5.18) suggests that we can determine the cell temperature T_D by measuring the temperature sensitivity of V_{mp}, a topic we will discuss in the following section.

Homework 5.4: Solar cell in a cold planet may not work at all

Compared to Earth, Pluto is very cold ($T_p = T_D = 50$ K). Compare the efficiencies of the solar cell deployed on Pluto vs. Earth.

Solution. The efficiency will be zero! Equation (4.1), from the previous chapter, shows that both terms of V_{mp}, namely, the Carnot factor and the angle entropy, improve at a low temperature. Recall that the angle entropy term is given by $k_B T_D \ln(\theta_D/\theta_S)$. From Eq. (1.3), $T_D \approx T_p \propto \sqrt{r_s/2d}$, while by Homework 1.1, $\theta_S = \pi(r_s/d)^2$. The decrease in $k_B T_D$ factor compensates for the increase in the angle ratio, i.e., θ_D/θ_S. (Earth is 8 light-minutes away, while Pluto is 4.5 light-hours away.) But all these considerations are irrelevant! We will see in Chapter 7 that a Si solar cell cannot operate properly at 50 K, because the dopants cannot be fully ionized at this temperature.

Homework 5.5: Thermodynamic limit of the temperature coefficient for c-Si

The temperature dependence of the bandgap of silicon is given by the empirical formula

$$E_g(T) = E_g(0) - \frac{\alpha_0 T_D^2}{T_D + \beta_0}. \tag{5.20}$$

(continued on the next page)

Homework 5.5 (*continued from the previous page*)

For silicon, $\alpha_0 = 4.77 \times 10^{-4}$ and $\beta_0 = 636$ K. Therefore, using Eq. (5.19), we find that $\frac{1}{I_{\text{mp}}}\frac{dI_{\text{mp}}}{dT_{\text{D}}} \sim 0.0006$. Also, from Eq. (5.18),

$$\frac{1}{V_{\text{mp}}}\frac{dV_{\text{mp}}}{dT_{\text{D}}} = \frac{(1.12/0.5 + 0.314/0.0258) \times 8.61 \times 10^{-5}}{(1.12 \times 0.95 - 0.314)} = -0.00172.$$

Therefore, $\beta = -0.00172 + 0.0006 = -0.0012$ or -0.12%. This limiting value applies to a variety of materials. Since thermodynamic limit calculations do not account for non-radiative recombination ($ERE = 1$), the temperature coefficient is somewhat smaller than the value observed in experiments. For silicon, V_{mp} can be ~280 mV lower than the thermodynamic limit due to non-radiative recombination and other series-resistance losses. This results in $\beta \approx -0.0048 + 0.0006 = -0.41\%$. (For these nonideal cases, the analysis is simplified by using the expressions derived in Homework 5.6.)

5.4 Determination of T_{D} by the temperature sensitivity of V_{oc}

In the discussion above, we have calculated T_{D} theoretically, but how do we know that our results are correct (without attaching a thermometer to the module)? An intensity-dependent measurement of V_{oc} can be used to directly estimate the operating temperature of a solar cell (see S. Ullbrich, "Turn Up the Heat: Absorption-Induced Open-Circuit Voltage-Turnover in Solar Cells," 2018). In particular, this approach can be useful in interpreting the self-heating in modules installed in a farm.

We saw in Homework 3.8 in Chapter 3 that the $J - V$ relationship of a solar cell (under concentrated c illumination) can be expressed in a classical form involving light and dark currents:

$$\begin{aligned} J &= c\,J_{\text{ph}} - J_0\left(\exp\left(\frac{qV}{nk_{\text{B}}T_{\text{D}}}\right) - 1\right) \\ &\sim c\,J_{\text{ph}} - J_{00}\,\exp\left(\frac{qV - E_{\text{g}}}{nk_{\text{B}}T_{\text{D}}}\right). \end{aligned}$$

Here, n is the ideality factor of the diode, and the bandgap-dependent reverse saturation current is given by $J_0 = J_{00}\exp\left(-E_{\text{g}}/nk_{\text{B}}T\right)$, where J_{00} is a device-dependent constant.

The temperature of the solar cell increases (approximately) linearly with c, i.e., $\Delta T = T_{\text{D}} - T_{\text{A}} = \zeta c$, where $\zeta c \approx (P_{\text{abs}} - \eta P_{\text{in}})/2h$ from Eq. (5.3).

Setting $J = 0$ at $V = V_{oc}$,

$$V_{oc} = (T_A + \zeta c)\frac{nk_B}{q}\ln\left(\frac{cJ_{ph}}{J_{00}}\right) + \frac{E_g}{q}. \tag{5.21}$$

The term $(T_A + \zeta c)$ is positive and increases linearly with c (see Fig. 5.5(a)). On the other hand, since $J_{ph}/J_{00} \ll 1$; therefore, the second term, $\ln(c\, J_{ph}/J_{00})$, is a large negative number for low c and its magnitude decreases with c. The product of the two terms, therefore, turns around at a specific c, as seen in Fig. 5.5(b). Fitting the experimental V_{oc} vs. the c curve, one can determine the device temperature, T_D.

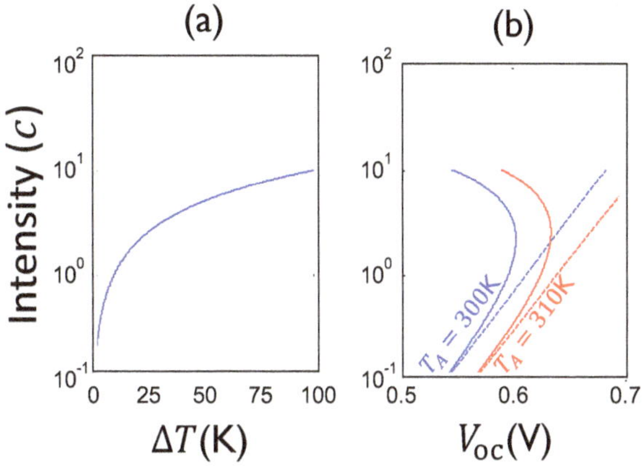

Figure 5.5: (a) Relationship between ΔT and the increasing intensity of sunlight, c. (b) The turnover of the V_{oc} curve due to self-heating can be fitted to obtain the cell temperature.

Homework 5.6: Temperature derivatives of V_{mp} and V_{oc}

Show that

$$\frac{dV_{mp}}{dT_D} = \frac{E_g/q - V_{mp}}{T_D} \quad \text{and} \quad \frac{dV_{oc}}{dT_D} = \frac{E_g/q - V_{oc} + \gamma^*(k_B T_D/q)}{T_D}.$$

In other words, the temperature dependence of V_{mp} and V_{oc} differs by a term proportional to γ^*.

Solution. Taking the derivative of Eq. (5.17) and substituting the expression for V_{mp}, we obtain the desired expression. For silicon, the derivative evaluates to

$$(1.12 - (1.12 \times (0.95) - 0.314))/300 = 1.23\ \text{mV/K}.$$

(continued on the next page)

Homework 5.6 (*continued from the previous page*)

In practice, V_{mp} of real cells is slightly smaller, so that the temperature coefficient is ~1.7–2 mV/K.

For the second part, recall that at the thermodynamic limit (Eq. (3.19))

$$\frac{V_{oc} - V_{mp}}{k_B T_D / q} = \ln \frac{\gamma(E_g, T_S)}{\gamma(E_g, T_D)} \quad \text{with} \quad \gamma(E, T) \equiv \frac{2k_B T}{c^2 h^3} \left(E^2 + 2k_B T E + 2k_B^2 T^2\right).$$

Taking derivative with respect to T_D, we find the expression for the coefficient γ^*. Typically, $\gamma^* \approx 1 - 3$.

For a panel installed outdoors, c is very low in the morning, then increases steadily and reaches its maximum at noon. Since the ambient temperature (T_A), light intensity (cP_{in}), and open-circuit voltage ($V_{oc}(T_A, c)$) are known, the data can be fitted into Eq. (5.21) to calculate the instantaneous device temperature T_D. We can use this approach to account for the bandgap narrowing with T_D.

Homework 5.7: V_{oc} turnaround for a silicon solar cell

Consider a silicon cell with $E_g = 1.1$ eV, $n = 1.2$, $J_0 = 10^{-10}$ mA/cm^2, $J_{ph} = 45$ mA/cm^2. Plot V_{oc} as a function of c to demonstrate the turnaround effect. Assume the wind velocity to be $v_w = 10.45$ m/s.

Solution. Homework 5.2 shows that for intensity $c = 1$ and wind speed $v_w = 10.45$ m/s, the module temperature is given by $\Delta T = 15$ K. Therefore, $c_T = \Delta T / c = 1 = 15$ K/sun. Now we can plot, in Fig. 5.5, ΔT and V_{oc} with varying intensity c. At low intensities, V_{oc} increases with intensity, as expected. However, the higher self-heating counters this increase and V_{oc} turns around beyond a critical intensity. If self-heating were absent, V_{oc} would keep increasing with intensity for a fixed T_A, as shown by the dashed lines in Fig. 5.5(b).

5.5 How to cool a solar cell

Since self-heating degrades both the efficiency and the lifetime of a solar module, several techniques have been developed to reduce its temperature. The techniques are effective, but the additional cost of implementing them is a concern. In this section, we will discuss some of these cooling approaches.

5.5.1 Active and passive convective cooling

There are several active and passive cooling schemes already in use. These include evaporative and fin cooling, submerged PV, heat pipe-based systems, and so on. These methods are effective, especially for concentrator solar cells, but the cost of integrating the additional components as well as their reliability in extreme weather conditions are important concerns.

5.5.2 Radiative cooling makes heat dissipation more effective

Modification of the module configuration based on the fundamental physics of self-heating of PV may create a simpler, yet more effective, cooling for modules. For example, recent proposals involving radiative cooling of solar cells have drawn much attention. The key idea of radiative cooling is to incorporate photonic crystals (PhC) so that the solar cell selectively radiates through the atmospheric window. PhC-enabled hemispherical emissivity to 0.9 has been demonstrated, which leads to considerable radiative cooling. Radiative cooling is particularly effective for extraterrestrial solar modules when air convective cooling is absent.

5.5.3 Spectrally selective cooling rejects sub-bandgap photons

In a different approach, spectrally selective filters can be used to reflect the sub-bandgap photons before they are absorbed by the solar cell. Optical filters with customized wavelength selectivity are commercially available and may be suitable for large-scale manufacturing. Including additional UV blocking in the filter can further prevent performance degradation from yellowing and delamination of encapsulants. The non-ideal wavelength cutoff of the filter can degrade short-circuit current and the trade-off between cutoff sharpness and pass-band transmissivity must be carefully engineered. The bandgap of solar absorbers depends on temperature (i.e., the bandgap of Si is ~1.12 eV and ~1.10 eV at 300 K and 400 K, respectively), which should also be considered for the filter design. Other methods of selective spectral cooling include de-texturing the front layer or nitridizing the back surface field in Si modules, both of which have been demonstrated experimentally.

5.5.4 Integrated spectral and radiative cooling

Integrating selective spectral and radiative cooling would increase the cooling efficiency of solar modules. As shown in Fig. 5.6, these strategies reduce the module temperature by 3 K, even in the absence of convective cooling. Increasing convective cooling (higher h) further reduces T_D (see Fig. 5.6(a)). For a fixed wind speed, radiative cooling is more effective in a hotter climate, as shown in Fig. 5.6(b), because thermal radiation power scales with temperature as $P \approx T_{PV}^4$. On the other hand, intrinsic power loss (e.g., carrier recombination) increases with temperature, leading to more heat being dumped from the above-bandgap irradiance. Hence, reflecting the heat power from sub-bandgap photons, i.e., selective cooling, is less effective with increasing T_A, as shown in Fig. 5.6(b). Interestingly, the degree of cooling by integrating these strategies is almost independent of T_A. Finally, water vapor and CO_2 reduce the transmittance between 8 μm and 13 μm of the atmosphere, directly suppressing thermal radiation from the glass encapsulation to the outer space. Consequently, radiative cooling is expected to be less useful in humid and cloudy climates.

5.5.5 Cooling by adding an energy harvester in series

Another option to cool the excess thermal energy is to harvest the energy produced. For example, one can use circulating fluid to cool the module and use the heated water for other applications.

Similarly, one can use a thermoelectric heater connected either in series or in parallel with the PV module. A thermoelectric (TE) converter operates between two reservoirs of T_D and T_A and converts thermal energy to useful electrical energy, namely,

$$V_{oc,TE} = S(T_D - T_A) \tag{5.22}$$

where S is the Seebeck coefficient. The Seebeck coefficient ($S \approx (k_B/q)[\eta_F + \Delta]$) depends on the relative Fermi energy $\eta_F = (E_C - E_F)/k_B T_D$ and $\Delta \sim 2$. It is easy to show that while TE does harvest energy from the PV, a series-connected PV-TE tandem cell cannot restore the loss of V_{mp} due to increased device temperature.

For a series-connected PV-TE system, $V_{mp} = V_{mp,PV} + V_{mp,TE}$, so that

$$V_{mp} = \left(1 - \frac{T_D}{T_S}\right)\frac{E_g}{q} - \frac{k_B T_D}{q}\ln\frac{\theta_D}{\theta_S} + S(T_D - T_A). \tag{5.23}$$

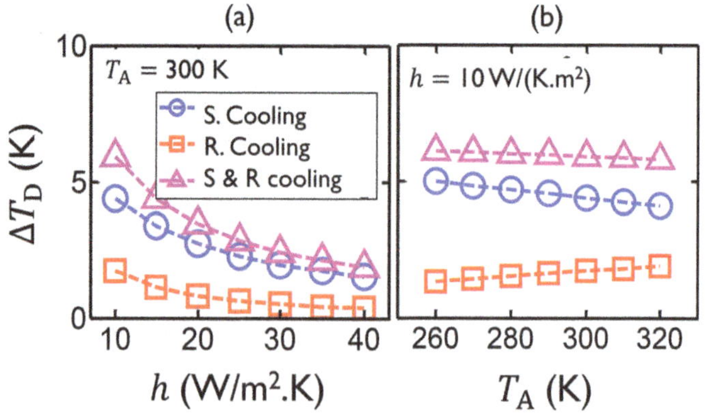

Figure 5.6: Temperature reduction of conventional Si modules as a function of (a) convective coefficient, h, and (b) the ambient temperature, T_A. The atmospheric transmittance used for this calculation is taken from Fig. 5.3.

Any increase in device temperature (T_D) reduces the first two terms due to PV, but increases the last term due to TE. The change is minimized by calculating $dV_{mp}/dT_D = 0$, i.e.,

$$S = \frac{E_g}{T_S} + k_B \ln \frac{\theta_D}{\theta_S} = \frac{E_g}{T_S} - \frac{0.31}{T_A} = 1250\ \mu V/K. \tag{5.24}$$

Therefore, unless $S \geq 1250\,\mu V/K$, the thermoelectric system will not be able to reach the isothermal Shockley–Queisser efficiency. Typically, $S = 100 - 200\,\mu V/K$; therefore, the harvested energy will only be a fraction of the total energy dissipated by a stand-alone PV system.

5.6 Conclusions

Self-heating is an intrinsic property of a solar cell that degrades its efficiency and lifetime. Self-heating due to above-bandgap photons cannot be prevented. The self-heating is further increased by the absorption of sub-bandgap photons (either in the contacts or in the backsheets). These conclusions hold across different solar technologies (i.e., GaAs, CIGS, Si, and CdTe). We used (a) the heat balance equation to explain why $T_D > T_A$ and (b) the fundamental properties of a solar cell to explain why $\beta \neq 0$. Taken together, one can calculate the intrinsic, material-specific limits of efficiency loss over time due to self-heating.

There are a number of ways to reduce the temperature of a solar cell. These include active and passive cooling, cooling by enhancing radiation, and reflecting sub-bandgap photons before they reach the cell. These strategies significantly improve the energy output (0.5%) and lifetime (80%) of conventional modules. The improvement is even more dramatic for concentrator solar cells (1.8% in efficiency, and 260% in lifetime).

The thermodynamic calculation so far presumed perfect optical absorption. In the next chapter, we will show that while absorption can be high, perfect absorption is fundamentally impossible!

References

[1] T.J. Silverman, M.G. Deceglie, Xingshu Sun, R.L. Garris, M.A. Alam, C. Deline, and S. Kurtz. Thermal and electrical effects of partial shade in monolithic thin-film photovoltaic modules. *IEEE Journal of Photovoltaics*, 5(6):1742–1747, November 2015.

[2] X. Sun, T. J. Silverman, Z. Zhou, M. R. Khan, P. Bermel, and M. A. Alam. Optics-Based Approach to Thermal Management of Photovoltaics: Selective-Spectral and Radiative Cooling. *IEEE Journal of Photovoltaics*, 7(2):566–574, March 2017.

[3] C. G. Granqvist and A. Hjortsberg. Radiative cooling to low temperatures: General considerations and application to selectively emitting SiO films. *Journal of Applied Physics*, 52(6):4205–4220, June 1981.

[4] Jun-long Kou, Zoila Jurado, Zhen Chen, Shanhui Fan, and Austin J. Minnich. Daytime Radiative Cooling Using Near-Black Infrared Emitters. *ACS Photonics*, 4(3):626–630, March 2017.

[5] Omemah Gliah, Boguslaw Kruczek, Seyed Gh Etemad, and Jules Thibault. The effective sky temperature: an enigmatic concept. *Heat and Mass Transfer*, 47(9):1171–1180, September 2011.

[6] David Faiman. Assessing the outdoor operating temperature of photovoltaic modules. *Progress in Photovoltaics: Research and Applications*, 16(4):307–315, June 2008.

[7] Sandia. PV Performance Modeling Collaborative | Sandia Module Temperature Model, 2016.

[8] Linxiao Zhu, Aaswath P. Raman, and Shanhui Fan. Radiative cooling of solar absorbers using a visibly transparent photonic crystal thermal blackbody. *Proceedings of the National Academy of Sciences*, 112(40):12282–12287, October 2015.

[9] T. J. Silverman, M. G. Deceglie, B. Marion, S. Cowley, B. Kayes, and S. Kurtz. Outdoor performance of a thin-film gallium-arsenide photovoltaic module. In *2013 IEEE 39th Photovoltaic Specialists Conference (PVSC)*, pages 0103–0108, June 2013.

[10] Priyanka Singh and N. M. Ravindra. Temperature dependence of solar cell performance — an analysis. *Solar Energy Materials and Solar Cells*, 101:36–45, June 2012.

[11] M. Chandrasekar and T. Senthilkumar. Passive thermal regulation of flat PV modules by coupling the mechanisms of evaporative and fin cooling. *Heat and Mass Transfer*, 52(7):1381–1391, August 2015.

[12] Anja Royne, Christopher J. Dey, and David R. Mills. Cooling of photovoltaic cells under concentrated illumination: A critical review. *Solar Energy Materials and Solar Cells*, 86(4):451–483, April 2005.

[13] A. Akbarzadeh and T. Wadowski. Heat pipe-based cooling systems for photovoltaic cells under concentrated solar radiation. *Applied Thermal Engineering*, 16(1):81–87, January 1996.

[14] Hans Christian Koehler. Cooling photovoltaic (PV) cells during concentrated solar radiation in specified arrangement in coolant with as low electric conductivity as possible, August 2000. International Classification H01L31/052;

Cooperative Classification Y02E10/52, H01L31/052; European Classification H01L31/052.

[15] Linxiao Zhu, Aaswath Raman, Ken Xingze Wang, Marc Abou Anoma, and Shanhui Fan. Radiative cooling of solar cells. *Optica*, 1(1):32, July 2014.

[16] A. R. Gentle and G. B. Smith. Is enhanced radiative cooling of solar cell modules worth pursuing? *Solar Energy Materials and Solar Cells*, 150:39–42, June 2016.

[17] Taqiyyah S. Safi and Jeremy N. Munday. Improving photovoltaic performance through radiative cooling in both terrestrial and extraterrestrial environments. *Optics Express*, 23(19):A1120, September 2015.

[18] W. H. Holley, S. C. Agro, J. P. Galica, L. A. Thoma, R. S. Yorgensen, M. Ezrin, P. Klemchuk, and G. Lavigne. Investigation into the causes of browning in EVA encapsulated flat plate PV modules. In *Proceedings of 1994 IEEE 1st World Conference on Photovoltaic Energy Conversion - WCPEC (A Joint Conference of PVSC, PVSEC and PSEC)*, volume 1, pages 893–896 vol. 1, December 1994.

[19] R. Santbergen and R. J. C. van Zolingen. The absorption factor of crystalline silicon PV cells: A numerical and experimental study. *Solar Energy Materials and Solar Cells*, 92(4):432–444, April 2008.

CHAPTER 6

Limits of Light Absorption

Chapter Summary

- A finite-sized solar cell cannot absorb all the photons incident on it due to wavelength dependent reflection, transmission, and radiative losses.
- Innovative light management techniques, including anti-reflection coating, surface texturing, photon recycling, etc. are critical design concepts to improve the short-circuit current.
- The classical light trapping limit, defined by the Yablonovitch formula, can be exceeded by using meta-surfaces, but interface recombination may compromise overall efficiency gain.
- It is possible to derive a general model of light absorption that accounts for short and long wavelength-dependent absorption, reflection, and transmission.

6.1 Introduction: A solar cell cannot absorb all incident photons

So far, we have assumed that all the photons (with $E \geq E_g$) incident on the solar cell are absorbed. In practice, the absorption in a finite-thickness absorber is always imperfect. In the 1980s, Yablonovitch showed that no matter how clever we are, the thermodynamic limit discussed in the previous chapters can be approached, but not achieved. In this chapter, we will use a simple geometrical argument based on Snell's law and elementary rules of probability to explain the Yablonovitch limit. Remarkably, the re-derivation suggests strategies for breaking the traditional absorption limit and improving PV efficiency (toward the Shockley–Queisser limit; see Sec. 4.6) by enhanced light absorption.

6.2 Absorption in a PV material: Overview

The thermodynamic argument proposed by Shockley–Queisser (S-Q) allows us to calculate the maximum efficiency of solar cells. In Chapter 2 (Sec. 2.2), we showed

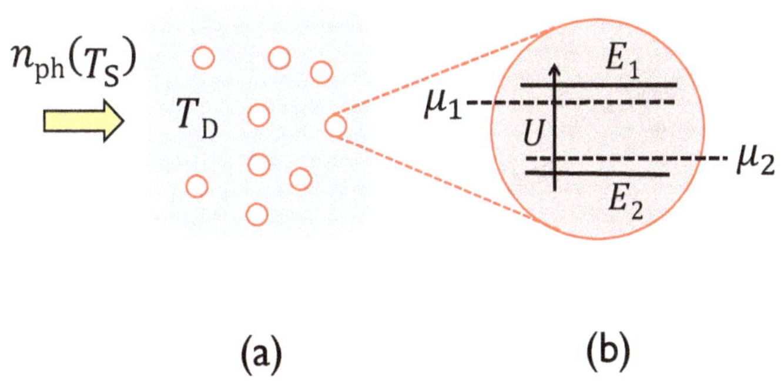

Figure 6.1: (a) A collection of two-level atoms is illuminated by photons from the sun. (b) Complete absorption of the photons is possible with a sufficiently large number of atoms.

that the essential features of the thermodynamic limit (as well as the strategies proposed to approach or exceed it) can be understood by shining sunlight onto a box of atoms characterized by two energy levels, E_1 and E_2 (see Fig. 6.1).

The S-Q limit can be theoretically achieved only if all the photons of the right energy ($\hbar\omega = E_1 - E_2$) entering the box of atoms are absorbed with a probability of 1. Our previous derivation of the S-Q limit for the two-level system presumed perfect absorption. For imperfect absorption, the upward and downward transition rates are rewritten as (see Eq. (2.24))

$$U = P \times \theta_S f_2(1 - f_1) n_{ph} \tag{6.1}$$
$$D = 1 \times \theta_D f_1(1 - f_2)(n_{ph} + 1) \tag{6.2}$$

allowing for the fact that some photons of the right energy may exit the box of atoms without being absorbed ($P < 1$). Here, n_{ph} is the B-E distribution related to the radiation from the sun (with appropriate $\Delta\mu \approx 0$ and $T = T_S$), and θ_S and θ_D are the input and output radiation angles, respectively. Equating Eqs. (6.1) and (6.2) and following the procedure in Sec. 2.2, we can recalculate the efficiency (η) of the simplified two-level model:

$$\eta = \left(1 - \frac{T_D}{T_S}\right) - \frac{k_B T_D}{E_g} \ln\left(\frac{\theta_D}{\theta_S}\right) - \frac{k_B T_D}{E_g} \ln\left(\frac{1}{P}\right). \tag{6.3}$$

Imperfect absorption ($P < 1$) reduces η below the thermodynamic limit.

More specifically, recall from Sec. 4.6.1 that in a practical, 3D system, imperfect absorption reduces the photocurrent $I^*_{mp} = \eta_Q I_{mp}$. Here, external quantum efficiency (EQE, η_Q) includes the combined effect of imperfect absorption and imperfect carrier collection. If carrier transport is idealized and carrier collection perfect, η_Q will be determined by the absorption P. The voltage V_{mp} is affected by both incident photon absorption and emitted photon reabsorption, as anticipated in Eq. (4.19):

$$
\begin{aligned}
V_{\mathrm{mp}}^{*} &= V_{\mathrm{mp}} - \frac{k_{\mathrm{B}}T_{\mathrm{D}}}{q}\ln\frac{1}{\eta_{\mathrm{R}}} - \frac{k_{\mathrm{B}}T_{\mathrm{D}}}{q}\ln\frac{1}{\eta_{\mathrm{Q}}} \\
&= \frac{E_{\mathrm{g}}}{q}\left(1-\frac{T_{\mathrm{D}}}{T_{\mathrm{S}}}\right) - \frac{k_{\mathrm{B}}T_{\mathrm{D}}}{q}\ln\left(\frac{\theta_{\mathrm{D}}}{\theta_{\mathrm{S}}}\right) - \frac{k_{\mathrm{B}}T_{\mathrm{D}}}{q}\ln\frac{1}{\eta_{\mathrm{R}}} - \frac{k_{\mathrm{B}}T_{\mathrm{D}}}{q}\ln\frac{1}{\eta_{\mathrm{Q}}}.
\end{aligned}
\tag{6.4}
$$

The external radiative efficiency (ERE, η_{R}) is determined by the emitted photon reabsorption. (As an aside, absorption or EQE is wavelength dependent, and is typically lower near the band-edge. For these cases, the generalized expression for V_{mp} is discussed in Khan *et al.*, PVSC 2013.) Let us now discuss why photon absorption is imperfect.

6.3 Finite absorption in a finite solar cell

6.3.1 Reflection from the air–cell interface

When sunlight is incident onto a semiconductor of complex refractive index, $n \equiv n_{\mathrm{r}} - in_{\mathrm{i}}$, the fraction of reflected light (R) is given by (see Fig. 6.2(a)):

$$
R = \frac{(n_{\mathrm{r}}-1)^2 + n_{\mathrm{i}}^2}{(n_{\mathrm{r}}+1)^2 + n_{\mathrm{i}}^2}. \tag{6.5}
$$

For silicon ($n_{\mathrm{r}} = n_{\mathrm{s}} = 3.5$), R increases from 0.3 (at 1.1 eV, $n_{\mathrm{i}} \sim 0$) to 0.5 (at 3.1 eV). In other words, without a clever design involving anti-reflection coating (ARC), at least one-third of the photons incident on a silicon cell will be lost to reflection. Placed between the glass cover ($n_{\mathrm{g}} = 1.5$) and Si substrate ($n_{\mathrm{s}} = 3.5$), a 60–90 nm thick ARC with index $n_{\mathrm{a}} = \sqrt{n_{\mathrm{s}}n_{\mathrm{g}}} = 2.3$ reduces the reflection loss to < 5–10% over the wavelength range of interest and most of the incident photons are transmitted into the substrate, i.e., $T \to 1$.

6.3.2 Entry does not guarantee absorption

The rate of absorption of a photon, after it enters a solar cell, depends on its energy (see Fig. 6.2(b)). A photon with energy E has equal probability of absorption per unit length ($\alpha(E) = 4\pi E n_{\mathrm{i}}/hc$) at any depth. The Beer–Lambert law describes the corresponding exponential decay of the photon density (and generation of electron–hole pairs):

$$
p(x,E) = e^{-\alpha(E)x}. \tag{6.6}
$$

For E near the bandedge E_{g}, $\alpha(E)$ is characterized differently for direct- vs. indirect-bandgap materials. For example, for direct-bandgap materials such as GaAs, $\alpha = A(E-E_{\mathrm{g}})^{0.5}$. On the other hand, for indirect-bandgap materials, e.g., Si, $\alpha = B_1(E-E_{\mathrm{g}})^2$. Equation (6.6) suggests that a complete absorption of the transmitted

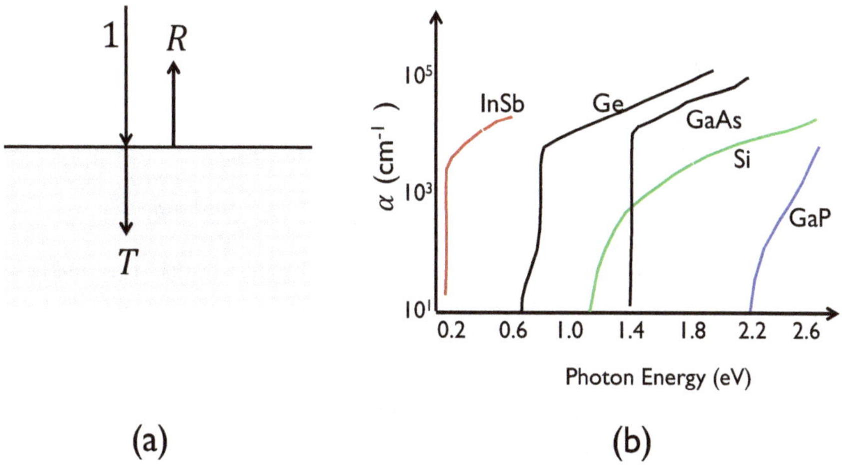

Figure 6.2: (a) Index mismatch between air and solar absorber leads to partial reflection (R) of the incident light. An ARC is needed to make $T \rightarrow 1$. (b) Energy-dependent absorption coefficient for various semiconductors.

photons (i.e., $P = 1 = \int_0^L p(x, E)dx = \int_0^\infty p(x, E)dx$ is only possible in an infinitely thick sample (i.e., $L \rightarrow \infty$)! In practice, as shown in Fig. 6.2(b), $\alpha(E \gg E_g)$ is large and these high-energy photons are absorbed within a few hundred nanometers. Therefore, cell thickness (L) is really an issue for lower-energy photons ($E \sim E_g$), with absorption so weak that $P \rightarrow \alpha L$. Since L must be as thin as possible (to efficiently collect photo-excited carriers before they are lost to recombination), and α is small (especially for an indirect-bandgap material), P would be negligibly small, with corresponding loss of photocurrent associated with the near-bandgap photons.

6.4 The Yablonovitch limit suggests strategies to improve absorption

Let us assume that there exists a clever arrangement of mirrors, reflectors, concentrators, photonic crystals, and metamaterials, which allow us to increase the absorptance of low-energy photons by the absorption enhancement factor, f_A. The absorption probability can then be expressed as follows:

$$P = \frac{f_A \, \alpha L}{f_A \, \alpha L + 1}. \tag{6.7}$$

With $f_A \rightarrow \infty$, we could achieve perfect absorption, i.e., $P = 1$. Is this possible?

In 1982, Yablonovitch used the theory of detailed balance of photons to show that no matter how clever or sophisticated the optical design, f_A cannot exceed $4n^2$ (n is the refractive index of the solar cell). With $f_A \rightarrow 4n^2$, however, $P(L) \rightarrow 1$, independent of E or L. In other words, although $P = 1$ is impossible, a clever

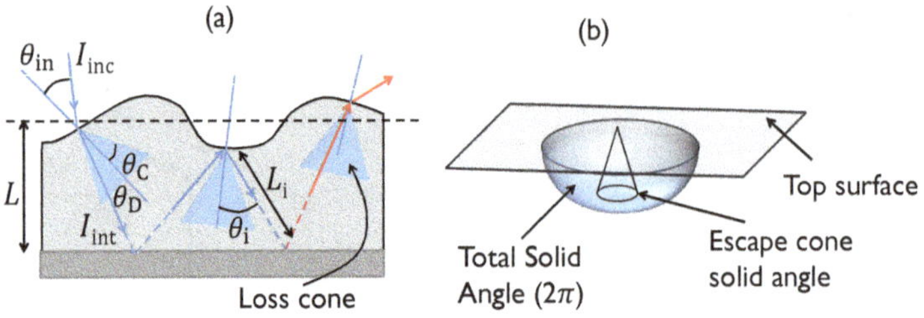

Figure 6.3: (a) Definition of loss cone (escape cone) and path lengths. (b) Illustration of escape cone solid angle in a 3D case.

design ensures nearly perfect absorption for low-energy photons even with a thin absorber! To find the optimum optical design, we must first understand the statistics of light rays within a solar cell.

6.4.1 Statistics of light rays within a solar cell

Consider the fate of a photon trapped within a finite dielectric slab as it tries to escape the dielectric region by repeatedly bouncing between the two (one reflecting, one random) surfaces, as in Fig. 6.3(a). Snell's law ($n_1 \sin\theta_1 = n_2 \sin\theta_2$) dictates that the maximum angle with which a photon incident onto dielectric/air interface can escape the dielectric is given by $\theta_C = \sin^{-1}(1/n)$. The probability that a ray will escape (P_{esc}) through the escape cone ($0 < \theta < \theta_C$) depends on the dimensionality of the confining surfaces.

If a ray is incident outside the escape cone, it will bounce back (total internal reflection) in a random angle defined by the local orientation of the top interface. The average number of bounces (β) a photon experiences before it escapes the dielectric is inversely proportional to the escape probability per bounce, (P_{esc}), i.e.,

$$\beta = {P_{esc}}^{-1}. \tag{6.8}$$

Moreover, note that the number of bounces before escape equals the enhancement of photon intensity inside the dielectric layer. Figure 6.4 explains why: if a photon bounces β times before it escapes, any arbitrary point A within the dielectric is visited by β (on average) number of photons, so that the intensity of the point goes up by the same factor. Therefore, the intensity enhancement is $f_I = \beta$ given that there is no mirror at the back. With a back mirror blocking the escaping photons, the net probability of escape is halved, and the intensity is doubled, i.e., $f_I = 2\beta$.

It should be understood that intensity enhancement ($f_I = 2\beta$) does not arise from an enhanced emission rate from atoms within the cell. The emission rate of photons depends only on the temperature of the dielectric (blackbody radiation). By reducing the probability of escape (P_{esc}) with suitably designed structures, we can increase the number of photons (N_{ph}) inside the dielectric, so that the number of photons escaping the cell ($P_{esc} \times N_{ph}$) remains independent of the escape angle.

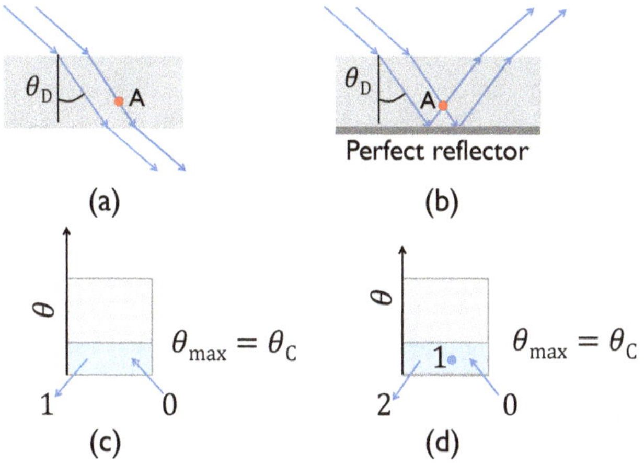

Figure 6.4: Angle statistics (c, d) of photons in a 1D object (a) without and (b) with a back reflector.

Within a given bounce, the rays are scattered at different angles ($0 < \theta_i < \pi/2$), see Fig. 6.3(a). On average, the path length per pass (up or down), $\langle L_i^{\text{eff}}(\theta_i)\rangle$, is greater than L, the thickness of the cell. Therefore, the absorption per round trip, with a back mirror, is $\alpha \times 2\langle L_{\text{i}}^{\text{eff}}\rangle \equiv 2f_{\text{L}}\alpha L$, where the path length enhancement is $f_{\text{L}} \equiv \langle L_{\text{i}}^{\text{eff}}\rangle/L$.

Now consider a solar cell in which photons bounce β times between top and bottom interfaces before exiting a dielectric slab of length L and absorption coefficient α. The probability of absorption per round trip is $\sim 2f_{\text{L}}\alpha L$, and in every round trip, a fraction $1/\beta$ of the photons escape through the top surface without being absorbed. Therefore, the absorption probability or absorptance is

$$
\begin{aligned}
P &= \frac{2f_{\text{L}}\alpha L}{2f_{\text{L}}\alpha L + \beta^{-1}} \\
&= \frac{2f_{\text{L}}\beta\alpha L}{2f_{\text{L}}\beta\alpha L + 1} \equiv \frac{f_{\text{L}}f_{\text{I}}\alpha L}{(f_{\text{L}}f_{\text{I}}\alpha L) + 1} \equiv \frac{f_{\text{A}} \times \alpha L}{(f_{\text{A}} \times \alpha L) + 1},
\end{aligned}
$$

which completes the derivation of Eq. (6.7). To find the optimum design, let us now calculate $\beta(= P_{\text{esc}}^{-1})$ and $f_{\text{L}}(L^{\text{eff}})$ for various optical configurations.

6.4.2 A dielectric slab confined by two parallel surfaces

Consider a dielectric defined by two parallel, planar surfaces. If a ray of sunlight refracts into the dielectric as in Fig. 6.4(a), it must enter the dielectric within the "escape cone" (see arrow labeled 0 in Fig. 6.4(c)). If the ray is neither absorbed nor scattered within the dielectric, it will be incident on the bottom surface within the escape cone and will escape into air with no further reflection (arrow 1, Fig. 6.4(c)). The path length enhancement depends on the angle of the refracted ray (θ_D) inside the dielectric, i.e., $f_L = 1/\cos\theta_D$. The number of trips through the dielectric is just 1 ($\beta = 1$). Only a single ray is associated with any point A, so that the density of photons at each point inside the dielectric is exactly equal to that in air ($f_I = \beta = 1$). Therefore, the absorption enhancement is $f_A = f_I f_L = \beta f_L = 1/\cos\theta_D$, an intuitive result. Note that if the photon is never absorbed nor scattered within the dielectric, its angle cannot change, and the lightly (gray) shaded region in Fig. 6.4(c) remains forever inaccessible. For practical dielectrics $\theta_D(< \theta_C) \to 0$ and $\cos\theta_D \to 1$.

If we make the back surface perfectly reflecting, as in Fig. 6.4(b), the ray still enters the dielectric within $\theta_{max} = \theta_C$ (Fig. 6.4(d)), bounces once on the back mirror (red dot, marked 1), and then escapes through the top interface (arrow marked 2), never once leaving the escape cone. The photon makes two trips ($f_I = 2$) through the dielectric before it escapes, so that $f_A = 2/\cos\theta_D = 2\beta f_L = f_I f_L$. By blocking off the exit from the bottom, the internal photon density has been raised by a factor of 2, because every point A is traversed by two rays: one on its way down to the mirror, the other after bouncing back, on its way to escape. Finally, using Eq. (6.7), we find that

$$P = \frac{2f_L\alpha L}{2f_L\alpha L + \beta^{-1}} = \frac{2\alpha L/\cos\theta_D}{2\alpha L/\cos\theta_D + 1} \approx \frac{2\alpha L}{2\alpha L + 1}. \tag{6.9}$$

Because without scattering, $f_L = 1/\cos\theta_D \to 1$. Clearly, a dielectric layer defined by parallel, planar surfaces makes a poor absorber, i.e., $P \sim 2\alpha L$ when $\alpha L \ll 1$ is small.

> **Homework 6.1: Absorption in a Si layer**
>
> Si has poor absorption close to the bandgap, e.g., $\alpha \approx 0.5$ cm^{-1} just above the band edge. For an $L = 200$ μm thick Si, calculate the fraction of normally incident photons absorbed by the Si-film with and without a back mirror.

6.4.3 Bottom mirror with 1D randomness

Figure 6.5(a) shows a dielectric with a planar surface on the top and a roughened (only along the x direction), fully reflective surface in the back. Let us assume that the incident ray is restricted to planes parallel to the $x - z$ plane. The incident ray enters the dielectric through the top planar surface within the escape cone ($\theta \leq \theta_C$) of the top surface. This is represented as state 0 in Fig. 6.5(c). The ray is scattered and reflected by the bottom rough surface. If the angle following the scattering is outside the escape cone ($\theta > \theta_C$), the state of the ray is characterized by a point outside the blue region in Fig. 6.5(c). Since the ray is outside the escape cone, it will be internally reflected from the top surface (total internal reflection). The bouncing between the surfaces will continue and the photon will remain trapped within the dielectric, as long as the ray occupies a state outside the escape cone

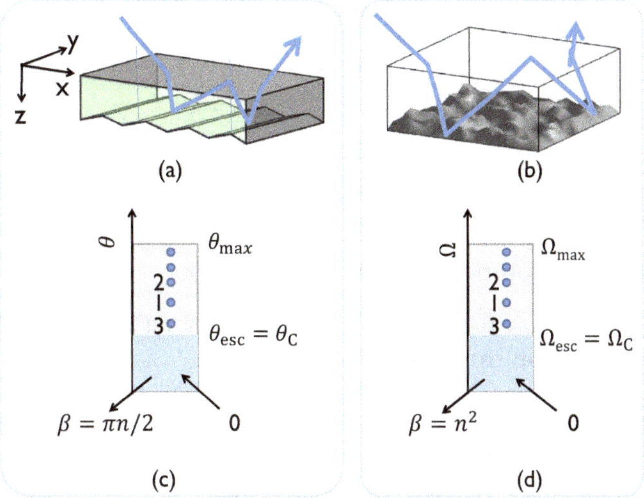

Figure 6.5: Angle statistics for photon scattering in (a) 1D and (c) 2D Lambertian reflectors at the back. The corresponding angle statistics are shown in (b) and (d). Here, θ and Ω represent angles and solid angles, respectively.

(the blue region in Fig. 6.5(c)). Statistically, on average, the photon will bounce β times (i.e., hop through β number of states in Fig. 6.5(c)) before it is randomly scattered into the escape cone and finally exits the structure (arrow β as shown in Fig. 6.5(c)). Note that β should be understood as an average, because some photons may escape after a single bounce, while others may be trapped for bounces larger than β.

The escape probability and β can be calculated as follows:

$$P_{\text{esc}} = \frac{\int_{-\theta_C}^{\theta_C} \cos\theta \, d\theta}{\int_{-\pi/2}^{\pi/2} \cos\theta \, d\theta} = \frac{2\sin\theta_C}{2} = \frac{1}{n}.$$

The intensity enhancement is $f_I = 2\beta = 2P_{\text{esc}}^{-1} = 2n$.

For a very weak absorbing dielectric, the absorption in a single pass by the randomly scattered light is

$$\begin{aligned} A_{\text{single-pass}} &= \frac{\int \left(1 - e^{(-\alpha L/\cos\theta)}\right) \cos\theta \, d\theta}{\int \cos\theta \, d\theta} \\ &\approx \frac{\int_{-\pi/2}^{\pi/2} \alpha L d\theta}{\int_{-\pi/2}^{\pi/2} \cos\theta \, d\theta} = \frac{\pi}{2}\alpha L. \end{aligned}$$

Therefore, the absorption path enhancement is

$$f_L = \frac{A_{\text{single-pass}}}{\alpha L} = \frac{\pi}{2}.$$

Taken together, $f_A = f_I f_L = \pi n = C_1 n$. Finally, from Eq. (6.7), we find that

$$P = \frac{2f_L \alpha L}{2f_L \alpha L + \beta^{-1}} = \frac{\pi n \alpha L}{\pi n \alpha L + 1} \tag{6.10}$$

for a reflective rough surface (in one direction). The appearance of π and $n(> 1)$ suggests an improved absorption even for a poor absorber like silicon.

Homework 6.2: Absorption in 1D textured Si cell

Let us assume that $\alpha \approx 0.5$ cm^{-1} and $L = 200\ \mu$m for a Si absorber layer with a 1D textured back mirror. The refractive index of Si is $n = 3.5$. Calculate the escape probability P_{esc}, absorption enhancement factor f_A, and absorption probability, P.

Hint. We expect to see that the 1D rough bottom reflector increases absorptance in Si by a factor of 11. This means that a 200 μm thick silicon layer will appear optically as a 2 mm thick film.

6.4.4 Two-dimensional random reflecting surface

It is possible to roughen the bottom reflector in both the x and y directions (see Fig. 6.5(b)). The light comes in through the top planar surface and gets scattered by the rough back reflector. The scattering of light and the trapping concept is the same as explained in the previous section. The light cannot escape from the dielectric if it is scattered into and then stays within the states beyond the blue region of Fig. 6.5(d). The light escapes after β bounces, when a random scattering by the bottom interface scatters the ray into the escape cone.

To calculate P_{esc} and β, we integrate over the solid angle (3D, represented by Ω in this chapter)[1] with $\theta_{esc} = \theta_C$, as follows:

$$\begin{aligned} P_{esc} &= \frac{\int \cos\theta \, d\Omega}{\int_{half-sphere} \cos\theta d\Omega} \\ &= \frac{\int_0^{2\pi}\int_0^{\theta_C} \cos\theta \sin\theta \, d\theta \, d\phi}{\int_0^{2\pi}\int_0^{\pi/2} \cos\theta \sin\theta \, d\theta \, d\phi} \\ &= \sin^2\theta_C \equiv 1/n^2, \end{aligned} \tag{6.11}$$

so that

$$\beta = \frac{1}{P_{esc}} = n^2.$$

[1]In other chapters, we used θ to indicate solid angles. For this chapter, to avoid confusion, we represent angles using θ, and solid angles using Ω.

The corresponding intensity enhancement is $f_{\mathrm{I}} = 2\beta = 2n^2$. In silicon, on average, the ray travels an astonishing ~25 times inside the layer before it can escape.

For a very weak absorbing dielectric, the single-pass absorption of the randomly scattered light is

$$\begin{aligned} A_{\text{single-pass}} &= \frac{\int \left(1 - e^{(-\alpha L/\cos\theta)}\right) \cos\theta \, d\Omega}{\int \cos\theta \, d\Omega} \\ &\approx \frac{\int_0^{\pi/2} \alpha L \sin\theta d\theta}{\int_0^{\pi/2} \cos\theta \sin\theta \, d\theta} = 2\alpha L. \end{aligned}$$

Therefore the absorption path enhancement is

$$f_{\mathrm{L}} = \frac{A_{\text{single-pass}}}{\alpha L} = 2.$$

The absorption enhancement factor is $f_{\mathrm{A}} = f_{\mathrm{I}} f_{\mathrm{L}} = 4n^2$, the Yablonovich result. For a weakly absorbing dielectric, the absorptance is given by

$$P = \frac{2 f_{\mathrm{L}} \alpha L}{2 f_{\mathrm{L}} \alpha L + \beta^{-1}} = \frac{4n^2 \alpha L}{4n^2 \alpha L + 1}. \qquad (6.12)$$

The formula implies that the sunlight entering a 200 μm thick silicon film will have an effective optical thickness of 1 cm for absorption! Even for very weakly absorbing light, $P \to 1$; such is the power of a single rough surface.

Homework 6.3: Absorption in a practical textured Si cell

Let us assume that $\alpha \approx 0.5\ \text{cm}^{-1}$ and $L = 200\ \mu$m for a Si layer with a 2D textured back mirror. The refractive index of Si is $n = 3.5$. Calculate the escape probability P_{esc}, absorption enhancement factor f_{A}, and absorption probability, P.

Hint. You should expect to see that the absorptance in Si is enhanced by a factor of 50. This means that a 200 μm thick silicon layer will appear optically as a 1 cm thick film! In addition, f_{I} tells you the number of times the rays bounce, on average, inside the Si layer.

We used a configuration with a planar, refracting surface facing the sun, and the roughened mirror at the back. The results are unchanged if the configuration is reversed by a roughened refracting surface on top and a planar reflecting surface at the back.

Homework 6.4: Absorption in a textured solar cell

Instead of a textured mirror, consider a planar mirror with a randomly textured interface between the dielectric (e.g., Si) and air.

1. Explain how the escape probability of light from the dielectric through the random surface into air is $P_{\text{esc}} = 1/n^2$.

2. Now consider a dielectric layer with random surfaces on both sides, but without a mirror at the back. Show that the absorption enhancement now $f_{\text{A1}} = 2n^2$.

3. Now assume that there is a planar mirror at the back of the layer — however, the reflectivity of the mirror is R. Recalculate f_{A2}. At $R < R_{\text{crit}}$, we expect $f_{A2} < f_{A1}$. For such a low reflectivity mirror, we are better off without a back mirror!

Solution.

1. The escape probability of isotropic light through a planar surface is $1/n^2$. The random surface essentially randomizes the angle between the light ray and the surface. Therefore, we can still expect the escape probability to be (approximately) $1/n^2$.

2. We can still set $f_{\text{I}} = 2$, $f_{\text{L}} = 2$. The net escape probability is now $1/n^2 + 1/n^2 = 2/n^2$ (i.e., $1/n^2$ for each top and bottom surface). Thus, $f_{\text{A1}} = 2 \times 2 \times (n^2/2) = 2n^2$.

3. Again, the net escape or loss probability is $(1/n^2) + (1 - R) = [1 + n^2(1 - R)]/n^2$, and $f_{\text{A2}} = 4n^2/[1 + n^2(1 - R)]$. In the limit $R = 1$, we have $f_{\text{A2}} = 4n^2$. Obviously, $1 - R_{\text{crit}} = 1/n^2$. For example, with $n = 3.5$ for Si, we have $R_{\text{crit}} \approx 92\%$. Clearly, having a mirror with $R < R_{\text{crit}}$ is as good as not having a back mirror at all.

These calculations assume low $1/n^2$ and high R under weak absorption conditions (i.e., αL is small).

Homework 6.5: Average number of bounces when absorption is strong

So far, we have calculated the escape probability, average number of bounces, and net enhanced absorption by assuming that the absorption is *weak* and the incident intensity is only weakly perturbed as the light travels through the absorber dielectric. This is obviously not correct if absorption is strong. Determine the number of bounces when the absorption is strong.

Solution. Assume a structure with a random texture at the front and a perfect mirror at the back. Remember that the escape probability from the dielectric to air is $P_{\text{esc}} = 1/n^2$ and the absorption per round trip is $A_0 = (1 - \exp(4\alpha L))$. After each round trip, we loose P_{esc} fraction of the rays intensity at the top surface. If we define
f_{i}^{-} = intensity (normalized) remaining after i round trips (before hitting the top surface),
f_{i} = intensity (normalized) remaining after i round trips (after fractional escape through the top surface),
E_i = escaped ray at i-th round trip, and
A_i = net absorption after i round trips,
then we can calculate $A_\infty = 1 - \sum E_i$. The following table summarizes the fraction of light absorbed in a strong absorber.

i	f_{i}^{-}	E_i	f_{i}
1	$1 - A_0$	$f_1^{-} P_{\text{esc}}$	$f_1^{-}(1 - P_{\text{esc}})$ $= (1 - A_0)(1 - P_{\text{esc}})$
$\vdots$			
j	$(1 - A_0)^j (1 - P_{\text{esc}})^{j-1}$	$f_j^{-} P_{\text{esc}}$	$f_j^{-}(1 - P_{\text{esc}})$ $= (1 - A_0)^j (1 - P_{\text{esc}})^j$

6.4.5 Planar surfaces and photon recycling

The angle diagram (Fig. 6.5(d)) suggests that it is not necessary to have a random surface to achieve a high photon intensity. Any process that scatters the photons away from the escape cone can achieve a similar amplification. For example, Fig. 6.6 shows that if a photon is absorbed and immediately re-emitted at a random angle (i.e., if a photon is recycled), the number of repeated bounces will be identical to those from scattering by rough surfaces. This process improves I_{sc} in luminescent solar cells, where incoming photons are first absorbed in a thin layer with embedded luminescent particles. These particles then immediately re-emit at random angles, thereby improving the absorption probability within the thin absorber layer.

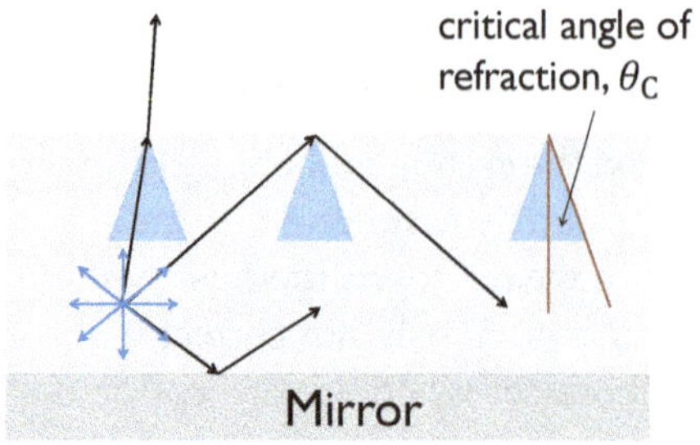

Figure 6.6: Trapping of the recycled photons between two planar surfaces.

The reabsorption of emitted photons through randomization of the angles is called **photon recycling**. This idea is mainly used to increase V_{mp} rather than improving I_{sc} (i.e., sunlight absorption) by enhancing external radiative efficiency (ERE); see Eq. (4.19) or Eq. (6.4). By reabsorbing the emitted photons, the steady state carrier density increases, which in turn increases V_{mp}. The photon recycling process has been used with great success in creating ultra-high-efficiency cells that do not require rough surfaces. Note that photon recycling is relevant and effective for materials with a high ratio of radiative to non-radiative recombination. For example, radiative recombination dominates in a high-quality GaAs cell — this allows us to utilize photon recycling to improve PV performance. However, Auger recombination (non-radiative) can dominate in indirect-bandgap materials such as Si — therefore, photon recycling is irrelevant in such cases.

6.5 Intensity enhancement: exceeding the Yablonovitch limit

For a very thin, poorly absorbing material, it may be necessary and possible to increase $f_A(\equiv f_I \times f_L) > 4n^2$. A first approach increases f_I by reducing the escape angle θ_{esc} below θ_C, so that P_{esc} is reduced, and the number of photons within the dielectric is enhanced. A second approach involves arranging the optics such that the scattering among various angles is no longer random. Rather, the probability of scattering outside the escape zone, i.e., outside the "blue region" in Fig. 6.5(d), is enhanced. In both cases, the Yablonovitch limit is exceeded, as discussed below.

6.5.1 Reducing the escape cone

In Homework 4.10, we discussed how angle-restricted solar cells reduce angle entropy loss and improve efficiency. If the angle of the output emission angle is reduced by a factor N_{out} by an angle-selective layer so that light can be emitted with an angle of ($\theta_{air} = 2\pi/N_{out}$), as in Fig. 6.7(a), the escape cone inside the dielectric will likewise be reduced by a factor N (where $\theta_{esc} = \theta_C/N$). From Snell's

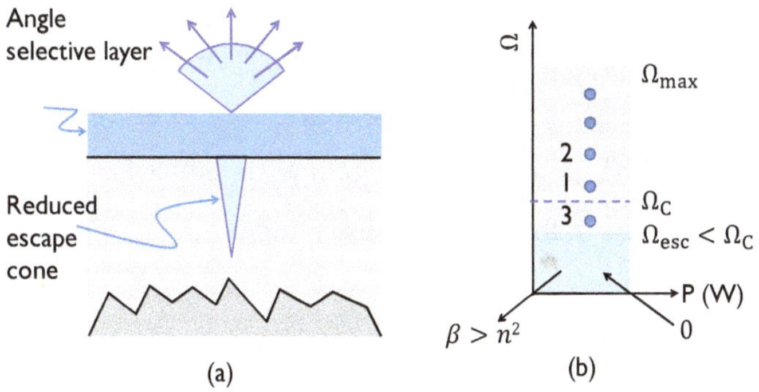

Figure 6.7: (a) Conventional structure with a Lambertian back reflector yielding $4n^2$ absorption enhancement. An extra angle-selective transmitter/reflector layer on the structure can yield an absorption enhancement of $> 4n^2$ as well as reduction in angle entropy loss. (b) The angle statistics shows that the suppressed escape angle allows for more states for photons inside the dielectric.

law,

$$\sin\left(\frac{\pi/2}{N_{\text{out}}}\right) = n \sin\left(\frac{\theta_{\text{C}}}{N}\right). \tag{6.13}$$

For practical dielectrics, $(\sin\theta_{\text{C}} \approx \theta_{\text{C}}) = (1/n)$, and for sufficiently large N_{out}, Eq. (6.13) can be rewritten as

$$\left(\frac{\pi/2}{N_{\text{out}}}\right) \approx n\left(\frac{\theta_{\text{C}}}{N}\right) \approx \frac{1}{\theta_{\text{C}}}\left(\frac{\theta_{\text{C}}}{N}\right) = \frac{1}{N}.$$

Therefore,

$$N \approx \frac{N_{\text{out}}}{\pi/2}.$$

The roughened back reflector continues to randomize the light angles inside the dielectric, so that the angle space in Fig. 6.7(b) is populated with equal probability. Following the derivation of Eq. (6.11), but now for the restricted escape angle, we find that

$$\begin{aligned} P_{\text{esc}} &= \sin^2(\theta_{\text{esc}}) \\ &= \sin^2(\theta_{\text{C}}/N) \\ &\approx (\theta_{\text{C}}/N)^2 = \frac{1}{N^2}\theta_C^2 \\ &= \frac{(\pi^2/4)}{N_{\text{out}}^2}\frac{1}{n^2}. \end{aligned}$$

Recall from Sec. 6.4.4 that for a 2D random surface, $f_{\text{L}} = 2$. Therefore, the absorption enhancement is

$$f_{\text{A}} = 2\beta f_{\text{L}} = \frac{2f_{\text{L}}}{P_{\text{esc}}} = N^2 \times 4n^2 = \frac{8N_{\text{out}}^2}{\pi^2} \times 4n^2.$$

Although the absorption path length enhancement is the same as before ($f_L = 2$), the intensity enhancement is very high: $\left(f_I = (8N_{\text{out}}^2/\pi^2) \times 2n^2\right)$.

Homework 6.6: Maximum absorption gain with reduced escape angle

Calculate the increase in the photon density and optical absorption with a reduced escape angle.

Solution. Ideally, we would want to restrict the emission solid angle so that $\Omega_{\text{emi}} \to 0$. However, sunlight comes in at a finite solid angle, Ω_S. We need $\Omega_{\text{emi}} \geq \Omega_S$, so that the sunlight is not blocked and the short-circuit current remains unaffected. Thus, if we reduce Ω_{emi} from 2π to ($\Omega_S \approx 6.5 \times 10^{-5}$sr), then the absorption gain will be maximized. Therefore, $N_{\text{out}} \approx (2\pi/\Omega_D) \approx 10^5$, and

$$f_A = N^2 \times 4n^2 \approx (4 \times 10^9) \times 4n^2,$$

which is significantly higher than the Yablonovich limit. In this case, the photons are virtually guaranteed absorption ($P \to 1$), because even the weakest absorbing materials have $\alpha(E) \approx 10^{-5}$/m.

Returning to Eq. (6.3) for a two-level system, we find that suppressing θ_{esc} not only improves P (reduces the third term on the right), but simultaneously suppresses the angle anisotropy and increases the open-circuit voltage close to the bandgap. The same idea holds for a practical 3D system as anticipated by Eq. (6.4). In practice, $N_{\text{out}} \to 10^5$ may be both impractical and unnecessary for absorption enhancement. The quadratic improvement of absorption with angle restriction ensures that even for moderate angle restriction, the absorbance enhancement is significant.

6.5.2 Anisotropic scattering by metasurfaces

The second approach to break the Yablonovitch limit involves anisotropic trapping. By clever optical design, we can trap the rays away from the escape cone, i.e., within the blue region in Fig. 6.5(d). The key idea is to design a planer metallic layer (a meta-surface) that can create a phase discontinuity in the optical path. Specifically, the meta-surface modifies the laws of refraction and reflection as follows:

$$n_t \sin\theta_t - n_i \sin\theta_i = \gamma \tag{6.14}$$

$$n_i \sin\theta_r - n_i \sin\theta_i = \gamma \tag{6.15}$$

where $\gamma = (\lambda_0/2\pi)(d\Phi/dx)$ is a measure of the discontinuity introduced by the new *interface layer* at wavelength λ_0. The subscripts "i," "r," and "t" denote the incident, reflected, and transmitted (refracted) rays, respectively. The well-known laws of refraction and reflection are restored for $\gamma = 0$. Let us assume that the meta-interface layer (see Fig. 6.8(a)) obeys the "new" laws of reflection and refraction for any polarization or wavelength of the incident light. With this generalized Snell's law, one can design an optical black hole — that traps forever an incident photon within the dielectric structure (see Figs. 6.8(a, b)).

For simplicity, let us assume that $n_A = 1$, $n_B = 3$, and $\gamma = 1$. In Fig. 6.8(a), the incident ray refracts according to Eq. (6.14). Note that the escape (critical) angle is asymmetric and the escape cone is skewed. Figure 6.8(a) shows that after the first bounce, the ray hits the top interface outside the escape cone and is internally reflected based on Eq. (6.15). The reflection angle increases at each successive bounce (with no possibility of escape) until eventually it propagates parallel to the interface as an evanescent mode.

In Fig. 6.8(c), we plot the reflection laws of the front surface for total internal reflection (black line) and the meta-surface (red line) for incident rays within the dielectric. The shaded region corresponds to the angles in the escape cone. The blue dashed line gives the refraction rule for the light going from air into

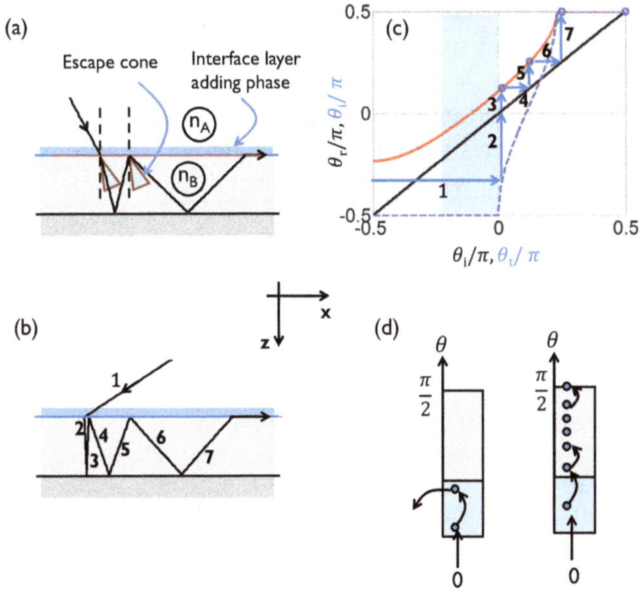

Figure 6.8: (a) Metasurface-based scheme for perfect light trapping. The top interface introduces phase discontinuities in the optical path. (b) Step-by-step study of light trapping in the structure. Light gets trapped into evanescent modes after a couple of bounces inside the structure. (c) The black and the red line present the reflection law for the mirror and the top interface. The blue dashed line describes the refraction law for rays entering the dielectric layer — for this case, the incident angle is along the y axis. (d) The angle statistics for the structure explain the light trapping scheme.

the dielectric — note that the incidence angle and transmitted angle are along the y and the x axis, respectively.

For a specific example of light trapping, consider the case shown in Fig. 6.8(b). The angles assumed by the rays in consecutive refraction and reflections are represented by the corresponding numbered arrows in Fig. 6.8(c). Ray 1 (Fig. 6.8(b)) is incident at an angle $-\pi/3$, as shown in Fig. 6.8(c). The refracted ray 2 bounces off the mirror. The corresponding angle is shown by arrow 2 on the black line (mirror reflection law). Ray 3 is outside the escape cone and reflects off the top interface (arrow 3 onto the red line shows the angle). The ray keeps on bouncing between the mirror and the top interface until ray 7 is reflected to $\pi/2$ (see arrow 7 on the red line). The light is now trapped into this mode forever.

Figure 6.8(d) shows angle statistics for the structure for a more general case of $0 < \gamma < 1$. If the ray stays inside the escape cone (blue region) after the first bounce, the ray escapes. On the other hand, the rays going outside the escape cone (blue region) keep bouncing away to larger angles and eventually get trapped.

Light trapping by meta-mirrors: Intensity vs. absorption enhancement In all the discussions before the meta-surface concept, we assumed the light to be scattered isotropically inside the dielectric. The intensity enhancement was $f_{\mathrm{I}} = 2\beta = 2n^2$ for a 2D random scattering surface. The average absorption path enhancement yields another factor of $f_{\mathrm{L}} = 2$. Note that the effective path length for light is much higher if it is scattered into large angles, allowing the rays to get absorbed. As shown in Fig. 6.9(a), rays with a smaller angle will require more number of bounces before absorption, with a higher probability of escape. A ray with an angle θ will undergo an absorption of $(1 - e^{-\alpha L/\cos\theta})$ for single pass through the dielectric. Now, assume that a surface scatters the rays according to a probability

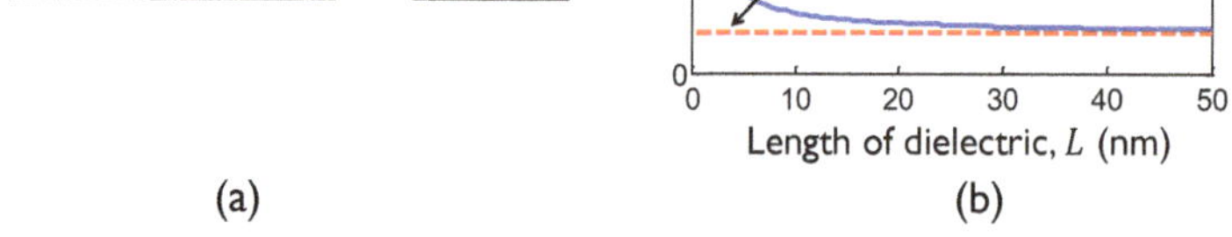

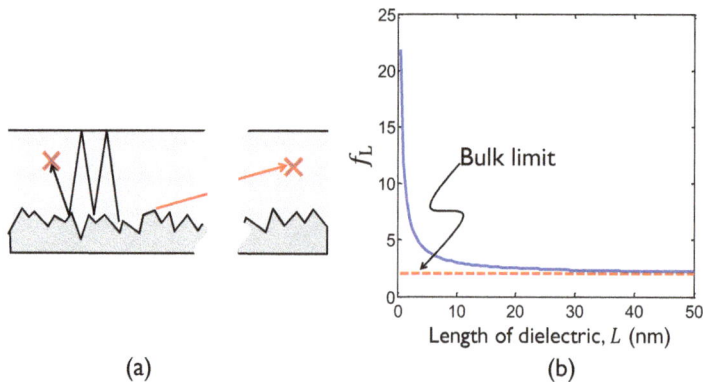

Figure 6.9: (a) Absorption of scattered light in the dielectric. (b) Absorption enhancement as a function of dielectric layer thickness L.

distribution of $P(\Omega)$. Therefore, the average absorption path length enhancement is

$$f_{\mathrm{L}} = \frac{1}{\alpha L} \int (1 - e^{-2\alpha L/\cos\theta}) P(\Omega) d\Omega. \tag{6.16}$$

If the surface is designed such that it preferentially scatters parallel ($\theta \to \pi/2$) to the surface, the absorption path enhancement f_{L} can be very high. The strong absorption implies fewer bounces ($\beta < n^2$); however, the overall absorption exceeds the Yablonovitch limit ($f_{\mathrm{A}} = 2\beta f_{\mathrm{L}} > 4n^2$).

Homework 6.7: Anisotropic meta-mirror

Focus on the structure in Fig. 6.8(b) where the back mirror has been modified to follow generalized Snell's law following Eq. (6.15). Assume that $n_{\mathrm{A}} = 1$ (air), and $n_{\mathrm{B}} = 3.5$ (Si).

1. Find the escape cone angle θ_{C} for Si.
2. Consider normal incidence from air. Light will refract into Si and still be at a normal angle. Thus, these rays will hit the back mirror at $\theta_{\mathrm{i}} = 0$. Now, if the mirror is designed such that $\theta_{\mathrm{r}} \geq \theta_{\mathrm{C}}$, then we can achieve perfect light trapping. Find the critical value of γ_{C} so that $\theta_{\mathrm{r}} = \theta_{\mathrm{C}}$. This should be the critical design for perfect light trapping for normally incident light.
3. For normally incident light, explain why (i) there is no light trapping for $\gamma < \gamma_{\mathrm{C}}$, and (ii) perfect light trapping for $\gamma > \gamma_{\mathrm{C}}$.

6.5.3 Anisotropic scattering into guided modes

Several recently proposed schemes beat the $4n^2$ limit of light absorption by innovative optical structures that scatters light predominantly into guided modes, as in Fig. 6.10(a). To understand this approach intuitively, we calculate the integral of f_{L}, shown above in Eq. (6.16), by partitioning the rays into two groups: one for very low angle evanescent wave absorption (A_{ev}) and the rest for bulk absorption (A_{bulk}). Taken together,

$$f_{\mathrm{L}} = \frac{A_{\mathrm{ev}} + A_{\mathrm{bulk}}}{\alpha L} \equiv f_{\mathrm{L(ev)}} + f_{\mathrm{L(bulk)}}. \tag{6.17}$$

Here, $f_{\mathrm{L(ev)}}$ and $f_{\mathrm{L(bulk)}}$ are path length enhancement due to evanescent mode and bulk absorption, respectively. We have already shown that for a 2D Lambertian surface, $f_{\mathrm{L(bulk)}} \approx 2$. The A_{ev} does not depend on dielectric thickness; therefore, $f_{\mathrm{L(ev)}} \sim 1/L$. For a moderately low absorption coefficient of $\alpha = 10^3\ \mathrm{m}^{-1}$, and low evanescent mode absorption $A_{\mathrm{ev}} = 10^{-5}$, Fig. 6.9(b) shows that we can obtain very high f_{L} as $L \to 0$, highlighting the importance of evanescent mode absorption.

The intensity enhancement f_{I} is calculated by turning off absorption. Now the ray of light goes into the dielectric (arrow 0 in Fig. 6.10(b)), bounces at the back (states shown by the set of dots marked 1), and then goes out (arrow 2). Therefore, $f_{\text{I}} = 2\beta = 2$. In summary, while f_{I} does not increase, f_{L} does, so that overall f_{A} exceeds the Yablonovitch limit.

Another form of such a slot waveguide structure has been proposed by Green *et al.* (see Fig. 6.10(b)). The cladding layers with higher refractive index increase the evanescent mode coupling to the active layer with a lower refractive index (see Fig. 6.10(c)), and increase A_{ev}. As the active layer is made thinner, $L \to 0$, $A_{\text{ev}} \gg A_{\text{bulk}}$. The increase in f_{L} compensates for the decrease in f_{I} to beat the Yablonovitch limit. The rays that enter into the active layer form the two cladding layers through the escape cones of the cladding-active layer interface as shown in the angle statistics diagram of Fig. 6.10(d). The escaped rays into the active layer are distributed such that evanescent modes are increased. Note that the evanescent mode coupling is purely a wave optics phenomenon and A_{ev} can only be calculated by solving Maxwell's equations.

It is important to realize that the absolute value of absorptance of these arrangements may not be high (i.e., not close to unity), although the absorption enhancement appears to be even orders of magnitude larger than the $4n^2$ limit.

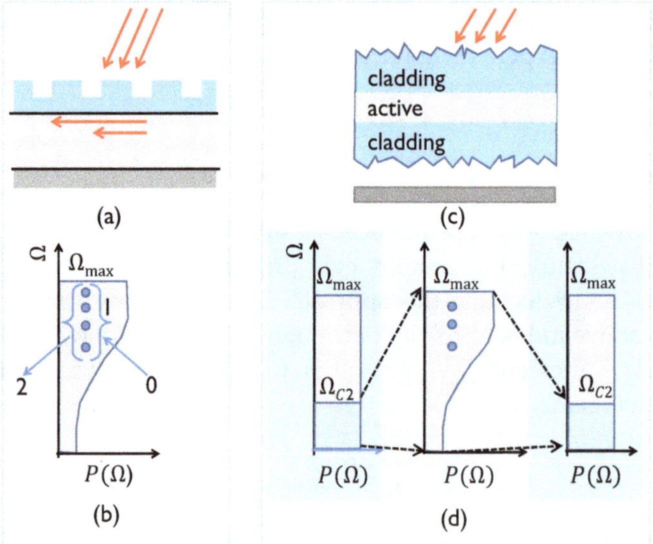

Figure 6.10: (a) Coupling of light into evanescent modes to reduce the probability of photon escape. (c) Thin active layer surrounded by high refractive index cladding for enhanced evanescent mode absorption. (b) and (d) show the angle statistics for (a) and (c), respectively. The two approaches are based on papers by Zongfu Yu *et al.*, *PNAS*, 107(41):17491, 2010, and M. A. Green, *Progress in Photovoltaics: Research and Applications*, 19(4):473, 2011.

6.5.4 Perfect absorption does not imply zero emission

The anisotropic meta-mirror structure discussed above is designed to perfectly trap the incoming light (an optical blackhole!), guiding it through a "path" within the material. Unfortunately, an the internally emitted light (from radiative recombination) can follow the same path backwards and escape the solar cell. Such reciprocity between absorption and emission is sometimes explained as a natural consequence of the time-reversal symmetry of the system. It is, however, possible to use magneto-optical materials to create metamaterials that break this symmetry and create alternate paths for incoming and outgoing lights. As discussed in Homework 4.15, this concept can be used to enhance solar cell efficiency.

6.6 A general model of absorption

So far, we have discussed light absorption in two limiting cases: (a) very weak absorption close to the band edge and (b) very strong absorption far above the band edge. In general, a solar cell with thickness L and perfectly randomizing front and rear interfaces is described by

$$P = \frac{\alpha}{\alpha_{\text{tot}}} \frac{1 - R_\text{b} T_\text{r}^2}{1 - R_\text{f} R_\text{b} T_\text{r}^2} \tag{6.18}$$

where α is the band-to-band absorption that generates electron–hole pairs, and α_{tot} is the absorption (with $E_{\text{ph}} > E_\text{g}$) from all sources, including those that do not generate electron–hole pairs (e.g., intraband free-carrier absorption). In addition, R_f and R_b are the reflection coefficients from the front and back interfaces, respectively, due to Lambertian light trapping for $\theta > \theta_\text{c}$ (with a small contribution from Fresnel reflection for $\theta \leq \theta_\text{c}$). Finally,

$$T_\text{r} = e^{-\alpha_{\text{tot}} L}(1 - \alpha_{\text{tot}} L) + (\alpha_{\text{tot}} L)^2 \int_{\alpha_{\text{tot}} L}^{\infty} dx \frac{e^{-x}}{x} \tag{6.19}$$

is the single-pass transmission probability between the top and the bottom interfaces. In Sec. 6.4.4, we would have derived Eq. (6.19), had we not approximated $A_{\text{single-pass}}$ integral for the "weak-absorption" limit.

Homework 6.8: Equation (6.18) **is easily derived and leads to a generalized expression for absorptance**

First read the paper by M. Green on Lambertian light trapping published in *Progress in Photovoltaics: Research and Applications*, 10:235–241, 2002. Then, derive Eq. (6.18) and show that it reproduces the limiting cases derived in this chapter.

(*continued on the next page*)

Homework 6.8 (*continued from the previous page*)

Solution. Using flux balance at the top surface among incident (I), reflected (R), and upward (I^+) and downward (I^-) fluxes, one writes $I - [RI + (1 - R_f)I^-] = I^+ - I^-$. Similarly, the reflection from the bottom surface leads to the expression $I^- = R_f R_b T_r^2 I^+$. Solving the equations, we can calculate the absorptance $P \equiv (I^+ - I^-)/I$. Assuming that $R \to 0$ (possibly with anti-reflection coating), we find Eq. (6.18).

Ignoring parasitic absorption ($\alpha_{tot} \to \alpha$), approximating Eq. (6.19) as $T_r \approx e^{-2\alpha L}$, and perfectly reflecting back surface ($R_b = 1$) and the reflection from the air–semiconductor front interface ($R_f = 1 - 1/n^2$), we find that

$$P = \frac{1 - e^{-4\alpha L}}{1 - (1 - 1/n^2)e^{-4\alpha L}}. \tag{6.20}$$

Equation (6.20) reduces to the correct limiting expressions. Expanding $e^{-x} \approx 1 - x$ and keeping only the linear terms leads to the Yablonovitch limit. With $e^{-x} \to 0$ in the strong absorption limit leads to $P \to 1$, as expected. Here, P increases monotonically with L. In practice, with the parasitic absorption included ($\alpha \neq \alpha_{tot}$), P initially increases with L due to improved absorption before turning round due to parasitic absorption in thicker films. The critical thickness depends on the absorber (e.g., L_{crit} ~60–90 μm).

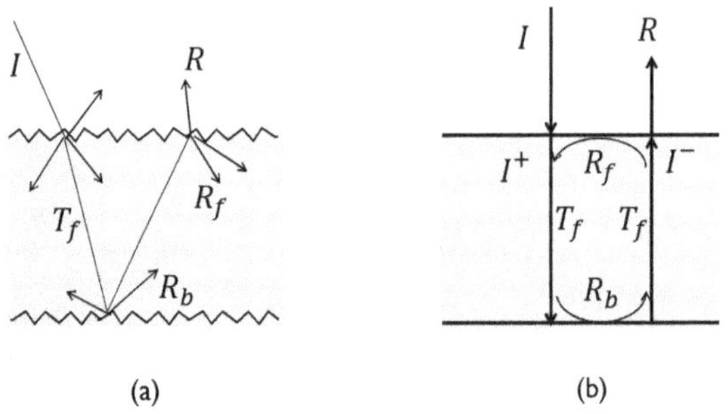

(a) Scattering of photons by the front and back surfaces. (b) At steady state, the net photon flux must be conserved throughout the solar cell.

6.7 Conclusions: absorption limit in PV

For a weakly absorbing layer with a back mirror, the intensity enhancement limit is found to be in the form $f_A = C_{DS} \times n^{D_S} \times (\theta_C/\theta_{esc})^{D_S}$, where n, C_{DS}, and θ_{esc} are the refractive index, proportionality constant, and maximum escape angle, respectively, as defined by the dimensionality D_S of the scattering surface. The derivation provides an intuitively simple explanation of the Yablonovich limit of $f_A = 4n^2$, as a special case with a random refracting surface. Additional gain beyond this limit is possible if we observe the following: the essence of light trapping and intensity enhancement is related to the reduction of escape probability of the photons. This can be achieved either by increasing the number of states occupied by the photons inside the dielectric or by decreasing the number of available states that allow photon escape. The example discussed in Sec. 6.5 suggests that angle restriction provides significant additional gain because it improves not only absorption but also the open-circuit voltage/efficiency of a solar cell. It is important to remember that this additional gain is achieved only for normal incidence of sunlight obtained by orienting the cell toward the sun throughout the day.

This chapter concludes Part I of the book, where we focused on fundamental limits of solar cells based on the balance of photon fluxes. The losses associated with transporting electrons and holes to their respective contacts and the design challenges of practical solar cells will be discussed in Part II.

References

[1] William Shockley and Hans J. Queisser. Detailed Balance Limit of Efficiency of p-n Junction Solar Cells. *Journal of Applied Physics*, 32(3):510, 1961.

[2] Muhammad A. Alam and M. Ryyan Khan. Fundamentals of PV efficiency interpreted by a two-level model. *American Journal of Physics*, 81(9):655–662, September 2013.

[3] E. Yablonovitch and G.D. Cody. Intensity enhancement in textured optical sheets for solar cells. *Electron Devices, IEEE Transactions on*, 29(2):300–305, 1982.

[4] Eli Yablonovitch. Statistical ray optics. *Journal of the Optical Society of America*, 72(7):899–907, July 1982.

[5] Antonio Luque. The confinement of light in solar cells. *Solar Energy Materials*, 23(2–4):152–163, December 1991.

[6] Owen D. Miller, Eli Yablonovitch, and Sarah R. Kurtz. Intense Internal and External Fluorescence as Solar Cells Approach the Shockley-Queisser Efficiency Limit. *arXiv:1106.1603v3*, June 2011.

[7] Xufeng Wang, M.R. Khan, M.A. Alam, and M. Lundstrom. Approaching the Shockley-Queisser limit in GaAs solar cells. In *2012 38th IEEE Photovoltaic Specialists Conference (PVSC)*, pages 002117–002121, June 2012.

[8] P. Campbell and M.A. Green. The limiting efficiency of silicon solar cells under concentrated sunlight. *Electron Devices, IEEE Transactions on*, 33(2):234–239, February 1986.

[9] Marius Peters, Jan Christoph Goldschmidt, and Benedikt Bläsi. Angular confinement and concentration in photovoltaic converters. *Solar Energy Materials and Solar Cells*, 94(8):1393–1398, August 2010.

[10] Jeremy N. Munday. The effect of photonic bandgap materials on the Shockley-Queisser limit. *Journal of Applied Physics*, 112(6):064501–064501–6, September 2012.

[11] M. Ryyan Khan, Xufeng Wang, Peter Bermel, and Muhammad A. Alam. Enhanced light trapping in solar cells with a meta-mirror following generalized Snell's law. *Optics Express*, 22(S3):A973–A985, May 2014.

[12] Nanfang Yu, Patrice Genevet, Mikhail A. Kats, Francesco Aieta, Jean-Philippe Tetienne, Federico Capasso, and Zeno Gaburro. Light Propagation with Phase Discontinuities: Generalized Laws of Reflection and Refraction. *Science*, 334(6054):333–337, October 2011.

[13] Francesco Aieta, Patrice Genevet, Nanfang Yu, Mikhail A. Kats, Zeno Gaburro, and Federico Capasso. Out-of-Plane Reflection and Refraction of Light by Anisotropic Optical Antenna Metasurfaces with Phase Discontinuities. *Nano Lett.*, 12(3):1702–1706, 2012.

[14] Evgenii E. Narimanov and Alexander V. Kildishev. Optical black hole: Broadband omnidirectional light absorber. *Applied Physics Letters*, 95(4):041106, 2009.

[15] Zongfu Yu, Aaswath Raman, and Shanhui Fan. Fundamental limit of nanophotonic light trapping in solar cells. *Proceedings of the National Academy of Sciences*, 107(41):17491–17496, October 2010.

[16] Martin A. Green. Enhanced evanescent mode light trapping in organic solar cells and other low index optoelectronic devices. *Progress in Photovoltaics: Research and Applications*, 19(4):473–477, 2011.

[17] Martin A. Green. Lambertian light trapping in textured solar cells and light-emitting diodes: Analytical solutions. *Progress in Photovoltaics: Research and Applications*, 10(4):235–241, 2002.

[18] M. Evstigneev and B. Arzhang. Effect of Fresnel Reflection on Limit Photoconversion Efficiency in Silicon Solar Cells. *IEEE Journal of Photovoltaics*, 10(5):1463–1464, September 2020. Conference Name: IEEE Journal of Photovoltaics.

[19] M. R. Khan, P. Bermel, and M. A. Alam, "Thermodynamic limits of solar cells with non-ideal optical response," in Photovoltaic Specialists Conference (PVSC), 2013 IEEE 39th, Jun. 2013, pp. 1036–1040. doi: 10.1109/PVSC.2013.6744318.

Part II

Transport Physics of Three Types of Cells

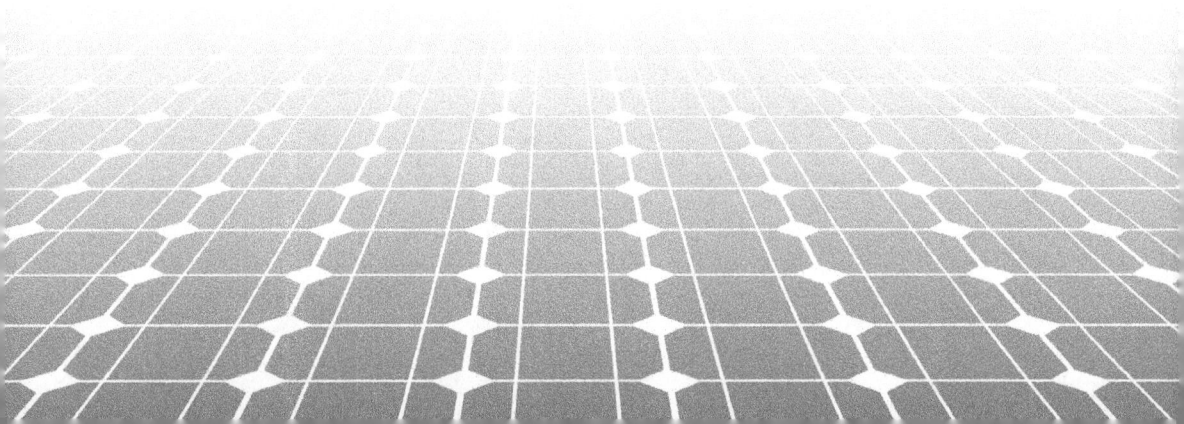

CHAPTER 7

Physics of Typical Solar Cells

Chapter Summary

- The current-voltage characteristics of an ideal solar cell can be obtained by a super-position of dark and light-currents. This I-V characteristics can be interpreted in terms of the thermodynamic considerations discussed in Part 1 of the book.
- The three types of solar cells (i.e. p-i-n, p-n, and heterojunction) approximate the ideal solar cell with field-independent charge collection, negligible wrong-contact recombination, and radiative recombination.
- The dark and light-currents of the three types of solar cells can be calculated by a simple flux-based approach.
- The presumption of the uncorrelated superposition of light and dark currents are often violated in heterojunction solar cells.
- Both the forward and reverse characteristics are important for the operation of a solar cells. The reverse characteristics are particularly important for various reliability issues, such as shadow degradation.

7.1 Introduction

In Part I of the book, we calculated the thermodynamic efficiency limit of an "ideal" solar cell. The performance of the cell is defined exclusively by radiative recombination and ultrafast charge collection. In practice, finite mobility and diffusion lengths as well as imperfect contacts suppress charge collection. The corresponding build-up of excess carriers increases total recombination even in the radiative limit. Moreover, defects and traps within the bulk and at the interfaces further add to the net recombination through non-radiative processes, such as defect-assisted Shockley–Read–Hall (SRH) recombination. In high-quality, direct-bandgap materials (e.g., GaAs), the radiative and non-radiative recombinations are comparable and both must be accounted for to calculate the practical efficiency of the solar cell. For other PV materials (e.g., Si, CIGS, etc.), solar cell performance is limited predominantly by non-radiative recombination. In Part II of this book, we will focus on calculating the efficiency of the solar cell by considering various practical considerations (e.g., non-radiative recombination, finite mobility, shunt resistance, and metallic grids for charge collection) and explain how this "practical" efficiency limit compares to the radiation-only thermodynamic limit calculated in the first part of the book.

7.2 Generation, recombination, and the $J - V$ relationship

We saw in Chapter 3 that the current–voltage characteristics of a solar cell operating at the thermodynamic limit can be calculated by the difference between the absorption (generation) and emission (radiative recombination) fluxes (see Eq. (3.1)). We also know from Chapter 6 that a finite-thickness practical solar cell shown in Fig. 7.1(a) cannot absorb all the incident photons, nor all the recombinations are radiative. Therefore, we should rewrite Eq. (3.1) as the difference between total generation ($G(x, V)$, upward arrows, Fig. 7.1(b)) and net recombination ($R(x, V)$, downward arrows, Fig. 7.1(b)), integrated over the cell, so that

$$J_{\text{total}}(V) = q \int_{\text{cell}} G(x, V)dx - q \int_{\text{cell}} R(x, V)dx. \tag{7.1}$$

This is shown schematically as the area under the position-resolved generation $G(x, V)$ and recombination $R(x, V)$ profiles in Fig. 7.1(c). The output current can be increased by reducing $R(x, V)$ through improved material processing and innovative device design. Improved processing reduces bulk and interface defects so that electron–hole pairs cannot recombine even if they are spatially collocated. Improved device design uses built-in electric fields or heterojunction to spatially separate the electrons and the holes so that they cannot recombine to begin with. In this chapter, we focus on three classes of device design concepts (i.e., p-i-n, p-n, and heterojunction) used to improve the efficiency of solar cells.

To calculate J_{total} using Eq. (7.1), we need to determine the generation and recombination fluxes, similar to Eqs. (3.2)–(3.4). Unfortunately, the fluxes are coupled (see Homework 7.9) and the calculation for a practical solar cell is non-trivial. Fortunately, approximate results can still be obtained analytically by calculating $R(x, V)$ and $G(x, V)$ separately. Specifically, we can calculate the dark current, $J_{\text{dark}}(V)$, by turning off the light (and the photogenerated carriers) as an approximate measure of $\int R(x, V)dx$. We can then calculate the photocurrent, $J_{\text{ph}}(G, V)$, by first turning off all electron–hole recombinations, as an approximate measure of $\int G(x, V)dx$. In other words,

$$J_{\text{total}} \simeq J_{\text{ph}}(G, V) - J_{\text{dark}}(V). \tag{7.2}$$

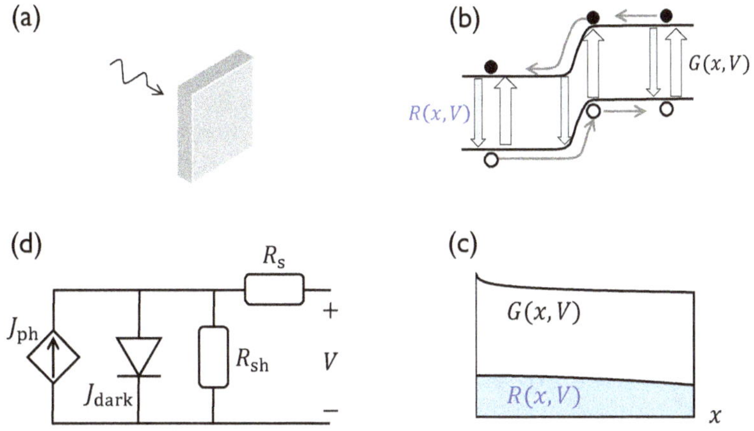

Figure 7.1: (a) A p-n-junction solar cell illuminated by sunlight. (b) Schematic band diagram of a typical p-n-junction solar cell, showing the photogeneration and recombination processes, along with the carrier transport. (c) Schematics of the voltage-dependent generation profile $G(x,V)$, and recombination profile $R(x,V)$ inside the solar cell shown in the area plot. The difference between the areas under these curves determines the output current of the solar cell. (d) An equivalent circuit for solar cells that obeys the superposition principle. The photocurrent is represented by a current source J_{ph}, in parallel to the dark diode current J_{dark}. The parasitic shunt R_{sh} and series R_{s} resistances are also shown.

The photogenerated carrier current J_{ph} depends on photogeneration rate G and the bias voltage across the cell, V. The dark current J_{dark} only depends on V, which is just the diode characteristic. Therefore, the equivalent circuit of the solar cell consists of a voltage-dependent current source with a diode connected in parallel (see Fig. 7.1(d)). In practice, there can be parasitic current paths between the two contacts of the device — this results in an equivalent shunt resistance R_{sh}. The external series resistance R_{s} is due to the contact sheet resistance. We will discuss the physics of R_{sh} and R_{s} in Chapters 9 and 10, respectively. For now, let us use Eq. (7.2) to calculate the $J-V$ characteristics of p-i-n, p-n, and heterojunction solar cells.

Homework 7.1: Energy diagram in Part I vs. band diagram in Part II

Explain the relation between the energy diagrams shown in Part I of the book to the band diagram for p-i-n, p-n, heterojunction solar cells discussed in this part of the book.

(continued on the next page)

Homework 7.1 *(continued from the previous page)*

Solution. In Part I of the book, we considered a set of atoms or a slab of semiconductor with two sets of energy levels (i.e., conduction band, E_c, and valence bands, E_v) separated by an energy bandgap, E_g (see Fig. 2.1). An electron-collecting contact was attached exclusively to one end of the conduction band, keeping the other end open (diamond). The same was true for the hole-collecting contact. The electron and hole separation by these carrier-selective contacts allowed us to reach the thermodynamic limit, because it ensured that electron and holes recombine only radiatively, but not accidentally either by exiting through the wrong contact or annihilating each other through non-radiative recombination. The only trouble with this great concept is that this scheme is impossible: no sooner do you touch a semiconductor with a metal electrode than both electrons and holes are free to move toward it and recombine with each other.

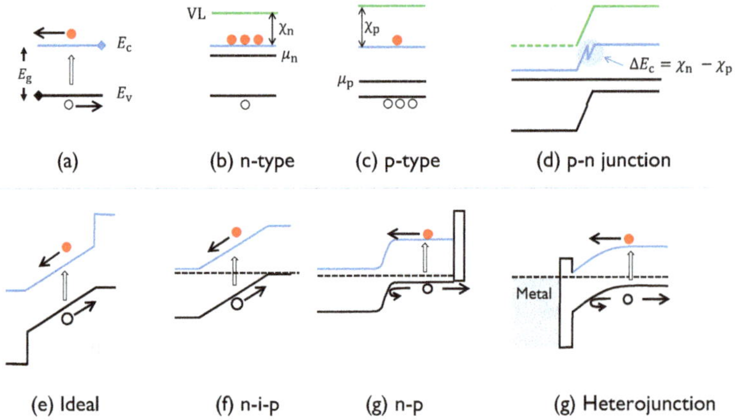

(a) Two-level solar cells attached to electron and hole contacts. (b) An n-doped semiconductor. (c) A p-doped semiconductor. (d) Formation of a junction between two dissimilar materials. (d) An n-p heterojunction solar cell. (e) An n-i-P heterojunction. (f) An n-i-p solar cells with staggered band diagrams. (g) An n-p junction solar cell has an effective hole blocking. (h) An n-p heterojunction solar cell.

It is easy to show that commercial p-i-n, p-n, and heterojunction solar cells use a combination of doping and bandgaps to approximate the ideal solar cell discussed above. To understand how it works, recall that several semiconductors, each defined by its bandgap E_g and vacuum level, χ, form the complex energy band profiles by following a few simple facts of semiconductor physics:

(continued on the next page)

Homework 7.1 (*continued from the previous page*)

- *Doping a semiconductor.* An n-type material, doped with electron donor atoms, has a lot of electrons ($n \sim N_{\mathrm{D}}$), so that in isolation its Fermi level μ_{n} is located close to E_{c} (Homework 7.1, Fig. (b)),

$$E_{\mathrm{c}} - \mu_{\mathrm{n}} = -(k_{\mathrm{B}}T_{\mathrm{D}})/q\ln(N_{\mathrm{D}}/N_{\mathrm{C}}), \tag{7.3}$$

where N_{C} is the effective density of states in the conduction band. Similarly, a p-type material, doped with electron acceptor atoms, has a lot of holes ($p \sim N_{\mathrm{A}}$), so that in isolation its Fermi level (μ_{p}) is located close to E_{v} (Homework 7.1, Fig. (c)),

$$E_{\mathrm{v}} - \mu_{\mathrm{p}} = (k_{\mathrm{B}}T_{\mathrm{D}})/q\ln(N_{\mathrm{A}}/N_{\mathrm{V}}), \tag{7.4}$$

where N_{V} is the effective density of states in the valence band. For an undoped (intrinsic) semiconductor, the Fermi level is located approximately in the middle of the bandgap, i.e., $\mu_i \sim (E_{\mathrm{c}} + E_{\mathrm{v}})/2$.

- *Band alignment and continuity of the vacuum level.* An external applied voltage, V, separates the Fermi levels, i.e., $qV = \mu_{\mathrm{p}} - \mu_{\mathrm{n}}$. Naturally, $\mu_{\mathrm{p}} = \mu_{\mathrm{n}}$ in equilibrium as well as at the short-circuit condition, as shown in Homework 7.1, Fig. (d). Moreover, the vacuum level must be continuous, so that the discontinuity in the conduction band at the interface between two materials is given by the work function difference between the two neighboring materials, i.e., $\Delta E_{\mathrm{c}} = \chi_n - \chi_p$ and $\Delta E_{\mathrm{c}} + \Delta E_{\mathrm{v}} = E_{g,n} - E_{\mathrm{g,p}}$. The conduction and valence band profiles control electron and hole movement: electrons roll down the conduction band, while holes float up the valence band.

- *Law of mass action.* The product of electron and hole concentrations is a voltage-dependent constant, i.e.,

$$n \times p \equiv n_{\mathrm{i}}^2 \exp(qV/k_{\mathrm{B}}T) \tag{7.5}$$

where $n_{\mathrm{i}}^2 = N_{\mathrm{C}}N_{\mathrm{V}}\exp(-E_{\mathrm{g}}/k_{\mathrm{B}}T)$ is defined by the intrinsic properties of a semiconductor. The law applies (a) at every point within the device in equilibrium ($G = 0, V = 0$) or (ii) within the junction, defined by the

(*continued on the next page*)

Homework 7.1 (*continued from the previous page*)

regions with $dE_c/dx \neq 0$ and/or $dE_v/dx \neq 0$, under non-equilibrium conditions ($G \neq 0$).

Homework 7.1, Figs. (e–g) show various types of solar cells to be discussed in Part II of this book. You should convince yourself that the n-i-p, n-p, and heterojunction band diagrams follow the four principles discussed above. Since electrons flow down the conduction band, $E_c(x)$, and holes float up the valence band, $E_v(x)$, an ideal solar cell involves a p-i-n heterojunction device where the intrinsic layer with bandgap $E_g \approx 1.34$ eV absorbs the photons, while the left and right heterojunctions perfectly block the wrong-contact escape and assist in unimpeded correct contact collection. A simpler homojunction n-i-p structure can reduce (but not block) the wrong-contact escape, and it does facilitate the correct-contact collection. For a p-n structure, the electron escape from the wrong contact is prevented by an additional layer on the right, called back-surface field. Finally, heterojunction solar cells (with a metal contact, large-E_g blocking layer, and the semiconductor absorber) create the correct energy band profile to prevent wrong-contact escape and correct-contact collection.

7.3 A versatile flux-based approach to calculate dark and photocurrents

To calculate the $J-V$ characteristics using Eq. (7.2), we need to determine $J_{\text{ph}}(G, V)$ and $J_{\text{dark}}(G, V)$. Traditionally, one solves a pair of drift–diffusion equations for electrons and holes, but an analytical solution is possible only for simple p-n-junction solar cells. An alternative (and intuitive) flux-based approach involves calculating carrier injection over an energy barrier. We do not need to solve any differential equation, and yet, we can calculate the dark and photocurrents analytically for a variety of solar cells.

The approach involves a simple idea based on the Landauer principle that initially appears unconnected to a solar cell: As shown in Fig. 7.2, electrons within the device can reach the left (or right) contact by going over the energy barrier $E_{\text{B,L}}$ (or $E_{\text{B,R}}$). Only electrons with sufficiently high energy can surmount the barriers, because the thermionic emission is suppressed exponentially with the barrier height. Therefore, the transmission coefficient of the electrons from the device to the left contact is $\gamma_{\text{RL}} = v_{\text{L}} \exp(-E_{\text{B,L}}/k_{\text{B}}T)$. Here, v_{L} is the collection velocity at the left contact. Similarly, $\gamma_{\text{LR}} = v_{\text{R}} \exp(-E_{\text{B,R}}/k_{\text{B}}T)$. We will now use this simple "injection-over-a-barrier" model to calculate the dark and photocurrents of a variety of solar cells.

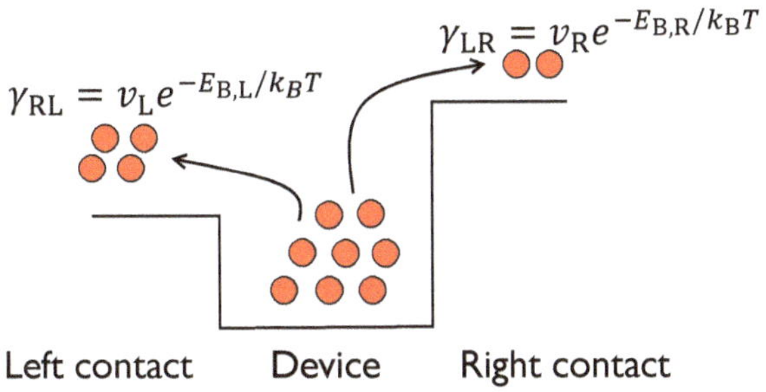

Figure 7.2: Illustration of over-the-barrier carrier transmission. The first and second indices of the transmission coefficient denote the relative positions of the injection point and the collection point, respectively, i.e., γ_{LR} implies injection from the left to the right.

7.4 The current–voltage characteristics of p-i-n solar cells

Let us begin by calculating the $J-V$ characteristics of a p-i-n solar cell by using the flux approach. A p-i-n solar cell includes an intrinsic absorption layer sandwiched between very thin n and p layers, as shown in Fig. 7.3(a). These n- and p-doped layers respectively collect the photogenerated electrons and holes and define the built-in potential ($qV_{bi} = E_{c,n} - E_{c,p}$). For a homojunction solar cell, $\Delta E_c = \chi_n - \chi_p = 0$, and therefore the wrong-contact collection cannot be prevented.

Practical examples of p-i-n thin-film solar cells (TFPV) include hydrogenated amorphous silicon (a-Si:H) or nanocrystalline silicon (nc-Si:H) solar cells. Unlike crystalline silicon (c-Si), a-Si:H is a direct-bandgap material (E_g ~1.7 eV) in which hydrogen passivates the defects in the amorphous silicon. In these cells, the i-layer thickness is defined by the depth necessary for photon absorption (~300–600 nm). We choose the a-Si:H cells as a model system, because we can gain deep insight

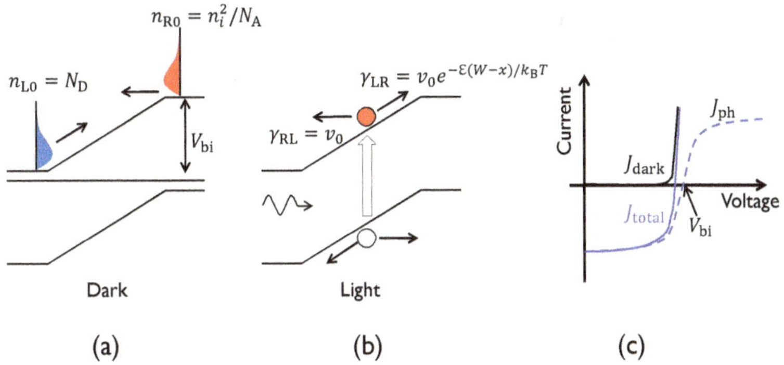

Figure 7.3: (a) Carrier injection from the contacts under dark conditions in a p-i-n device. (b) Flow of photogenerated carriers. (c) $J-V$ characteristics of a p-i-n solar cell. Details are shown in Secs. 7.4.1 and 7.4.2.

into the current collection in a solar cell by solving the flux equations analytically under the following (very reasonable) assumptions:

(a) the i-layer is much thicker than the p/n layers, i.e. $W \approx d$, where W and d are the cell and i-layer thicknesses, respectively;

(b) the electric field inside the i-layer is constant and is given by $\mathcal{E} = (V - V_{\text{bi}})/W$, where V_{bi} is the built-in potential;

(c) the generation rate G is approximately constant across the ultra-thin i-layer; and finally,

(d) the i-layer is so thin and the built-in field so high that the carriers are collected before they recombine, i.e., bulk recombination is absent.

With these assumptions, the transport for electrons and holes are decoupled, making it easy to calculate the photocurrent (J_{ph}) and the "ideal" diode current (J_{dark}).

7.4.1 Dark current: A p-i-n solar cell

Let us first focus on electron transport in the p-i-n cell under dark conditions ($G = 0$). Figure 7.3(a) shows that at equilibrium ($G = 0$, $V = 0$), the electron density is $n_{\text{L0}} = N_{\text{D}}$ at the left (n/i) interface and $p_{\text{R0}} = N_{\text{A}}$ (or equivalently by Eq. (7.5), $n_{\text{R0}} = n_{\text{i}}^2/N_{\text{A}}$) at the right (i/p) interface). Assuming that bulk recombination is absent, the carrier velocity is equal to the thermal velocity v_0 in all layers. Now, the electrons at any position within the cell can move/partition to the right or to the left with transmission probabilities γ_{LR} or γ_{RL}, respectively, as discussed in Sec. 7.3. Therefore, the number of electrons moving from the left n/i interface toward the right i/p interface is $n_{\text{L0}}v_0 \times \gamma_{\text{LR}}/(\gamma_{\text{LR}} + \gamma_{\text{RL}})$, while the corresponding current from the right i/p-contact to the left n/i contact is $n_{\text{R0}}v_0 \times \gamma_{\text{RL}}/(\gamma_{\text{LR}} + \gamma_{\text{RL}})$. The net electron current is

$$J_{\text{n}} = qn_{\text{L0}}v_0 \frac{\gamma_{\text{LR}}}{\gamma_{\text{LR}} + \gamma_{\text{RL}}} - qn_{\text{R0}}v_0 \frac{\gamma_{\text{RL}}}{\gamma_{\text{LR}} + \gamma_{\text{RL}}}. \tag{7.6}$$

At equilibrium ($G = 0$, $V = 0$), the built-in potential barrier qV_{bi} exponentially suppresses the number of electrons injected from the left n/i contact that can reach the right i/p-contact. Thus, $\gamma_{\text{LR,0}} = v_0 \exp(-q\mathcal{E}_{\text{B,0}}\, d/k_{\text{B}}T) = v_0 \exp(-qV_{\text{bi}}/k_{\text{B}}T)$. The electrons injected from the right i/p junction, however, face no energy barrier whatsoever; therefore, $\gamma_{\text{RL,0}} = v_0 \times 1$. Electron flux balance at $V = 0$ ensures that $J_{\text{n}}(V = 0) = 0$, and we find that

$$\frac{n_{\text{L,0}}}{n_{\text{R,0}}} = \frac{\gamma_{\text{RL,0}}}{\gamma_{\text{LR,0}}} = e^{+qV_{\text{bi}}/k_{\text{B}}T}. \tag{7.7}$$

Since $n_{\text{R,0}} = n_{\text{i}}^2/N_{\text{A}}$, we can write $n_{\text{L,0}} = (n_{\text{i}}^2/N_{\text{A}})e^{+qV_{\text{bi}}/k_{\text{B}}T}$.

Homework 7.2: Boltzmann distribution and transmission over a barrier

Show that the thermionic emission probability is given by $\gamma_{\mathrm{RL}} = v_{\mathrm{L}} \exp\left(-E_{\mathrm{B,L}}/k_{\mathrm{B}}T\right)$.

Solution. We saw in Chapter 2 that the electron density at energy E, with the density of states $D_c(E)$, is given by

$$n(E) = D_c(E)\, f_c(E - \mu_{\mathrm{n}}) \sim D_c(E)\, e^{-(E-\mu_{\mathrm{n}})/k_{\mathrm{B}}T}.$$

Here, $f_c(E - \mu_{\mathrm{n}}) = \left[1 + \exp\left((E - \mu_{\mathrm{n}})/k_{\mathrm{B}}T\right)\right]^{-1}$ is the Fermi–Dirac probability of a state being occupied. The Boltzmann approximation (i.e., occupation probability falling exponentially with energy) holds for $(E_{\mathrm{c}} - \mu_{\mathrm{n}}) > 3k_{\mathrm{B}}T$. With the Boltzmann approximation and $D_c(E)$ being a weak function of energy, Eq. (7.3) is easily derived:

$$N_{\mathrm{D}} \sim n = \int_{E_{\mathrm{c}}}^{\infty} n(E)dE = N_{\mathrm{C}}\, e^{-(E_{\mathrm{c}}-\mu_{\mathrm{n}})/k_{\mathrm{B}}T}.$$

Here, N_{C} is the effective density of states in the conduction bands that are reasonably occupied by electrons. The average injection probability over the barrier is given by

$$\gamma_{\mathrm{RL}} = \frac{\int_{E_{\mathrm{B,L}}+E_{\mathrm{c}}}^{\infty} v(E)\, n(E)dE}{\int_{E_{\mathrm{c}}}^{\infty} n(E)dE} \approx v_0 \frac{N_{\mathrm{C}} e^{-(E_{\mathrm{B,L}}+E_{\mathrm{c}}-\mu_{\mathrm{n}})/k_{\mathrm{B}}T}}{N_{\mathrm{C}} e^{-(E_{\mathrm{c}}-\mu_{\mathrm{n}})/k_{\mathrm{B}}T}} = v_0 e^{-E_{\mathrm{B,L}}/k_{\mathrm{B}}T}.$$

Here, $v_0 \approx \sqrt{k_{\mathrm{B}}T/m_n}$ is the average velocity of the electrons dictated by the effective mass (m_n) of the electrons.

If we now apply a voltage across this p-i-n diode ($G = 0, V > 0$), the electronic barrier height in the i-layer is reduced to $q(V_{\mathrm{bi}} - V)$, so that $\gamma_{\mathrm{LR}} = v_0 \exp(-q(V_{\mathrm{bi}} - V)/k_{\mathrm{B}}T)$ and $\gamma_{\mathrm{RL}} = v_0$. The voltage-dependent electron current is given by

$$\begin{aligned} J_{\mathrm{n}}(G=0,V) &= q n_{\mathrm{L},0} v_0 \frac{\gamma_{\mathrm{LR}}}{\gamma_{\mathrm{LR}} + \gamma_{\mathrm{RL}}} - q n_{\mathrm{R},0} v_0 \frac{\gamma_{\mathrm{RL}}}{\gamma_{\mathrm{LR}} + \gamma_{\mathrm{RL}}} \qquad (7.8) \\ &= \frac{q v_0}{\gamma_{\mathrm{LR}} + \gamma_{\mathrm{RL}}} \left(n_{\mathrm{L},0}\gamma_{\mathrm{LR}} - n_{\mathrm{R},0}\gamma_{\mathrm{RL}}\right) \\ &= \frac{q v_0}{\exp(-q(V_{\mathrm{bi}} - V)/k_{\mathrm{B}}T) + 1} \frac{n_{\mathrm{i}}^2}{N_{\mathrm{A}}} \left[e^{qV_{\mathrm{bi}}/k_{\mathrm{B}}T} e^{-q(V_{\mathrm{bi}}-V)/k_{\mathrm{B}}T} - 1\right] \\ &= q\frac{n_{\mathrm{i}}^2}{N_{\mathrm{A}}} \left[\frac{\mu_{\mathrm{n}}(V - V_{\mathrm{bi}})/d}{\exp(q(V - V_{\mathrm{bi}})/k_{\mathrm{B}}T) + 1}\right] \left[e^{qV/k_{\mathrm{B}}T} - 1\right]. \qquad (7.9) \end{aligned}$$

In deriving the last line, we assumed that $v_0 \equiv \mu_{\mathrm{n}}\mathcal{E} = \mu_{\mathrm{n}}(V - V_{\mathrm{bi}})/d$. It is easy to see that the hole current, $J_{\mathrm{p}}(G = 0, V)$, has a similar form, except that N_{A} is replaced by N_D and μ_n by μ_p. Finally, the total dark current given by the sum of electron and hole currents is

$$\begin{aligned} J_{\text{dark}} &= J_n(G=0,V) + J_p(G=0,V) \\ &= q\left[\mu_n \frac{n_i^2}{N_A} + \mu_p \frac{n_i^2}{N_D}\right]\left[\frac{(V-V_{bi})/d}{\exp(q(V-V_{bi})/k_B T)+1}\right]\left[e^{qV/k_B T}-1\right] \quad (7.10) \\ &\equiv J_{01}\left[e^{qV/k_B T}-1\right]. \quad (7.11) \end{aligned}$$

Here, q is electron charge, d is the thickness of the intrinsic region, N_D and N_A are the doping densities of the n and p layers, respectively, μ_n and μ_p denote the electron and hole mobilities, respectively, n_i is the intrinsic carrier density in a-Si:H, k_B is the Boltzmann constant, and T is absolute temperature. A plot of J_{dark}, with a characteristic exponential increase at $V \approx V_{bi}$, is shown in Fig. 7.3(c). The dark current actually saturates above V_{bi}, as discussed in Homework 7.3.

Homework 7.3: Approximate expressions for dark current

Based on Eq. (7.11), derive an approximate expression for J_{dark} for $V < V_{bi}$ and $V > V_{bi}$.

Solution. For $V < V_{bi}$, we can simplify the middle term of Eq. (7.10) by first linearizing the exponential term, rationalizing the fraction, and using the Einstein relation ($D_n/\mu_n \equiv k_B T/q$). The simplified expression $(k_B T/q)(1/d)$ reduces Eq. (7.10) to

$$J_{\text{dark}} \approx \underbrace{q\left(\frac{D_p}{d}\frac{n_i^2}{N_D} + \frac{D_n}{d}\frac{n_i^2}{N_A}\right)}_{J_{01}}\left[e^{qV/k_B T}-1\right]. \quad (7.12)$$

We will see later that this dark current expression for a p-i-n solar cell is identical to that of an ideal p-n diode solar cell, except that the electron and hole diffusion lengths are replaced by the device thickness, $W \approx d$.

For $V > V_{bi}$, the middle term of Eq. (7.10) can be simplified because $\exp{(V - V_{bi})/k_B T} \gg 1$. Moreover, from Homework 7.1, we find $qV_{bi} \equiv E_{c,p} - E_{c,n} = (k_B T/q)\ln{(N_D N_A/n_i^2)}$. Taken together, Eq. (7.10) reduces to:

$$J_{\text{dark}} \approx \underbrace{q(\mu_n N_D + \mu_p N_A)}_{\rho^{-1}}\underbrace{\frac{V - V_{bi}}{d}}_{\mathcal{E}}. \quad (7.13)$$

(*continued on the next page*)

Homework 7.3 (*continued from the previous page*)

The built-in barrier disappears at $V > V_{\mathrm{bi}}$ and the i-layer serves as a simple ohmic resistor described by $J_{\mathrm{dark}} = \rho^{-1}\mathcal{E}$. Here, ρ is intrinsic resistivity of the a-Si:H layer. Any external series resistance must be accounted for separately.

7.4.2 Photocurrent: p-i-n cells

Now let us consider the p-i-n cell under illumination, as shown in Fig. 7.3(b). For simplicity, we assume that the generation rate G is constant throughout the device. At depth x, the photogenerated electrons (or holes) partition into fluxes to right and left. The electrons (or holes) reach the contact and recombine with holes (or electrons) to yield the photocurrent J_{ph}. Since the current is continuous, we can calculate the net recombination or collection at either contact. Therefore, we calculate electron flux toward the right p-contact to find electron contribution to the photocurrent, i.e.,

$$J_{\mathrm{ph(n)}} = -q \int_0^W G\,dx \frac{\gamma_{\mathrm{LR,n}}}{\gamma_{\mathrm{RL,n}} + \gamma_{\mathrm{LR,n}}}. \tag{7.14}$$

The hole current to the right contact is similarly given by

$$J_{\mathrm{ph(p)}} = q \int_0^W G\,dx \frac{\gamma_{\mathrm{LR,p}}}{\gamma_{\mathrm{RL,p}} + \gamma_{\mathrm{LR,p}}}. \tag{7.15}$$

Homework 7.4: Thin-film solar cell: Dark current

Consider a thin-film p-i-n solar cell with the following parameters: $W \approx d =$ 300 nm, $N_{\mathrm{A}} = N_{\mathrm{D}} = 10^{18}$ cm^{-3}, and $n_{\mathrm{i}} = 10^{10}$ cm^{-3}. For the i-region, $\mu_{\mathrm{n}} = 1$ cm^2/Vs, and $\mu_{\mathrm{p}} = 1$ cm^2/Vs. Use Eq. (7.12) to calculate $J_{\mathrm{dark}}(V)$ in terms of the reverse saturation current, J_{01}. Also determine the built-in voltage of the solar cell.

Solution. Using the Einstein relationship ($D/\mu = k_{\mathrm{B}}T/q$) and $T = 300$ K, we find that $D_{\mathrm{n}} = D_{\mathrm{p}} = 0.258$ cm^2/s. The corresponding saturation current is $J_{01} = 1.52 \times 10^{-11}$ A/cm^2. Therefore,

$$J_{\mathrm{dark}}(V) = 1.52 \times 10^{-11} \left(e^{qV/k_{\mathrm{B}}T} - 1\right).$$

The built-in voltage for this homojunction diode is

$$V_{\mathrm{bi}} = (k_{\mathrm{B}}T/q)\ln\left(N_{\mathrm{D}}N_{\mathrm{A}}/n_{\mathrm{i}}^2\right) = 0.0258\ \ln\left(10^{18} \times 10^{18}/10^{20}\right) = 0.952\ \mathrm{V}.$$

Finally, the total photocurrent is given by

$$\begin{aligned} J_{\mathrm{ph}} &= J_{\mathrm{ph(p)}} + J_{\mathrm{ph(n)}} \\ &= q\int_0^W G\,dx \left[\frac{\gamma_{\mathrm{LR,p}}}{\gamma_{\mathrm{RL,p}} + \gamma_{\mathrm{LR,p}}} - \frac{\gamma_{\mathrm{LR,n}}}{\gamma_{\mathrm{RL,n}} + \gamma_{\mathrm{LR,n}}}\right] \quad (7.16) \\ &= q\int_0^W G\,dx \left[\frac{\gamma_{\mathrm{RL,n}}}{\gamma_{\mathrm{LR,n}} + \gamma_{\mathrm{RL,n}}} - \frac{\gamma_{\mathrm{LR,n}}}{\gamma_{\mathrm{RL,n}} + \gamma_{\mathrm{LR,n}}}\right]. \quad (7.17) \end{aligned}$$

Here, we have used the fact that $\gamma_{\mathrm{LR,p}} = \gamma_{\mathrm{RL,n}}$ by symmetry. Although photogenerated carriers involve both electrons and holes, Eq. (7.17) suggests that J_{ph} can be calculated from the difference of left- and right-moving fluxes associated with a single type of carrier (electron or hole). Using the expressions of γ's at depth x, we find that

$$\begin{aligned} J_{\mathrm{ph}} &= q\int_0^W G\,dx \left[\frac{v_0}{v_0 + v_0 e^{-\mathcal{E}(W-x)/k_{\mathrm{B}}T}} - \frac{v_0 e^{-q\mathcal{E}(W-x)/k_{\mathrm{B}}T}}{v_0 + v_0 e^{-q\mathcal{E}(W-x)/k_{\mathrm{B}}T}}\right] \quad (7.18) \\ &= qGW \times \frac{2L_{\mathrm{D}}}{W} \ln\cosh\frac{W}{2L_{\mathrm{D}}} \quad (7.19) \\ &\approx qGW \left[\coth\frac{W}{2L_{\mathrm{D}}} - \frac{2L_{\mathrm{D}}}{W}\right]. \quad (7.20) \end{aligned}$$

Here $L_{\mathrm{D}} \equiv k_{\mathrm{B}}T/q\mathcal{E} = (k_{\mathrm{B}}T/q)(W/(V - V_{\mathrm{bi}}))$ defines the length of the region (e.g., Price length) close to the contacts over which wrong-contact injection reduces the photocurrent.

The plot of Eq. (7.20) in Fig. 7.3(c) shows that the photocurrent of a p-i-n cell is voltage dependent. Figure 7.3(b) suggests an intuitive explanation: the photogenerated carriers can be swept downstream by the electric field (γ_{RL}) and contribute to the useful current, or jump over the built-in barrier upstream and recombine at the wrong contact (γ_{LR}). With V increasing toward V_{bi}, the reduction in the electric field ($\mathcal{E} \approx (V - V_{\mathrm{bi}})/W$) makes it difficult to collect charges, while the reduction in the barrier height ($q(V_{\mathrm{bi}} - V)$) makes wrong-contact recombination increasingly likely. At $V \approx V_{\mathrm{bi}}$, J_{ph} vanishes, because without an electric field to separate the photogenerated carriers, they can escape through either contact. For $V > V_{\mathrm{bi}}$, the electric field changes sign, and the photogenerated carriers flow to opposite contacts, resulting in a *positive* photocurrent. This voltage-dependent reduction (and eventual reversal) of J_{ph}, shown in Fig. 7.3(c), is a characteristic feature of a thin-film p-i-n solar cell. As a result, the open-circuit voltage (V_{oc}) and maximum power point voltage (V_{mp}) are reduced below the thermodynamic limit. Moreover, it explains why $J_{\mathrm{total}}(V)$ is seen to cross $J_{\mathrm{dark}}(V)$ at $V \approx V_{\mathrm{bi}}$.

Although the closed form expression in Eq. (7.20) is derived for a p-i-n cell, the photocurrent will always be voltage dependent for any solar cell that relies on the built-in electric field to collect photogenerated carriers. Since the electric field region in all TFPV cells is either equal to or comparable to absorber layer thickness, all TFPV cells exhibit an intrinsically voltage-dependent photocurrent.

This voltage dependence violates a widely used and often-invoked superposition principle for solar cells, to be discussed in Sec. 7.7.

Homework 7.5: Photocurrent in a thin-film solar cell

Reconsider the solar cell in Homework 7.4 and assume that the total photogenerated flux is given by $qGW = 15\ \mathrm{mA/cm^2}$. Calculate $J_{\mathrm{ph}}(V)$ to determine $J_{\mathrm{sc}} \equiv J_{\mathrm{ph}}(G, V = 0)$. What fraction of the photogenerated flux is actually collected as J_{sc}? Can you explain the collection efficiency in terms of assumptions implicit in deriving the equations? Also, calculate V_{oc} by setting $J_{\mathrm{total}} = J_{\mathrm{ph}} - J_{\mathrm{dark}} = 0$. Note that V_{oc} is hard to calculate analytically, but we can estimate it by setting $qGW - J_{\mathrm{dark}} = 0$ because $V_{\mathrm{oc}} < V_{\mathrm{bi}}$.

Solution. The short-circuit current and open-circuit voltage are $J_{\mathrm{sc}} = 14.4\ \mathrm{mA/cm^2}$ and $V_{\mathrm{oc}} = 0.892$ V, respectively.

Homework 7.6: The p-i-n solar cell with carrier-selective contacts

Derive the $J - V$ characteristics of a p-i-n solar cell, assuming a perfect electron blocking layer on the right and a perfect hole-blocking layer on the left.

Solution. The blocking layers make $\gamma_{\mathrm{LR,n}} = 0$ and $\gamma_{\mathrm{RL,p}} = 0$. We can sum the electron and hole fluxes in either the left or the right contact. For example, the electron current collected at the left contact is given by

$$J_{\mathrm{n}} = q \int G dx \frac{\gamma_{\mathrm{RL,n}}}{\gamma_{\mathrm{LR,n}} + \gamma_{\mathrm{RL,n}}} = q \int G dx = qGW,$$

whereas the hole current collected at the left contact is given by

$$J_{\mathrm{p}} = -q \int G dx \frac{\gamma_{\mathrm{RL,p}}}{\gamma_{\mathrm{LR,p}} + \gamma_{\mathrm{RL,p}}} = 0.$$

We would get the same answer if we evaluated the fluxes on the right contact, except with $J_n = 0$ and $J_{\mathrm{p}} = qGW$. Regardless, the photocurrent, $J_{\mathrm{ph}} = J_{\mathrm{n}} + J_{\mathrm{p}} = qGW$, is independent of the applied voltage, because recombination is absent and wrong-contact collection has been suppressed. The total current of this ideal solar cell is easily calculated as a simple superposition of the dark and photocurrents (Eq. (7.2)):

$$J = J_{\mathrm{ph}} - J_{01}\left(e^{qV/k_{\mathrm{B}}T} - 1\right).$$

7.5 Current–voltage characteristics of p-n-junction solar cells

Many commercially important solar cell technologies use a simple p-n junction to collect the photocurrent and suppress the dark current (see Fig. 7.4). A simple p-n-junction solar cell cannot prevent wrong-carrier recombination; therefore, innovative modifications (e.g., back-surface field, point-contact technology, etc.) are necessary to increase the efficiency of the cells. In this chapter, we will discuss the fundamental charge collection issues of a p-n-junction solar cell. We will discuss the technology-relevant topics in Part III of the book.

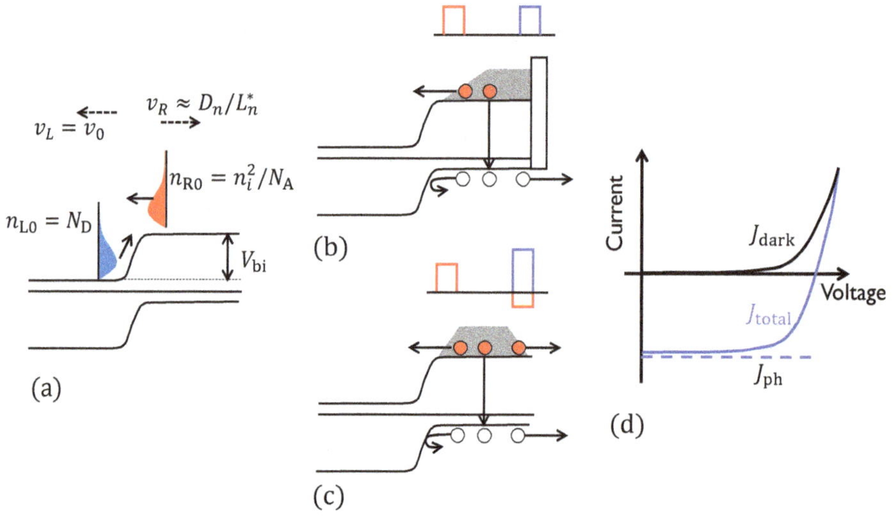

Figure 7.4: (a) Carrier injection from the contacts under dark conditions in a p-n device. (b) and (c) show the flow of photogenerated carriers. The light is incident from the p-side, and most of the carrier generation is in the p-region. (b) Shows the structure with a blocking layer for electrons at the p-contact; (c) shows a case when there is no blocking layer. (d) $J-V$ characteristics of a p-n solar cell.

7.5.1 Dark current: p-n-junction solar cells

Let us first derive an expression of dark current for a p-n-junction cell by using the flux balance approach explained in Sec. 7.3. Figure 7.4(a) shows that at equilibrium ($G = 0$, $V = 0$), the electron density is $n_{L0} = N_D$ in the n-doped region on the left and $n_{R0} = n_i^2/N_A$ in the p-doped region on the right. These densities hold in the quasi-neutral regions (i.e., $\mathcal{E} \sim 0$) up to the edges of the space charge region, with $\mathcal{E} \neq 0$. We assume that there is no recombination in the space charge region.

Now, the effective velocity of minority electrons injected from the left n-region into the right p-region (see Fig. 7.4(a)) is limited by diffusion, so that $v_R = D_n/L_n^*$. Here, $L_n^* = L_n || W_p$ is the effective diffusion length of minority electrons. The velocity of majority carriers (electrons) in left n-region is equal to the thermal velocity, i.e., $v_L = v_0$. We can now use the previously introduced idea of electron partitioning and transmission coefficients to write the net electron current as

$$J_n = qn_{L,0}v_L \frac{\gamma_{LR}}{\gamma_{RL} + \gamma_{LR}} - qn_{R,0}v_R \frac{\gamma_{RL}}{\gamma_{RL} + \gamma_{LR}}. \quad (7.21)$$

Homework 7.7: The i-region recombination in a p-i-n solar cell

The defects in the i-region (time constant, τ) lead to an electron–hole recombination of the photogenerated and contact-injected carriers. If the recombination is weak, the increase in the dark current and the reduction in the photocurrent are given by

$$\Delta J_{ph} = \frac{1}{4} \frac{qGW^3}{\mu\tau(V_{bi} - V)},$$
$$\Delta J_{dark} = \frac{qn_i W}{\tau} \frac{2k_B T/q}{V_{bi} - V} \left(e^{qV/2k_B T} - 1\right),$$
$$J_{total} = (J_{dark} + \Delta J_{dark}) - (J_{ph} - \Delta J_{ph}).$$

Here, J_{dark} and J_{ph} are dark and photocurrents for an idealized defect-free p-i-n solar cell, given by Eqs. (7.10) and (7.20), respectively. Provide an intuitive interpretation of the recombination current components.

Solution. Let us rewrite ΔJ_{ph} in terms of the electric field $\mathcal{E} = |V_{bi} - V|/W$ and the drift distance before recombination (i.e., $L_2 \equiv \mu\mathcal{E}\tau$), so that $\Delta J_{ph} = qGW(W/4L_2)$. The equation is interpreted as the fraction (i.e., $W/4L_2$) of photo-carriers (qGW) lost to recombination. Similarly, ΔJ_{dark} is understood as follows. The recombination is maximized when electron and hole concentrations are equal at the middle of the i-layer, i.e., $n(W/2) \approx p(W/2) = n_i \exp(qV/2k_B T)$, from Eq. (7.5), and the recombination width of the recombination region is given by ($L_1 \equiv 2k_B T/q\mathcal{E}$); therefore, $\Delta J_{dark} = qn(W/2)L_1/\tau$.

Here, transmission coefficients: $\gamma_{LR} = v_R \exp(-q(V_{bi} - V)/k_B T)$ and $\gamma_{RL} = v_L = v_0$. Since $V_L \gg v_R$, we expect (for a range of V) $\gamma_{RL} \gg \gamma_{LR}$. Therefore:

$$J_n = qn_{L,0}V_L \frac{\gamma_{LR}}{\gamma_{RL} + \gamma_{LR}} - qn_{R,0}v_R \frac{\gamma_{RL}}{\gamma_{RL} + \gamma_{LR}} \quad (7.22)$$
$$\approx qn_{L,0}V_L \frac{\gamma_{LR}}{\gamma_{RL}} - qn_{R,0}v_R \frac{\gamma_{RL}}{\gamma_{RL}} \quad (7.23)$$
$$= qn_{L,0}v_0 \frac{\frac{D_n}{L_n^*} e^{-q(V_{bi}-V)/k_B T}}{v_0} - qn_{R,0} \frac{D_n}{L_n^*} \quad (7.24)$$
$$= q\frac{n_i^2}{N_A} \frac{D_n}{L_n^*} \left(e^{qV/k_B T} - 1\right). \quad (7.25)$$

Note that $J_n(V = 0) \equiv 0$ by definition. Equation (7.24) can satisfy this zero-current condition only if $V_{bi} \equiv k_B T/q \ln(n_{L,0}/n_{R,0})$. Equation (7.25) is derived by substituting the expression for V_{bi} and $n_{r,0} = n_i^2/N_A$. One obtains a similar expression for J_p by replacing D_n, L_n^*, N_A by D_p, L_p^*, N_D, respectively. The dark current involves a sum of electron and hole currents, i.e.

$$
\begin{aligned}
J_{dark}(V) &= J_n(G = 0, V) + J_p(G = 0, V) \\
&= q\left(\frac{n_i^2}{N_A}\frac{D_n}{L_n^*} + \frac{n_i^2}{N_D}\frac{D_p}{L_p^*}\right)\left(e^{qV/k_B T} - 1\right) \qquad (7.26) \\
&\equiv J_{02}\left(e^{qV/k_B T} - 1\right). \qquad (7.27)
\end{aligned}
$$

This well-known diode equation for p-n junction is similar to Eq. (7.12), except the diffusion lengths have replaced the width of the intrinsic layer of p-i-n solar cell. Note that, by assuming finite diffusion lengths L_n, L_p, we have implicitly assumed that the recombination is restricted to the quasi-neutral n and p regions. Figure 7.4(c) plots the dark current of a typical p-n-junction solar cell.

7.5.2 Photocurrent: p-n-junction solar cells

Next, let us assume that sunlight illuminates the solar cell from the right (p-side), and that the photogeneration of electron–hole pairs is confined to the p-region (width W_p). A potential barrier may (Fig. 7.4(b)) or may not (Fig. 7.4(c)) block the photogenerated electrons from recombining at the wrong contact. The photocarriers generated close to the p-n junction are quickly separated (i.e., electrons relax to the n-side, while holes are blocked and reflected). The photogenerated carriers away from the junction, however, diffuse randomly in the field-free quasi-neutral region until they recombine with each other or are separated once they reach the junction. Naturally the probability of reaching the junction decreases exponentially with distance from the junction (i.e. $\gamma_{RL} \sim \exp(-x/L_n)$ where L_n is the diffusion length for electrons). Therefore, the electron flux exiting the left contact is given by

$$
\begin{aligned}
J_{RL} &= q\int_0^{W_p} G\, e^{-x/L_n}\, dx \\
&= qGL_n\left(1 - e^{-W_p/L_n}\right). \qquad (7.28)
\end{aligned}
$$

Now with a large-bandgap blocking layer shown at the far right of Fig. 7.4(b), the electrons cannot escape through the (wrong) p-contact, i.e. $J_{LR} = 0$. Therefore, *with* a blocking layer.

$$
J_{ph} = J_{RL} - J_{LR} = \begin{cases} qGW_p, & \text{for } L_n \gg W_p \\ qGL_n, & \text{for } L_n \ll W_p. \end{cases} \qquad (7.29)
$$

In other words, one collects all the photo-carriers from the p-region if the recombination is weak, and therefore the diffusion length is larger than the thickness of the p-region (i.e., $L_n \gg W_p$). On the other hand, for highly defective region with relatively small diffusion length (i.e., $L_n \ll W_p$), only carriers generated within the diffusion length L_n from the junction can be collected by the left contact. The rest of the electrons are lost to recombination with holes in the p-layer.

Without the blocking layer, however, the efficiency of photocurrent collection degrade significantly, see Fig. 7.4(c). For $L_n \ll W_p$, only carriers within the diffusion length L_n from the junction can be collected. It does not matter if the rest of the electrons either recombine in the p-layer or at the (wrong) p-contact. Thus, we end up with the same result: $J_{ph} = qGL_n$ for $L_n \ll W_p$. On the other hand, for large diffusion length (i.e. $L_n \gg W_p$), one half the carriers goes toward the junction, while the remaining half diffuses toward the p-contact and recombine. The charge collection reduces by half compared to the case with the blocking layer. Therefore, the photocurrent *without* blocking layer is

$$J_{ph} = \begin{cases} qGW_p/2, & \text{for } L_n \gg W_p \\ qGL_n, & \text{for } L_n \ll W_p. \end{cases} \tag{7.30}$$

Unlike the n-i-p solar cells, the photocurrent in a p-n-junction solar cell is (essentially) voltage independent (see Fig. 7.4(c)). Moreover, photocurrent (and the efficiency) would reduce dramatically for high-quality, defect-free semiconductors if wrong-contact collection is not suppressed by introducing a blocking layer (e.g., back-surface field) or by reducing the contact area (e.g., point-contact technology).

7.5.3 Device design for improved photocurrent collection

An ideal blocking (or passivation layer) improves charge collection by preventing electron escape, but allowing hole collection. As shown in Fig. 7.5(a), some solar cell technologies use a heavily p-doped region near the bottom p-contact to create a back-surface barrier or back-surface field (BSF) to reflect the electrons, but allow hole collection. More modern passivated emitter and rear contact (PERC) cells first surround the localized p-contacts by back-surface field (as in BSF), and then passivate the rest of the back surface by a high-bandgap material (e.g., SiO_2) to prevent electron and hole escape (see Fig. 7.5(c)). The acronym PERC is often used as shorthand for a variety of PV technologies, such as PERL, PERT, and PERF. In this section, we will focus on the physics of point-contact charge collection.

The physics of BSF is obvious, but how do the localized PERC contacts work? Wouldn't smaller contacts make *both* electron and hole collection difficult? Not really! An electron diffusing on the p-region has two escape routes: the p-n junction (correct-contact collection) vs. the p-contact (wrong-contact collection). The passivation layer reduces the wrong-contact escape, thereby increasing the *relative* probability of correct-contact collection. For photogenerated holes, the p-n

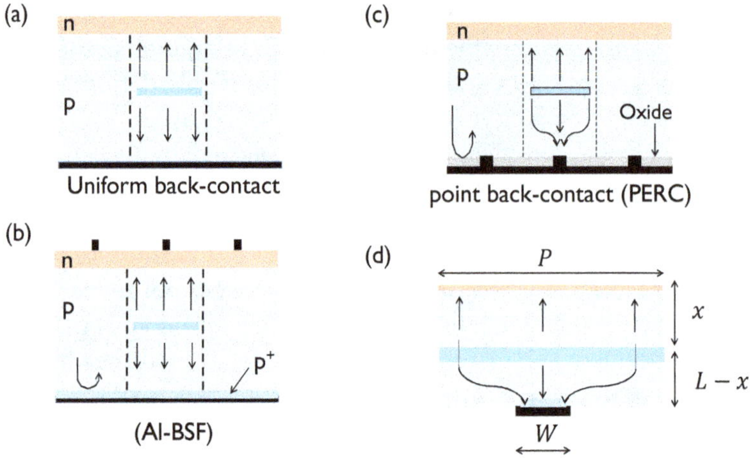

Figure 7.5: (a) A uniform metallic back-contact electrode, (b) Al-BSF, (c) PERC, and (d) a unit cell of various types of solar cells.

junction prevents its wrong-contact escape altogether; therefore, they must still exit through (some more resistive) p-contact.

Let us consider electrons photogenerated within the p-region at a depth x from the p-n junction, indicated by the blue horizontal stripe in Fig. 7.5(d). The probability to escape through the p-n junction on the top (light brown) by diffusion is given by $\gamma_{\mathrm{RL}} = D_{\mathrm{n}}/x$, and the escape-probability through the p-contact at the bottom (black stripe) is given by $\gamma_{\mathrm{LR}} = s\, D_{\mathrm{n}}/(L-x)$, where the suppression efficiency

$$s \equiv (W/P)\exp(-\Delta E_{\mathrm{c}}/k_{\mathrm{B}}T)$$

depends on the relative geometric constriction, W/P, and the back-surface potential defined by the p-p$^+$ doping, $\Delta E_{\mathrm{c}} = k_{\mathrm{B}}T\ln\left(N_{A,p}/N_{A,p^+}\right)$, as shown in Fig. 7.5(d).

The photocurrent of a PERC cell can be calculated by summing the electron and hole currents exiting n-contact on the top, i.e., $J_{\mathrm{ph}} = J_{\mathrm{ph(n)}} + J_{\mathrm{ph(p)}}$, see Eq. (7.17). Since $J_{\mathrm{ph(p)}} = 0$ due to the perfect blocking of holes by the p-n junction; therefore,

$$J_{\mathrm{ph}} = q\int_0^{W_{\mathrm{p}}} G\,dx\left[\frac{\gamma_{\mathrm{RL,n}}}{\gamma_{\mathrm{RL,n}} + \gamma_{\mathrm{LR,n}}}\right] \tag{7.31}$$

$$= q\int_0^{W} G\,dx\left[1 - \frac{s\,x}{W - x + s\,x}\right]. \tag{7.32}$$

Without an electron blocking layer (i.e., $s = 1$; see Fig. 7.5(a)), $J_{\mathrm{ph}} = qGW_{\mathrm{p}}/2$, as expected. With perfect blocking ($s = 0$), $J_{\mathrm{ph}} = qGW_{\mathrm{p}}$, defines complete charge collection. The factor s is improved by making the contacts smaller ($W \ll P$), but this must be balanced by the increasing difficulty of hole collection (i.e., series resistance). The higher p doping increases ΔE_c (good), while simultaneously increasing optical loss (bad). A good blocking layer is an important part of the design of a p-n-junction solar cell.

We mentioned above that half of the photocurrent would be lost without blocking. Actually, things are not so horrible! Recall from Sec. 6.3 that photogeneration depends on energy and position, i.e., $G = P_{in}(1 - R(E))\exp(-\alpha(E)x) = G_0 \exp(-\alpha(E)x)$. Inserting this in Eq. (7.32), we see that high energy photons with large α are absorbed close to the junction and the p-contact design is irrelevant (the second term does not contribute). Unfortunately, the low-energy photons with poor α must bounce repeatedly between the top and bottom surfaces before absorption. The blocking layer design improves the collection of charges generated by low-energy photons, especially those generated close to the p-contact. In short, high-efficiency solar design must care about the whereabouts of every electron and hole generated within the absorption layer. Getting a reasonably good solar cell is not hard, but approaching the thermodynamic limit is really difficult!

Homework 7.8: The design of a PERC cell

Assume that 15% of the photons are uniformly generated within the absorber region. What fraction of these carriers will be collected by the correct contact, if the p and p^+ doping differ by a factor of 10 and if $W/P = 1, 0.5, 0.1$. Show that the result is essentially identical even if we replace $\gamma_{\mathrm{RL,n}} = s\,D_{\mathrm{n}}/(L - x)$ with the more accurate expression:

$$\gamma_{\mathrm{LR,n}} = \frac{2\pi D_{\mathrm{n}}}{\ln \frac{\sinh(\pi P/2(L-x))}{\sinh(\pi W/(4(L-x))}}.$$

7.6 Current–voltage characteristics of heterojunction solar cells

A third type of solar cells involves innovative use of a heterojunction (HJ) to enhance correct-contact charge collection and suppress wrong-contact charge escape. A variety of solar cell technologies, such as HIT (heterojunction with intrinsic thin layer), CIGS (copper indium gallium selenide), CZTS (copper zinc tin sulfide), OPV (organic solar cells), and perovskite (depending on transport layers), use heterojunctions to reduce dark current and increase photocurrent. In this section, we will consider a simple HJ solar cell (e.g., CIGS) where a very thin highly n-doped buffer layer forms a heterojunction with the p-type photon absorption layer, as shown in Fig. 7.6(a). The buffer layer has a large bandgap; thus, it blocks hole transport.

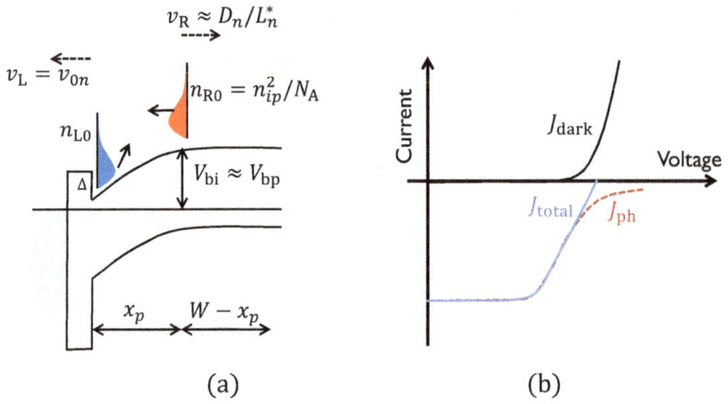

Figure 7.6: (a) Carrier injection from the contacts under dark conditions in a heterojunction solar cell. (b) $J - V$ characteristics of the HJ solar cell.

7.6.1 Dark current: Heterojunction solar cells

For the typical HJ structure in Fig. 7.6(a), let us assume that the equilibrium minority carrier electron density at the left edge of the p-layer at the HJ (blue carriers) is $n_{\mathrm{L},0}$, while the electron density at the right edge of the space charge region (red carriers) is $n_{\mathrm{R},0} = n_{ip}^2/N_{\mathrm{A}}$. The electron velocity in the n-type buffer layer is $v_{\mathrm{L}} = v_{0n}$, and that in the quasi-neutral region of p-layer is $v_R = D_{\mathrm{n}}/L_{\mathrm{n}}^* = v_{Dp}$. The electron transmission γ_{LR} from the p- to the n-layer is exponentially suppressed by the conduction band discontinuity (Δ) at the HJ interface, i.e., $\gamma_{\mathrm{RL}} = v_{\mathrm{L}}e^{-\Delta/k_{\mathrm{B}}T} = v_{0n}e^{-\Delta/k_{\mathrm{B}}T}$. Similarly, the electron injection to the right γ_{RL} is suppressed by the built-in potential of the p-region, i.e., $\gamma_{\mathrm{LR}} = v_R e^{-q(V_{bp}-V)/k_{\mathrm{B}}T}$. The net electron current is given by

$$J_{\mathrm{n}} = qn_{\mathrm{L},0}v_{\mathrm{L}}\frac{\gamma_{\mathrm{LR}}}{\gamma_{\mathrm{RL}}+\gamma_{\mathrm{LR}}} - qn_{\mathrm{R},0}v_R\frac{\gamma_{\mathrm{RL}}}{\gamma_{\mathrm{RL}}+\gamma_{\mathrm{LR}}}. \tag{7.33}$$

In equilibrium, the electron fluxes cancel (i.e., $J_{\mathrm{n}}(G=0, V=0) = 0$), so that

$$qn_{\mathrm{L},0}v_{\mathrm{L}}\frac{\gamma_{\mathrm{LR}}}{\gamma_{\mathrm{RL}}+\gamma_{\mathrm{LR}}} - qn_{\mathrm{R},0}v_{\mathrm{R}}\frac{\gamma_{\mathrm{RL}}}{\gamma_{\mathrm{RL}}+\gamma_{\mathrm{LR}}} = 0 \tag{7.34}$$

$$\text{or, } n_{\mathrm{L},0}v_{0n}v_{Dp}e^{(-qV_{bp}+\Delta)/k_{\mathrm{B}}T} - \frac{n_{ip}^2}{N_{\mathrm{A}}}v_{Dp}v_{0n} = 0 \tag{7.35}$$

$$\text{or, } n_{\mathrm{L},0} = \frac{n_{ip}^2}{N_{\mathrm{A}}}e^{-(-qV_{bp}+\Delta)/k_{\mathrm{B}}T}. \tag{7.36}$$

Using the expression for $n_{\mathrm{L},0}$, we calculate the dark current at any bias voltage V,

$$J_{\mathrm{n}} = qn_{\mathrm{L},0}V_{\mathrm{L}}\frac{\gamma_{\mathrm{LR}}}{\gamma_{\mathrm{RL}}+\gamma_{\mathrm{LR}}} - qn_{\mathrm{R},0}v_{R}\frac{\gamma_{\mathrm{RL}}}{\gamma_{\mathrm{RL}}+\gamma_{\mathrm{LR}}} \tag{7.37}$$

$$= q\frac{n_{ip}^2}{N_{\mathrm{A}}}v_{\mathrm{L}}v_{R}\left(e^{qV/k_{\mathrm{B}}T}-1\right)\frac{e^{-\Delta/k_{\mathrm{B}}T}}{\gamma_{\mathrm{LR}}+\gamma_{\mathrm{RL}}} \tag{7.38}$$

$$= q\frac{n_{ip}^2}{N_{\mathrm{A}}}\left[\frac{1}{\dfrac{1}{v_{n0}e^{(q(V_{bp}-V)-\Delta)/k_{\mathrm{B}}T}}+\dfrac{1}{v_{Dp}}}\right]\left(e^{qV/k_{\mathrm{B}}T}-1\right) \tag{7.39}$$

$$= q\frac{n_{ip}^2}{N_{\mathrm{A}}}\left[\frac{1}{\dfrac{1}{v_{ems}}+\dfrac{1}{v_{Dp}}}\right]\left(e^{qV/k_{\mathrm{B}}T}-1\right). \tag{7.40}$$

Here, $v_{ems} \equiv v_{n0}\exp[(q(V_{bp}-V)-\Delta)/k_{\mathrm{B}}T]$ is the HJ limited emission velocity, $v_{Dp} \equiv D_{\mathrm{n}}/L_{\mathrm{n}}^*$ is the diffusion velocity, and $v_{\mathrm{eff}}^{-1} \equiv v_{ems}^{-1} + v_{Dp}^{-1}$ is defined by the smaller of the diffusion and thermionic emission velocities. In the absence of any recombination in the p-region, there is no hole current in this device ($J_{\mathrm{p}} = 0$), because the blocking layer prevents charge transfer between contacts. The dark current is given by the sum of the electron and hole currents:

$$J_{\mathrm{dark}} = J_{\mathrm{n}} + J_{\mathrm{p}} = q\frac{n_{\mathrm{i}}^2}{N_{\mathrm{A}}}v_{\mathrm{eff}}\left(e^{qV/k_{\mathrm{B}}T}-1\right) \equiv J_{03}\left(e^{qV/k_{\mathrm{B}}T}-1\right). \tag{7.41}$$

A plot of the dark current is shown in Fig. 7.6(b). Note that, as $\Delta \to 0$, $v_{\mathrm{eff}} \approx D_{\mathrm{n}}/L_{\mathrm{n}}^*$. In such a case the structure reduces to a one-sided p-N$^+$ junction.

7.6.2 Photocurrent: Heterojunction solar cells

Total photocurrent is obtained by integrating the fractions of electrons and holes collected by their respective contacts. Focusing on the left contact, $\gamma_{\mathrm{RL,p}} = 0$ because hole escape is blocked by the HJ. The photocurrent consists of two contributions from the electrons: the carriers generated within the space charge region ($0 < x < x_p$) are separated by the electric field in a manner similar to that of an n-i-p solar cell shown in Fig. 7.3(b), while the carriers generated within the field-free quasi-neutral region ($W > x > x_p$) are separated by diffusing to the junction in a manner similar to a p-n junction shown in Fig. 7.4(c). Based on the expressions already derived, the two components are given by

$$
\begin{aligned}
J_{\text{ph}} &= q\int_0^{x_p} G\,dx\left[\frac{\gamma_{\text{RL,n}}}{\gamma_{\text{RL,n}}+\gamma_{\text{LR,n}}}-\frac{\gamma_{\text{RL,p}}}{\gamma_{\text{RL,p}}+\gamma_{\text{LR,p}}}\right]\\
&\quad + q\int_{x_p}^{W} G\,e^{-(x-x_p)/L_{\text{n}}}\,dx \qquad (7.42)\\
&= q\int_0^{x_p} G\,dx\left[\frac{v_{0n}e^{-\Delta/k_{\text{B}}T}}{v_{0n}e^{-\Delta/k_{\text{B}}T}+(D_{\text{n}}/L_{\text{n}}^{*})e^{-\mathcal{E}(x_p-x)/k_{\text{B}}T}}\right]\\
&\quad + q\int_0^{W-x_p} G\,e^{-x/L_{\text{n}}}\,dx\\
&= qG(x_p+L_{\text{n}}) - qG\,\frac{k_{\text{B}}T}{\mathcal{E}}\cdot\alpha.
\end{aligned}
$$

The second term indicates that the wrong-contact escape over the built-in potential increases with decreasing barrier height (i.e., $\mathcal{E} = (V_{bp} - V)/x_p$). Here, $\alpha \equiv \ln(N/D)$, where $N \equiv (v_{Dp} + v_{0n}\exp(-\Delta/k_{\text{B}}T)$ and $D \equiv (v_{Dp}\exp(-q(V_{bp} - V)/k_{\text{B}}T) + v_{0n}\exp(-\Delta/k_{\text{B}}T))$. As shown in Fig. 7.6(b), the photocurrent eventually approaches zero as $V \to V_{bp}$ because the electron and hole collection are suppressed by the respective barriers of the left contact. A comparison of the photocurrent in Figs. 7.3, 7.4, and 7.6 shows that the essential physics of a solar cell can be recognized by the characteristic voltage dependence of their photocurrent, namely, voltage independence of a homojunction p-n cell, current crossover in a homojunction p-i-n cell, and suppression of charge collection at a high voltage for an HJ cell.

7.7 Principle of (current) superposition

The conventionally known principle of superposition states that the net current out of an illuminated solar cell at bias V is the sum of the short-circuit photocurrent and the dark (recombination) current, namely,

$$J_{\text{total}}(V) = J_{\text{sc}}(G) - J_{\text{dark}}(G = 0, V). \qquad (7.43)$$

In other words, the superposition principle assumes the following:

1. The photocurrent collection is voltage independent and equals the short-circuit current, namely, $J_{\text{ph}}(G, V = 0) = J_{\text{sc}}(G)$.

2. Current injected from the contacts contributing to recombination is equal to the dark current (i.e., $J_{\text{inj}} = J_{\text{dark}}$).

The classical superposition principle thus relates the solar cell net current to that under dark. J_{total}, J_{sc} and J_{dark} can be found through $I - V$ measurements. In classical c-Si solar cells (p-n devices), Eq. (7.43) works reasonably well, because the photocurrent is indeed voltage independent and the photogenerated carriers do not accumulate at heterojunctions to substantially change the energy band profile compared to dark conditions (see Fig. 7.4). However, it is generally not valid for thin-film solar cells (typically p-i-n or HJ cell), because $J_{\rm ph}$ is voltage dependent and $J_{\rm dark}$ does not equal the injection flux from the contact under illumination, as shown in Fig. 7.6(b).

Regarding the first assumption of the superposition principle, our analysis of p-i-n and HJ cells has shown that it is relatively easy to include the voltage dependence of $J_{\rm ph}$:

$$J_{\rm total}(V) = J_{\rm ph}(G, V) - J_{\rm dark}(V). \tag{7.44}$$

This is the same basic equation (7.2) we have used to discuss $J-V$ characteristics in this chapter.

The second assumption, namely, $J_{\rm inj}(G, V) \simeq J_{\rm dark}(V)$ is less intuitive and more difficult to justify. It assumes that the injection current under illumination is the same as the dark current, independent of generation. This approximation fails in solar cells when the diffusion length is small, or photogeneration is significant. The high photogeneration modifies the band profile and relative carrier distribution compared to dark. This results in a generation-dependent injection current (at given bias V) to be substantially different from the dark current.

In particular, HIT solar cells show generation-dependent diode injection, as well as bias-dependent photo-collection — thereby violating both constraints of the superposition principle. For these solar cells, the J-V is more appropriately written

Homework 7.9: Frozen potential approach to isolate photocurrent and Injection Currents

The following documents explains how the failure of the superposition principle makes the injection current different from the dark current of a solar cell.

1. J. Moore, "Sentaurus Resources for Frozen Potential Analysis of Solar Cells," 2014, `https://nanohub.org/resources/21561`.
2. J.E. Moore, S. Dongaonkar, R.V.K. Chivali, M.A. Alam, M.S. Lundstrom, "Correlation of Built-In Potential and I–V Crossover in Thin-Film Solar Cells," *IEEE Journal of Photovoltaics*, July 2014.
3. R. V. Chavali *et al.*, "The Frozen Potential Approach to Separate the Photocurrent and Diode Injection Current in Solar Cells," *IEEE Journal of Photovoltaics*, 865–873, 2015.

in the following form:

$$J_{\text{total}}(G, V) = J_{\text{ph}}(G, V) - J_{\text{inj}}(G, V). \tag{7.45}$$

In other words, the solar cell current is the 'superposition' of photo-collection and injection current found at the specific bias and illumination.

7.8 Performance parameters of a practical solar Cell

We recall from Sec. 3.5 that once the complete J-V characteristics of a solar cell is obtained, one wishes to calculate the key parameters of interest, such as the maximum cell efficiency (η), and the voltage (V_{mp}) and current (I_{mp}) at which the efficiency is maximized. In turn, these quantities are related to open-circuit voltage (V_{oc}) and short-circuit current (I_{sc}), and the fill factor, $\text{FF} \equiv V_{\text{mp}} I_{\text{mp}} / (V_{\text{oc}} I_{\text{sc}})$. Note that $I(V) = A J_{\text{total}}(V)$, where A is the cross-sectional area of the solar cells and the expressions for J_{total} for p-i-n, p-n, and HJ solar cells have been derived in the preceding three sections.

The maximum efficiency is obtained by taking the voltage derivative of $P_{\text{out}} = V \times I(V)$ and setting the resultant expression to 0. Since $I(V)$ is a complex function of V, in practice it may be easier to plot $P_{\text{out}}(V)$ and determine the maximum power point graphically or numerically. The efficiency of a practical solar cell is always lower than the thermodynamic efficiency calculated in Part 1 of the book. For a p-n-junction solar cell that obeys the superposition principle (Sec. 7.7), however, J_{ph} and J_0 are voltage independent, making it easier to derive expressions for maximum power output, as discussed in the following Homework problem.

Homework 7.10: Maximum power point voltage is related to open-circuit voltage and short-circuit current

The power delivered by a solar cell, $P_{\text{out}} = VI$, is maximized at $I = I_{\text{mp}}$ and $V = V_{\text{mp}}$. For a p-n-junction solar cell, show that

$$v_{\text{mp}} = v_{\text{oc}} - \ln(v_{\text{mp}} + 1) \tag{7.46}$$

$$i_{\text{mp}} = 1 - \exp(-(v_{\text{oc}} - v_{\text{mp}})), \tag{7.47}$$

where normalized voltage and current are given by $v_{\text{mp}} \equiv qV_{\text{mp}}/nk_{\text{B}}T$, $v_{\text{oc}} = qV_{\text{oc}}/nk_{\text{B}}T$ and $i_{\text{mp}} \equiv I_{\text{mp}}/I_{\text{sc}}$. How do the results compare to the thermodynamic expressions for V_{mp}, V_{oc}, and I_{mp} derived in Sec. 3.5, namely, Eqs. (3.16), (3.19), and (3.21)?

Solution. Recall that $P_{\text{out}} = I \times V = A\, J_{\text{total}} \times V = A\left[J_{\text{ph}} - J_{02}\left[\exp\left(qV/nk_{\text{B}}T\right) - 1\right)\right]$. Calculate dP/dV and set it to 0 derive the pair of relationships.

Homework 7.11: Calculation of the solar cell parameters

A c-Si solar cell has $E_g = 1.1$ eV, $n = 1.58$, $V_{oc} = 0.747$ V, and $I_{ph} = 38$ mA with an area $A = 1$ cm^2. Assuming that superposition principle holds, find the maximum power point voltage, maximum power point current, and the fill factor of the solar cell. How do these quantities compare to the thermodynamic limits derived in Homework 3.5?

Solution. Insert the numerical values to the expressions in Homework 7.10 to calculate the relevant quantities. These parameters are $V_{mp} = 0.62$ V, $I_{mp} = 36.47$ mA, and FF $= 0.79$. The corresponding thermodynamic limits are, $V_{mp} = 0.75$ V, $V_{oc} = 0.84$ V, and FF $= 0.87$. There is room for significant improvement.

7.9 The puzzle of the reverse saturation current

We explained in Homework 3.8 that the thermodynamic current–voltage characteristics can be partitioned into dark and photocurrent components, namely, $J(V) = J_{ph} - J_{dark} = J_{ph} - J_0(e^{qV/k_BT} - 1)$. The reverse saturation current was expressed in terms of the internal flux due to radiative recombination and the emittance of the solar absorber, i.e.,

$$J_0 \equiv q \int_{E_g}^{\infty} n_D(V=0)\, A_{emi}(E) dE.$$

Remarkably, the expression does not involve any geometrical or material parameters of a solar cell. In contrast, the reverse saturation currents derived in this chapter (e.g., Eq. (7.11) for p-i-n solar cells, Eq. (7.27) for p-n solar cells, or Eq. (7.41) for heterojunction solar cells) all depend on material parameters, such as W, D_n, N_D, n_i, W_n, Δ and so on. Is there something wrong with the derivations in this chapter?

Not really! The thermodynamic limit is defined by the *radiation limit* and is based on photon fluxes observed from *outside* the device. In contrast, the expressions derived in this chapter are based on non-radiative recombination and calculated *inside* the device. For example, electrons injected over V_{bi} in a forward-biased p-i-n junction eventually dissipates the energy as heat at the contacts. Similarly, the trap-assisted recombination within the semiconductor dissipates non-radiatively via lattice vibration. In many practical cells, non-radiative processes dominate, leading to the dark current expressions derived in this chapter.

We will see below that if we suppress all non-radiative processes (by heterojunction blocking and defect-free semiconductor), the residual radiative recombination within the solar cell produces the radiation flux outside the solar cell that exactly matches the thermodynamic expression for the dark current.

7.9.1 Radiative photon flux inside the solar cell

Consider a forward-biased p-n-junction solar cell. In the absence of other non-radiative processes, the dark current is produced by electrons radiatively recombining with the holes, so that

$$J_{\text{dark}} = J_0^* \left(e^{qV/k_\text{B}T} - 1\right), \text{ where } J_0^* = qBn_\text{i}^2 W.$$

In other words, J_0^* in the radiative limit depends on the bimolecular electron–hole recombination coefficient B, the device thickness $W(= W_\text{p} + W_\text{n})$, and the intrinsic carrier density n_i. Unlike the thermodynamic expression, apparently the reverse saturation current, J_0^*, still depends on the thickness of the solar cell and the material parameters, because $n_\text{i}^2 = N_\text{C}N_\text{V}\exp(-E_\text{g}/k_\text{B}T)$ by Eq. (7.5).

The coefficient B, however, is defined as an integral over solid angle and the energy-resolved recombination of electron hole pairs (see Eq. (3.34)), namely,

$$B = 4\pi\frac{\hat{n}^2}{n_\text{i}^2}\int_0^{\pi/2}\left(\int_\text{g}^{\infty}\alpha(\theta,E)b(E)dE\right)\cos\theta\sin\theta\,d\theta \tag{7.48}$$

$$\approx 4\pi\frac{\hat{n}^2}{n_\text{i}^2}\alpha_g b_g \Delta E_\text{g}. \tag{7.49}$$

Here, $\hat{n}$ is the refractive index, n_i is the intrinsic carrier concentration, $\alpha(\theta, E)$ is the absorption coefficient, and $b(E) \equiv (2E^2/h^3c^2)\, f_{BE}(E, T_\text{D})$ is the emission rate per solid angle per energy. While Eq. (7.48) is exact, Eq. (7.49) involves two assumptions: First, the Boltzmann distributions of electron and holes make the emission sharply peaked at the bandgap E_g, so we can approximate the energy integral by $\alpha(E_\text{g})b(E_\text{g})\Delta E_\text{g} = \alpha_g b_g \Delta E_\text{g}$. Here, ΔE_g is the effective width of the sharply peaked $\alpha(E)b(E)$ function. Second, the integration over the solid angle produce (4π) because internal emissions occur in random directions. Taken together,

$$J_0^* = qBn_\text{i}^2 W = qW \times 4\pi\hat{n}^2\alpha_g b_g \Delta E_\text{g}. \tag{7.50}$$

In other words, J_0^* (obtained by integrating over the internal emission) scales with the device dimension, W.

7.9.2 Radiative photon flux outside the solar cell

As discussed in Chapter 6 and shown in Fig. 7.7(a), only a fraction of photons produced by electron–hole radiative recombination escapes the solar cell through the escape cone and contribute to the dark current flux used in the thermodynamic calculation. Photons with larger angles experience total internal reflection and are

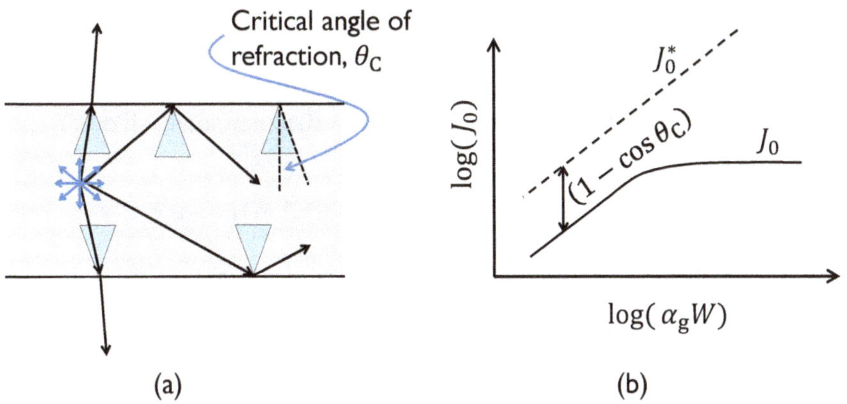

Figure 7.7: (a) Schematic of photon recycling. (b) Classical and thermodynamic reverse saturation current (J_0 and J_0^*) as a function of effective reabsorption length $\alpha_g W$.

eventually reabsorbed (i.e., recycled) — this process is called photon recycling. Simply put, an isotopically generated ray due to radiative recombination, traveling inside the device layer at an angle θ_i, will be reabsorbed with the probability:

$$a(\theta, E) = 1 - \exp(-\alpha(E)W/\cos\theta_i). \tag{7.51}$$

We can now calculate J_0 by recalling that only the photons within the escape cone (defined by Snell's law: $\sin\theta = \hat{n}\sin\theta_i$) contribute to dark current. In other words, the photons within $\theta_i < \theta_c$ will escape outside at an angle θ and contribute to the overall recombination current J_0 (the rest of the photons are reabsorbed or recycled). Therefore,

$$\begin{aligned} J_0 &= 4\pi q \int_0^{\theta_c} (1 - \exp(-\alpha(E_g)W/\cos\theta_i))b_g \Delta E_g \hat{n}^2 \cos\theta_i \sin\theta_i d\theta_i \\ &= q \times 4\pi\hat{n}^2(1 - \exp(-\alpha_g W))b_g \Delta E_g(1 - \cos\theta_c). \end{aligned} \tag{7.52}$$

This exponential term accounts for photon recycling and the cosine term accounts for the photon escape.

7.9.3 Internal and external fluxes compared

Now, let us use Eqs. (7.52) and (7.50) to compare the dark currents computed within and outside the device, i.e.,

$$J_0 = \left[\frac{1 - \exp(-\alpha_g W)}{\alpha_g W}\right](1 - \cos\theta_c)\, J_0^*. \tag{7.53}$$

The result is plotted in Fig. 7.7(b). The key differences between J_0^* and J_0 are

- For the classical limit of $\alpha_g W < 1$: $J_0 \to (1 - \cos\theta_c)\, J_0^*$ is determined by the fraction of internally emitted photons that are not recycled, but rather allowed to escape to the surrounding. Here, $J_0 < J_0^*$, implying that the recombination predicted by the thermodynamic model (by considering photon recycling) is lower than that predicted from the classical model. This also indicates that the classical model underestimates V_{oc} in a high-efficiency solar cell dominated by radiative recombination.

- In the thermodynamic limit of complete absorption (i.e., $\alpha_g W \gg 1$),

$$J_0 = ((1 - \cos\theta_c)/\alpha_g W)\ J_0^* = 4\pi q\hat{n}^2 b_g \Delta E_g (1 - \cos\theta_c)$$

  is independent of the device dimension W, as expected. Although J_0^* keeps increasingly linearly with W, the emitted photons are repeatedly recycled, so that the net emission from the edges becomes independent of the device dimension.

7.10 Conclusions

In this chapter, we calculated the J-V characteristics of three types of solar cells (i.e., p-i-n, p-n, and heterojunction) based on an intuitively simple "transmission over a barrier" approach. This approach explains the essential difference among the three types of solar cells. For example, we find that the dark currents for the solar cells all depend exponentially on the applied voltage, but the prefactors are significantly different. Moreover, the photocurrents in p-i-n and heterojunction cells are voltage dependent, but the photocurrent of p-n-junction cells are voltage independent. We have also emphasized that the widely used analytical formula for V_{oc}, V_{mp}, FF derived in Homework 7.10 rely on the validity of the classical superposition formula. The classical superposition formula must be generalized or numerical simulators such as ADEPT or PC1D must be used to interpret the I-V characteristics of thin-film and heterojunction solar cells. In the final section, we related the dark current expression calculated in this chapter with its corresponding thermodynamic limit derived Part I of the book.

In this chapter, we focused on three types of solar cells made of inorganic semiconductors (i.e., Si, GaAs, CIGS). In fact, there is a fourth type of solar cell called organic photovoltaics (OPV) that operates on a slightly different principle. In the next chapter, we will discuss the physics of carrier transport in organic solar cells.

References

[1] N. G. Tarr and D. L. Pulfrey. The superposition principle for homojunction solar cells. *IEEE Transactions on Electron Devices*, 27(4):771–776, April 1980.

[2] S. M. Sze. *Physics of Semiconductor Devices*. John Wiley & Sons, September 1981.

[3] M. Burgelman, P. Nollet, S. Degrave, and J. Beier. Modeling the cross-over of the i-v characteristics of thin film cdte solar cells. In *Conference Record of the Twenty-Eighth IEEE Photovoltaic Specialists Conference - 2000 (Cat. No. 00CH37036)*, pages 551–554, 2000.

[4] M. Igalson and C. Platzer-Björkman. The influence of buffer layer on the transient behavior of thin film chalcopyrite devices. *Solar Energy Materials and Solar Cells*, 84(1):93–103, October 2004.

[5] M. Lundstrom. Elementary scattering theory of the Si MOSFET. *IEEE Electron Device Letters*, 18(7):361–363, July 1997. Conference Name: IEEE Electron Device Letters.

[6] Yoseph Imry and Rolf Landauer. Conductance viewed as transmission. *Reviews of Modern Physics*, 71(2):S306–S312, March 1999. Publisher: American Physical Society.

[7] Pietro P. Altermatt. Models for numerical device simulations of crystalline silicon solar cells — A review. *Journal of Computational Electronics*, 10(3):314, July 2011.

[8] H. Wagner-Mohnsen and P. P. Altermatt. A Combined Numerical Modeling and Machine Learning Approach for Optimization of Mass-Produced Industrial Solar Cells. *IEEE Journal of Photovoltaics*, 10(5):1441–1447, September 2020. Conference Name: IEEE Journal of Photovoltaics.

[9] A. Fell, K. R. McIntosh, P. P. Altermatt, G. J. M. Janssen, R. Stangl, A. Ho-Baillie, H. Steinkemper, J. Greulich, M. Müller, B. Min, K. C. Fong, M. Hermle, I. G. Romijn, and M. D. Abbott. Input Parameters for the Simulation of Silicon Solar Cells in 2014. *IEEE Journal of Photovoltaics*, 5(4):1250–1263, July 2015. Conference Name: IEEE Journal of Photovoltaics.

[10] A. Martí, J. L. Balenzategui, and R. F. Reyna. Photon recycling and Shockley's diode equation. *Journal of Applied Physics*, 82(8):4067–4075, October 1997.

[11] Julian Mattheis, Jürgen H. Werner, and Uwe Rau. Finite mobility effects on the radiative efficiency limit of pn-junction solar cells. *Phys. Rev. B*, 77:085203, February 2008.

[12] Robert Street, editor. *Technology and Applications of Amorphous Silicon*. Springer Series in Materials Science. Springer-Verlag, Berlin Heidelberg, 2000.

[13] Robert A. Street. *Hydrogenated amorphous silicon*. Cambridge solid state science series. Cambridge Univ. Press, Cambridge u.a, new edition, 2005. OCLC: 845372037.

[14] Steven S. Hegedus and William N. Shafarman. Thin-film solar cells: Device measurements and analysis. *Progress in Photovoltaics: Research and Applications*, 12(23):155–176, March 2004.

[15] Steven Hegedus, Darshini Desai, and Chris Thompson. Voltage dependent photocurrent collection in CdTe/CdS solar cells. *Progress in Photovoltaics: Research and Applications*, 15(7):587–602, 2007. _eprint: https://onlinelibrary.wiley.com/doi/pdf/10.1002/pip.767.

CHAPTER 8

Organic Solar Cells

Chapter Summary

- There are three types of organic solar cells (OPV): planar heterojunction, vertical heterojunction, and bulk heterojunction.
- The electron-holes produced in OPVs are so tightly coupled that they would recombine within the diffusion length unless sliced apart by a heterojunction.
- Simple flux-balance equations can be used to calculate the light and dark currents of various types of organic solar cells, and explain various anomalous behaviors, such as, very low fill-factor in planar heterojunction solar cells.
- Vertical hetrojunction solar cells offers most efficient charge collection because heterojunctions can immediately separate electron-hole pairs.
- A bulk heterojunction solar cell mimics a vertical heterojunction solar cell, with a simpler fabrication process.

8.1 Introduction: Bulk-heterojunction solar cells

The fourth and final type of solar cell to be discussed in this book is called a bulk-heterojunction (BHJ) solar cell. In Sec. 2.3.1 we have discussed the thermodynamic efficiency limit of this type of solar cells. In this chapter, we will focus on device-specific issues that reduces the efficiency below the thermodynamic limit. The key distinguishing feature of a BHJ solar cell is this: Instead of having a heterojunction only at the front or back contacts as wrong-contact blocking layers, as discussed in Chapter 7, the heterojunctions are inserted horizontally, vertically, or randomly within the bulk of the film, as shown in Fig. 8.1. This BHJ design strategy is particularly important in solar cells that uses organic semiconductors (e.g., PPV, P3HT, DTBT, PCP, PTB7, etc.) as the photon absorber. An organic semiconductor is typically composed of carbon atoms in rings or long chains, to which are attached other atoms such as hydrogen, oxygen, and nitrogen. The solution-processed

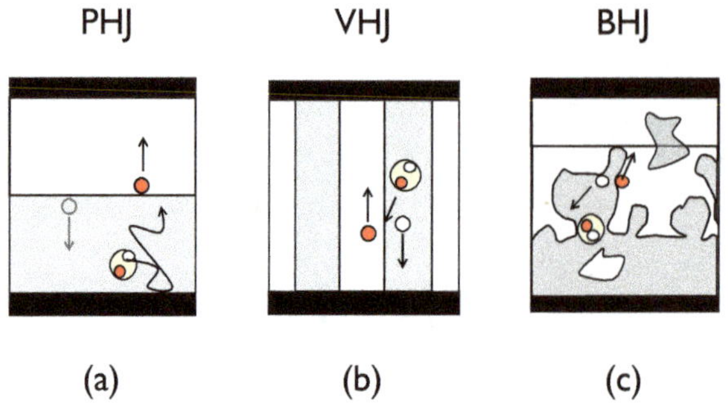

Figure 8.1: Different types of excitonic (organic) solar cells: (a) Planar heterojunction solar cell. (b) Vertical-heterojunction solar cell. (c) Distributed bulk-heterojunction solar cell. The transport of electrons (red circle), holes (white circle), and excitons (yellow circle) within the donor (gray) and acceptor (white) regions are indicated by arrows.

organic solar cells (OPV) are cheaper and easier to make than vapor-deposited or epitaxially grown inorganic solar cells, such as Si or GaAs. In order to understand why a BHJ principle is essential for OPVs, but not for Si solar cells, for example, we will need to understand the concept of excitons and how a heterojunction can slice open an exciton into a pair of free electron and hole.

8.2 Exciton involves an electron–hole diffusing together

Once a photon is absorbed, it creates a pair of electron and hole, as shown in see Fig. 8.2(a). The negatively charged electron and the positively charged hole are attracted to each other through Coulomb force. In fact, the Coulomb force makes the pair orbit each other just like gravitational force makes the planet go around the sun. The energy that binds the pair (E_1) and the radius of the orbiting pair (r_{B}) are given by

$$E_1 = \frac{m\,q^4}{32\pi^2\hbar^2\epsilon_0^2\kappa^2} = 13.6 \times \frac{1}{\kappa^2}\frac{m}{m_0}\ \mathrm{eV} \tag{8.1}$$

$$r_{\mathrm{B}} = \frac{4\pi\epsilon_0\hbar^2}{m\,q^2}\kappa = 0.53 \times \kappa\frac{m_0}{m}\,\mathrm{\AA}. \tag{8.2}$$

Here, q is the electronic charge, κ is the relative dielectric constant of the material, m_0 is the free electron mass. The second part of each equation is written in terms of an hydrogen atom, where a free electron orbits a proton in the free space ($\kappa = 1$) with free electron mass (m_0), just as in exciton a free electron orbits a hole in a solid ($\kappa > 1$) with an effective mass m (see Fig. 8.2(c)).

First, let us consider a "hypothetical" silicon solar cell with $\kappa = 10$ and $m = 0.1\,m_0$. The excitons in silicon would have $E_1 \sim 13.6$ meV and radius $r_{\mathrm{B}} = 5.3$ nm. The atoms themselves vibrate on average with the energy $k_{\mathrm{B}}T = 25$ meV at room

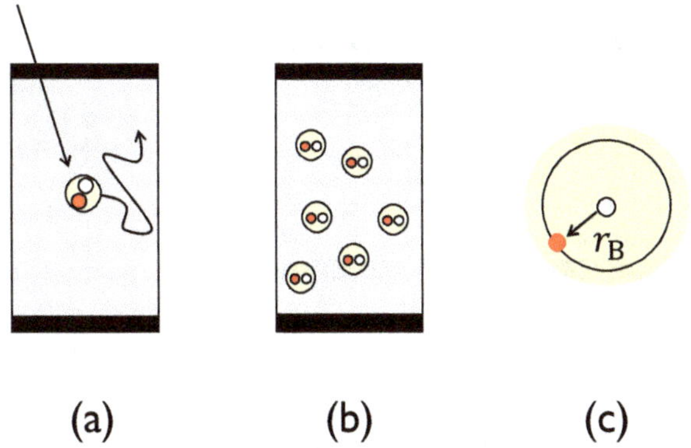

Figure 8.2: (a) A single semiconductor excitonic cell is geometrically similar to a p-i-n solar cell. (b) The excitons are generated throughout the volume of the cell. (c) An exciton can be viewed as a "hydrogen atom" with positive and negative charges orbiting each other.

temperature ($T = 300$ K). This energy is sufficient to immediately dissociate the exciton into a pair of free electron and hole. Therefore, excitons are not important for the three types of solar cells discussed in Chapter 7.

For a solar cell made of an organic material, with $\kappa = 3$ and $m = 0.1\,m_0$, the binding energy $E_1^{org} = 151$ meV and the $r_\mathrm{B} = 1.5$ nm. The electron and the hole move in a tight orbit, and the binding energy is too strong ($E_1 \gg k_\mathrm{B}T$) to be dissociated at room temperature. These strongly bound excitons diffuse around as a charge neutral particle (see Fig. 8.2(b)). The potential energy associated with the electric field in a solar cell, $q\mathcal{E}_{max} \times r_\mathrm{B} = V_\mathrm{bi} r_\mathrm{B}/W$ is too small to perturb the pair. As they wander around, sooner or later the electron will recombine with the hole, and the photon absorbed will have produced no external current. A single semiconductor OPV, therefore, is characterized by low J_sc and poor cell efficiency η.

8.3 Physics of a planar heterojunction cell

Let us consider an absorber consisting of two organic semiconductors, called *donor* and *acceptor*, stacked on top of each other (see Fig. 8.3(a)). The materials have similar bandgaps, but their electron affinities (χ_1, χ_2) are different. Recall from Homework 7.1 that a pair of material with different affinities produces an energy band diagram shown in Fig. 8.4(a), with a conduction band discontinuity of $\Delta E_c \equiv \chi_1 - \chi_2$. The ΔE_c transition occurs within a fraction of a nm ($\ll r_\mathrm{B}$), which is smaller than the size of the exciton. If $\Delta E_\mathrm{c} > E_1$, any exciton that diffuses to and reaches the junction will be pulled apart as free electron and hole (see Fig. 8.3(a)). The electron transfers to the acceptor (A), while the hole transfers to donor (D). Since the electrons and holes are spatially separated into different materials, they

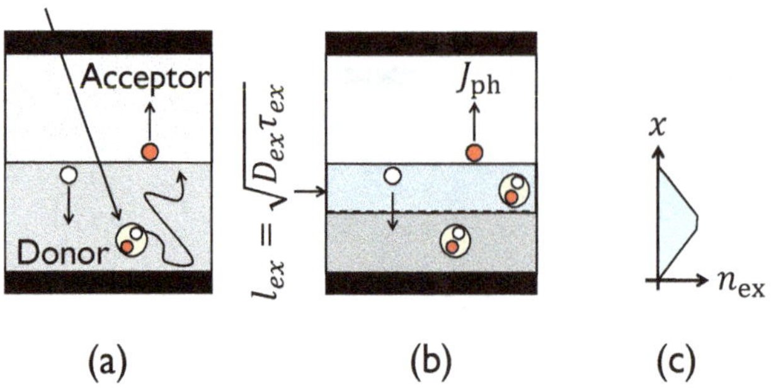

Figure 8.3: (a) Donor and acceptor semiconductors create an heterojunction at the interface. (b) Excitons created within one diffusion length of the interface can be dissociated by the heterojunction. (c) Spatial profile of excitons generated within the donor semiconductor.

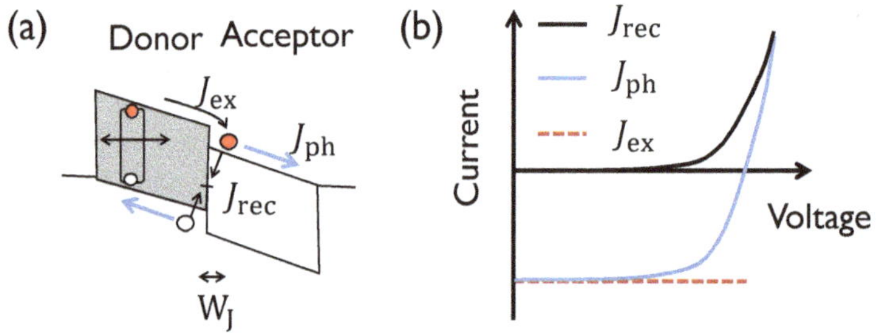

Figure 8.4: (a) Band diagram and flux components of a planar heterojunction solar cell. (b) voltage dependence of the current components.

cannot recombine, even if the material is defective. In short, OPV uses BHJ to dissociate excitons and suppress recombination to produce high J_{sc}.

If the excitons are generated uniformly in one of the two semiconductors (see Fig. 8.3(a)), their density profile (n_{ex}) shown in Fig. 8.3(c) is reminiscent of the electron profile of a p-n junction shown in Fig. 7.4(c), because they describe the same transport phenomena involving diffusion-recombination in the semiconductor followed by dissociation at the junction. Therefore, in analogy to Eq. (7.28), the rate of exciton dissociation is given by

$$J_{ex} = qG\, l_{ex} \tanh\left(\frac{W}{2l_{ex}}\right) \sim qG\, l_{ex}\left[1 - e^{-\frac{W}{4l_{ex}}}\right] \tag{8.3}$$

where $l_{ex} = \sqrt{D_{ex}\tau_{ex}}$ is the diffusion distance of an exciton travels before recombination in a material characterized by the diffusion coefficient, D_{ex} and the recombination time constant τ_{ex}. The factor $(W/4)$ in the exponent reflects the fact that photogeneration occurs only in the acceptor (i.e., $W/2$) and approximately half the excitons (i.e., $W/4$) are collected by the donor–acceptor heterojunction.

For a thick absorber ($W/2 \gg l_{ex}$), $J_{ex} = qGl_{ex}$, i.e., only the excitons generated within the diffusion distance away from the junction are converted into free electron and hole pairs, the rest recombine without producing any current. A thin absorber ($W/2 \ll l_{\text{ex}}$) can harvest half the excitons, $J_{\text{ex}} = qGW/4$, while the other half is lost to the bottom (wrong) contact. Unfortunately, typical processing condition require $W \gg l_{\text{ex}}$; therefore, only a small fraction ($\eta_{\text{ex}} \sim l_{\text{ex}}/W$) of the exciton generated ($qGW/2$) are harvested by a planar heterojunction solar cell. We will see later in Secs. 8.4 and 8.5 how vertical-heterojunction and bulk-heterojunction solar cells solve this exciton harvest bottleneck. For now, let us calculate the photo and dark current of a planar heterojunction solar cell.

8.3.1 Photocurrent can be calculated by flux balance argument

To calculate the electron- (or hole) current produced by exciton dissociation by the donor–acceptor heterojunction, we can use the flux balance approach:

$$\frac{J_{\text{ph}}}{J_{\text{ex}}} = \left[\frac{\gamma_{R,n}}{\gamma_{L,n} + \gamma_{R,n} + \gamma_{rec}} - \frac{\gamma_{R,p}}{\gamma_{L,n} + \gamma_{R,n} + \gamma_{rec}}\right] \tag{8.4}$$

Fig. 8.4 shows that an electron generated at the junction has two options. First, electrons can drift down the potential energy profile to be collected by the contact to contribute to J_{ph}:

$$\gamma_{R,n} = \mu_n \mathcal{E}(0) = \mu_n \frac{V_{\text{bi}} - V}{W_n}. \tag{8.5}$$

Here we *assume* that the electric field $\mathcal{E}(0)$ drops linearly across the device. The second option for the electron is to recombine with the holes right across the barrier. In Fig. 8.4(a), the direct recombination is proportional to the product of electron and hole concentrations at the interface, $W_J B n(0) p(0)$; while the indirect recombination is proportional to $W_J n(0)/\tau$ where W_J is the width of the junction, and τ is the interface recombination time. Taken together

$$\gamma_{rec} = W_J B n(0) p(0) + W_J n(0)/\tau \sim W_J (B p(0) + \tau^{-1}) n(0) \equiv S\, n(0). \tag{8.6}$$

If $W_n \sim W_p$ and $\mu_{\text{n}} \sim \mu_{\text{p}}$, then $n(0) \sim p(0)$, by symmetry. Therefore, the interface recombination velocity, $S = W_J(Bn(0) + \tau^{-1})$, depends on $n(0)$, the electron (or hole) concentration at the interface. Finally, we note that $\gamma_{L,n} = 0$, because $\Delta E_c \gg k_{\text{B}}T$ is designed to prevent electron escape to the left contact. Similarly, the holes cannot be collected by the right contact, i.e., $\gamma_{R,p} = 0$. Inserting Eqs. (8.5) and (8.6) into Eq. (8.4), we find

$$\frac{J_{\text{ph}}}{J_{\text{ex}}} = \frac{\mu_n \mathcal{E}(0)}{\mu_n \mathcal{E}(0) + S}. \tag{8.7}$$

In other words, the J_{ph} depends on the efficiency of exciton dissociation J_{ex}, the surface recombination velocity S, and the electric field at the junction, $\mathcal{E}(0) = (V_{\text{bi}} - V)/W_n$. Like a p-i-n or heterojunction solar cell, the photocurrent of an planar organic solar cell is voltage-dependent, because the reduced ($\mathcal{E}(0)(V)$ cannot pull the carriers away from the junction; the pile-up of carriers at the junction increases γ_{rec}.

Equation (8.7) must be solved iteratively, because J_{ph} determines $n(0)$ which in turn determines S. We can obtain an explicit solution by writing the flux balance equation, namely,

$$J_{\text{ex}} = q\left[W_J\left(Bn(0)^2 + n(0)\tau^{-1}\right) + n(0)\mu_n\mathcal{E}\right] \tag{8.8}$$

which states that the electrons produced by the excitons will either be lost by direct (first term) or trap-assisted recombination (second term), or drift to the contacts. Solving for $n(0)$ and realizing that $J_{\text{ph}} = qn(0)\mu\mathcal{E}(0)$, we find

$$J_{\text{ph}} = \frac{-v_{\text{T}} + \sqrt{v_T^2 + 4BJ_{\text{ex}}}}{2B}v_T \tag{8.9}$$

where $v_{\text{T}} \equiv \mu\mathcal{E} + W_J\tau^{-1}$. Figure 8.4(b) plots the three current components.

The analytical theory just described captures several essential aspects of the photocurrent in an OPV. For example, the theory explains the increase in the photocurrent as a consequence of charge separation that suppresses bulk recombination. The voltage-dependent photocurrent is also explained as a competition between interface recombination and drift-dominated charge collection.

8.3.2 Dark current is small and is calculated by elementary arguments

Let us now calculate the dark current. When the diode is forward-biased, electrons and holes are injected from the respective contacts. Since the electrons and holes co-exist only at the junction, therefore

$$J_{\text{dark}} = qW_JB\left[n(0)p(0) - n_{i,0}^2\right], \tag{8.10}$$

where B is the bimolecular recombination rate constant at the donor–acceptor interface, $n_{i,0}$ is the intrinsic carrier concentration at the interface and W_J is the width of the interface. The voltage-dependent interface concentrations $n(0)$ and $p(0)$ are respectively related to the corresponding concentrations at the contacts (i.e. n_{R} and p_{L}) as follows: $n(0) = n_{\text{R}}\exp\left(-qV_n/k_{\text{B}}T\right) = n_{\text{R}}\exp\left(-q\mathcal{E}W_{\text{n}}/k_{\text{B}}T\right)$ and $p(0) = p_{\text{L}}\exp\left(-qV_p/k_{\text{B}}T\right) = p_{\text{L}}\exp\left(-q\mathcal{E}W_{\text{p}}/k_{\text{B}}T\right)$. Inserting these relations in Eq. (8.10) and realizing that $\mathcal{E}(W_n + W_p) = V_{\text{bi}} - V$ we find

$$J_{\text{dark}} = qBW_J\left[n_{\text{R}}p_{\text{L}}e^{-q(V_{\text{bi}}-V)/k_{\text{B}}T} - n_{i,0}^2\right]. \tag{8.11}$$

Since $J_{\text{dark}}(V = 0) = 0$ in equilibrium, therefore $n(0, V = 0) \cdot p(0, V = 0) = n_{\text{R}}p_L\exp\left[-q(V_n(V=0) + V_{\text{p}}(V=0))/k_{\text{B}}T\right] = n_{\text{R}}p_{\text{L}}\exp\left(-qV_{\text{bi}}/k_{\text{B}}T\right) = n_{i,0}^2$. The dark current expression is now simplified to

$$J_{\text{dark}} = qBW_J\,n_{i,0}^2\left[e^{qV/k_{\text{B}}T} - 1\right] \equiv J_{03}\left[e^{qV/k_{\text{B}}T} - 1\right]. \tag{8.12}$$

The dark current has an expression similar to the p-n, p-i-n, or HJ solar cells (see Fig. 8.5(b)). Unlike J_{ph}, the dark current is not affected by charge pile-up either.

Therefore, Eq. (8.12) holds even for low-mobility organic semiconductor. Finally, we have focused on bimolecular recombination of electrons and holes at the interface. For trap-assisted interfacial recombination, the equation is modified to

$$J_{\text{dark}} = qW_J \frac{n_{i,0}}{\tau} \left[e^{qV/2k_{\text{B}}T} - 1\right] \equiv J_{04} \left[e^{qV/2k_{\text{B}}T} - 1\right]. \tag{8.13}$$

The total current, given by the sum of the dark (Eqs. (8.13) and (8.12)) and photo- (Eq. (8.9)) currents, is plotted in Fig. 8.5.

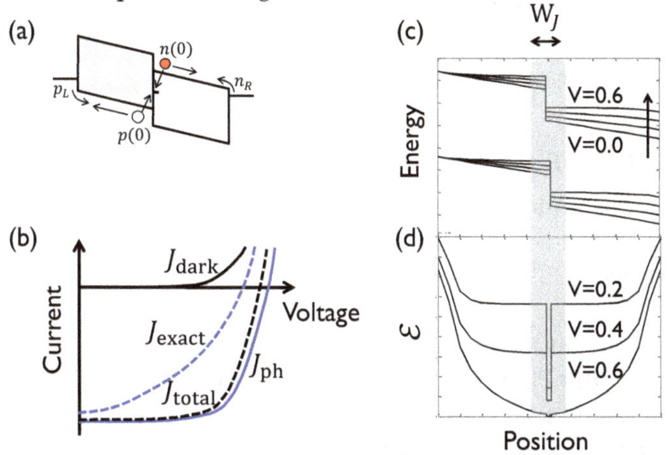

Figure 8.5: (a) The energy band diagram to calculate the light current. (b) The exact numerical result (J_{exact}) differ from the analytical result (J_{total}) derived above by assuming that the potential drop across the absorber layer is linear and the electric field is constant. (c) Non-linear drop of potential across the junction. (d) The junction electric field is reduced because low-mobility organic semiconductors cannot sweep away the charges sufficiently fast.

8.3.3 The puzzle of very low fill factor

Unfortunately, the theory appears not to capture the low FF observed in the experiments, as shown in the $J_{\text{exact}}(V)$ plot in Fig. 8.5(b). For low-mobility, intrinsic organic semiconductors, we cannot calculate $J_{\text{ph}} = qn(0)\mathcal{E}$ without accounting for the contribution to $n(0)$ from dark current. This limitation is easily corrected. Since the PHJ cell can be viewed as a pair of n-i-p and p-i-n cells in series, we can use Eq. (7.11) by replacing $n_{\text{i}}^2/N_{\text{A}}$ and $n_{\text{i}}^2/N_{\text{D}}$ by the generation and voltage-dependent electron and hole concentrations at the interface, namely, $n(0)(G,V)$ and $p(0)(G,V)$.

$$J_n(G,V) = q\mu_{\text{n}} n(0)\,\mathcal{E} \left[\frac{1}{\exp(-q|\mathcal{E}|W_{\text{n}}/k_{\text{B}}T) - 1}\right] \left[\Delta\, e^{qV/k_{\text{B}}T} - 1\right]$$
$$J_p(G,V) = q\mu_{\text{p}} p(0)\,\mathcal{E} \left[\frac{1}{\exp(-q|\mathcal{E}|W_{\text{p}}/k_{\text{B}}T) - 1}\right] \left[\Delta\, e^{qV/k_{\text{B}}T} - 1\right] \tag{8.14}$$

with $\Delta \equiv n(0)(G, V = 0)/n(0)(G, V)$. The ratio of this exact expression and the approximate expression $J_e = qn(0)\mu_n\mathcal{E}$ (second term on the right of Eq. (8.8)) identifies the voltage-dependent correction factor $c(\mathcal{E})$ given by the two terms in the bracket of Eq. (8.14). With this correction, we can recalculate the photocurrent as follows: Fig. 8.4 shows that $J_e = J_{\text{ph}} = J_{\text{ex}} - J_{\text{rec}}$ and $J_e = J_h$. Inserting Eqs. (8.6) and (8.14), we calculate $n(0)(G, V)$ and $p(0)(G, V)$ explicitly. These interface carrier densities allow us to calculate the $J(V)$. This (somewhat messy) expression explains the experimentally observed reduced FF as a consequence of carrier pile-up at the interface due to poor mobility and dark current injection from the contact and the associated electron–hole recombination.

Homework 8.1: Explaining reduced fill factor

Find the exact solution of the PHJ current–voltage characteristics. Assume that the donor and acceptor have identical parameters. Plot $J(V)$ to explain the dependence of FF on μ.

Solution. The exact expression has been derived in Appendix E of Biswajit Ray's 2013 PhD thesis, "Redefining the operation and design considerations of organic solar cells: Role of morphology and defect states." Specifically, assuming that donor and acceptors have identical parameters, we can write $J_n = qc(\mathcal{E})n(0)\mu\mathcal{E}$ where $c(\mathcal{E})$ is the field-dependent "correction" factor formed by the second and third terms of the right Eq. (8.14). Solve the flux balance equations, $J_e = J_{\text{ph}} = J_{\text{ex}} - J_{\text{rec}}$ and $J_e = J_{\text{n}}$, to find an exact expression for $J_e(E)$ where $E = (V_{\text{bi}} - V)/2W_{\text{n}}$.

Homework 8.2: Numerical simulation explicitly accounts for space charge effects

A planar HJ solar cell is composed of two organic semiconductors, P3HT (donor) and PCBM (acceptor). Both semiconductors are 100 nm thick, and the cathode and anode are also 100 nm thick. The exciton diffusion length (l_{ex}) is 10 nm in both materials. The electron and hole mobilities are $\mu_{\text{n}} = 10^{-5}$ cm^2/V.s and $\mu_{\text{p}} = 10^{-6}$ cm^2/V.s.

Use the web-enabled device simulator **OPVLabs** posted at `https://nanohub.org/resources/opv/about` to calculate the following: light $J - V$ characteristics, dark $J - V$ characteristics, electron and hole concentration profiles, and electric field distribution within the device. Do the $I - V$ characteristics appear to have a very different shape compared to the cases discussed in this chapter? Explain.

8.3.4 Organic solar cells may not obey superposition

Some organic semiconductors have very low mobility, which limits charge collection. As shown in Fig. 8.5(c), the electrons begin to pile up close to the junction, which reduces the field and suppresses charge collection and increases recombination (see Homework 8.2). Since the accumulated carriers changes the band profile, the superposition principle cannot hold. The analytical formula can be improved either by calculating the J_{ph} using the space charge limited theory (to be discussed in Chapter 9), or by using a numerical simulator, such as OPVLab.

8.4 Physics of a vertical-heterojunction cell

In this section, we will see how a vertical-heterojunction solar cell (VHJ) increases the efficiency of exciton collection, η_{ex}. Since the excitons are charge neutral, it makes no difference whether the heterojunctions used to separate them are parallel or perpendicular to the built-in field. In a VHJ cell, shown in Figs. 8.6(b)–(d), the excitons are dissociated laterally at the heterojunction. Once dissociated, the electrons and holes are confined to the channel defined by their respective energy barriers, without any possibility of bulk recombination with the other charge carrier traveling in the neighboring channel. The only recombination occurs at the interface.

Although the physics of exciton dissociation and charge transport are nominally similar, the VHJ cells have an important (geometrical) advantage over PHJ cells. Recall that for PHJ, efficient exciton dissociation requires that $W_n \sim W_p \sim l_{\text{ex}}$. It is impossible to bring the contacts so close to the junction (few nm) without shorting it electrically, especially for typical large-area solar cell ($\sim \text{m}^2$). Instead, in the vertical array shown in Fig. 8.6(b), the excitons are dissociated horizontally parallel to the contacts, but the free electron and holes are transported vertically, perpendicular to the contact. The film thickness is no longer determined by the exciton diffusion length.

8.4.1 Exciton harvesting in a VHJ solar cell

Consider a set of vertical cells arranged in a checkerboard pattern. The dimensions of the cells are shown in Figs. 8.6(b, c). The donor (or acceptor) finger density per unit area of the solar cell is $N_{\text{F}} = (1/2)(1/\text{S}^2)$, the finger volume is $V_{\text{f}} = W\,\text{S}^2$, and the finger surface area is $S_{\text{F}} = 4W\text{S}$.

Assuming a relatively coarse morphology (i.e., $l_{\text{ex}} \ll \text{S}$), we use Eq. (8.3) to calculate the total number of excitons generated *per finger*: $qG\,l_{\text{ex}} \times S_{\text{F}}$ and *per unit area*:

$$J_{\text{ex}}^{\text{VHJ}} = qG\,l_{\text{ex}} \times S_{\text{F}} \times N_{\text{F}} = qG\,l_{\text{ex}}(2W/\text{S}) = J_{\text{ex}}^{\text{PHJ}}\,(2W/\text{S}). \tag{8.15}$$

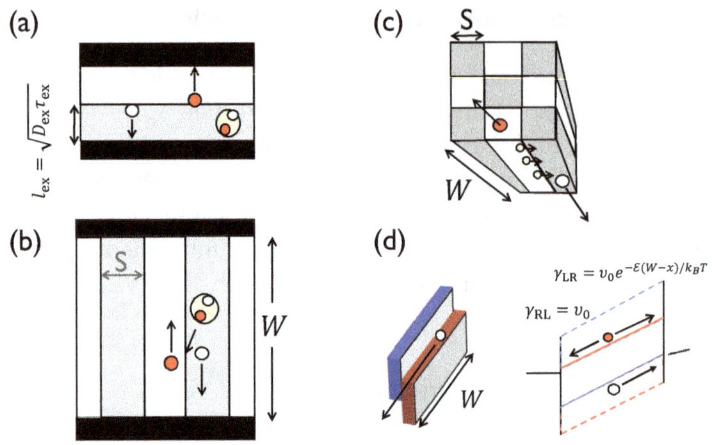

Figure 8.6: (a) Ultra-thin PHJ solar cell. (b) Vertical-heterojunction solar cells. (c) A 3D cartoon of the checkerboard-like structure of VHJ cells. (d) The energy band diagram explains how electrons and holes are transported in parallel channels.

The geometrical enhancement factor (i.e., $2W/\mathrm{S} \gg 1$) explains the more efficient exciton harvest by a distributed (checkerboard) VHJ, seen in Fig. 8.7.

Homework 8.3: Limiting exciton harvest of a VHJ solar cell

Show that $J_{\mathrm{ex}}^{\mathrm{VHJ}} \to qGW$ for $2W/S \gg 1$.

Solution. Equation (8.15) may imply that J_{ex} can be increased indefinitely by reducing S and increasing the geometrical enhancement factor $2W/\mathrm{S}$. This cannot be true: after all, the harvest cannot exceed the photogenerated excitons, i.e., $J_{\mathrm{ex}}^{max} = qGW/2$. (The factor $1/2$ reflects our assumption that only the donor region photogenerates excitons.)

The puzzle is resolved by returning to Eq. (8.3). As S is reduced and becomes comparable to l_{ex} (i.e. $\mathrm{S} \sim l_{\mathrm{ex}}$), the harvested excitons per unit area of a finger are obtained by linearizing Eq. (8.3), i.e., $J_{\mathrm{ex}}^{\mathrm{VHJ}} \sim qG\,l_{\mathrm{ex}} \times (\mathrm{S}/4l_{\mathrm{ex}}) = qGS/4$. The number of harvested excitons per unit area saturates to: $J_{\mathrm{ex}}^{\mathrm{VHJ}} = qGS \times N_{\mathrm{F}} \times S_{\mathrm{F}} = qGW/2$, as expected. In fact, one can plot

$$J_{\mathrm{ex}}^{\mathrm{VHJ}}(S) = qG\,l_{\mathrm{ex}} \tanh(\mathrm{S}/2l_{\mathrm{ex}}) \times N_{\mathrm{F}} \times S_{\mathrm{F}} = qGl_{\mathrm{ex}}\,(2W/\mathrm{S})\,\tanh(\mathrm{S}/2l_{\mathrm{ex}})$$

to explain the performance gain of a VHJ cell with respect to PHJ cell as a function of donor–acceptor morphology.

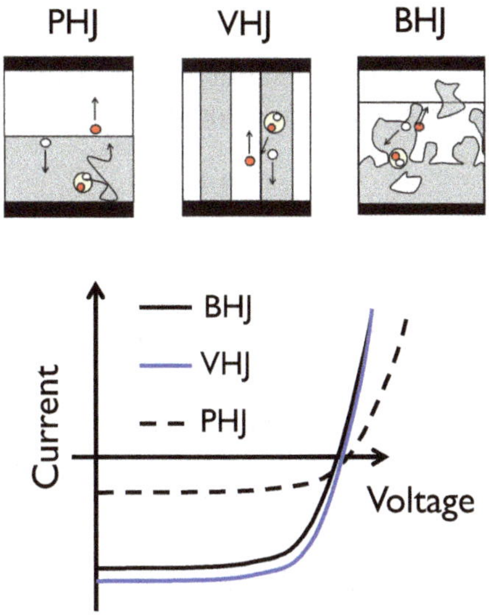

Figure 8.7: Comparison of light $I-V$ characteristics of PHJ, VHJ (ordered), and bulk-HJ excitonic solar cells.

8.4.2 Photocurrent from harvested excitons

As the electron and holes are transferred to the respective channels, and as they drift toward their respective contacts as shown in Fig. 8.6, we can write the flux equations as follows

$$\frac{J_{\mathrm{ph}}}{J_{\mathrm{ex}}} = \int_0^W dx \left[\frac{\gamma_{L,n}}{\gamma_{L,n} + \gamma_{R,n} + \gamma_{rec}} - \frac{\gamma_{L,p}}{\gamma_{L,n} + \gamma_{R,n} + \gamma_{rec}} \right]. \tag{8.16}$$

If we ignore the interfacial recombination $\gamma_{rec} = Bn_{\mathrm{i}}(x)p_i(x)4SW_J$, where n_{i} and p_i are the interface concentration of electrons and holes, and realize that the energy band profile of VHJ devices is essentially identical to the p-i-n solar cell, then we can use Eq. (7.20), to write an expression for photocurrent in a VHJ solar cell.

$$\frac{J_{\mathrm{ph}}}{J_{\mathrm{ex}}} = \frac{2L_D}{W} \ln\cosh\left(\frac{W}{2L_{\mathrm{D}}}\right) \tag{8.17}$$

where $L_{\mathrm{D}} = k_BT/(q(\mathcal{E}))$ is the Price length. Just as in a p-i-n cell, we anticipate voltage-dependent charge collection and light-dark current crossover in a VHJ solar cell. See Fig. 8.7.

8.4.3 Dark current of a VHJ solar cell

Once again, ignoring the interface recombination, we can use Eq. (7.10) to write an expression for J_{dark} associated with the parallel p-i-n channels,

$$J_{\text{dark}} = J_{01}\left(e^{qV/k_{\text{B}}T} - 1\right).$$

The ideal p-i-n diode current is relatively small and may not be important.

Unfortunately, one cannot ignore interface recombination completely in an organic solar cell. In Homework 7.7, we discussed the importance of recombination in a p-i-n solar cell. Actually, this is even more important for VHJ cell, because geometrical enhancement factor ($2W/\text{S}$) increases the surface area available for recombination both for photogenerated carriers as well as dark current, namely,

$$\Delta J_{\text{ph}}^{\text{VHJ}} = \Delta J_{\text{ph}} \times (2W/\text{S}), \tag{8.18}$$

$$\Delta J_{\text{dark}}^{\text{VHJ}} = \Delta J_{\text{dark}} \times (2W/\text{S}). \tag{8.19}$$

The dark current increases and photocurrent is suppressed as $\text{S} \rightarrow l_{\text{ex}}$. Therefore, although $J_{\text{sc}}^{\text{VHJ}} \gg J_{\text{sc}}^{\text{PHJ}}$, as shown in Fig. 8.7, the voltage-dependent increase in the recombination current reduces $V_{\text{oc}}^{\text{BHJ}} < V_{\text{oc}}^{\text{PHJ}}$! This would not be expected in a p-n-junction solar cell, where $V_{\text{oc}} \sim (k_{\text{B}}T/q)\ln(1 + J_{\text{sc}}/J_0)$, increases with J_{sc}.

Overall, the performance of VHJ cells is defined by a close relationship between geometry and performance. Excitons are harvested more efficiently in cells with higher surface-to-volume ratio, namely, cells with a smaller S. On the other hand, interfacial recombination increases with a decrease in S. Therefore, an optimum dimension S, defined by the exciton dissociation distance and the free electron–hole diffusion distance, maximizes performance.

8.5 Physics of bulk-heterojunction cells

The PHJ cells are inefficient and the nanoscale patterning of VHJ cell is difficult, if not impossible. The commercially available bulk-heterojunction cells (BHJ) approximate the essence of VHJ cells without resorting to expensive nanoscale patterning (see Fig. 8.7). The BHJ cells are prepared by mixing two special types of organic semiconductors in a solvent. The self-affinity of the organic semiconductors is higher compared to their affinity to each other; therefore, when the solution is heated, the polymer begins to segregate into donor-rich and acceptor-rich regions. Once the polymer columns reach the desired thickness, heating is removed, the contacts are attached. This completes the cell fabrication!

Although the morphology of BHJ cells shown in Fig. 8.7 appears very complicated and resembles the well-known picture of a "plumber's nightmare," the essential physics is exactly the same as in planar or vertical cells. The complete numerical solutions show that typical BHJ cells perform as well as the vertical bulk-heterojunction cells (see Fig. 8.7). If fact, we only need to replace S by the average width of donor–acceptor filaments of BHJ cells in Eqs. (8.15), (8.17), and (8.19) to find an analytical expression for their current–voltage characteristics.

8.6 Conclusions

In this chapter, we have derived the $J - V$ characteristics of an organic solar cell. Although the cell functions differently compared to inorganic solar cells, we can use the same "transmission over the barrier" formulation to calculate the dark and the light currents. Once again, we find that the dark current depends exponentially on the applied voltage, but with a prefactor defined by the physics of OPV. We also find that the photocurrent collection depends on the applied voltage. We assumed that the superposition principle holds, and that the total current can be calculated by adding the dark and light currents at each voltage. This is not always true, because the photogenerated carriers may make the potential profile very different under light compared to dark. If so, web-enabled numerical simulation tools (e.g., OPVLabs) can be used to calculate the $I - V$ characteristics.

In deriving the $I - V$ relationship, we assumed that the electrons and holes must drift/diffuse against a potential barrier. This is indeed the case for an idealized diode. In practice, the cell fabrication process may sometimes locally destroy the junction and the carriers can flow through these localized spots called *shunts*. Shunts is a universally present non-ideal effect that reduces the efficiency of a solar cell by increasing the dark current. We will discuss the physics of shunts in the next chapter.

References

[1] R. Sokel and R. C. Hughes. Numerical analysis of transient photoconductivity in insulators. *Journal of Applied Physics*, 53(11):7414–7424, November 1982.

[2] Brian A. Gregg. Excitonic Solar Cells. *The Journal of Physical Chemistry B*, 107(20):4688–4698, May 2003. Publisher: American Chemical Society.

[3] G. Yu, J. Gao, J. C. Hummelen, F. Wudl, and A. J. Heeger. Polymer Photovoltaic Cells: Enhanced Efficiencies via a Network of Internal Donor-Acceptor Heterojunctions. *Science*, 270(5243):1789–1791, December 1995.

[4] Alan J. Heeger. 25th Anniversary Article: Bulk Heterojunction Solar Cells: Understanding the Mechanism of Operation. *Advanced Materials*, 26(1):10–28, 2014. _eprint:
https://onlinelibrary.wiley.com/doi/pdf/10.1002/adma.201304373.

[5] Sung Heum Park, Anshuman Roy, Serge Beaupré, Shinuk Cho, Nelson Coates, Ji Sun Moon, Daniel Moses, Mario Leclerc, Kwanghee Lee, and Alan

J. Heeger. Bulk heterojunction solar cells with internal quantum efficiency approaching 100%. *Nature Photonics*, 3(5):297–302, May 2009.

[6] Thomas Kirchartz, Kurt Taretto, and Uwe Rau. Efficiency Limits of Organic Bulk Heterojunction Solar Cells. *The Journal of Physical Chemistry C*, 113(41):17958–17966, October 2009.

[7] Ye Huang, Edward J. Kramer, Alan J. Heeger, and Guillermo C. Bazan. Bulk Heterojunction Solar Cells: Morphology and Performance Relationships. *Chemical Reviews*, 114(14):7006–7043, July 2014. Publisher: American Chemical Society.

[8] L. J. A. Koster, E. C. P. Smits, V. D. Mihailetchi, and P. W. M. Blom. Device model for the operation of polymer/fullerene bulk heterojunction solar cells. *Physical Review B*, 72(8):085205, August 2005.

[9] Biswajit Ray, Pradeep R. Nair, and Muhammad A. Alam. Annealing dependent performance of organic bulk-heterojunction solar cells: A theoretical perspective. *Solar Energy Materials and Solar Cells*, 95(12):3287–3294, December 2011.

[10] Biswajit Ray and Muhammad A. Alam. Random vs regularized OPV: Limits of performance gain of organic bulk heterojunction solar cells by morphology engineering. *Solar Energy Materials and Solar Cells*, 99:204–212, April 2012.

[11] Biswajit Ray, Mark S. Lundstrom, and Muhammad A. Alam. Can morphology tailoring improve the open circuit voltage of organic solar cells? *Applied Physics Letters*, 100(1):013307, January 2012.

[12] Biswajit Ray, Aditya G. Baradwaj, Mohammad Ryyan Khan, Bryan W. Boudouris, and Muhammad Ashraful Alam. Collection-limited theory interprets the extraordinary response of single semiconductor organic solar cells. *Proceedings of the National Academy of Sciences*, page 201506699, August 2015.

CHAPTER 9

Physics and Universality of Shunt Distribution

Chapter Summary

- All solar cells are characterized by a localized non-ideal junction conduction, defined by the shunt resistance.
- Shunt resistance have symmetric power-law voltage dependence, with low thermal activation, and statistically distributed with a log-normal distribution.
- The voltage-symmetry of the shunt conduction provides a simple characterization technique to determine the I-V characteristics of a shunt-free solar cell.
- The shunt-resistance can be in interpreted by space-charge limited conduction given by the generalized Mott–Gurney law.
- Various reliability mechanisms, such as PID, reduce the shunt resistance and degrades the power conversion efficiency of a solar cell.

9.1 Introduction: Parasitic shunt current reduces solar cell efficiency

In the previous two chapters, we explained the physics of idealized p-n, p-i-n, heterojunction, and organic solar cells and calculated their dark and light currents. Experiments consistently show, however, that the *measured* dark current (J_{dark}) deviates significantly from the theoretical dark current (i.e., $J_{\mathrm{dark}}^{theory} \equiv J_0[\exp(qV/k_BT) - 1]$) both at reverse bias and low forward bias, as seen in Fig. 9.1(a). This deviation is attributed to the voltage-, temperature-, and thickness-dependent shunt current, $I_{\mathrm{sh}}(V, L, T)$ (or equivalently to the shunt resistance, R_{sh}, shown in Fig. 7.1):

$$I_{\mathrm{dark}} = I_{\mathrm{dark}}^{theory} + I_{\mathrm{sh}}(V, T, L). \tag{9.1}$$

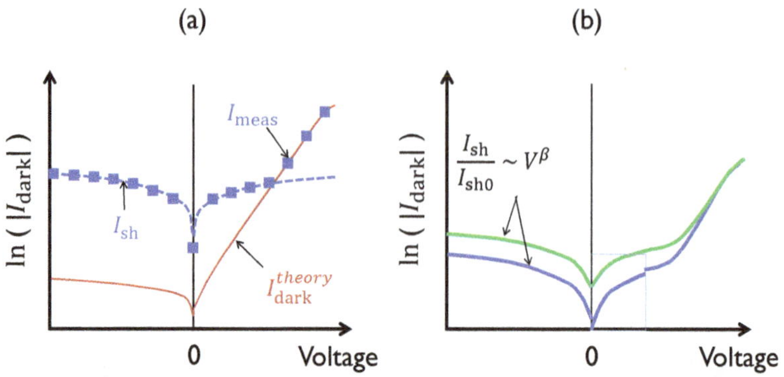

Figure 9.1: (a) The dark diode current ($|I_{dark}|$) is plotted as a function of the forward ($V > 0$) and the reserve bias ($V < 0$). The shunt current (dashed line) is obtained by subtracting the dark current (solid line) from the experimentally measured diode current (symbols). The shunt current dominates in the reserve and low forward bias. (b) The shunt current varies significantly from one device to the next (green vs. blue curves) and has a power law voltage dependence, $I_{sh} \sim V^\beta$.

The increase in the dark current (or "shunting" of the photocurrent) reduces cell efficiency; therefore, it is an important design consideration for practical solar cells. In this chapter, we will explain the physics and statistics of shunt conduction.

9.2 Unique voltage, temperature, and thickness dependence of $I_{sh}(V, T, L)$

9.2.1 Four features of the shunt current

The shunt current is defined by four characteristic features: voltage symmetry, significant device-to-device fluctuation, power law voltage and thickness dependencies, and weak temperature dependence. First, Fig. 9.1(a) shows that once we subtract I_{dark}^{theory} (solid red line) from the measured $J - V$ characteristics (blue squares), the shunt current (dashed line) is symmetric with respect to voltage, i.e., $|I_{sh}(V)| = \alpha|I_{sh}(-V)|$, with $\alpha \sim 1$. In Homework 9.1, we will use this voltage symmetry to extract shunt current from a collection of measured $I - V$ characteristics. Second, unlike I_{dark}^{theory}, the shunt current at low biases fluctuates significantly from cell to cell, as shown in Fig. 9.1(b), even when all cells are processed identically. Third, Fig. 9.1(b) also shows that shunt conduction has a power law dependence on voltage, i.e., $I_{sh}(V) \sim |V|^\beta$, with $\beta = 1 - 3$. If $\beta = 1$, the I_{sh} can be modeled using a parallel parasitic ohmic shunt resistance R_{sh}, as in Fig. 7.1. Since $R_{sh} = dI_{sh}/dV \sim V^{\beta-1}$, therefore shunt resistance is generally voltage dependent (non-ohmic). Finally, one can measure I_{dark} at low and high temperatures (e.g., $T_L \sim 200$ K and $T_H \sim 350$ K) and then determine the voltage-symmetric I_{sh} to show that $\alpha_T = I_{sh}(T_H)/I_{sh}(T_L) \sim 1 - 3$. Compared to the exponential voltage and temperature sensitivity of I_{dark}^{theory}, shunt current is characterized by a weak temperature dependence. An analysis of hundreds of solar cells demonstrates the validity of the four properties of shunt current: $\alpha \sim 1$ for voltage symmetry; device-to-device I_{sh} variability by 2–3 orders of magnitude; $\beta = 1-3$ for nonlinear voltage dependence, $\alpha_T = 1-3$ for weak temperature sensitivity.

Homework 9.1: PV Analyzer can extract relevant shunt parameters

Determine I_{sh} from the measured dark current I_{meas} by using the principle of voltage symmetry of the shunt current.

Solution. To solve this problem, we will use a web-enabled software called PV Analyzer to extract the idea diode and the parasitic shunt current parameters from the measured dark $I-V$ characteristics.

1. Login to nanohub.org ... Go to the link:
   `https://nanohub.org/resources/PVAnalyzer`
2. Download the user manual posted at
   https://nanohub.org/resources/11075/download/
   PV AnalyzerManual.pdf
3. Generate a set of dark J-V data using this simple formula: $J(V) = J_0\,(\exp(qV/k_BT) - 1) + V/R_{sh}$. You can use Excel, Matlab, or any program of your liking to generate the data.
4. Now launch the PV Analyzer and follow the instructions to run analysis on your data. Try different sets of values for R_{sh} — check the plots and the output log. How accurately does the PV Analyzer estimate the shunt resistance?

9.2.2 Interpreting shunt current by the space charge-limited theory

The four characteristic features of shunt conduction is a consequence of space charge-limited (SCL) current through parasitic current conduction paths (see Fig. 9.2). Figure 9.2(a) shows the cross-section of a typical p-i-n solar cell discussed in Chapter 7, except that the $n^+ - i$ junction on the top-left corner has been breached by a p^+ doped region. (The p-doping of this region is often attributed to metal ions diffusing from the transparent metal due to contact annealing, although other impurities may contribute as well.) Even a small incursion of p-doped region dramatically changes the diode $I-V$ characteristics, because this $p^+ - i - p^+$ region (shunted section) eliminates the energy barrier associate with $p^+ - i - n^+$ diode. For the reverse-biased diode shown in Fig. 9.2(a), the ideal diode current (Eq. (7.10)) should have been negligibly small (not shown), but the shunt current (blue arrows) through the shunted region adds to total current. Similarly

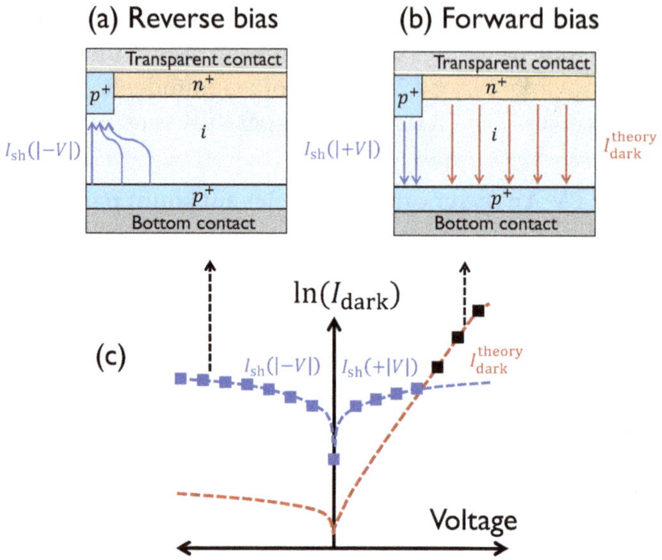

Figure 9.2: Shunt current is associated with local incursion of metal ions from the electrode into the grain boundaries of a solar absorber (p^+ region, top-left corner). The parasitic current path modifies the diode $I-V$ characteristics. (a) Localized $I_{sh}(|-V|)$, indicated by blue arrows, dominates the dark current of a reverse-biased diode. (b) In a forward-biased diode, localized $I_{sh}(|+V|)$ is dominant at low bias, but is superseded by the spatially uniform diode current (red arrows) at higher voltages. The approximate symmetry of the $I_{sh}(|-V|) \sim I_{sh}(|+V|)$, indicated by blue arrows, is a characteristic feature of shunt conduction. (c) The measured dark $I-V$ characteristics with various current components indicated.

for the forward-biased diode shown in Fig. 9.2(b), the ideal diode current at the low bias is very small and the shunt current dominates. Once the diode turns on fully at higher bias (red arrows), the shunt current (blue arrows) becomes insignificant. In short, we can think of the experimentally observed diode current as an independent sum of the ideal diode current through the $\mathrm{p}^+ - i - \mathrm{n}^+$ region and the shunt current through $\mathrm{p}^+ - i - \mathrm{p}^+$ region, and we can calculate the shunt current simply by focusing on shunted region alone.

The current through the $\mathrm{p}^+ - i - \mathrm{p}^+$ region shown on the left of Fig. 9.3(a) can be calculated by space charge theory. We generally use SCL theory to calculate current in a metal–semiconductor–metal (MSM) device. Typically, these semiconductors have poor carrier mobility, and the metal work functions are close to the conduction or valence band of the semiconductor, so that electron or holes (but not both) can be injected into the low-mobility semiconductor. For example, in Fig. 9.3(b), the metal workfunction supports hole injection from the contact, but electron injection is suppressed by the energy barrier. In a low-mobility semiconductor, the buildup of slow-moving injected charges modifies the internal

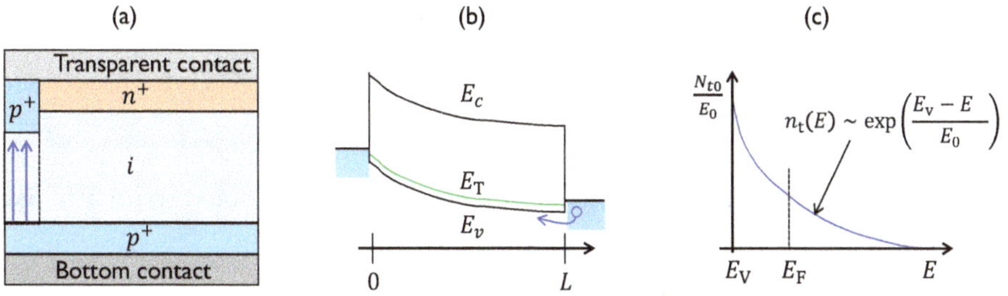

Figure 9.3: (a) One calculates the shunt current by focusing on the localized conduction through the shunted region on the left. (b) The band diagram of the shunted region is similar to the MSM structure with single-carrier (hole) injection. The charge buildup due to slow-moving carriers is reflected in the nonlinear valence band profile. (b) A schematic diagram showing the exponential trap distribution (band tails) near the valence band edge.

electric field to produce voltage-dependent non-ohmic space charge-limited current, e.g., $I_{\text{scl}}^{ideal} \propto V^2$. For a defective material, the voltage exponent of the SCL current depends on the energy-dependent density of shallow traps with trap-energy distributed exponentially within the bandgap, as in Fig. 9.3(c):

$$I_{\text{scl}} = \epsilon \mu_c(\gamma) A \frac{V^{\gamma+1}}{L^{2\gamma+1}}. \tag{9.2}$$

Here, ϵ is the material permittivity, μ_c is the carrier mobility, A is the cross-sectional area, and L is the film thickness. The parameter γ is a function of trap distribution inside the bandgap, and is usually between 1 and 2. It is easy to see that Eq. (9.2) can qualitatively capture the four features of I_{sh} discussed earlier. The expression is symmetric in voltage, and the measured power exponent $\beta \equiv \gamma + 1$ is also in the expected range. Finally, μ_c is the only temperature dependent term in Eq. (9.2): the weak temperature dependence of μ_c explains the weak temperature dependence of I_{sh}. Therefore, we can write the shunt current in thin-film cells, using the non-ohmic relation

$$I_{\text{sh}} = I_{sh0}(L,T) \frac{V^{\gamma+1}}{L^{2\gamma+1}}; \tag{9.3}$$

where I_{sh0} denotes the shunt current magnitude depending on shunt defect area, and L is the absorber layer thickness of the solar cells.

9.3 Understanding the space charge-limited current

Let us now consider the physics of SCL current in a 1D metal/semiconductor/metal (MSM) structure, shown in Fig. 9.3. This will essentially describe the shunt current of a p-i-p shunt path discussed in the previous section.

9.3.1 Ideal, trap-free semiconductor: Mott–Gurney law

The derivation of SCL current in an idealized MSM structure is based on the following assumptions:

1. The semiconductor is intrinsic and defect-free: it does not contain any fixed or trapped charges.

2. The contacts inject/extract only one type of carrier (holes in this case).

3. The electric field at the injecting contact is zero, $\mathcal{E}(0) = 0$.

4. With a large applied voltage, the drift current is much larger than the diffusion current.

Given the assumptions, we may consider the MSM device as a parallel plate capacitor separated by an intrinsic semiconductor. The total bound charge on the electrode is

$$Q_b = CV, \tag{9.4}$$

where, the capacitance $C = \epsilon A/L$ is characterized by the permittivity ϵ, thickness L, and cross-sectional area A of the MSM capacitor. Since $\mathcal{E}(0) = 0$ at the contact, the free carriers (i.e., holes) injected are equal to (but have the opposite polarity of) the bound charges in the electrodes (i.e., $Q_f \approx Q_b$). Their average velocity of the free charges with mobility μ_{p} is given by $v = \mu_{\text{p}}\mathcal{E}_{av}$, with $\mathcal{E}_{av} = V/L$.

The space charge current is determined by calculating the number of charges traversing the semiconductor per unit time. The travel time of holes between the contacts is

$$\tau = \frac{L}{v} = \frac{L^2}{\mu_{\text{p}} V}. \tag{9.5}$$

Taken together, the steady state SCL current density is

$$J_{scl} = \frac{1}{A}\frac{Q_f}{\tau} = \epsilon\mu_{\text{p}}\frac{V^2}{L^3}. \tag{9.6}$$

The band diagram in Fig. 9.3(b) shows that $\mathcal{E} \equiv (1/q)dE_v/dx$ is small at the right injecting contact, but increases rapidly toward the collecting contact on the left. Therefore, our use of $\mathcal{E}_{av} = V/L$ is not completely accurate. A more detailed derivation (see Homework 9.2) corrects the numerical prefactor:

$$J_{scl} = \frac{9}{8}\epsilon\mu_{\text{p}}\frac{V^2}{L^3}, \tag{9.7}$$

which is known as the **Mott–Gurney law**.

Homework 9.2: Exact derivation of the Mott–Gurney law

Calculate the exact expression for the Mott–Gurney law.

Solution. With the assumption set for trap-free semiconductor, the steady state semiconductor current is

$$J_{scl} = q\mu_{\rm p} p\mathcal{E}, \tag{9.8}$$

and the Poisson equation is given by

$$\frac{d\mathcal{E}}{dx} = \frac{qp}{\epsilon}. \tag{9.9}$$

Here, J_{scl} is the steady state SCL current, $\mu_{\rm p}$ is hole mobility, $\mathcal{E}$ is the electric field, $p(x)$ is the hole concentration, ϵ is the semiconductor permittivity, and q is electron charge. Using Eq. (9.8), we can write $p = J_{scl}/q\mu_{\rm p}\mathcal{E}$, and substitute it in Eq. (9.9) to obtain

$$\frac{d\mathcal{E}}{dx} = \frac{J_{scl}}{\epsilon\mu_{\rm p}}\frac{1}{\mathcal{E}}. \tag{9.10}$$

This can be rearranged, and integrated from the injecting contact to an arbitrary point x inside the semiconductor, using the assumption (3) above, to get

$$\int_0^{\mathcal{E}} \mathcal{E}d\mathcal{E} = \frac{J_{scl}}{\epsilon\mu_{\rm p}}\int_0^x dx, \tag{9.11}$$

which gives the electric field inside the semiconductor as

$$\mathcal{E}(x) = \sqrt{\frac{2J_{scl}}{\epsilon\mu_{\rm p}}}\sqrt{x}. \tag{9.12}$$

The band diagram in this situation is shown in Fig. 9.3(b), which demonstrates the situation under single-carrier (hole) injection. Finally, we write the electric field in terms of the potential inside the semiconductor as $\mathcal{E} = -d\phi/dx$ and integrate across the semiconductor layer thickness L to write

$$-\int_0^V d\phi = \sqrt{\frac{2J_{scl}}{\epsilon\mu_{\rm p}}}\int_0^L \sqrt{x}dx, \tag{9.13}$$

which when integrated and rearranged yields the SCL current expression as

$$J_{scl} = \frac{9}{8}\epsilon\mu_{\rm p}\frac{V^2}{L^3}. \tag{9.14}$$

This detailed derivation determines the exact prefactor of the Mott–Gurney expression.

9.3.2 SCL current in a semiconductor with shallow traps

Most amorphous, low-mobility semiconductors are characterized by shallow traps close to the conduction and valence bands. As shown in Fig. 9.3(c), these shallow traps are distributed exponentially in energy below the conduction band, i.e.,

$$n_{\mathrm{t}}(E) = \frac{N_{t0}}{E_0} \exp\left(\frac{E_{\mathrm{v}} - E}{E_0}\right), \tag{9.15}$$

where E_0 is the characteristic energy of the trap distribution. To calculate the SCL, we need to account for trapped as well as free charges. An analytical solution can be obtained for the following additional assumptions:

1. For the hole traps, the states above E_F are positively charged.
2. The number of immobilized trapped charges is much larger than the mobile, free charges.

The number of charged traps at a given location is

$$p_{\mathrm{t}}(x) = \int_{E_F}^{\infty} \frac{N_{t0}}{E_0} \exp\left(\frac{E_{\mathrm{v}} - E}{E_0}\right) dE = N_{t0} \exp\left(\frac{E_{\mathrm{v}} - E_F}{E_0}\right). \tag{9.16}$$

The upper integration limit is set to infinity, because for most materials, $E_0 \ll E_{\mathrm{g}}$, where E_{g} is the semiconductor band gap. Similarly, we recall from Homework 7.1 that the free carrier density is given by

$$p_f(x) = N_{\mathrm{V}} \exp\left(\frac{E_{\mathrm{v}} - E_F}{k_{\mathrm{B}}T}\right). \tag{9.17}$$

Using Eq. (9.16), we can write

$$p_f(x) = N_{\mathrm{V}} \left(\frac{p_{\mathrm{t}}(x)}{N_{t0}}\right)^{\gamma}, \tag{9.18}$$

where $\gamma \equiv E_0/k_{\mathrm{B}}T$. For now, let us assume that p_t, p_f are independent of position.

To calculate the space charge current, we can assume (as we did in the last section) that the holes are injected from the electrode with bound charge concentration $Q_b = CV$ associated with the parallel plate capacitor. Assuming $p_t \gg p_f$, the trapped and free carrier densities can be written as:

$$\begin{aligned} p_t &= \frac{1}{q}\frac{Q_b}{LA} = \frac{CV}{qLA} = \frac{\epsilon V}{qL^2}, \\ p_f &= N_{\mathrm{V}} \left(\frac{p_t}{N_{t0}}\right)^{\gamma} = N_{\mathrm{V}} \left(\frac{\epsilon V}{qN_{t0}L^2}\right)^{\gamma}. \end{aligned} \tag{9.19}$$

Now, by setting the hole velocity to $v = \mu_p \mathcal{E} = \mu_p V/L$, we have

$$J_{scl} = qp_f \times v = q\mu_p N_V \left(\frac{\epsilon}{qN_{t0}}\right)^{\gamma} \frac{V^{\gamma+1}}{L^{2\gamma+1}}. \tag{9.20}$$

A more detailed analysis in Homework 9.3 derives the following exact expression (Mark–Helfrich law) for the SCL current in a semiconductor with shallow traps

$$J_{scl} = q\mu_p N_V \left(\frac{\epsilon\gamma}{qN_{t0}}\right)^{\gamma} \left(\frac{2\gamma+1}{\gamma+1}\right)^{\gamma+1} \frac{V^{\gamma+1}}{L^{2\gamma+1}}. \tag{9.21}$$

Our approximate result (Eq. (9.20)) differs from the exact solution (Eq. (9.21)) by a constant: $\gamma^{\gamma}\ [(2\gamma+1)/(\gamma+1)]^{\gamma+1}$!

9.3.3 A scaling theory for SCL conduction

In the last section, we neglected any spatial distribution of the traps. For spatially distributed traps, we can write

$$p_t(x) = \int_{E_v}^{\infty} n_t(E)\mathcal{S}(x)dE \tag{9.22}$$

where $n_t(E)$ and $\mathcal{S}(x)$ represent the energy and the spatial distribution functions of the traps. In such cases L will be replaced by effective L_{eff}, which accounts for inhomogeneous spatial distribution of free and trapped carriers. Note that L_{eff} carries information from both $n_t(E)$ and $\mathcal{S}(x)$. It can be shown that the $J_{scl} - V$ relation always follow the following scaling law:

$$\frac{J_{scl}}{L_{\text{eff}}} = f\left(\frac{V}{L_{\text{eff}}^2}\right). \tag{9.23}$$

For example, referring to Eq. (9.21) or (9.2), we see that trap-free semiconductors, semiconductor with single-level trap, or exponentially distributed traps, or Gaussian distributed traps can all be described by the following general formula:

$$\frac{J_{scl}}{L_{\text{eff}}} = C^* \left(\frac{V}{L_{\text{eff}}^2}\right)^{\gamma+1}. \tag{9.24}$$

For uniformly distributed traps, however, we have a slightly different functional form,

$$\frac{J_{scl}}{L_{\text{eff}}} = C_1^* \left(\frac{V}{L_{\text{eff}}^2}\right) \exp\left(C_2^* \frac{V}{L_{\text{eff}}^2}\right). \tag{9.25}$$

The universal scaling law (Eq. (9.23)) is valid if the mobility is field independent and the diffusion is negligible. If the carrier diffusion is important, a modified formula is given by

$$V \approx \frac{2}{3}\left(\frac{2J}{\epsilon q\mu}\right)^{1/2} L^{3/2} - 4.68\left(\frac{k_B T}{q}\right)^{2/3} \left(\frac{J}{2\epsilon q\mu}\right)^{1/6} L^{1/2}. \tag{9.26}$$

The first term yields the Mott–Gurney equation (Eq. (9.7)), while the second term incorporates the diffusion effect.

Homework 9.3: SCL current in presence of shallow traps

Derive the exact expression for the SCL current with shallow traps and compare the results to Eq. (9.21).

Solution. Now using assumption (2) above, we can write the Poisson and the continuity equations as

$$\frac{d\mathcal{E}}{dx} \approx \frac{qp_t(x)}{\epsilon}, \tag{9.27}$$

and

$$J_{scl} = q\mu_{\text{p}}p_f\mathcal{E}. \tag{9.28}$$

Then, using the relation between trapped and free charge, we get

$$\frac{d\mathcal{E}}{dx} \approx \frac{qN_{t0}}{\epsilon}\left(\frac{N_{\text{V}}}{p_f}\right)^{1/\gamma} = \underbrace{\frac{qN_{t0}}{\epsilon}\left(\frac{J_{scl}}{q\mu_{\text{p}}N_{\text{V}}}\right)^{1/\gamma}}_{\Theta}\frac{1}{\mathcal{E}^{1/\gamma}}, \tag{9.29}$$

where the constants are collected as Θ. Rearranging the terms and integrating from 0 to arbitrary point x inside the semiconductor we get

$$\int_0^{\mathcal{E}} \mathcal{E}^{1/\gamma}d\mathcal{E} = \Theta\int_0^x dx, \tag{9.30}$$

which gives the electric field as

$$\mathcal{E} = -\frac{d\phi}{dx} = \left(\frac{\gamma+1}{\gamma}\Theta\right)^{\frac{\gamma}{\gamma+1}} x^{\frac{\gamma}{\gamma+1}}. \tag{9.31}$$

This is then integrated over the semiconductor layer thickness L to obtain the applied voltage as

$$V = \left(\frac{\gamma+1}{\gamma}\Theta\right)^{\frac{\gamma}{\gamma+1}} \frac{\gamma+1}{2\gamma+1}L^{\frac{2\gamma+1}{\gamma+1}}. \tag{9.32}$$

Finally, substituting the value of Θ from above and rearranging, we obtain the full expression of the SCL current in the presence of shallow traps as

$$J_{scl} = q\mu_{\text{p}}N_{\text{V}}\left(\frac{\epsilon\gamma}{qN_{t0}}\right)^{\gamma}\left(\frac{2\gamma+1}{\gamma+1}\right)^{\gamma+1}\frac{V^{\gamma+1}}{L^{2\gamma+1}}. \tag{9.33}$$

The result is essentially identical to Eq. (9.21).

9.4 Experimental validation of the universal non-ohmic shunt conduction

The theory of shunt conduction not only explains the voltage symmetry, power law exponent, and weak temperature dependence of shunt current, but it also makes a series of important predictions that can be verified by experiments.

9.4.1 SCL current involves either electrons or holes

SCL conduction is a result of single-carrier injection in the semiconductor layer. In case of a-Si:H cells, shunt formation is correlated to metal incursion from the top contact, which creates a p-type counter-doping. Consequently, a p-i-p shunt, as shown in Figs. 9.2(a, b). Therefore, in a-Si:H cells, holes injected through the shunt region determine I_{sh}. We can indirectly establish the dominance of hole transport in I_{sh} by examining the voltage power exponent $\beta = \gamma + 1$. In materials with exponentially distributed shallow traps (i.e., $n_t = (N_{t0}/E_0)\exp((E_v - E)/E_0)$, Eq. (9.15)), we know that $\gamma \equiv E_0/k_BT$, where E_0 is the characteristic energy of the exponential trap distribution, and E_v is the valence band edge. In other words, the characteristics energy of shallow traps can be obtained from the experimentally measured voltage power law exponent β as follows:

$$E_0 = (\beta - 1)k_BT. \tag{9.34}$$

Once we determine the power exponent β from the slope of $\ln(I_{sh}) - \ln V$ plot of a cell characterized at temperature T, Eq. (9.34) determines E_0 for the cell. Figure 9.4 plots E_0 from a large collection of solar cells. On average, $E_0(\sim 40\text{ meV})$, comparable to the slope of the *valence* band tail in a-Si:H, affirming that I_{sh} is indeed dominated by hole current.

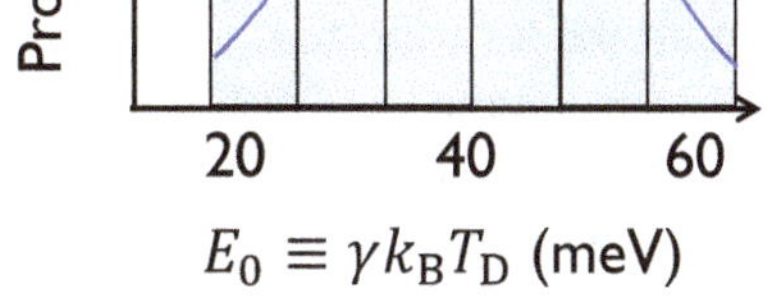

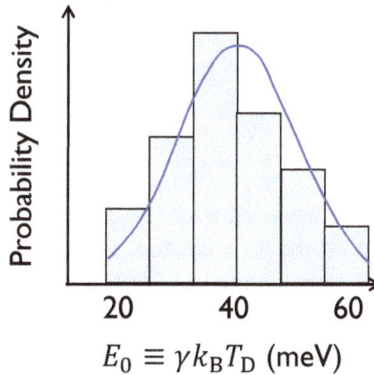

Figure 9.4: A histogram of E_0 is obtained by analyzing the voltage exponent β of the shunt currents of a large number of solar cells.

9.4.2 Shunt conduction has an inverse power law thickness dependence

Another important prediction of SCL shunt current model discussed in Sec. 9.2 is the power law thickness dependence, i.e., $I_{\rm sh} \propto L^{-2\gamma-1}$ (see Eq. (9.33) or (9.21)). Figure 9.5 compares two different cell technologies: (i) a-Si:H devices with i-layer thicknesses 140 nm to 1120 nm, and (ii) OPV cells with absorber thicknesses ranging from 50 nm to 250 nm. For both technologies, the log–log plots confirm the power law thickness dependence, with the slope given by $\gamma+1$. Since $\gamma = E_0/k_{\rm B}T$, the higher slope of OPV solar cell implies a broader trap distribution, as expected.

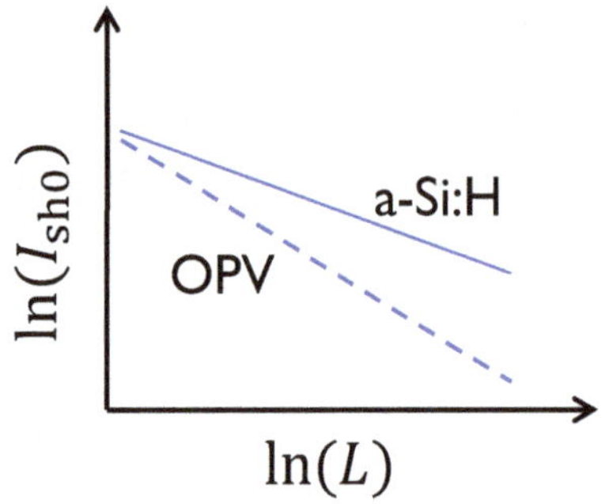

Figure 9.5: I_{sh0} vs. L plots for (a) a-Si:H and (b) OPV cells. The shunt current has a broad scatter; only median current values are shown. The inverse power law for thickness dependence holds for both technologies.

9.4.3 Shunt current is a universal feature of solar cells

In the last section, we saw that the space charge-limited shunt current in a-Si:H cells is caused by local metal incursion through the thin n-layer. In fact, all heterojunction solar cells share similar device geometry and manufacturing techniques. They all use thin (~10–100 nm) emitter layers deposited on the absorber to create the solar cells (see Fig. 9.6). Moreover, absorbers and emitters are deposited at relatively low temperatures over large-area substrates leading to surface nonuniformity, voids, and grain boundaries. As a result, local shunt formation through metal incorporation through pinholes in emitter (or through contact metal diffusion) is observed in a variety of technologies. This is true even for c-Si solar cells with thicker emitter layers, because the contacts are formed at higher temperature, and high voltages force Na atoms through the grain boundaries. We will discuss the physics of potential-induced degradation (PID) in Chapter 22 of the book.

Figure 9.6(a) shows the schematic representation of local shunt paths formed due to metal incorporation through thin emitter layers for three different thin-film solar cell technologies. The shunt current is spatially localized; therefore, one can separate the diode and shunt current paths, as shown in Fig. 9.6(b), and use the equations for diode current developed in the Chapters 7 and 8 and shunt current

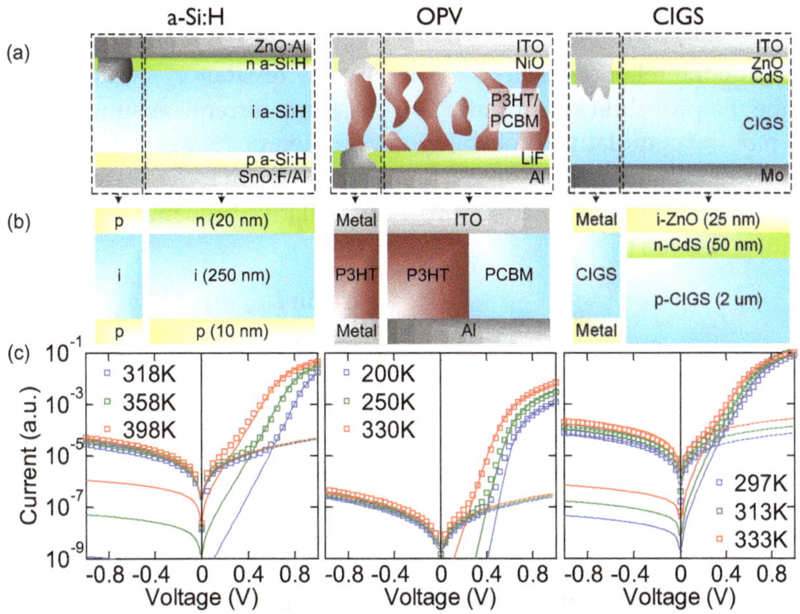

Figure 9.6: (a) Schematics showing the likely shunt formation in a-Si:H, OPV and CIGS cells, due to contact metal diffusion through emitter. (b) From the localization of shunt current, we can separate the shunt and diode current paths, simplifying the 2D picture in part (a). (c) Simulations combining the 1D shunt in parallel to a diode for all three cell types can reproduce the measured voltage and temperature dependencies, using typical material parameters.

developed in this chapter to interpret the measured dark current characteristics shown in Fig. 9.6(c). Regardless the technology, the shunt currents are characterized by voltage symmetry, power law voltage dependence, and relatively low temperature sensitivity.

9.5 Shunt magnitude distribution is log-normal, but its spatial distribution is random

In Sec. 9.1, we saw that shunt current varies significantly from one cell to the next. Understanding and accounting for this variability in shunt magnitude is essential for understanding the physics of shunt current. We will see in Part III of the book that a relatively few bad shunts (large I_{sh}) located close to the electrical contacts can significantly reduce the efficiency of a solar cell. Therefore, shunt formation in solar cells is a universal challenge for achieving reproducible, good quality, high-performance solar cells.

9.5.1 Statistical analysis of shunt current magnitude

To determine the relative fraction of bad (i.e., low resistance) shunts, we need to determine the probability distribution of the shunt current. A simple way to do so is to plot the cumulative probability distribution of I_{sh} against the theoretical probability distribution "presumed" to describe the data. For example, if I_{sh} is described by a log-normal distribution, the cumulative probability for log-normal distribution is given by

$$F_X(I_{sh0}; \mu, \sigma) = \frac{1}{2}\left[1 + \text{erf}\left(-\frac{\ln I_{sh0} - \hat{\mu}}{\sqrt{2}\,\hat{\sigma}}\right)\right]. \tag{9.35}$$

Or, equivalently,

$$\underbrace{-\sqrt{2}\,\text{erf}^{-1}(2F_X - 1)}_{Y} = \underbrace{\frac{\ln I_{\text{sh}} - \hat{\mu}}{\hat{\sigma}}}_{\hat{X}}. \tag{9.36}$$

Here, $\hat{\mu}$ and $\hat{\sigma}$ are the mean and standard deviations of $\ln(I_{sh0})$, respectively. If I_{sh0} follows a log-normal distribution, a $\hat{X} - \hat{Y}$ plot of the shunt current based on Eq. (9.36) should be described by a straight line. Alternatively, the scale variable $\hat{X}$ of a log-normal distribution should exhibit standard normal ($\mathcal{N}(0,1)$) behavior. This can be checked using a quantile–quantile (QQ) plot, which shows the quantiles of data vs. theoretical quantiles from a hypothesized distribution. This means that if I_{sh0} is indeed distributed log-normally, then the plot with quantiles of the scaled variable $\hat{X}$ (see Eq. (9.36)) vs. theoretical quantiles of $\mathcal{N}(0,1)$ distribution would give a straight line.

Figure 9.7 shows that the QQ plot of $\hat{X}$ vs. theoretical quantiles of $\mathcal{N}(0,1)$ shows reasonable agreement with the hypothesized log-normal behavior, for multiple technologies. The line shows that the best fit to the scaled variable $\hat{X}$ for all technologies is a $\mathcal{N}(0,1)$ distribution, as expected for a log-normal distribution of I_{sh0}. The analysis shows the magnitude of shunt current indeed follows a log-normal distribution. It is a remarkably universal trend, especially considering the substantial differences between these technologies, ranging from cell structures, materials, fabrication processes, and even the variety of sources of these data sets.

Once we determine the $\hat{\mu}$ and $\hat{\sigma}$ by analyzing the shunt current distribution associated with a PV technology, we can use the log-normal distribution to calculate the probability of developing very large shunt that will dramatically reduce the efficiency of modules based on the solar cell technology.

9.5.2 Spatial size and position distribution of shunts

Given the hierarchical nature of photocurrent collection in a solar module, we will see in Part III that some areas of a solar cell (e.g., close to the grids) are more vulnerable to shunt conduction compared to others. In particular, clustering of shunts could be particular detrimental.

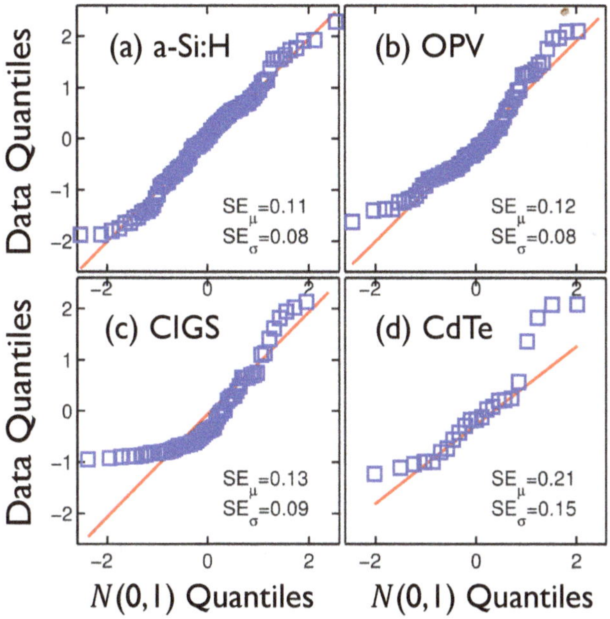

Figure 9.7: QQ plots for $\hat{X}$ values for four different thin-film solar cells (symbols) show very good fit to the theoretical standard normal quantiles (red line) for (a) a-Si:H, (b) OPV, (c) CIGS, and (d) CdTe technologies. For standard errors associated with $\hat{mu}$ and $\hat{\sigma}$ are shown in insets.

It is possible to determine the spatial distribution of shunts through various imaging techniques, such as dark lock-in thermography (DLIT), electroluminescence (EL), and photoluminescence (PL). These methods allow us to identify the physical location of shunt defects on the cell surface. For example, the DLIT technique takes thermal images of solar cells, when subjected to pulsed reverse bias under dark conditions. As bulk diodes are turned "off" in reverse bias, only the shunt paths conduct significant current, and the resultant local temperature rise due to Joule heating can be imaged using thermal imaging.

The schematic in Fig. 9.8(a) uses bright blue spots to indicate the randomly distributed shunt locations of an a:Si-H solar cell. A DLIT image, associated with the top view of the solar cell, is shown in Fig. 9.8(b). The image allows us to determine both the pairwise distances among the neighboring spots (see Fig. 9.8(c)) and the distribution of the spot sizes (see Fig. 9.8(d)). The pairwise distance distribution identifies clustering of shunts, if any. The size distribution should correlate to the shunt current distribution given by Eq. (9.36) and shown in Fig. 9.7.

Spot sizes are log-normally distributed. Figure 9.9(a) shows an identical log-normal distribution (Eq. (9.36)) of spot sizes obtained from the DLIT images of a-Si:H, CIGS, and multi-crystalline Si. Remarkably, the spot size distribution shows

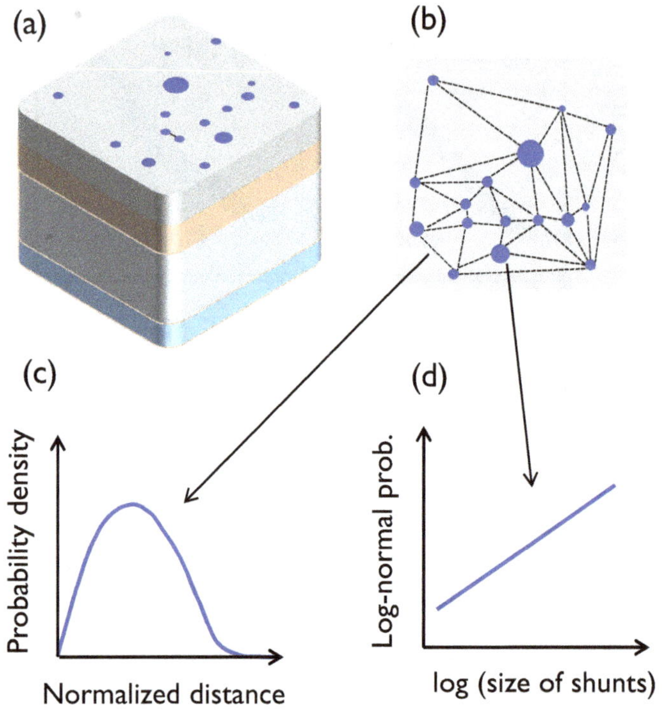

Figure 9.8: (a) Parasitic shunt formation due to local metal incursion from the contact into an a-Si:H solar cell. (b) Top view of solar cell with local shunts as seen in dark lock-in thermography (DLIT) images, with the spots distributed on the surface. The centroid and the radius of each bright spot is determined by processing the DLIT image. (c) Given the location (centroid) of each spot, one can determine the probability distribution of pairwise distances. (d) The size distribution of the spots can also be checked against various probability distributions.

identical log-normal behavior for all technologies (Eq. (9.36)), despite the fact that these are fabricated using entirely different materials, processes, and structures. The spot sizes are proportional to the local heat generation, which in turn is proportional to the local shunt current. Therefore, this log-normal spot size distribution correlates, as it must, to the log-normal magnitude of shunt current observed in electrical measurements of a-Si:H cells.

Shunt formation is spatially uncorrelated. Figure 9.9(b) shows the probability density of the pairwise distances between the centroids (Fig. 9.8(c)), for all three cell technologies. This position statistics are similar for various cell types and follow the PDF of a complete spatial randomness (CSR) process. The probability

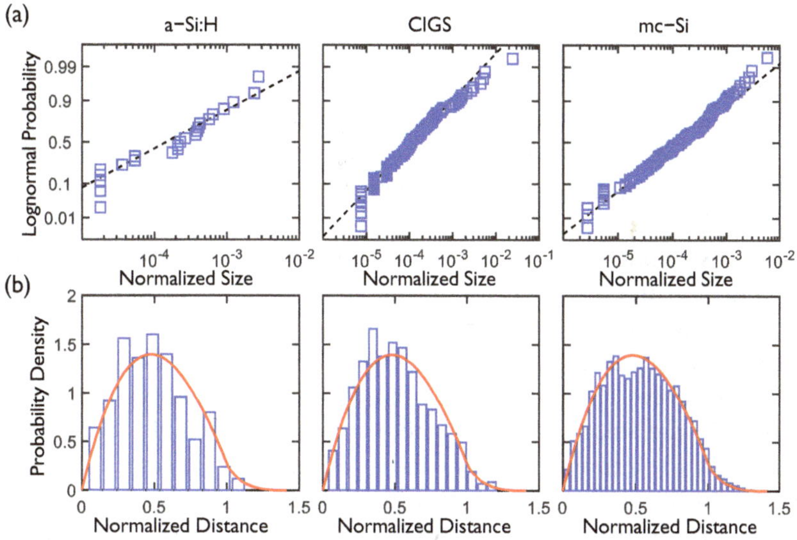

Figure 9.9: (a) Normalized spot size distributions for DLIT images of a-Si:H, CIGS, and mc-Si solar cells show the same log-normal behavior. Normalized spot size is defined as $N_{pixel,spot}/N_{pixel,image}$. (b) The distribution of pairwise distances between spot locations, for the same three images normalized to unit square (bars), closely follows the PDF obtained from a CSR process in a unit square (line).

distribution is given by

$$P(l) \simeq \begin{cases} 4l\left[1 - \left(\frac{1}{2} + \frac{\pi}{4}\right)l + \frac{1}{3}l^2\right], & 0 \le l \le 1 \\ \frac{5}{2}(\sqrt{2} - l)^3, & 1 < l < \sqrt{2}. \end{cases} \tag{9.37}$$

The spatial randomness implies that the formation of an individual shunt path does not affect the probability of shunt formation in its immediate vicinity. This spatially uncorrelated shunt formation has important implications on the efficiency of solar cells.

Why is the shunt formation uncorrelated, log-normally distributed, and universal? The short answer is: we do not know, but we can make a reasonable guess. The universality of shunt formation implies a common mechanism, possibly related to the metal contact and the TCO layer (see Fig. 9.8(a)). It is likely that the universal log-normal shunt statistics is linked to the universal log-normal grain size distribution of metal/TCO layers. This hypothesis is supported by the fact that interlayers, most notably in OPVs and CIGS, suppress parasitic shunting. In multi-crystalline silicon, shunt formation in mc-Si cells has been attributed to impurity precipitates. If the precipitate concentration is random and log-normally distributed, so will be shunt distribution.

9.6 Conclusions

In this chapter, we discussed the importance of parasitic shunts in determining the efficiency of practical solar cells. We showed that the shunts arise as a natural consequence of solar cell geometry (ultra-thin emitter layer) and processing conditions (homogeneity associated with moderate temperature processing). The space charge-limited model describes the electrical characteristics (e.g., voltage symmetry, power law voltage and thickness dependence, temperature insensitivity) of shunt conduction and provides insights into the log–normal distribution of shunt current magnitudes for a variety of PV technologies.

The second important parasitic component in solar cells, shown in Fig. 7.1, is the series resistance R_s, which arises from distributed photogeneration of localized current in the solar cells. The photocurrent must travel a certain distance before reaching the terminals, and the finite resistance of the contact layers causes additional power and efficiency loss in the solar cell. This series resistance includes both the intrinsic ohmic resistivity of the semiconductor layers as well as the extrinsic resistivity associated with the front contact grids and busbars. In the following chapter, we will discuss this loss mechanism and the designs to minimize the loss.

References

[1] Simon Min Sze and Kwok Kwok Ng. *Physics of semiconductor devices*. Wiley-Interscience, Hoboken, N.J., 2007.

[2] N. F. Mott and Ronald W. Gurney. *Electronic processes in ionic crystals*. Dover Publications, New York, 1964.

[3] A. Rose. Space-charge-limited currents in solids. *Physical Review*, 97(6):1538–1544, March 1955.

[4] Gernot Paasch and Susanne Scheinert. Space-charge-limited currents in organics with trap distributions: Analytical approximations versus numerical simulation. *Journal of Applied Physics*, 106(8):084502, 2009.

[5] Kwan-Chi Kao. *Dielectric phenomena in solids: With emphasis on physical concepts of electronic processes*. Academic Press, Amsterdam ; Boston, 2004.

[6] S. Dongaonkar, J. D. Servaites, G. M. Ford, S. Loser, J. Moore, R. M. Gelfand, H. Mohseni, H. W. Hillhouse, R. Agrawal, M. A. Ratner, T. J. Marks, M. S. Lundstrom, and M. A. Alam. Universality of non-Ohmic shunt leakage in thin-film solar cells. *Journal of Applied Physics*, 108(12):124509, 2010. Publisher: AIP.

[7] Ken Durose, Jon D. Major, and Yuri Y. Proskuryakov. Nucleation and grain boundaries in CdTe/CdS solar cells. In *Materials research society symposium proceedings*, volume 1165. Materials Research Society, 2009.

[8] Steve Johnston, Thomas Unold, Ingrid Repins, Rajalakshmi Sundaramoorthy, Kim M. Jones, Bobby To, Nathan Call, and Richard Ahrenkiel. Imaging characterization techniques applied to Cu(In,Ga)Se[sub 2] solar cells. *Journal of Vacuum Science & Technology A: Vacuum, Surfaces, and Films*, 28(4):665, 2010.

[9] E. Miranda, E. O'Connor, and P.K. Hurley. Analysis of the breakdown spots spatial distribution in large area MOS structures. In *2010 IEEE international reliability physics symposium*, pages 775–777. IEEE, 2010.

[10] M. D. Irwin, D. B. Buchholz, A. W. Hains, R. P. H. Chang, and T. J. Marks. p-Type semiconducting nickel oxide as an efficiency-enhancing anode interfacial layer in polymer bulk-heterojunction solar cells. *Proceedings of the National Academy of Sciences*, 105(8):2783–2787, February 2008.

[11] Karin Ottosson. *The role of i-ZnO for shunt prevention in Cu (In, Ga) Se2-based solar cells*. PhD thesis, Uppsala University, 2006. Number: April.

[12] M. D. Abbott, T. Trupke, H. P. Hartmann, R. Gupta, and O. Breitenstein. Laser isolation of shunted regions in industrial solar cells. *Progress in Photovoltaics: Research and Applications*, 15(7):613–620, 2007. _eprint: https://onlinelibrary.wiley.com/doi/pdf/10.1002/pip.766.

[13] Otwin Breitenstein, Jan Bauer, Thorsten Trupke, and Robert A. Bardos. On the detection of shunts in silicon solar cells by photo- and electroluminescence imaging. *Progress in Photovoltaics: Research and Applications*, 16(4):325–330, 2008. _eprint: https://onlinelibrary.wiley.com/doi/pdf/10.1002/pip.803.

[14] X. Saura, J. Suñé, S. Monaghan, P. K. Hurley, and E. Miranda. Analysis of the breakdown spot spatial distribution in Pt/HfO2/Pt capacitors using nearest neighbor statistics. *Journal of Applied Physics*, 114(15):154112, October 2013. Publisher: American Institute of Physics.

[15] Silke Steingrube, Otwin Breitenstein, Klaus Ramspeck, Stefan Glunz, Andreas Schenk, and Pietro P. Altermatt. Explanation of commonly observed shunt currents in c-Si solar cells by means of recombination statistics beyond the Shockley-Read-Hall approximation. *Journal of Applied Physics*, 110(1):014515, July 2011. Publisher: American Institute of Physics.

CHAPTER 10

Physics of Series Resistance of Solar Cells and Modules

Chapter Summary

- Series resistance is another parasitic component of a solar cell that reduces the efficiency of a solar cell below the thermodynamic limit.
- The series resistance arises from the bulk material, finger and busbars that collect the current, and the wires that connect the modules to the load. The shadow of the fingers and busbars reduces short-circuit current.
- The series resistance can be optimized by a judicious choice of front contact (grid) design.
- Thin film PV and thick film solar cells have very different optimization principles.
- The topic is defined by many innovations and rapid commercialization.

10.1 Introduction

In the previous chapters, we discussed the semiconductor physics of a solar cell and calculated the photocurrent, dark current, and shunt current. Eventually, the electrons and holes must be transferred to the outside world by metallic contacts. You may be surprised to see that, unlike a typical electronic device, an intricate pattern of metal grids decorates the top and/or bottom surfaces of a solar module (see Fig. 10.1). For a thin-film solar cell, the grids run parallel to each other, across the width of the module, as in Figs. 10.1(a, b). The grids divide the module into a series of subcells. For c-Si modules, each rectangular wafer is patterned with horizontal and vertical metal lines. The wafers are then arranged in series to complete the module, see Figs. 10.1(c, d). When these subcells in thin-film PV or wafers in c-Si PV are connected in series, they can be viewed as a network of series-connected diodes, as in Fig. 10.1(e).

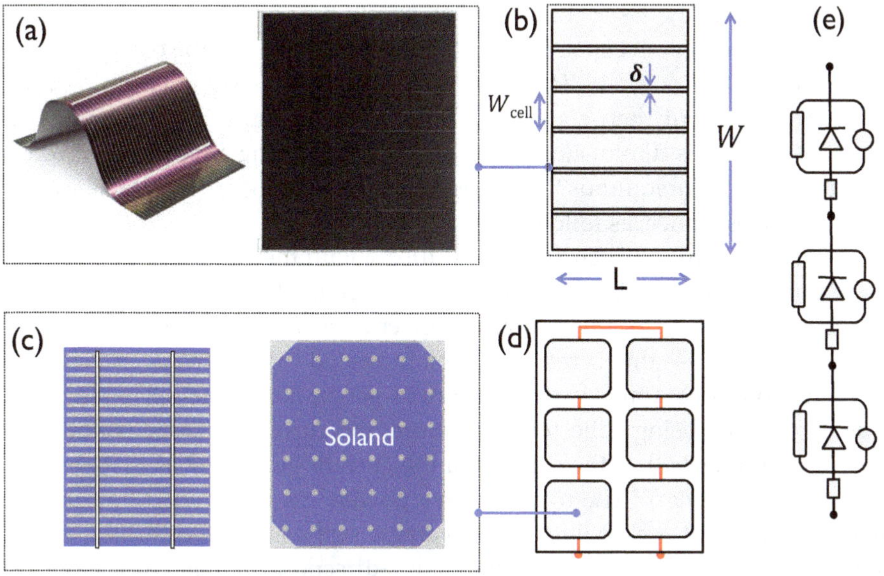

Figure 10.1: Solar cells have a variety of specially designed grids to collect the photocurrent: (a) Thin film solar cells use thin lateral grids. This is shown schematically in (b). (c) c-Si solar cells use a combination of lateral grids (fingers) and vertical grids (busbar). (d) The cells are connected in series to create a module. (e) The series connection of the cells can be represented by equivalent circuit associated with each cell.

After examining the patterns closely, you will realize that someone must have thought very carefully about the layout. You may even wonder if the grids of a c-Si solar cell are similar to the veins of a leaf. In this chapter, we will show that the grid patterns indeed involve a combination of geometrical and optoelectronic optimization to reduce the series resistance loss of a solar cell. We will begin with the grid layout and cell topology of thin-film solar cells. Once you are comfortable with the basics, it will not be difficult to understand the slightly more complex physics of c-Si gridding.

10.2 You cannot avoid the losses due to series resistance

Consider a thin-film solar cell (*length* W and *width* L, area of $A_T = W \times L$) being illuminated by sunlight. If the output voltage is V and the current I, then the power output is $P_{out} = V_{oc} \times I_{sc} \times FF \equiv C_0$. If you cut (typically by laser) this large-area cell into N segments and reconnect them in series (Fig. 10.1(b)), then the current of each subcell will reduce to I_{sc}/N, while the voltage produced will increase to $V_{oc} \times N$, so that the net power output remains the same, i.e., $P_{out} = (I_{sc}/N) \times (V_{oc} \times N) \times FF \equiv C_0$. Since FF does not depend on N in the thermodynamic limit, it is easy to see that the power output is independent of N. In practice, the laser cut needed to isolate the subcells has a width δ. The fraction of the total active area (A_T) lost to laser cuts reduces the output power by $P_{area} = (N - 1)(\delta \times L/A_T)P_{out}$. This calculation suggests that cutting a large cell in N pieces and then connecting them does not offer any advantage — if anything, the P_{out} is diminished. Then why is subdividing a solar cell into subcell (i.e., stripping) so ubiquitous? The answer is that dividing a cell reduces power lost to series resistance, as follows.

Assume a non-rectangular cell in Fig. 10.2(c) with the contact at the top shown by the thick green line. Electron hole pairs are generated at each point of the solar cell, resulting in a locally generated current density J_0 (A/m^2) at every point (x_0, y_0) of the cell — this is the small box indicated by current J_0 in Fig. 10.2(c). These electrons must travel vertically from its generation point (y_0) to the top electrode $g(x)$ before being collected. Therefore, the current at (x, y) is not just the local current $J_0 dxdy$, but the sum of all the current generated at $y_0 \leq y$, so that $J(y) = J_0 \times y$ (A/m). The total resistive power loss is obtained by calculating the Joule heating at each point along a vertical stripe in the y-direction, and then summing over the vertical stripes along the x-direction, namely,

$$\begin{aligned} P_\rho &= \int_0^L dx \int_0^{g(x)} dy (J_0 \times y)^2 \rho \\ &= J_0^2 \rho \int_0^L dx g^3(x)/3, \end{aligned} \tag{10.1}$$

where the area of each cell is defined by $A = \int_0^L g(x)dx$. This equation shows, as long as there is a finite 'ρ', there will be a resistive loss within each cell.

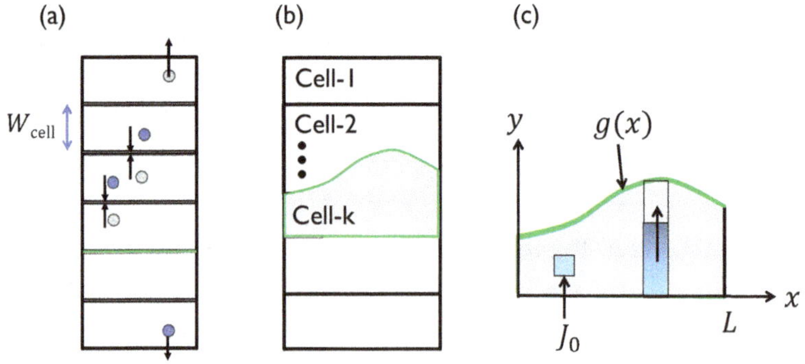

Figure 10.2: (a) The thin-film module is divided into segments, and reconnected in series. This striping reduces the distance traveled by the electrons and holes before they are collected by their respective contacts. (b) The thin-film module is shown to be divided into arbitrarily shaped cells. (c) Current flow within one of the segments. The shape of a subcell need not be rectangular, but for a fixed L and subcell area A, a rectangular cell minimizes the resistive loss, P_ρ.

10.3 For typical subcells, a rectangular grid minimizes power dissipation

The equation above applies to any arbitrarily shaped subcells constrained by the cell area, A. For example, $g(x) \equiv ax^n$, with $a = (n+1)A/L^{(n+1)}$ ensures that the cell area is preserved for all n so that the amount of sunlight absorption is the same. By inserting this $g(x)$ in Eq. (10.1), we find

$$P_{\rho,cell}(n) = J_0^2 \rho \times \frac{(n+1)^3}{3(3n+1)} \times \frac{A^3}{L^2}. \tag{10.2}$$

As shown in Fig. 10.3, if we plot $P_1 = P_{\rho,cell}(n)/(J_0^2\rho)$ as a function of n, the energy loss is minimized for rectangular cells with $n = 0$, or $g(x) = A/L = W_{cell}$. No wonder we find the solar cells are laser-scribed into rectangular subcells. Let us conclude this section with three observations: First, the analysis above assumes that the length of the subcell, L, equals the width (L) of the module — this is typical, but not universal. We will see the implication of violating this rule later. Second, we have focused only on the top electrode–current. The current will flow through the back electrode as well, and therefore, there is a similar power loss at the back (with resistivity ρ_b) that must be added to the front-contact power loss discussed above. For a traditional monofacial solar cell, $\rho_b \ll \rho$, we can therefore neglect the back-contact resistive loss. However, it may not be the case for interdigitated back-contact (IBC) or bifacial solar cells. Finally, the derivation does not tell us what the value of W is, because to determine $A = A_T/N$, we must find N by minimizing the power dissipated for the entire module. Here, A_T is the area of the module. We will discuss the topic in the next section.

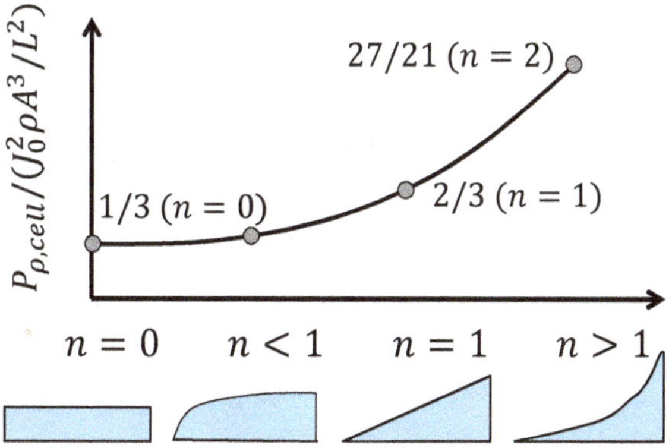

Figure 10.3: Resistive power loss plotted as a function of the shape of a subcell characterized by exponent n.

Homework 10.1: Lagrange optimization and optimum shape of a subcell

In the discussion above, the conclusion that the rectangular cells minimize power dissipation was based on special shapes defined by a power law, i.e., $g(x) = ax^n$. Show that the conclusion is general, as long as the current paths can be presumed approximately vertical.

Solution. Since Eq. (10.1) applies to arbitrary shaped cells, we can use Lagrange ($\mathbf{L}$) optimization to find $g(x)$ that minimizes P_ρ subject to the area constraint:

$$\mathbf{L} = J_0^2 \rho \int_0^L dx \frac{g^3(x)}{3} - \lambda^2 \left(\int_0^L g(x)dx - A \right) \tag{10.3}$$

where λ is the Lagrange multiplier. Setting $d\mathbf{L}/dg = 0$, we find $g(x) = 2\lambda/(J_0^2\rho)$, so that $A = \int_0^L g(x)dx = 2\lambda L/(J_0^2\rho)$, implies that $\lambda = (A/L) \times (J_0^2\rho/2)$ is a constant. Since J_0 and ρ are constants, $A/L = W$ must be a constant. The result implies that rectangular cells minimize power dissipation — no wonder all the cells we see in practice are rectangular.

10.4 Determining the number of subcells in an optimization problem

We now know that the rectangular cell minimizes the loss in each cell. How many of such cells should we have in a given module? If the module is divided into N cells, the total power P_{out} can be written as follows:

$$P_{\text{out}} = P_{\text{ideal}} - P_{\rho,T}(N) - P_{\text{scribe}}(N), \tag{10.4}$$

where, $P_{\text{ideal}} = C_0$ is the power delivered to the load if the power lost to scribe lines P_{scribe} and total series resistance of the module $P_{\rho,T}$ are negligible. In practice,

$$\begin{aligned} P_{\rho,T}(N) &= P_{\rho,cell}(n) \times N \\ &= (J_0^2\rho) \times \frac{(n+1)^3}{3(3n+1)} \times \frac{A^3}{L^2} \times N. \end{aligned} \tag{10.5}$$

Since $A = A_{\text{T}}/N$, therefore, $P_\rho = C_1/N^2$, where

$$C_1 \equiv (J_0^2\rho) \times \frac{(n+1)^3}{3(3n+1)} \times \frac{A_{\text{T}}^3}{L^2}. \tag{10.6}$$

Similarly,

$$P_{\text{scribe}} = C_2(N-1) \tag{10.7}$$

where $C_2 \equiv (L\delta/A_{\text{T}})C_0$ with a width δ lost to each scribe line. Inserting these expressions in Eq. (10.4) and maximizing the power as a function of N, we find that

$$
\begin{aligned}
N_{opt} &= (2C_1/C_2)^{1/3} \\
&\equiv (n+1)\frac{A_\mathrm{T}}{L}\left(\frac{2/3}{3n+1}\right)^{1/3}\left[\frac{J_0^2\rho A_\mathrm{T}}{C_0\delta}\right]^{1/3}. \qquad (10.8)
\end{aligned}
$$

The formula explains why the number of subcells depends on the solar cell technology (e.g., Si, CdTe, etc.) as well as the efficiency ($\eta \propto C_0/A_\mathrm{T}$) of the cell. Remember that the module is subdivided into N cells along the module length W, i.e., $W_{cell}N = W$. Thus,

$$
\frac{N_{opt,1}}{N_{opt,2}} = \frac{W_{2,cell}}{W_{1,cell}} = \left[\frac{(J_{0,1}^2\,\eta_2)}{(J_{0,2}^2\,\eta_1)}\right]^{1/3}.
$$

In other words, cells with higher short-circuit current (i.e., smaller bandgap) will lose less energy with wider subcells. Table 10.1 summarizes the optimum for several technologies (assuming that the cells have the same ρ and same shape, i.e., characterized by the same n).

Table 10.1: Cell width W_{cell} of various thin-film solar cells.

Technology	E_g	J_sc	W_{cell} (cm)
c-Si	1.1	40	0.52
a-Si	1.7	16	1.18
CIGS	1.5	35	0.75

Thick solar cells, such as c-Si, are difficult to subdivide into subcells. Therefore, we will see later that we will need to grid the c-Si solar cells differently.

10.5 Rectangular cells are typical, but non-rectangular cells can reduce module power dissipation

Modern solar modules are rectangular. The analysis in the preceding section assumed that the length of a subcell equals the width of the module L. In that case, a rectangular subcell minimized power dissipation and maximized the power output. In practice, a rectangle can accommodate N equal-area subcells of a variety of geometrical shapes. The area of the subcells must be equal to ensure current continuity from one cell to the next.

Figures 10.4 and 10.5 illustrate three examples: rectangular cells, cells with triangles, and cells with discrete Cantor transform. All the cells have the same area (can you convince yourself that this is indeed the case?), but the resistive power dissipation is not the same. Let us calculate the power dissipation to see if these nontraditional cells would actually perform better.

Homework 10.2: Using a module simulator to estimate the power output of a thin-film solar module

1. Login to nanohub.org ... Go to the link: `https://nanohub.org/resources/pvpanelsim`
2. Download the user manual posted at https://nanohub.org/resources/12259/download/Users_Guide.pdf
3. Work out the default example to understand the power output and 2D power generation profile.
4. Change the grid spacing and the width of the metal lines to find the optimum power output. How does the results compare with the analytical results discussed above?

We have already calculated power dissipation of a module divided into N cells of "power law" shapes, characterized by exponent n, and given by Eq. (10.5). Going back to a single-cell or subcell power dissipation in Eq. (10.2), for rectangular subcells ($n = 0$), we find that

$$P_{\rho,cell}(n = 0) = J_0^2 \rho \frac{A_{\mathrm{T}}^3}{3L^2} \tag{10.9}$$

where L is the width of the cell (and the module).

For a triangular subcell shown in Fig. 10.4(b), the widths of the subcells are different, but the area of the cells are exactly the same! The N triangular subcells were created as follows. First, mark the outer edge of the module by $2N$ points, separated by $\Delta \equiv (W + L)/N$, and connect these points to the center with straight lines. The triangles created (half-cells) have equal area, $A = \Delta L/2$ because they share the same base ($L/2$) and height ($2W/N$), provided $L = W$. A pair of half-cells (reflected over the diagonal) forms a single subcell, and the N subcells so formed are connected in series and carry the current in two parallel branches, as shown in Fig. 10.4(b). Unlike typical rectangular cells, the "electrical edges" (marked by dark black lines in Fig. 10.4) do not align with the physical edges of the module. Rather, the current enters through the edge that connects the center to the bottom-right corner and exits through the edge that connects the center to the top-right corner. The edges must be separated by a hole in the center that is large enough to prevent any arcing between the electrodes.

Unlike rectangular cells, power dissipated in the triangular subcells depends on their orientations. For subcells with bases along the x or y axes have a length

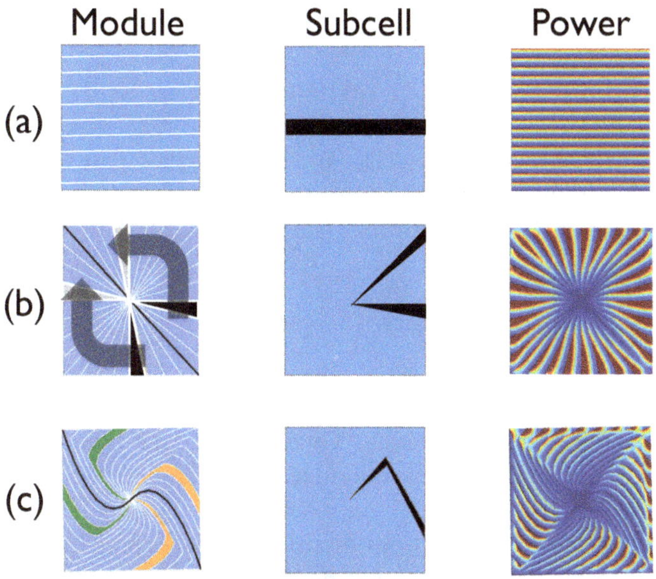

Figure 10.4: Power dissipation in non-rectangular cells. Rows marked (a), (b), and (c) depict thin-film modules with rectangular, triangular, and spiral subcells, respectively. The first column shows the module, the second column shows individual subcells within the module, and the third column plots the power dissipation in each subcell within the module. The darker regions define areas with higher power dissipation. While all the rectangular subcells have the same power dissipation, this is not true for triangular and spiral cells.

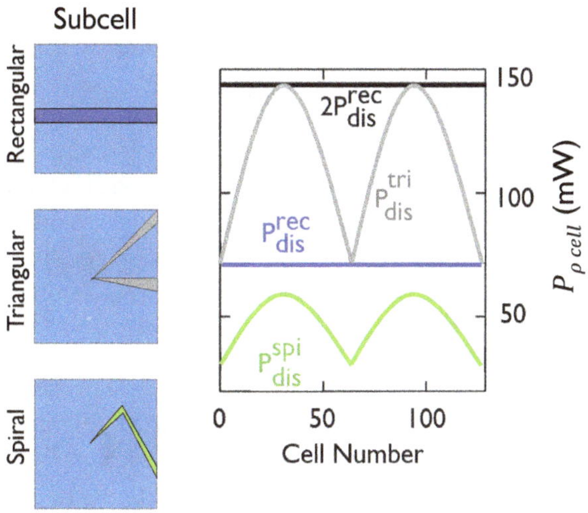

Figure 10.5: Power dissipation associated with various subcell shapes. Rectangular cells (blue solid line), triangular subcells (sinusoidal light black line), and spiral subcells (sinusoidal green line).

$L_{\text{cell}} = L$ and $n = 1$. Therefore,

$$P_{\rho,cell}(n = 1) = J_0^2 \rho \frac{2A_{\text{T}}^2}{3L^2},$$

roughly twice that of rectangular cells. We already knew that this would be the case for $n = 1$ based on Fig. 10.3.

For subcells along the 45 degree axis (see Fig. 10.4(b)), however, $L_{\text{cell}} = \sqrt{2}L$ and $n = 1$, so that

$$P_{\rho,cell}(n = 1, L_{\text{cell}} \to \sqrt{2}L) = J_0^2 \rho \frac{A_{\text{T}}^3}{3L^2},$$

same as the rectangular subcells. Other triangular subcells will dissipate power between the two limits (see Fig. 10.5). Overall, however, the power dissipation will be 25–50% more than that in rectangular cells. Although we did relax the rule $L_{\text{cell}} = L$ for the subcells, it did not help the situation much!

Does it mean that rectangular cells dissipate the least amount of power? Not really! You can design the subcells by using discrete Archimedean spirals. We will describe the algorithm to generate the cells a little later, but first let us examine how the cells look and how much power is dissipated in these cells.

Figure 10.5 shows that, similar to the triangular subcells, the spiral subcells have identical area, but their lengths and widths are unequal. On average, $L_{\text{cell}} = (\pi/2)L$ and shape exponent is $n = 2/3$. Therefore, you can use Eq. (10.2) to show that

$$\frac{P_{\rho,cell}(n \sim 2/3, L_{\text{cell}} \to (\pi/2)L)}{P_{\rho,cell}(n = 0, L_{\text{cell}} = L)} < 1.$$

Homework 10.3: Creating a discrete Archimedean grid

Figure 10.6 compares the methodology of generating a standard vs. discrete Archimedean spiral. Explain the procedure clearly with reference to a traditional Archimedean spiral. Show that you can apply the same algorithm to a rectangle to create a discrete Archimedean spiral. Also convince yourself that the area of the classical and discrete spirals produce subcells with an equal area.

Solution. For the standard Archimedean spiral, a circle of radius R is divided into a set of concentric circles, where the radius of the i-th circle is iR/N, as in Fig. 10.6(a). Next, the circle is divided into $2M$ sectors, by running M diagonals through the center. The intersection points of the radial lines and concentric circuits are used to form the spiral by connecting successive intersection points, as shown in Fig. 10.6(b). As M and N approach infinity, the piecewise line becomes a smooth spiral curve.

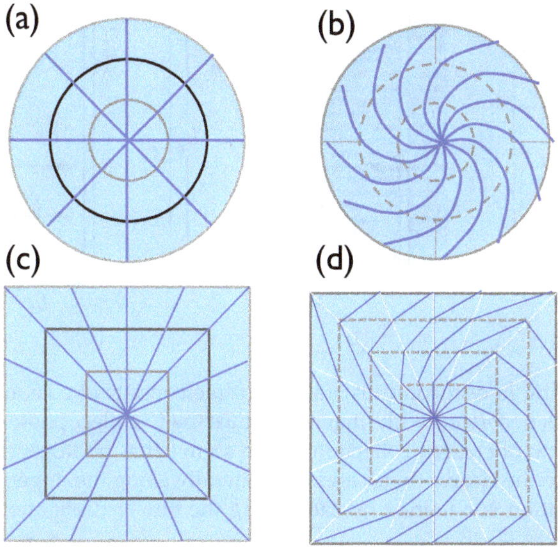

Figure 10.6: (a) A traditional Archimedean spiral is constructed by first dividing an area into concentric circles. (b) Next, radial lines are drawn from the center of the circles. The intersection points are joined to form the final spirals (blue line). (c) Likewise, a rectangular spiral is generated by dividing the squares into concentric squares and drawing the radial lines through the center. (d) The intersection points are joined by the same algorithm used to generate the traditional spirals.

In other words, if a module is scribed by a discrete Archimedean spiral, it will dissipate roughly half the power of a rectangular cell (see the oscillating green line in Fig. 10.5). These cells are more shadow-tolerant, a concept we will return to in the next chapter.

10.6 Crystalline Si solar cells are gridded differently than thin-film solar cells

The grids in c-Si and thin-film cells appear similar and they both serve the same purpose of reducing the overall resistive losses in the top and bottom electrodes. On closer inspection, however, we find that they are very different, see Fig. 10.7. First, the subcells of a c-Si module are slices of c-Si wafers, and the subcells are connected in series to obtain the desired output voltage, as in Fig. 10.7(a). Each subcell has a set of gridlines Fig. 10.7(b). These grids in the silicon cell do not divide the modules into a set of subcells; therefore, unlike thin-film cells, the output voltage and current do not scale with the grid number. Rather, similar to the veins in a leaf or small roads merging into highways, the gridlines are arranged hierarchically: the photogenerated carriers first reach the metal fingers vertically, then they flow along the finger until they reach the thicker metal lines, sometimes

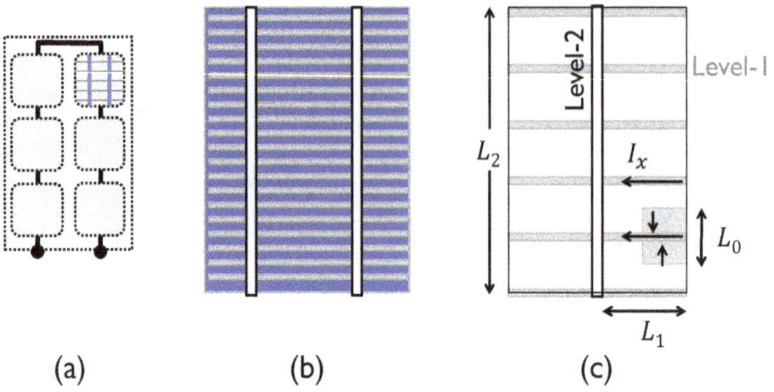

Figure 10.7: Gridding in c-silicon cells. (a) A module consists of a set of series-connected cells. (b) Each cell is decorated with fingers and busbars to collect photogenerated current. (c) A schematic view of current collection which shows that photogenerated carriers at a given point move to the closest finger and flow laterally along the finger. By symmetry, the total power dissipation is obtained by calculating the power dissipation in each unit cell (dash colored box).

called busbars (see Fig. 10.7(c)). In principle, these second-level metal lines could merge into third-level metal lines, and so on. In calculating the c-Si resistive loss, we will need to add up the losses of current flow along the metal lines.

Let us begin by calculating the losses as the electrons travel to the level 1 electrodes, see Fig. 10.7(c). Equation (10.1) calculates the loss of a region of width L and height $g(x)$ as being $P_\rho = J_0^2\rho\int_0^L dxg^3(x)/3$. Therefore, for a rectangle of width $L = L_1$ and height $g(x) = L_0/2$,

$$P_0 \equiv 2P_\rho = 2 \times J_0^2\rho\frac{L_1L_0^3}{24} = J_0^2\rho\frac{L_1L_0^3}{12}. \tag{10.10}$$

The prefactor 2 in Eq. (10.10) arises from the fact that the current in metal 1 is supplied by the two neighboring subcells for current collection. The total loss over the module is $P_{0,T} = MP_0$, where the M is the number of cells with size $L_0 \times L_1$, namely $M = \frac{L \times W}{L_0L_1}$.

Once the electrons are collected by level 1 metal contacts, they will travel along it. The level 1 subcell is shown in Fig. 10.8. The current at any point x involves all the electrons collected up to that point, namely,

$$I_x = 2 \times \int_0^x dx \int_0^{g(x)} J_0 dy = J_0 h(x), \tag{10.11}$$

where $h(x) \equiv 2 \times \int_0^x dx \int_0^{g(x)} dy = xL_0$. In general, the level 1 metal can have a non-uniform cross section $f(x)$. For now, if we assume the cross section of level 1 metal is $f(x) = a_1$ and its resistivity is ρ_1, the power dissipated along level 1 metal

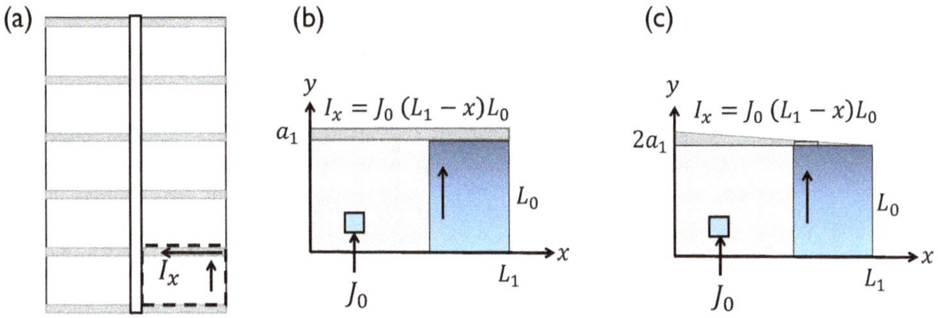

Figure 10.8: (a) A schematic of a gridded c-Si cell. Power dissipation in c-Si grid can be calculated first by calculating energy dissipation for vertical current flow and then lateral current flow along the fingers. (b) Shows fingers with constant width, and (c) shows fingers width increasing width.

is given by

$$P_1 = \int_0^{L_1} I_x^2 \rho_1 dx/f(x) \tag{10.12}$$

$$= \int_0^{L_1} J_0^2 \rho_1 h(x)^2 dx/f(x) \tag{10.13}$$

$$= \frac{J_0^2 \rho_1}{a_1} \times \frac{L_0^2 L_1^3}{3} \tag{10.14}$$

$$= (J_0 A_1)^2 \times \frac{R_1}{3}. \tag{10.15}$$

In deriving Eq. (10.15), as we have assumed a wire with constant cross section, i.e., $f(x) = a_1$, the wire resistance is given by $R_1 = \rho_1 L_1/a_1$. There are $M(\equiv A_{\text{module}}/A_1)$ metal lines, each associated with $A_1 = L_1 \times L_0$ unit cell. Therefore, $P_{1,T} = P_1 A_{\text{module}}/A_1$.

In general, in the k-th level metal, the power dissipation is

$$P_k = \frac{1}{3}(J_0 A_k)^2 R_k, \tag{10.16}$$

where $A_k = L_k \times L_{k-1}$, and R_k is the resistance of the k-th metal level.

Now that we know the power dissipation in grids with rectangular fingers and busbars, there are two ways to design the grids, namely, (i) not considering and (ii) considering the shading loss due to the metal lines.

Homework 10.4: The finger shapes can be optimized to reduce resistive losses

For the triangle-shaped fingers shown in Fig. 10.8(b), show that the power dissipated is given by

$$P_1 = \frac{1}{4}(J_0 A_1)^2 R_1. \tag{10.17}$$

Assume that the finger thickness ($a_1 \ll L_0$). How does the result compare to Eq. (10.15)?

Solution. Since the fingers are thin, let us assume that the current collection is not affected. Therefore, the current at a distance x from the busbar is given by $I(x) = J_0(L_1 - x)L_0$. Since the finger height is h_1 and the finder width is increasing linearly toward the busbar, the cross section at x is $a(x) = 2a_1 h_1(L_1 - x)/L_1$. Therefore,

$$\begin{aligned} P_1 &= \int_0^{L_1} dx \frac{\rho_1}{a(x)} I^2(x) \\ &= \frac{1}{4}\frac{\rho_1 L_1}{a_1 h_1}(J_0 L_1 L_0)^2 \\ x &= \frac{1}{4} R_1 (J_0 A_1)^2. \end{aligned} \tag{10.18}$$

If k-th level metal grid is also triangular, then we can write in general that

$$P_k = \frac{1}{4}(J_0 A_k)^2 R_k. \tag{10.19}$$

In general, for the same wire volume, triangular fingers have a smaller prefactor (1/4) compared to that of rectangular fingers (1/3). One must balance the performance improvement against the challenges in reliably manufacturing the thin edges of the triangular grids.

10.6.1 A simple constant-dissipation approach specifies the ratio of metal grids

The first approach simply requires that each metal level dissipate the same amount of energy. In other words, the $(i + 1)$ metal level (e.g., busbar) must dissipate the same amount of energy as all the i-th metal levels (e.g., fingers) that feed the $(i+1)$ layer, i.e.,

$$P_{i+1} = n_i P_i.$$

Since $n_{\mathrm{i}} \equiv L_{i+1}/L_i$ by inspection, therefore

$$\frac{L_{i+1}}{L_i} \times \frac{1}{3} r_{\mathrm{i}} \times (J_0 A_i)^2 = \frac{1}{3} R_{i+1} \times (J_0 A_{i+1})^2 .$$

Using Eq. (10.15), we find that

$$\frac{L_{i+1}}{L_i} \times \frac{1}{3} \frac{\rho_i L_i}{a_i} \times (J_0 L_i L_{i-1})^2 = \frac{1}{3} \frac{\rho_{i+1} L_{i+1}}{a_{i+1}} \times (J_0 L_{i+1} L_i)^2$$

which simplifies to

$$\frac{a_{i+1}}{a_i} = \left(\frac{L_{i+1}}{L_{i-1}}\right)^2 . \tag{10.20}$$

For example, if we know L_0 and L_2, we can determine the ratio of the cross-sectional area for the fingers (a_i) and busbars (a_{i+1}). Simply put, the busbars are wider than the fingers because they carry more current.

10.6.2 An optimized grid design must balance module-wide shading and power dissipation

We have already seen that the total power dissipation for a specific metal level in the gridding hierarchy is given by $P_{i,T} = P_i \times A_{\text{module}}/A_i$. Summing up the power dissipated in the semiconductor and the metal levels 1 through K, we find that

$$P_{\rho,T} = \sum_{i=0}^{K} P_i A_{\text{module}}/A_i . \tag{10.21}$$

Remember that, $A_i = L_i L_{i-1}$. A typical grid consists of fingers and busbars, so that $K = 2$.

From our discussion about thin-film solar cells, we know that the shadow of the grid reduces the short-circuit current. Therefore, we can optimize the total power output by writing an equation similar to Eq. (10.4). The optimization is considerably simplified if one requires the power dissipated at each metal level to be the same (ensured by appropriately increasing the width of the metal lines at each successive level), so that for a wafer gridded to the k-th level,

$$P_{\rho,T} = K P_0 \frac{A_{\text{module}}}{A_0} \tag{10.22}$$

so that the problem reduces exactly to that of Eq. (10.4), except that $P_{\rho,T}$ increases by a factor of K, and P_{scribe} increases by the additional shadow loss associated with the increasing width of the successive metal layers, i.e., $\sum_1^k a_i L_i$. It is a good exercise to work out the optimization problem.

10.7 c-Si mimics a thin-film cell: The Q-cell strategy

We now appreciate that, compared to a thin-film PV, the c-Si cell has additional resistive loss as electrons travel through the network of grids to the exit point of the substrate. The loss increases with area. Therefore, if each of the N cells in a module is segmented into Q smaller pieces and these segmented cells are directly connected to each other by highly conducting ribbons (see Fig. 10.9), the busbars will be shortened by a factor of Q. Moreover, the smaller cells will reduce the photocurrent that flows along the busbar. Taken together, the busbar resistive loss will reduce significantly.

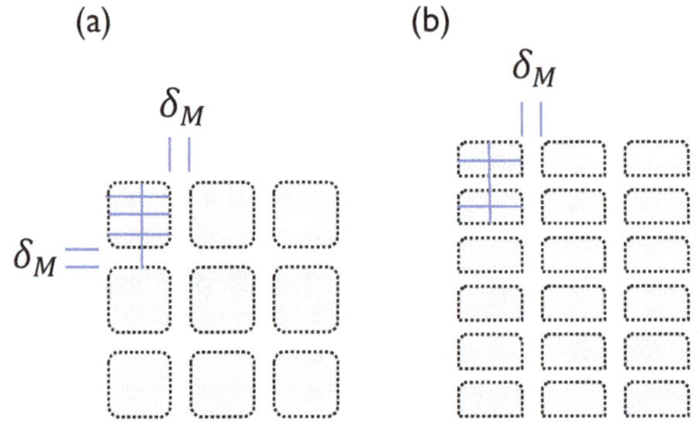

Figure 10.9: Halving the cells and reconnecting them by high conducting ribbons reduces series resistance loss. (a) The original design with $N = 9$ and $Q = 1$. (b) The half-cell design with $N = 9$ and $Q = 2$. Any gap between the halved cells, however, will increase the dead-area loss, with the corresponding loss of photocurrent.

We can write the module output power by subtracting the various resistive losses. Note that the resistive losses due to semiconductor series resistance (P_0) and the resistive loss (P_1) and shadow loss ($P_{s,1}$) due to the level 1 metal grid remain unchanged because segmentation does not affect these loss components. The segmentation approach, however, will reduce the level 2 metal (busbar) loss $P_2(N, Q = 1)$ by a factor of Q^2, but increase the shadowing loss by a factor of Q. Therefore, we can write

$$P(M, N, Q) = C_0(M) - C_1 Q - C_2/Q^2. \tag{10.23}$$

Here, C_0 accounts for the net output power, including the semiconductor and level 1 metal losses, $C_1 = C_0 \times L \times \delta_M / A_T$ accounts for the additional shadowing loss, and $C_2 = P_2(N, Q = 1)$ accounts for the resistive loss in the original module (before segmentation). As in the case in thin-film solar cells, the loss is minimized for $Q_{opt} = (2C_1^*/C_2^*)^{\frac{1}{3}}$. The relatively large spacing encourages low Q. In practice, each wafer is divided into 2 or 3 pieces and then arranged in 60 or 72 fragments.

Homework 10.5: Design of metal grids for bifacial solar cells

A bifacial solar cell accepts sunlight from the top directly (1-sun, say) and from the bottom through the albedo reflected light (R-sun, say). Read the paper by Yubo Sun *et al.*, "Tailoring Interdigitated Back Contacts for High-Performance Bifacial Silicon Solar Cells," published in *Applied Physics Letters*, 114.10 (2019):103901, to demonstrate that the shadowing and series resistance losses define a fundamental design constraint for an interdigitated back-contacted (IBC) solar cell designed for bifacial solar cell operation. In particular, given an albedo R_A, the ratio of the power output to power loss due to shading and resistive losses (c), the gap-to-period ratio (w) of the metal grid is given by $w = 1 - (1 + R_A)\,(R_A(R_A + c))^{-1/2}$, with the corresponding power output of $P \propto (1 + R_A)(1 - 2\sqrt{R_A/c})$.

10.8 More complex grids must be designed by computer simulation

The analytical theory discussed above explains the motivation and defines the principle of gridding a module. Unlike thin films, c-Si grids cannot be optimized analytically without some additional constraints. Moreover, the optimization assumes that the two electrodes are placed at the top and the bottom, respectively. In general, simulators such as Griddler allows one to explore a wide variety of grid geometries and find the structures appropriate for a particular application.

Some of the modern c-Si technologies are fabricated differently and therefore require a different gridding strategy. For example, interdigitated back-contact (IBC) solar cells have both n- and p-contacts at the back, obviating the need to consider front-grid shading. Two other technologies (i.e., emitter wrap-through (EWT) and metal wrap-through (MWT)) have been introduced. Both technologies

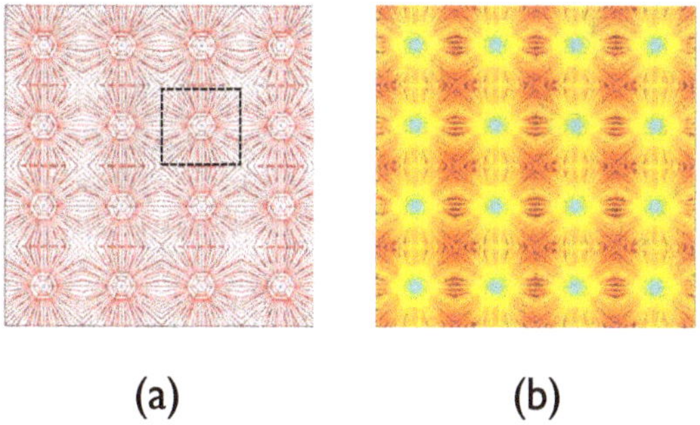

Figure 10.10: Numerical simulation tools, such as Griddler, can simulate complex grids. (a) Grid layout is provided as an input to Griddler. (b) Griddler returns a spatially resolved map of power dissipation at various regions of the cell.

eliminate front-grid busbars. For example, Fig. 10.10(a) shows the top view of a solar cells containing a collection of laser-drilled through-holes (indicated by the circles) and a set of fingers radiating from the holes. The photogenerated carriers are collected by the radial fingers, funneled to the nearest through-hole, and transported to the back of the solar cells. The shorter fingers reduce power dissipation (see Fig. 10.10(b)). Moreover, the elimination of the busbars increases short-circuit current by reducing shadowing, although we must account for the dead-area loss (and the processing cost) associated with laser-drilled through-holes.

Homework 10.6: Grid analysis in 'Griddler2.0'

Download Griddler2.0 from the web to simulate the performance of the following gridding scheme. Consider a c-Si cell with dimensions 156 mm by 156 mm.

- The front side has 60 fingers (width 100 μm) and two busbars (width 3 mm). Simulate the system to find the total energy dissipated.
- Change the number of fingers to 100, but reduce its width to 60 μm. Increase the number of busbars to 3, but reduce their width to 2 mm each. The change keeps the amount of metal needed and the shadowing of the grid essentially the same. Recalculate the performance of the cell.
- Redo both the calculations above by choosing the "half-cell" option.
- Of the three cells simulated, which one would have the lowest loss?

10.9 A summary of gridding principles of c-Si solar cells

- An optimized grid must balance the energy loss due to shadowing of the grids (similar to the area lost due to laser scribing for thin-film solar cells), the cost/benefit of hierarchical grids, and the shapes and sizes of the grids themselves.
- Analytical solutions exist for rectangular grids, but one must use numerical optimization for more complex grids. Indeed, since the problem is analogous to the generalized Poisson problem in two dimensions, the modern theory of complex variables can be used to solve them.
- For comparable systems, c-Si has higher resistive losses compared to thin-film solar cells.
- One must optimize for resistivity and transparency of the front electrodes. Although the resistive losses can be reduced by heavy doping and wider grid lines, the optical absorption and shadowing negate any benefit gained.

10.10 Emerging trends and the future of gridding

In addition to intriguing mathematics, the technology of gridding offers many interesting trends, summarized as follows:

- Typical c-Si grids are made of Ag, applied to the electrode surface by screen printing. Given the cost of Ag (we sometime use them for jewelery), there have been efforts to replace Ag with Cu. It remains to be seen if the cost benefit is significant enough to justify the switch.

- A grid overlay approach pre-embeds the grid lines into a thin transparent polymer host material. Once the solar cell is processed, the film is overlayed — the process obviates screen printing and provides a better control of the grid shapes, which may reduce the shadowing effects. For example, one may be able to use circular gridlines to scatter light into the system or use a reflector to trap light through internal reflection as shown in Fig. 10.11. A grid that casts no shadow — now that is really amazing!

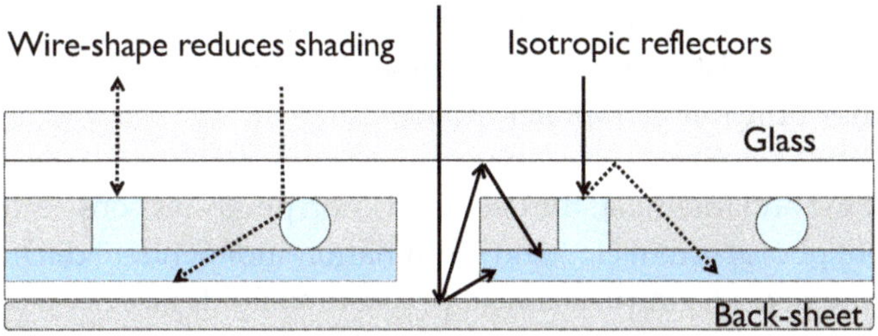

Figure 10.11: Light scattering by the wires with circular cross section and various internal reflections improve the light collection by the cell.

- Another recent trend to reduce metallic gridding and the associated shadow is to use high-conductivity transparent electrodes, such as ITO. Recently, graphene-wrapped Ag nanowire networks have been reported as a high-conductivity transparent conductors that do not require structured metal grids. The commercial adoption of these techniques is not guaranteed, but it is a topic of active research.

- Although a substantial amount of current flows through the grid, the current density is typically low; therefore, electromigration is seldom an issue. However, a metallic grid does pose significant reliability challenges. For example, the exposure to moisture that seeps through the polymer encapsulant and the corrosion of metal lines is a significant concern. We will discuss these reliability issues of solar cells in Part IV of the book. Second, broken metal lines or inhomogeneity in photocurrent causes significant redistribution of current in the grid and increases resistive losses. For c-Si cells, the inhomogeneity can be addressed by increasing the number of busbars. For thin-film solar cells, vertical laser stripes can isolate heavily shunted cells from contaminating adjacent regions within the same subcells. We will discuss the physics of reliability-aware gridding in Part IV of the book.

Homework 10.7: A c-Si solar cell can be described by a five-parameter model

We have discussed in detail the physics of dark and photocurrent (Chapter 7), shunt resistance (Chapter 9), and series resistance (Chapter 10). This information is sometimes summarized by a five-parameter model given by the following equation:

$$I = I_{\rm sc} - I_0(e^{q(V+IR_{\rm s})/nk_{\rm B}T(\rm D-1)} - (V + IR_{\rm s})/R_{\rm sh},$$

where $I_{\rm sc}, I_0, n, R_{\rm s}, \text{and} R_{\rm sh}$ are the five fitting parameters used to described the $I - V$ characteristics of a cell or a module (see Fig. 7.1). Explain the key assumptions of the five-parameter model.

Solution. First and foremost, the decomposition photo-response into a voltage-independent photocurrent ($I_{\rm sc}$) and illumination-independent dark diode current (with prefactor I_0) implies that superposition holds (see Sec. 7.7). This assumption is typically valid only for c-Si and other high-efficiency solar cells. The non-ideality factor $1 < n < 2$ effectively describes the combination of "injection-over-the-barrier" and the junction recombination current components of the diode. Sometimes the diode current is divided into two parts, with $n = 1$ and $n = 2$ factors. In this case, the five-parameter model includes the prefactor of the $n = 2$ term, instead of the fractional non-identity factor n. Incidentally, $n > 2$ is sometimes observed for heterojunction diodes with high injection barrier.

Second, the $R_{\rm sh}$ describes a linear, voltage-independent shunt resistance. We know from Eq. (9.3) that this is generally not the case. The temperature and voltage dependence of $R_{\rm sh} = dI_{\rm sh}/dV$ is not well represented by a linear resistance. Finally, the appearance of $R_{\rm s}$ in the exponential diode term is appropriate for solder bond failure, but does not adequately capture the distributed voltage drop in fingers (see Chapter 23, Homework 23.6). Nevertheless, we will see in Part IV of the book that the five-parameter model is widely used to describe the system and farm-level performance of solar cells. A variety of fitting algorithms to extract the five parameters have been reported in the literature.

10.11 Conclusions

In this chapter, we have discussed the elegant physics of stripping a solar cell to reduce the series resistance loss and to maximize power output. For mature technologies (e.g., c-Si, CdTe, etc.), innovation in module shapes, size, and stripping continues to play an important role. Now that the module design is complete, we are ready to install it in solar farms to harvest solar energy, the topic of the next chapter.

References

[1] Martin A. Green. *Solar cells: Operating principles, technology, and system applications.* Prentice-Hall, Englewood Cliffs, NJ, 1982.

[2] Sourabh Dongaonkar and Muhammad A. Alam. Geometrical design of thin film photovoltaic modules for improved shade tolerance and performance. *Progress in Photovoltaics: Research and Applications*, 23(2):170–181, 2015. _eprint: https://onlinelibrary.wiley.com/doi/pdf/10.1002/pip.2410.

[3] S. Dongaonkar and M. A. Alam. End to end modeling for variability and reliability analysis of thin film photovoltaics. In *2012 IEEE International Reliability Physics Symposium (IRPS)*, pages 4A.4.1–4A.4.6, April 2012.

[4] Deepak K. Gupta, Marco Barink, Yulia Galagan, and Matthijs Langelaar. Integrated Front–Rear-Grid Optimization of Free-Form Solar Cells. *IEEE Journal of Photovoltaics*, 7(1):294–302, January 2017. Conference Name: IEEE Journal of Photovoltaics.

[5] Henning Schulte-Huxel, Robert Witteck, Malte Ruben Vogt, Hendrik Holst, Susanne Blankemeyer, David Hinken, Till Brendemühl, Thorsten Dullweber, Karsten Bothe, Marc Köntges, and Rolf Brendel. Reducing the electrical and optical losses of PV modules incorporating PERC solar cells. *Photovoltaics International Papers, Module*, October 2016.

Part III

Design of a PV System: Panels, Farms, and Storage

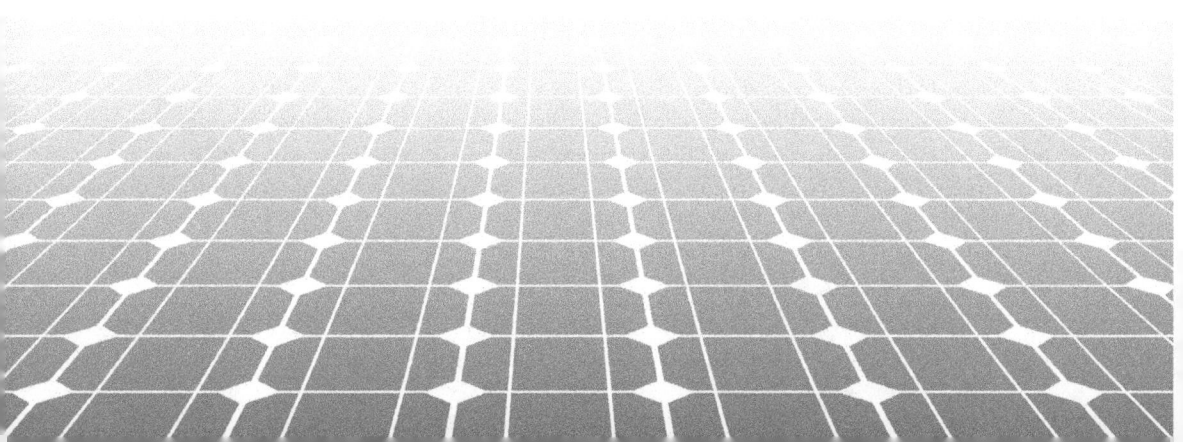

CHAPTER 11

System Integration of Solar Modules

Chapter Summary

- A PV system contains a variety of components, such as, solar modules, inverters, storage elements, maximum-power point tracker, electrical loads.
- Stand-alone systems are configured very differently compared to grid-connected systems.
- Inverters convert DC solar power to AC power suitable for appliances and power grids. Different types of inverters support different applications.
- The efficiency of the integrated system can be much lower than the individual components. The system must be optimized for the end applications.

11.1 Introduction

In the last chapter of Part II, we discussed how modules are designed to maximize light collection and minimize resistive losses to maximize the output power. In the simplest case, we would connect a single module to run a DC load. In general, however, solar cells must be able to run AC loads and be compatible with other sources of electricity. In this chapter, we will see how a collection of modules are configured along with additional electrical components to efficiently supply electricity to the consumer loads. The design and configuration vary depending on the scale of the system and connectivity to the power grid.

11.2 A PV system consists of a variety of electronic components

It is important to understand the basic functionality of the components of a PV system (see Fig. 11.1) before we integrate them into a system.

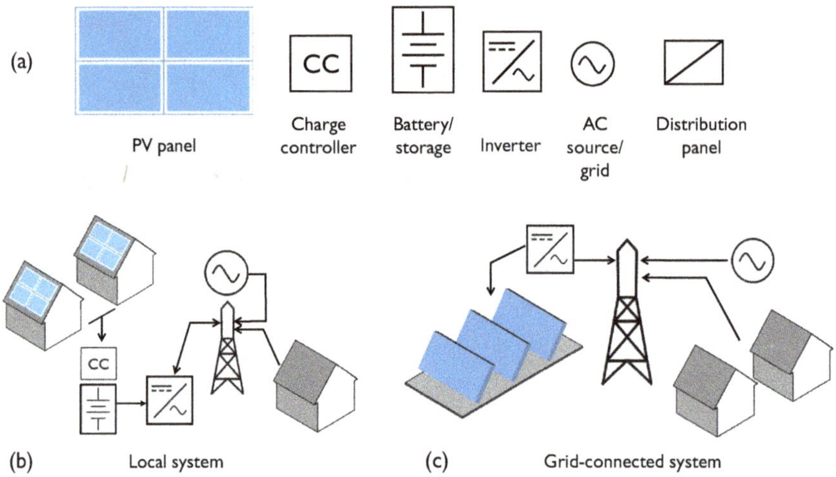

Figure 11.1: (a) Key components in a PV-connected system. (b) and (c) show the placement of these components within typical local and grid-connected systems.

1. **Cells, Modules, and Panels:** Wafer based solar cells are individually fabricated and then connected in series to form a module. Thin film cells are monolithically fabricated in a module configuration. Several modules are then connected to form a panel, see Fig. 11.2. These large panels are appropriately tilted, and mounted on fixtures for installations in roof-tops or solar farms.

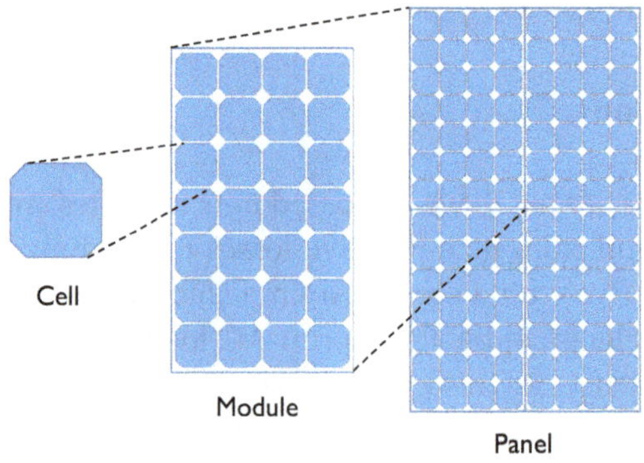

Figure 11.2: A cell to module to panel configuration.

Table 11.1: Efficiency of various components in a PV system.

Derate Category	Typical [2, 3]	Range [2]	Micro-inverter Systems [3]	See Chapter
PV module nameplate DC rating	0.950	0.85–1.05	0.950	
Inverter and transformer	0.920	0.89–0.94	0.965	
Mismatch	0.980	0.97–0.985	1.000	
Diodes and connections	0.995	0.99–0.997	0.995	
DC wiring	0.980	0.97–0.99	0.995	18
AC wiring	0.990	0.98–0.993	0.980	
Soiling	0.950	0.75–0.98	0.970	17
System availability	0.980	0.0–0.995	0.998	
Shading	1.000	0.0–1.00	1.000	18
Sun tracking	1.000	0.98–1.00	1.000	
Aging and degradation (corrosion, LID, PID, etc.)	1.000		1.000	20, 21, 22, 23, 24
Total derate factor (system efficiency)	0.770		0.861	
Storage efficiency:				15
(a) Li-ion battery		0.8–0.9		
(b) Electrochemical cell		0.4–0.8		
(c) Water pump		0.75–0.93		

2. **Charge controller:** A DC-DC converter is a power electronic device that transforms the DC voltage generated by the solar module to a voltage level that can be used by various DC loads. This is usually called a charge controller.

3. **Inverters:** An inverter converts the PV-generated DC power to AC either for local AC-loads such as air-conditioner, and/or to connect to the power grid. An inverter is an expensive component, second only to the module. There are several configurations for inverter connection to the PV panels for optimum output. We will discuss these inverter connections later in Sec. 11.4. A distributor allows switching between solar energy and the energy from standard power grid.

4. **Maximum Power Point Tracker:** A maximum power point tracker (MPPT) is an electronic device connected to the leads of a panel. It monitors the time-varying voltage and current output of a solar panel and adjusts the load to ensure that power output is maximized. Depending on the system configuration (discussed in the next section), MPPT can be a part of the charge controller or the inverter.

5. **Storage Components:** Depending on the PV-system, the panels may be connected to a storage/battery through charge controller. The charge controller prevents overcharging the battery, and discharging through PV panels at night. We will discuss various storage options in more detail in Chapter 15.

11.3 Types of systems

11.3.1 Standalone PV system

Standalone or off-grid systems supply power to loads exclusively from the PV source, see Fig. 11.3.

(a) Standalone PV without storage: These systems directly connect the DC loads to the PV source. For example, small devices such as calculators, watches, small water pumps, medical storage, bus stop lighting, etc., can be directly powered by a single PV module.

(b) Standalone PV with a storage: The PV panels can be connected to the battery (storage) and to the DC loads through a charge controller. AC loads can also run via an inverter connected to the battery. The battery stores energy when the net load demand is lower than the PV output and then supplies energy to the load while PV output is low, e.g., at night. Small houses and farms secluded from the main power grid can use such systems.

Figure 11.3: A standalone PV-powered system (a) without and (b) with energy storage.

11.3.2 Grid-connected solar modules

Larger number of PV panel arrays can be directly connected to the grid to contribute to the net grid power, see Fig. 11.4. Two types of configurations are of commercial interest:

(a) Home/rooftop and utility installations
In utility scale grid-connected systems, PV panel arrays are connected to an inverter and a distribution panel to either supply on-site AC loads or contribute to the power-grid. Unlike standalone systems, these inverters must

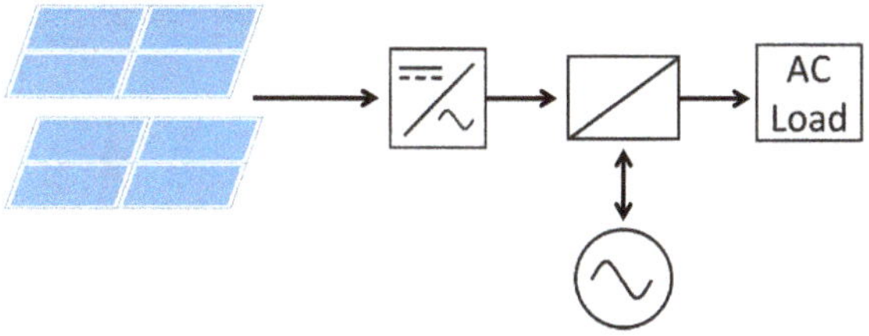

Figure 11.4: A PV system connected to the power grid.

be synchronized with the AC grid before connection. The distribution panel in this system is bidirectional so that for large on-site load-demand, power from both the PV and the grid can be fed to the on-site load. For example, systems with home/rooftop installations allow loads to run on combined PV and grid power during peak on-site demand when PV power generation is insufficient. At off-peak hours, excess power from the PV panels can be sold to the grid. For home/rooftop systems, shading from adjacent walls, trees, and electric poles must be avoided (see Chapter 18). That is why site analysis is required to determine panel tilt and shading from nearby structures before home/rooftop installations. Utility scale installations, such as at airport parking lots, electric car charging docks etc., operate in a similar fashion.

(b) Commercial power-grid connected systems

Large area PV panel arrays (solar farms) can work as power plants supplying energy to the grid. The first solar power plant Lugo in San Bernardino, California produced 1 MW electricity. The commercial solar farms have come a long way since then — currently, the largest farms can produce GWs of power. The website https://en.wikipedia.org/wiki/List_of_photovoltaic_power_stations contains a list of very large solar farms. We will see below that, despite the size of the farm involving millions of solar panels, the basic principle of operation and grid connection of these farms is nominally similar to home/rooftop systems. We will discuss in detail the design principles of solar farms in Chapters 12–14.

11.3.3 Hybrid connection

The standalone solar panels with a battery storage system can be combined with a backup electricity generation source to form a hybrid system, see Fig. 11.5. The backup source (e.g., diesel, gas, wind generator, or grid) is used to recharge the battery or run the AC loads when supply from the PV is low. The backup AC source is particularly useful when PV energy is inadequate in absence of grid connection.

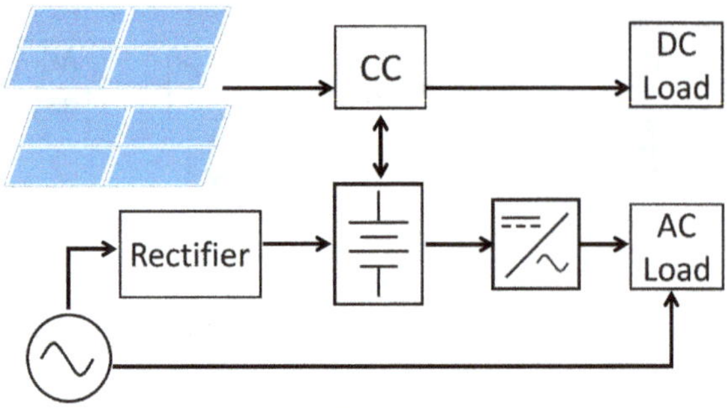

Figure 11.5: A hybrid power system configured with PV and another source running both AC and DC loads.

11.3.4 Microgrids

While standalone solar panels with storage systems are good for powering small households, it may not be appropriate for multiple homes. For locations (e.g., small village) with no grid connection, a PV plant may be designed with a local power grid and backup generator to form a 'micro-grid' to supply the local power demand, see Fig. 11.6. The micro grid system installation and maintenance should cost less than multiple standalone systems. The capacity of these systems is typically a few hundred kWs.

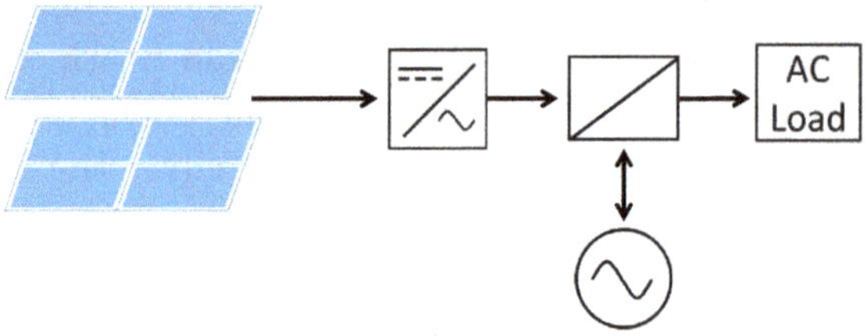

Figure 11.6: A micro-grid includes PV and another source powering a grid for a small community.

11.4 Inverter connection topologies

One of the most important power electronic components in the PV system is the inverter. Typically, inverters convert DC electricity to AC. However, the inverters used in PV systems often have a built-in MPP tracker. In this section, we will focus on inverter connection schemes, mainly for larger systems installed in buildings and solar farms.

11.4.1 Central inverter

In a large system, several modules are connected in series (i.e., in a string) to obtain sufficiently high voltages to operate various loads or to connect to the grid. The strings are then connected in parallel to increase the current. This complete array of series–parallel panels is then connected to a "central" inverter as shown in Fig. 11.7(a). In this configuration, bypass diodes and fuses are used within the strings and arrays so that the strings that are not operational can be disconnected from the system. However, the partial power generated from the operational modules in those disconnected strings cannot be recovered.

11.4.2 Micro-inverter

In an extreme case, each module (or a few modules) can be connected to a smaller inverter as shown in Fig. 11.7(b). In this configuration, a large number of micro-inverters are required, and the associated losses can accumulate. Obviously, micro-inverters operate at a much lower voltage compared to a central or string inverter. This system has the benefit of disconnecting only the specific modules that are non-operational. With the development of more efficient inverters, the micro-inverter system is gaining popularity.

11.4.3 String inverter

This approach combines the benefits of central and micro-inverter configurations by allowing inverters for each string (at $\sim$ 1 kV), see Fig. 11.7(c). String inverters allows each string to operate at its own MPPT — thus, this system can be more efficient than the central inverter configuration.

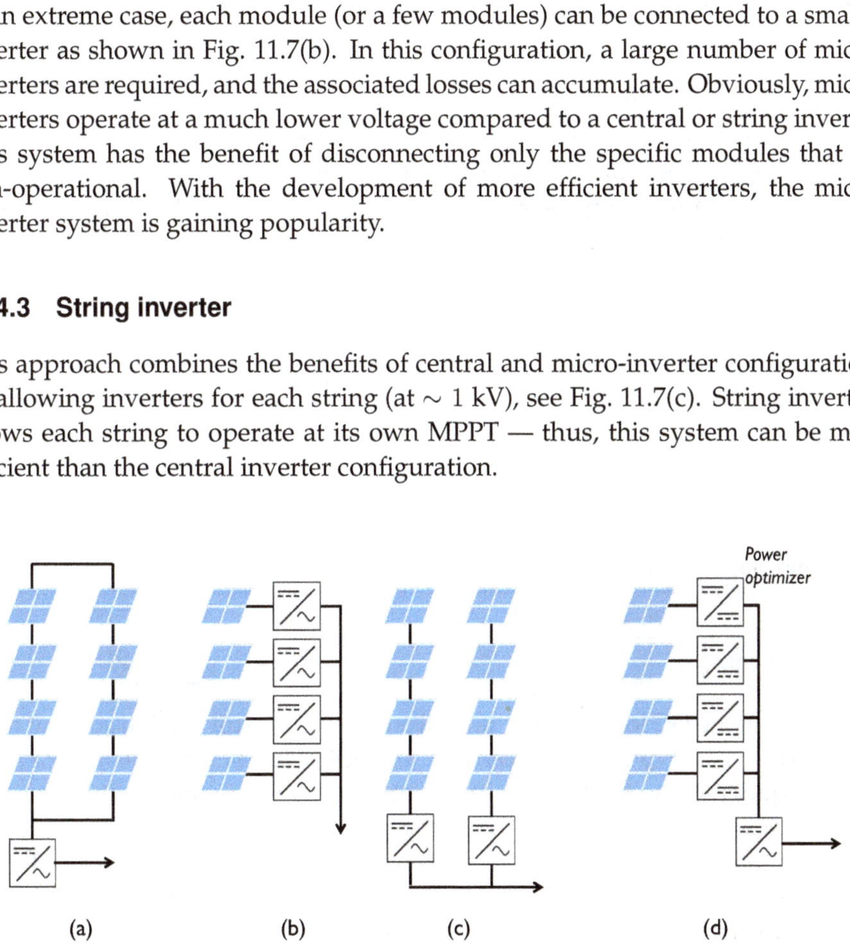

Figure 11.7: Inverter topologies for PV panel arrays: (a) central inverter, (b) micro-inverter, (c) string inverter, and (d) inverter with power optimizer.

Homework 11.1: Designing a standalone PV system requires a few simple considerations

We would follow three basic steps to design a standalone PV system:

1. Determine the load requirement (system voltage, Wh, Ah).
2. Battery size based on load and backup requirement.
3. Panel array current requirement and arrangement based on load requirement and local insolation conditions.

Here, we assume for simplicity that all equipments run with DC input with their own inverters when necessary.

1. Load requirement

Typical nominal voltages for a DC system are 12 V, 24 V, or 48 V. For our system, let us choose a 24 V system and use the equipments listed in the following table. Their ratings and daily usage are known. Therefore, we can estimate the daily energy usage (power rating × usage duration) and the corresponding ampere hour (energy usage/system voltage) as shown in the table below.

	Load information		Calculated requirement	
Load	Rating	Usage/day (h)	Wh/day	Ah/day @24 V
Lights ×5	8 W	4	160	6.67
Fan ×3	12 W	8	288	12
TV	24 W	10	240	10
Phone chargers ×3	1 Ah/day @ 5 V		15	0.625
Sum			703	29.295
Total = Sum×1.2			843.6	35.15

The sum of the DC loads is not the final answer — we still need to account for the system loss (typically between 20% and 30%). To calculate the total or net load requirement in the last row of the table, we have assumed a 20% system loss.

Note that, in this example, we have assumed that all the loads operate at DC. In case of an AC load, the calculated load (Wh or Ah) should be divided by the inverter efficiency (typically ∼95%).

(*continued on the next page*)

Homework 11.1 (*continued from the previous page*)

2. Battery size
If we assume a 3-day reserve capacity, then battery size should be $35.15 \times 3 = 105.45$ Ah. But then, should the panel be three times larger as well? No, because the panels are sized slightly larger than the daily power required so that the battery is gradually charged for the reserve days. With 25% additional capacity, the DC load requirement, including the reserve capacity, is $35.15 \times 1.25 = 43.94$ Ah.

3. Panel size and arrangement
We know that the standard 1-sun irradiance $I_{AM1.5}$ is 1 kW/m^2. For any locality, we can find the mean irradiance (from weather stations or databases) called GHI (I_{GHI} in kWh/m^2/day). We can therefore estimate the average or equivalent sun hours as $I_{\mathrm{GHI}}/I_{\mathrm{AM1.5}}$, i.e., essentially the value of GHI.

For example, with an equivalent sun hours of 4 h/day in a location, and a required DC load of 43.94 Ah/day (including charging for reserve capacity, as found earlier), the solar panel is required to produce 43.94/4 = 10.98 A.

Now, we would need to choose a module to set up the panel arrangement — say, we have a module with 18 V and 6.1 A at maximum power point. Therefore, to supply the required 10.98 A current, we need 10.98/6.1≈2 panels in parallel, and to attain 24 V system voltage, we need 24/18 = 1.33, i.e., 2 modules in series. Therefore, the panel arrangement is 2×2 of the chosen modules. If we had an MPP charge controller or smart junction box with buck-boost converters, then the 18 V panels would not have to operate at 12 V each, and the system would be more efficient.

11.4.4 Inverter with power optimizer

In recent years, power optimizers have been used at each module before utilizing string inverters. The power optimizers (also called module-level power electronics [MLPE]) pre-process the DC electricity at each module, reducing the effect of non-uniform power among modules, and then send it to the string inverter as shown in Fig. 11.7(d). Therefore, this system is highly efficient compared to the string inverter alone. The power optimizer works similar to the micro-inverter system; however, it is more affordable. Currently, manufacturers are integrating the power optimizer in the modules. These "smart modules" are more expensive, but they may reduce the installation time and cost significantly.

Typical efficiencies of the various components in the PV system are summarized in Table 11.1. Conventional inverter-based systems can be ~77% efficient, while micro-inverter systems can boost the system efficacy to ~86%. These do not include the soiling losses and module degradation (aging), which can further reduce the system efficiency.

11.5 Conclusions

In this chapter, we have described how a PV system is integrated to other sources and the power grid to support consumer-level loads. Depending on the application and scale of the overall power system, the PV system can be configured as a standalone, or grid-connected, or hybrid system. A micro-grid system is more common for small communities disconnected from the main power grid. We also briefly discussed the inverter connection schemes in panel arrays (e.g., for rooftops or solar farms). In the next three chapters, we will study how panels are arranged to form solar farms to maximize its power output.

References

[1] J. C. Arnett, L. A. Schaffer, J. P. Rumberg, and R. E. L. Tolbert. Design, installation and performance of the ARCO Solar one-megawatt power plant, pages 314–320, 1984.

[2] B. Marion, J. Adelstein, K. Boyle, H. Hayden, B. Hammond, T. Fletcher, B. Canada, D. Narang, A. Kimber, L. Mitchell, G. Rich, and T. Townsend. Performance parameters for grid-connected PV systems. In *Conference Record of the Thirty-first IEEE Photovoltaic Specialists Conference, 2005*, pages 1601–1606, January 2005. ISSN: 0160-8371.

[3] Enphase. Technical Brief: Guide to PVWatts Derate Factors for Enphase Systems (EN-US). Technical report.

[4] Klaus-Dieter Jäger, Olindo Isabella, Arno H.M. Smets, René A.C.M.M. van Swaaij, and Miro Zeman. *Solar energy: fundamentals, technology and systems*. 2016. OCLC: 1013815023.

[5] Antonio Luque and Steven Hegedus, editors. *Handbook of Photovoltaic Science and Engineering, Second Edition*. March 2011.

[6] Arno Hendrikus Marie Smets, Klaus Jäger, Olindo Isabella, René Adrianus Christianus Marinus Maria van Swaaij, and Miro Zeman. *Solar energy: The physics and engineering of photovoltaic conversion, technologies and systems*. UIT, Cambridge, England, 2016. OCLC: 960694728.

CHAPTER 12

Design of Solar Farms

Chapter Summary

- The sunpath determined by the geographical location and the season.
- It is easy to derive an empirical rule that relates the sunpath to the tilt of a single panel.
- No-shadowing condition on the longest day determines the row-spacing between the panels.
- The yearly energy yield depends on the amount of direct, diffuse, and albedo light incident on a solar cell. The row-to-row shadowing of a solar farm reduces per-module energy collection.
- A calculation technique called the view-factor method allows simple derivation of complex light incidence problems.

12.1 Introduction

In the previous chapter, we discussed the electrical configuration of various types of PV systems, and how they are connected to the electric grid and various types of storage schemes. The solar modules provided from manufacturers are arranged in larger panels or panel arrays in these PV installations. In the next three chapters, we will focus on the physical installation of the panels in a farm. We will answer questions such as: What should be location-specific tilt angle (β) of a stand-alone solar panel? How would the tilt angle change if the panels are installed in a farm? How far apart should the rows be? What is the maximum energy yield of such a farm? Is the farm cost-effective, given the prices of the panels, land, etc.? In this chapter, we will focus on simple, qualitative understanding of the basic considerations of a farm design. The next two chapters will focus on quantitative calculations.

12.2 The sun path depends on the geographic location of the farm

We need to know where the sun is (i.e., sun path) to orient the solar panel toward it. As we discussed in Chapter 1, insolation and sun path depends on the geographical location on the earth (e.g., latitude, L), and the day of the year (D_{n}). The sun path for a location on latitude L is defined by two angles: the elevation angle (θ_e), measured from the horizon, and the azimuth angle (γ_s), measured from the north towards the east (see Fig. 12.1(a)). From the figure, we see that the zenith angle is related to the elevation angle through $\theta_Z = 90° - \theta_e$.

The elevation angle is given by

$$\sin(\theta_e) = \sin(\delta)\sin(L) + \cos(\delta)\cos(L)\cos(\omega), \tag{12.1}$$

and the azimuth angle is given by

$$\cos(\gamma_s) = \frac{\sin(\delta) - \cos(\theta_e)\sin(L)}{\sin(\theta_e)\cos(L)}. \tag{12.2}$$

Here, δ is the declination angle (different for different day of the year), and ω is the hour angle. For example, on December 21, 3 hours before solar noon, $\delta = -23.5°$ and $\omega = -3 \times 360°/24 = -45°$. The location-specific instantaneous sun path along with the irradiance allows us to calculate the instantaneous energy yield of a solar panel throughout the year.

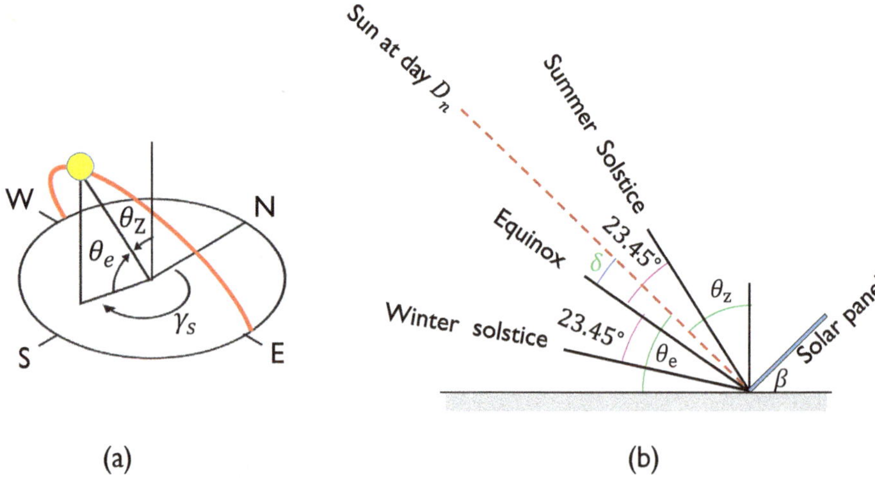

Figure 12.1: (a) The sun path is defined by azimuth and elevation (or zenith) angles of the sun as marked here. (b) The sun has the highest and lowest elevation on the summer and winter solstice, respectively. The declination angle δ (i.e., deviation from an equinox) can be used to determine the maximum elevation for any given day of the year.

12.2.1 Location of the sun at the peak of the sun path

Fortunately, it is easy to calculate maximum θ_e (at $\gamma_s = 180°$ or $0°$ for the Northern or the Southern Hemisphere) for a given day of the year, as follows. The day and night becomes equal (spring equinox) on the Northern Hemisphere on March 21, the 80th day of the year. On this day, the zenith of the sun equals the latitude, i.e., $\theta_Z = L$. Or, equivalently, the elevation angle is given by its complement, $\theta_e = 90° - L$. For any other day D_n, the zenith angle, shown in Fig. 12.1(b), is given by

$$\theta_z(D_n) = |L| \mp \delta(D_n). \tag{12.3}$$

Here, "$-$" and "$+$" should be used for the Northern and the Southern Hemisphere, respectively. The declination angle of the sun ($\delta(D_n)$) is given by

$$\delta(D_n) = 23.45° \sin(2\pi(D_n - 80)/365).$$

As we will discuss in the following sections, given θ_Z, we can calculate the tilt angle of a panel (β) for maximum energy collection, determine the separation between the rows based on "no shadow on the longest day" condition, and so on.

12.3 Energy yield is determined by the panel tilt

In the following two sections, we will discuss the tilt of the stand-alone panel vs. the panels in a solar farm.

12.3.1 An empirical rule for the tilt of a stand-alone panel

We see from Fig. 12.1 that the peak solar elevation varies from L (spring equinox) to $L + 23.45°$ (Summer solstice). Therefore, we can increase the yearly yield in energy by favoring a tilt angle that maximizes energy collection during the summer by pointing approximately halfway between $0°$ and $23.45°$ (say, $\delta = 10°$), i.e.,

$$\beta_1 = \theta_z(@\delta = 10°) = L - 10°. \tag{12.4}$$

A proper integral over the solar intensity over the year would have given a slightly more complicated formula:

$$\beta_2 = 0.69L + 3.7°. \tag{12.5}$$

A solar farm optimizes for yearly energy yield, but a small, stand-alone home installation cannot sacrifice the yield during the winter, when the days are shorter and the need for energy is greater. A two-tilt strategy, with $\beta_{3,s} = L - 23.45°$ during summer and $\beta_{3,w} = L + 23.45°$ during winter, addresses the need of a stand-alone system.

Homework 12.1: Determine the optimum tilt angles for West Lafayette, IN; Chennai, India; and Shanghai, China.

The table below summarizes the tilt angles predicted by Eqs. (12.4) and (12.5) for Lafayette, IN ($L = 40.27°$ N), Chennai, IN ($L = 13.5°$ N), and Shanghai, China ($L = 31.23°$ N).

	Lafayette	Chennai	Shanghai
L	40.27	13.5	31.23
β_1	30.27	3.50	21.23
β_2	31.79	12.92	24.83
$\beta_{3,s}$	16.82	−9.95	7.78
$\beta_{3,w}$	63.72	36.95	54.78

Note that while $\beta_1 \approx \beta_2$ for high L, β_2 provides a better estimate for low L. The negative value for $\beta_{3,s}$ implies that the panel will be looking to the south during the summer even though Chennai is located in the Northern Hemisphere. Finally, the tilt formula focuses on idealized energy yield. A panel may have to be tilted higher in soiling-prone regions in deserts close to the equator, to improve practical energy yield of the system.

12.3.2 No-shadowing constraint determines the row spacing of a solar farm

A solar farm consists of multiple rows of solar panels, each with height h. Analogous to a stand-alone solar panel, we need to determine the tilt angle (β) and row spacing (p) of the farm, and then integrate the direct, diffuse, and albedo light components to determine the energy yield $E(h, L)$. In practice, this involves an iterative optimization: given an initial β, the no-shadowing constraint discussed below determines the row spacing, $p(\beta, h)$. The integrated insolation incident on the panel determines $E(p(\beta), h, L)$. The design goal is to iteratively determine β to maximize energy yield per unit area of the farm, or cost of energy (COE) for a given installation, etc. We will discuss the no-shadowing constraint in Chapter 14. For now, let us assume that we know the optimized parameters p and β.

12.4 Calculation of the energy yield of a panel

Now that the panel has been oriented properly toward the sun, we will calculate the energy output by summing the contributions from the direct (I_{DNI}), diffuse (I_{DHI}), and albedo (I_{alb}) light components.

Homework 12.2: Approximate row spacing of a solar farm

Determine the setback ratio (SBR) of a solar farm to be installed in West Lafayette, IN (latitude 40.27° N). The module of height h is tilted at an angle β.

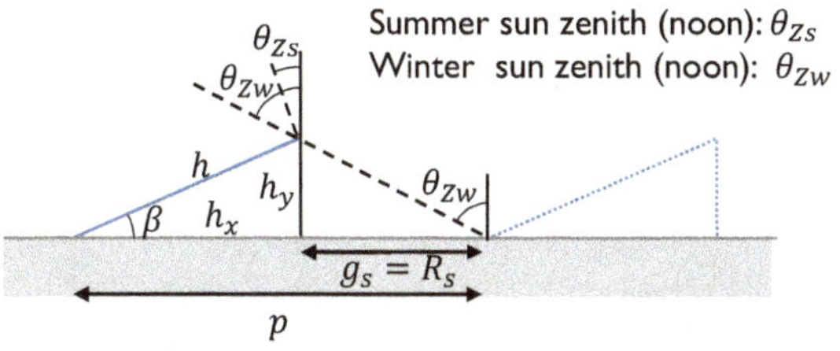

Solution. From Eq. (12.3), the zenith and azimuth of the sun at solar noon on December 21 are, respectively, $\theta_{Zw} = 40.27 + 23.45 = 63.72°$ and $\gamma_s = 180°$. The setback ratio of a farm is defined by

$$\text{SBR} = \frac{R_s}{h_y} = \tan(\theta_{Zw}),$$

where R_s is the empty space between the rows obtained from ground project and h_y is the vertical height measured from the top of the panel to the ground. The SBR for W. Lafayette, IN, is SBR $= \tan(63.72) = 2.03$.

We can now calculate the relative pitch between the panels (Eq. (12.4) determines β):

$$\begin{aligned} p/h &= \cos(\beta) + \sin(\beta) \cdot \text{SBR} \\ &= \cos(40.27 - 10) + \sin(40.27 - 10) \times 2.03 = 1.88. \end{aligned}$$

We will see in Chapter 14 that this is an excellent approximation to the exact result.

12.4.1 Components of the sun's illumination

Let us recall from Chapter 1 that the sunlight reaching the surface of the earth is composed of several components. These components arise due to scattering of the extraterrestrial photons by the atmosphere (diffuse light) and the ground (albedo light).

Extraterrestrial illumination The solar illumination reaching the earth's outer atmosphere (extraterrestrial) vs. the earth's surface is not equal. The extraterrestrial illumination is I_0 ~1350–1400 W/m^2 — this can be estimated from the sun's radiation (i.e., blackbody radiation at ~6000 K) and the distance from the sun to the earth. This, however, is partially absorbed and scattered before reaching the earth's surface, leading to a beam and diffuse component of insolation.

Global horizontal illumination The total sunlight reaching the earth's surface normal to the beam direction, i.e., the global illumination, can be approximated as $I_G \sim c_1 I_0 \tau^{c \times AM}$. Here, τ is the transmittance of the atmosphere, and $AM = (1/\cos\theta_Z)$ represents the air mass the sunlight has to travel through at a given zenith. c_1 and c are fitting parameters. The global insolation on the horizontal plane, termed Global Horizontal Irradiance (GHI), $I_{\mathrm{GHI}} = I_G \cos\theta_Z$, is a more commonly used and reported value.

Direct and diffuse insolation The GHI is composed of a direct (beam) and a diffuse horizontal illumination (DHI) component (see Fig. 12.2):

$$I_{\mathrm{GHI}} = I_{\mathrm{DNI}} \cos\theta_z + I_{\mathrm{DHI}}, \tag{12.6}$$

where I_{DHI} is the DHI and $I_{\mathrm{DNI}} \cos\theta_z$ is the beam/direct horizontal insolation. Here, I_{DNI} is the direct normal illumination (DNI): the beam normal to the sun's rays. Clearly, insolation is low close to sunrise and sunset — at these times the illumination is predominantly diffuse. The insolation peaks at solar noon. There are various databases which provide insolation information (both direct and diffuse) for any date, time, and location, see Tables 26.1 and 26.2.

Albedo light Every point on an open ground is illuminated by GHI. If R_A is the isotropic reflectivity of the ground surface (e.g., grass, concrete, etc.), then albedo light can act as a secondary source for the panel with intensity $R_A \times I_{\mathrm{GHI}}$.

In the next section, we will discuss how a panel intercepts these three components of light to convert sunlight to electricity.

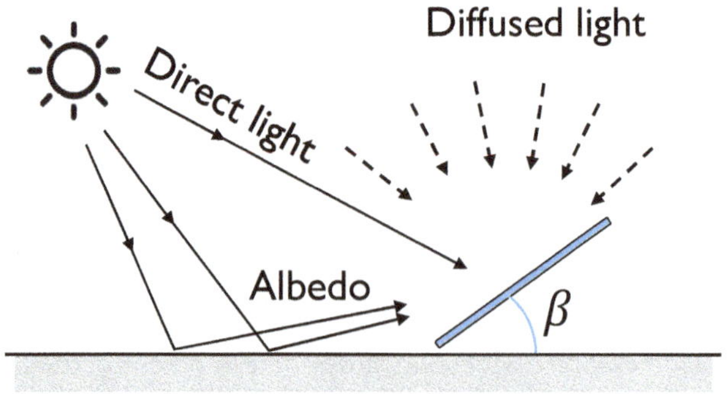

Figure 12.2: Definition of the three components of sunlight arriving on the panel surface.

12.4.2 Light collection by solar panels

The total light collection is obtained by summing the contributions from the three light components discussed above, namely,

$$I_T = I_{\text{dir}} + I_{\text{diff}} + I_{\text{alb}}. \tag{12.7}$$

Here, I_{dir}, I_{diff}, and I_{alb} are the light collection on the panel surface from DNI, DHI, and albedo, respectively. We will now calculate these contributions for a stand-alone panel and panels installed in a solar farm.

1. **Direct light collection**

   Stand-alone panel
   The panel is tilted at an angle β with respect to the ground, and the sun is elevated at an angle $\theta_e = 90° - \theta_z$. The light collection by the panel at solar noon due to the beam (direct) component of the light is given by (see Fig. 12.3(a)) taking the projection of DNI onto the normal to the panel surface as follows:

$$I_{\text{dir}} = I_{\text{DNI}} \cos(90° - (\theta_e + \beta)) = I_{\text{DNI}} \sin(\theta_e + \beta) = I_{\text{DNI}} \cos(\theta_z - \beta). \tag{12.8}$$

   Since θ_z is known from Eq. (12.3), it is easy to calculate the direct light collection during the solar noon of any day at a given location.

   Panels in a solar farm
   In a periodic array of panels spaced p apart, the direct light collection by each panel will be exactly the same as that found in the stand-alone case if there is no shading between rows, see Fig. 12.3(b).

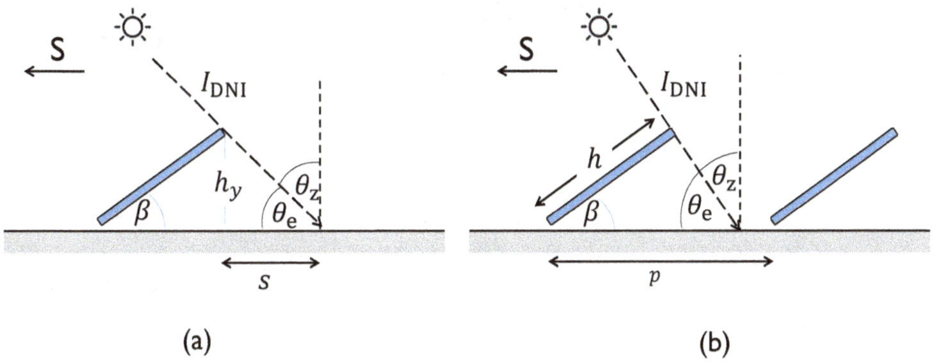

Figure 12.3: Collection of direct light by the (a) stand-alone panel and (b) rows of panels in a farm.

Homework 12.3: View factor is a very useful concept

Before we discuss collection of scattered or diffuse light (from the sky or the ground to the panel), we will introduce the concept of view factor. The view factor allows us to find the fractional diffuse radiation collection from one surface segment to another. It is defined as follows:

$$VF = F_{i-j} = \frac{\text{diffuse energy received by surface j from i}}{\text{total diffuse energy leaving surface i}}. \tag{12.9}$$

Therefore, the view factor (VF) quantifies the ability of a receiving surface to intercept isotropic emission from an uniformly illuminated surface. The evaluation of view factors involves several complicated integrals. Luckily, the "cross-string" method simplifies the view factor calculations for the two-dimensional cases considered in our current analysis. For infinitely long surfaces (along one dimension), we can calculate the view factor from i to j as follows (see Fig. 12.5(c)):

$$VF = F_{i\text{-}j} = \frac{\sum \text{crossed string} - \sum \text{uncrossed string}}{2 \times \textit{string on surface i}} \tag{12.10}$$

$$= \frac{(S_1 + S_2) - (S_3 + S_4)}{2L_i}. \tag{12.11}$$

2. **Diffuse light collection**

Stand-alone panel

The diffuse light I_{DHI} is isotropic — the photons scattered by the atmosphere and the clouds arrive from all angles. For a panel tilted at β, Fig. 12.4(a) shows that the panel will intercept the photons arriving between $0°$ and $(180° - \beta)$. The rays arriving from $0°$ makes an angle (β) with respect to the plane of the panel, while those arriving from $(180° - \beta)$ arrives tangentially to the surface. The diffuse light collection on the panel is

$$I_{\text{diff}} = I_{\text{DHI}} \times F_{PV\text{-}sky} = I_{\text{DHI}} \times \frac{(1 + \cos\beta)}{2}. \tag{12.12}$$

Here, $F_{PV\text{-}sky} = (1 + \cos\beta)/2$ is the view factor from the panel to the sky. When $\beta = 0$, the horizontal panel sees the full sky, and thus $F_{PV\text{-}sky} = 1$. The view factor decreases as the panel is tilted (i.e., β is increased), because the panel sees a smaller part of the sky.

Panels in a solar farm

Since the height of the panel is much smaller compared to the distance to the horizon, the diffuse light arriving at each point on a stand-alone panel is identical. In contrast, in a solar farm, the neighboring panels restrict the sky seen by various points on the panel (Fig. 12.4(b)) and make the diffuse illumination dependent on position. For a position ξ, from the bottom of the panel,

$$I_{\text{diff}}(\xi) = I_{\text{DHI}} \times \frac{(1+\cos(\beta+\alpha(\xi)))}{2}. \tag{12.13}$$

Here, $(1+\cos(\beta+\alpha(\xi)))/2$ is the view factor from the position ξ to the unrestricted part of the sky. The formula suggests that diffuse light intensity at the bottom of the panel (which sees a smaller fraction of the sky) is smaller than that at the top (which sees a larger part of the sky). The average diffuse sunlight collected per unit area of the panel is found by integrating $I_{\text{diff}}(\xi)$ along the panel as follows:

$$I_{\text{diff}} = \frac{1}{h}\int_0^h I_{\text{diff}}(\xi)\, d\xi. \tag{12.14}$$

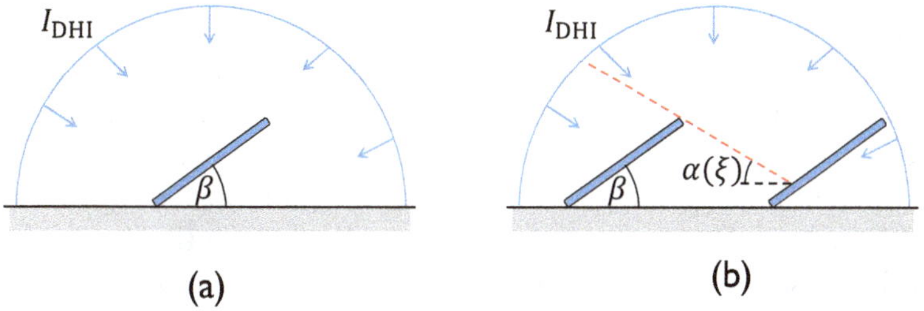

Figure 12.4: Collection of diffuse light by (a) a stand-alone panel (b) rows of panels in a farm.

3. **Albedo light collection**

Stand-alone panel

If we assume that the light incident on the ground (I_{GHI}) is scattered randomly with an albedo coefficient R_A, we can think of the ground as a secondary light source with intensity $I_{\text{GHI}} \times R_A$. The fraction of this secondary light intercepted by a panel of height h and tilted at an angle β is given by (see Fig. 12.5(a))

$$I_{\text{alb}} = I_{\text{GHI}}\, R_A\, F_{PV\text{-}gnd} = I_{\text{GHI}}\, R_A\, (1-\cos\beta)/2. \tag{12.15}$$

Here, $F_{PV\text{-}gnd} = (1-\cos\beta)/2$ is the view factor from the tilted panel to the infinite ground.

Homework 12.4: View factor from panel to ground

Let us explain how to find $F_{PV\text{-}gnd}$. (A similar calculation determines $F_{PV\text{-}sky}$.) Assume that the reflecting ground has length s, as shown in Fig. 12.5(a). The view factor can be calculated by the cross-string method:

$$\begin{aligned} F_{PV\text{-}gnd} &= \frac{(h+s) - (0 + \sqrt{s^2 + h^2 + 2sh\cos\beta})}{2h} \\ &= \frac{1}{2}(1 + r - \sqrt{1 + r + 2r\cos\beta}). \end{aligned}$$

Now, the ground is large compared to the panel, i.e., $r = s/h \to \infty$. In that limit,

$$\begin{aligned} F_{PV\text{-}gnd} &= \lim_{r\to\infty} \frac{1}{2}(1 + r - \sqrt{1 + r + 2r\cos\beta}) \\ &= \frac{1}{2}(1 - \cos\beta). \end{aligned}$$

Panels in a solar farm

The calculations for albedo light collection in a panel array (say, the period is p) of a solar farm are more complicated due to shadows on the ground from adjacent rows. Figure 12.5(b) shows that the ground between the panels has two different kinds of illumination: region s_1 is illuminated by direct and partial diffuse light, but region s_2 is illuminated only by a fraction of the diffuse light. Panel shading blocks direct illumination in region s_2. Summing up the contributions from these two regions, we find that

$$I_{\text{alb}} = I_1\, R_A\, F_{PV\text{-}s1} + I_2\, R_A\, F_{PV\text{-}s2}. \tag{12.16}$$

Here, I_1 and I_2 should be calculated using their respective view factors defined from the ground to the sky opening (*sky(p)*) seen through the row spacing:

$$\begin{aligned} I_1 &= I_{\text{DNI}} + I_{\text{DHI}}\, F_{s1\text{-}sky(p)} \\ I_2 &= I_{\text{DHI}}\, F_{s2\text{-}sky(p)}. \end{aligned}$$

All the view factors can be easily calculated using the cross-string method. The results, however, depend on the shadow length s_2, and the opening $s_1 = p - s_2$. The shadow s_2 will depend on the sun's position (which is defined by the location (latitude), date, and time of day), panel size h, and panel tilt β.

Total light collection

The sum of the three components

$$I_T = I_{\text{dir}} + I_{\text{diff}} + I_{\text{alb}} \tag{12.17}$$

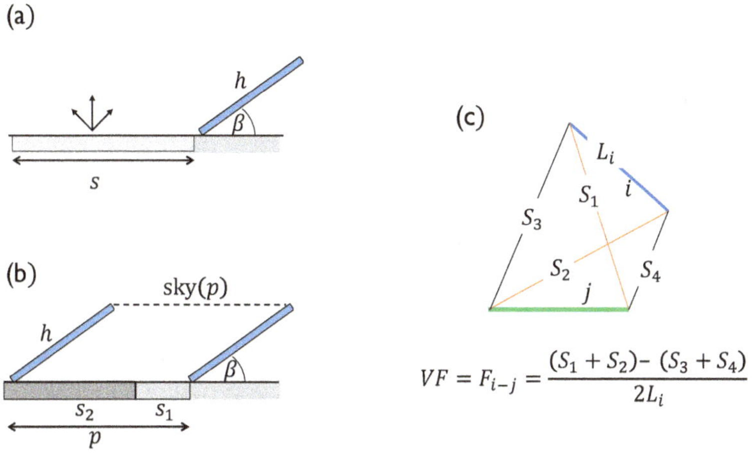

Figure 12.5: Collection of albedo light by (a) a stand-alone panel and (b) rows of panels in a farm. The part marked *sky(p)* shows the open sky as seen by a single period of the array. (c) Definition of the cross-string method for calculating the view factor between two surfaces.

determines the insolation onto unit area of a panel. The energy yield is calculated by accounting for the angle-dependent reflectivity and the efficiency of the panel. For solar farms, it may be required to estimate the panel light collection per unit area of the farm instead of per unit panel area. In that case, we can simply scale I_T by h/p.

12.5 Conclusions

In this chapter, we qualitatively discussed the design considerations of optimally tilted, stand-alone panels and the panels installed in a solar farm. The energy yield of a stand-alone panel defines the upper limit of energy yield per panel and, therefore, can be used to calculate the optimistic upper bound of the energy yield of a solar farm. In practice, we have seen that panel-to-ground shading and panel-to-panel shading reduces the energy yield of a solar farm. The general problem with non-uniform diffuse/albedo illumination can only be solved numerically. Fortunately, we can gain considerable insight into the physics of farm design through closed-form analysis of a vertical bifacial farm. This will be the topic of discussion in the next chapter.

References

[1] Antonio Luque and Steven Hegedus, editors. *Handbook of Photovoltaic Science and Engineering, Second Edition*. March 2011.

[2] POWER. Surface meteorology and Solar Energy: A renewable energy resource web site (release 6.0), 2017.

[3] M. F. Modest. *Radiative heat transfer*. Academic Press, New York, third edition edition, 2013. OCLC: ocn813855549.

[4] M. R. Khan, M. T. Patel, R. Asadpour, H. Imran, N. Z. Butt, and M. A. Alam, "A review of next generation bifacial solar farms: predictive modeling of energy yield, economics, and reliability," *J. Phys. D: Appl. Phys.*, vol. 54, no. 32, p. 323001, May 2021, doi: 10.1088/1361-6463/abfce5.

[5] M. T. Patel, R. A. Vijayan, R. Asadpour, M. Varadharajaperumal, M. R. Khan, and M. A. Alam, "Temperature-dependent energy gain of bifacial PV farms: A global perspective," *Applied Energy*, vol. 276, p. 115405, Oct. 2020, doi: 10.1016/j.apenergy.2020.115405.

CHAPTER 13

Design of a Vertical Solar Farm

Chapter Summary

- Bifacial solar panels accept light from both front and back faces.
- Installed in a vertical configuration, the bifacial panels provide more uniform energy output compared to the monofacial panels.
- The optimal elevation of a vertical solar panel depend on the latitude.
- The albedo collected by a vertical farm is significantly lower compared to a standalone vertical solar panel.

13.1 Introduction

In Chapter 12, we integrated the contributions from direct, diffuse, and albedo light to calculate the total irradiance onto a panel. The calculation is relatively simple for a stand-alone, optimally tilted panel. For a solar farm, however, the shading related to neighboring panels makes the farm design (e.g., optimum orientation, tilt, etc.) difficult. Although the panel or farm analysis at noon, as discussed in Chapter 12, gives a zeroth-order estimation, we need a time-resolved calculation followed by integration over time for accurate results.

In this chapter, we will discuss a simplified situation assuming an east-to-west sun path, and a farm consisting of only vertical east–west-facing, ground-mounted bifacial panels. This will exemplify the time-resolved calculation method for a standalone panel or an array of panels (i.e., solar farms). This will be the basis for extending the concepts to more complex and realistic farm designs, to be discussed in the next chapter.

13.2 Basics of a solar farm output: A simplified case study

Let us assume that the sun moves from the east (sunrise) to the west (sunset), and there is no tilt of the sun path toward the north or south. This is depicted in Fig. 13.1(a). At any time of day, the position of the sun is characterized by the zenith angle $\theta_Z(t)$ (or, the elevation angle θ_e). Here, we define $\theta_Z(t)$ to be negative when the sun is in the east (before noon), and positive when the sun is in the west (after noon). Note that θ_e attains the same values before and after noon, i.e., θ_e (or θ_Z) is a double-valued function of time. Therefore, it is easier to see the time progression of the incident and collected light as a function of θ_Z.

13.2.1 Vertical bifacial panel: Stand-alone

Direct insolation collection. A vertically mounted bifacial panel of height h is shown in Fig. 13.1(i)(a) facing E-W — it runs infinitely along the N-S direction.

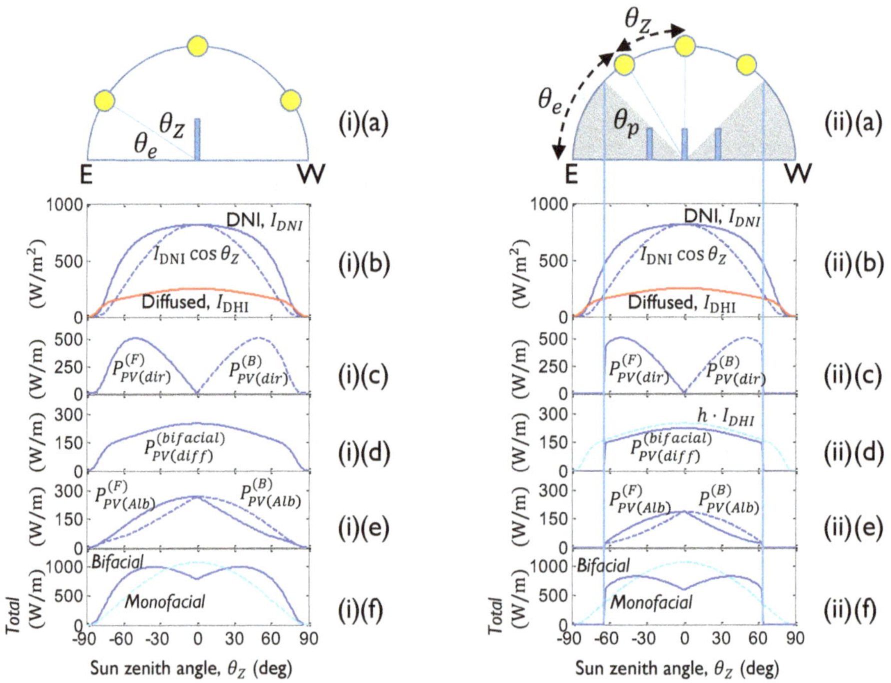

Figure 13.1: The insolation and the panel light collection components are shown as a function of the sun's zenith angle $\theta_Z = [-90°, 90°]$. The angle θ_Z and time change proportionally as the day progresses. The plots are shown for (i) a stand-alone panel and (ii) a panel array. The corresponding panel output can be found by multiplying the light collection with the panel efficiency. The dashed lines marked 'Monofacial' in figure (f) represent light collected on horizontal monofacial panel.

The front face (east-facing) of the panel sees the sun from sunrise till noon. The back face (west-facing) of the panel sees the sun from noon till sunset. The angle between the sun's beams and the normal of the panel is equal to the elevation angle, $\theta_e = 90° - \theta_Z$. Illumination collected at the front face of the panel is

$$P^{(F)}_{PV(dir)} = \begin{cases} h \times I_{\text{DNI}} \cos\theta_e, & \text{for } \theta_Z \leq 0 \text{ (i.e., till noon)} \\ 0, & \text{for } \theta_Z > 0. \end{cases} \tag{13.1}$$

Here, $P^{(F)}_{PV(dir)}$ is the light collected per unit width of the panel (front face). The solid line in Fig. 13.1(i)(c) shows $P^{(F)}_{PV(dir)}$ as the day progresses. After the solar noon ($\theta_Z > 0$), the front face will not directly see the sun, and $P^{(F)}_{PV(dir)} = 0$ for the later part of the day. Similarly, the back face shows a mirrored characteristic for $P^{(B)}_{PV(dir)}$, as shown by the dashed line. For the bifacial panel,

$$P^{(bifacial)}_{PV(dir)} = h \times I_{\text{DNI}} \cos\theta_e = h \times I_{\text{DNI}} |\sin\theta_Z|. \tag{13.2}$$

Diffuse insolation collection. The diffuse sunlight is isotropic and covers all zenith angles $[-90°, 90°]$. Only half of the diffuse rays reach the vertical front panel face. Thus, $I^{(F)}_{PV(diff)} = I_{\text{DHI}}/2$. The back face also collects half of the diffuse light. Therefore, the total diffuse light collected is

$$P^{(bifacial)}_{PV(diff)} = h \times I_{\text{DHI}}. \tag{13.3}$$

Up to this point, it does seem as if the bifacial panel would underperform compared to the horizontal monofacial panel, as the direct light collection is lower for the vertical bifacial panel. However, the vertical panel has the opportunity to collect light scattered from the ground (albedo light). We will need to treat the albedo light originating from diffuse and direct insolation separately.

Albedo light collection. Let us assume that all the diffuse illumination I_{DHI} reaches the ground, i.e., we neglect the partial shading from the panel. The ground isotropically scatters the light with albedo R_A (reflectance). Similar to the discussion for

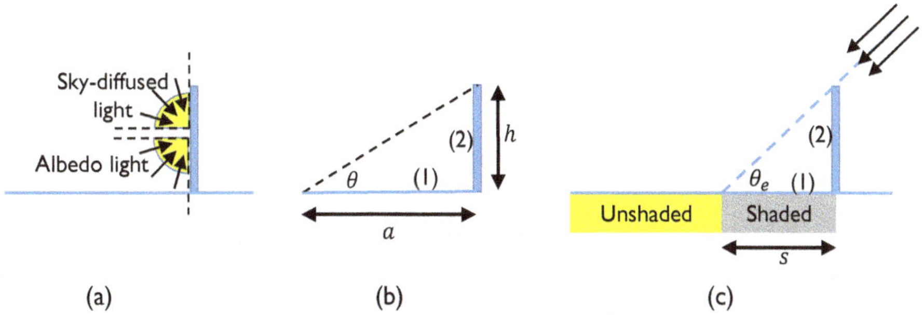

Figure 13.2: (a) Collection of diffuse light from sky and ground. (b) We want to find the view factor from surface (1) to surface (2). (c) Panel shadow on the ground from direct light.

the sky-diffuse light, only half of the isotropic light ($1/2 \times I_{\mathrm{DHI}} R_A$) will be collected at the front (or back) face:

$$P^{(F)}_{\mathit{PV(Alb:diff)}} = P^{(B)}_{\mathit{PV(Alb:diff)}} = h \times \frac{1}{2} I_{\mathit{diff}} R_A. \tag{13.4}$$

Next, to understand the albedo light collection originating from the direct light, we would need to use the concept of view factors.

Homework 13.1: View factor defined

Explain the concept and the calculation procedure of view factors.

Solution. The concept of view factor allows us to find the fractional diffuse radiation collection from one surface segment to another. The view factor is defined as follows:

$$F_{i-j} = \frac{\text{diffuse energy received by surface j from i}}{\text{total diffuse energy leaving surface i}}. \tag{13.5}$$

Consider the setup shown in Fig. 13.2(b), where light is isotropically scattered from surface (1) (reflector; length a) and a fraction of this scattered light is intercepted by surface (2) (collector; height h). It can be shown that the view factor from the reflector to the collector is

$$F_{1-2} = \frac{1}{2}\left(1 + H - \sqrt{1+H^2}\right) \tag{13.6}$$

$$= \frac{1}{2\cos\theta}\left(\sin\theta + \cos\theta - 1\right). \tag{13.7}$$

Here, $H = h/a = \tan\theta$. If light of intensity I_{inc} is incident on surface (1) of reflectance R_A, then total diffuse emission from the reflector is $a \times I_{inc} R_A$. By definition of the view factor, the diffuse light collected by surface (2) is

$$P_{1-2} = F_{1-2} \times a \times I_{inc} R_A$$

$$= F_{1-2} \times \frac{h}{\tan\theta} \times I_{inc} R_A$$

$$= h I_{inc} R_A \times \frac{1}{2}\left(1 - \tan\frac{\theta}{2}\right) \tag{13.8}$$

$$= h I_{inc} R_A \times F_{2-1}. \tag{13.9}$$

Let us first consider a case where there is no shading on the ground from the direct beam and the panel. Therefore, just as in the case of albedo from the sky-diffuse light, the power collected at the panel front-face is

$$P^{(F)}_{\mathit{noShade}} = h \times \frac{1}{2} I_{dir} R_A. \tag{13.10}$$

Here, $I_{dir} = I_{\text{DNI}} \cos\theta_Z$ is the direct insolation on the horizontal surface[1] — which of course changes as the day progresses. In the afternoon, as the sun moves toward the west, we expect shading near the front face of the panel (as shown in Fig. 13.2(c)). How much albedo light contribution do we lose due to a shade of length s? We immediately get the answer from Eq. (13.9) by setting s, θ_e, and I_{dir} in place of a, θ, and I_{inc}:

$$P_{lost}^{(F)} = hI_{dir}R_A \times \frac{1}{2}\left(1 - \tan\frac{\theta_e}{2}\right).$$

The net direct beam activated albedo light collection in the front face is therefore

$$P_{\text{Alb-afternoon}}^{(F)} = P_{\text{noShade}}^{(F)} - P_{\text{lost}}^{(F)} = h \times \frac{1}{2} I_{dir} R_A \tan\frac{\theta_e}{2}. \quad (13.11)$$

Finally,

$$P_{\text{PV(Alb:dir)}}^{(F)} = \begin{cases} h \times \frac{1}{2} I_{dir} R_A, & \text{for } \theta_Z \le 0 \text{ (i.e., till noon)} \\ h \times \frac{1}{2} I_{dir} R_A \tan\frac{\theta_e}{2}, & \text{for } \theta_Z > 0. \end{cases} \quad (13.12)$$

The net albedo light contribution on the front face ($P_{\text{PV(Alb)}}^{(F)} = P_{\text{PV(Alb:dir)}}^{(F)} + P_{\text{PV(Alb:diff)}}^{(F)}$) is shown by the solid line in Fig. 13.1(i)(e). For the back face, $P_{\text{PV(Alb:dir)}}^{(B)}$ is just the flipped version of $P_{\text{PV (Alb:dir)}}^{(F)}$ around noon, i.e.,

$$P_{\text{PV(Alb:dir)}}^{(B)} = \begin{cases} h \times \frac{1}{2} I_{dir} R_A \tan\frac{\theta_e}{2}, & \text{for } \theta_Z \le 0 \text{ (i.e., till noon)} \\ h \times \frac{1}{2} I_{dir} R_A, & \text{for } \theta_Z > 0. \end{cases} \quad (13.13)$$

By adding the previous two equations, we find that

$$P_{\text{PV(Alb:dir)}}^{(F)} + P_{\text{PV(Alb:dir)}}^{(B)} = \begin{cases} h \times \frac{1}{2} I_{dir} R_A + h \times \frac{1}{2} I_{dir} R_A \tan\frac{\theta_e}{2} & \text{for } \theta_Z \le 0 \\ h \times \frac{1}{2} I_{dir} R_A \tan\frac{\theta_e}{2} + h \times \frac{1}{2} I_{dir} R_A & \text{for } \theta_Z > 0 \end{cases} \quad (13.14)$$

$$= h \times \frac{1}{2} I_{dir} R_A + h \times \frac{1}{2} I_{dir} R_A \tan\frac{\theta_e}{2}, \quad \text{for all } \theta_Z. \quad (13.15)$$

The overall albedo light collection by the bifacial panel is

$$\begin{aligned} P_{\text{PV(Alb)}}^{(\text{bifacial})} &= \left[P_{\text{PV(Alb:diff)}}^{(F)} + P_{\text{PV (Alb:diff)}}^{(B)}\right] + \left[P_{\text{PV(Alb:dir)}}^{(F)} + P_{\text{PV(Alb:dir)}}^{(B)}\right] \\ &= \left[h \times I_{\text{diff}} R_A\right] + h \times \frac{1}{2} I_{dir} R_A \left[1 + \tan\frac{\theta_e}{2}\right] \\ &= h \times I_{\text{diff}} R_A + h \times \frac{1}{2} I_{dir} R_A \left[1 + \tan\left(\frac{\pi}{4} - \frac{|\theta_Z|}{2}\right)\right]. \end{aligned} \quad (13.16)$$

Remember that, sunlight on the horizontal ground is $I_{dir} = I_{\text{DNI}} \cos\theta_Z = I_b \sin\theta_e$.

[1]Note that I_{dir} is defined here as the light on the horizontal surface, unlike in the previous chapter, where it was assumed to be the light incident on a tilted panel.

Total light collection. Finally, we can write the total light collection by the stand-alone bifacial panel as

$$\begin{aligned} P_{PV}^{(bifacial)} &= P_{PV(dir)}^{(bifacial)} + P_{PV(diff)}^{(bifacial)} + P_{PV(Alb)}^{(bifacial)} \\ &= h \times I_{\text{DNI}} \cos\theta_e + h \times I_{\text{DHI}} + h \times I_{\text{DHI}} R_A + h \times \frac{1}{2} I_{\text{DHI}} R_A \left[1 + \tan\frac{\theta_e}{2}\right] \\ &= h \times I_{\text{DNI}} \left[\cos\theta_e + \frac{1}{2}\sin\theta_e R_A \left(1 + \tan\frac{\theta_e}{2}\right)\right] + h \times I_{\text{DHI}}(1 + R_A). \end{aligned} \tag{13.17}$$

We recall that $\theta_e = \pi/2 - |\theta_Z|$. The total bifacial collection is shown in Fig. 13.1(i)(f). With a panel efficiency of η, the energy output (per row length) from the panel would be $= \eta \times P_{PV}^{(bifacial)}$.

13.2.2 Vertical bifacial panels: Array

Now, consider that the vertical bifacial panels of height h are arranged in an array with a period of p as shown in Fig. 13.3(a). Similar to the previously discussed stand-alone case, each of the panels faces E-W and runs infinitely along the N-S direction.

Direct insolation collection. Due to the array configuration, the front face of a panel is partially illuminated (partially shaded) until the sun elevation θ_e reaches

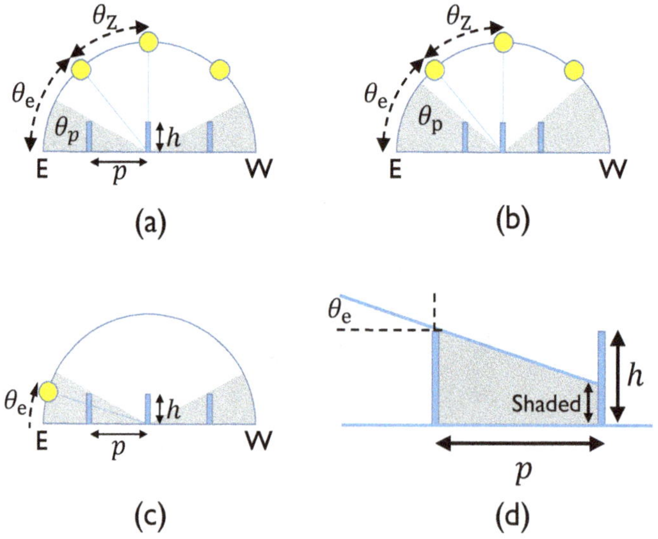

Figure 13.3: (a) Mutual shading of panels for different times of the day. There is no shade on the panel for $\theta_e > \theta_p$. Increased and reduced shading times with a shorter and a longer array period p in (b) and (c), respectively. (d) A partially shaded panel.

$\theta_p = \tan^{-1}(h/p)$. The angle θ_p is shown by gray shading in Figs. 13.3(a–c). In Fig. 13.3(c), we show $\theta_e < \theta_p$. In such a case, the bottom part of the panel is shadowed, as illustrated in Fig. 13.3(d). As discussed in Chapter 18, partial shading in conventional panels can cause excessive heating and degrade the lifetime. Therefore, we assume that the panels are turned off while there is shading — the panels operate only when $\theta_e > \theta_p$. A similar situation occurs for the back face of the panel before sunset. The non-operating conditions are shown as the gray-shaded region in Fig. 13.3(a). The operating time of the arrays are in a narrower range of θ_e (i.e., shorter time span) when the panels are positioned closer — compare (a) and (b) in Fig. 13.3.

The collection of direct illumination throughout the day by a bifacial panel in the array is the same as before:

$$P_{PV(dir)}^{(b\text{-}array)} = h \times I_{\rm DNI} \cos\theta_e = h \times I_{\rm DNI} |\sin\theta_Z|, \tag{13.18}$$

within the time range $\theta_e > \theta_p$. This truncated plot is shown in Fig. 13.1(ii)(c).

Diffuse insolation collection. Ideally, when the panels are far apart, as we have explained earlier (for the stand-alone case), half of the diffuse rays reach the front face of the panel. These rays cover a zenith angle range of $[-\pi/2, 0]$. However, a fraction of these angles are obstructed/shaded when the panels are arranged in an array, as shown by the shaded circles in Fig. 13.4(a). From this illustration, we can see that the top portion of the vertical panel receives more diffuse light than the bottom. We assume here that the net current from the panel is limited by the cell with the least light collection, i.e., the one at the bottom. The bottom cell only collects light from the zenith angle range of $[-(\pi/2 - \theta_p), 0]$. This means the front face of the panel receives only $(1/2)(1 - \sin\theta_p)$ fraction of the net diffuse light within $[-\pi/2, \pi/2]$. The same is true for the back face. Thus,

$$P_{PV(diff)}^{(F)} = P_{PV(diff)}^{(B)} = h \times I_{\rm DHI} \times \frac{1}{2}(1 - \sin\theta_p). \tag{13.19}$$

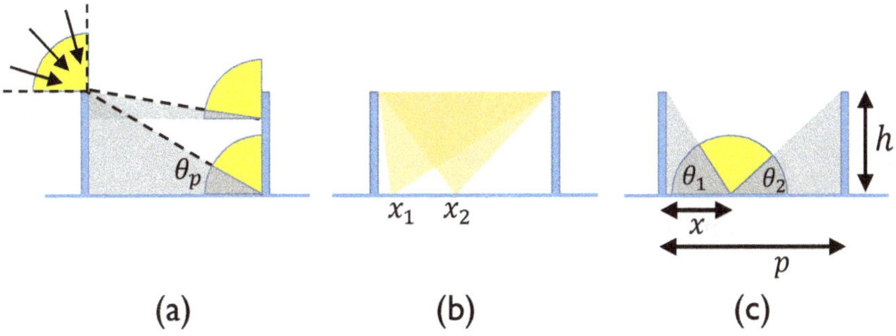

Figure 13.4: (a) Partial masking of DHI on the face of the panel. (b, c) Partial masking of DHI on the ground. The fractional DHI reaching the ground is a source of albedo light.

The total diffuse insolation collection:

$$P^{(b\text{-}array)}_{PV(diff)} = P^{(F)}_{PV(diff)} + P^{(B)}_{PV(diff)} = h \times I_{\mathrm{DHI}} \times (1 - \sin\theta_p). \tag{13.20}$$

Homework 13.2: Calculating the view factor of a bifacial panel in a PV farm

The view factor from a point at height z on any face of the vertical panel (in the array) to the sky is $F_{dz-sky} = 1/2 \times (1 - \sin\psi(z))$. Here, $\psi(z) = (h - z)/p$ is the masking angle of the diffuse light (see Fig. 13.4(a)).

1. Find the total view factor $F_{h-sky} = 1/h \times \int_0^h F_{dz-sky} dz$.
2. Use this to find the total diffuse light collection onto the panel. How does it compare to diffuse light $P_{PV(diff)}$ collected at the bottom of the panel?

Solution. $F_{h-sky} = 1/h \times \int_0^h F_{dz-sky} dz = 1/2 \times (1 - \tan\theta_p/2)$. Therefore, $P^{(F)}_{PV(diff)} = hI_{diff}(1 - \tan\theta_p/2)/2$. Note that the corresponding power generated is limited by the collection at the bottom cell of the panel (due to the series connection):

$$\frac{\text{collection}^{(F)}}{P^{(F)}_{PV(diff)}} = \frac{1 - \tan\theta_p/2}{1 - \sin\theta_p} \geq 1. \tag{13.21}$$

This loss of power can be partially recovered by using bypass diodes and DC-DC converters.

Albedo light collection. Let us first describe the effect of diffuse insolation on albedo. As explained in the preceding discussion, there is a fractional shadowing or masking of the diffuse light reaching the panel. A similar scenario is true for the diffuse light reaching the ground. Depending on the position of the panels, the amount of diffuse sunlight reaching the ground is different. For example, as shown in Fig. 13.4(b), the masking angles are different at x_1 and x_2.

Consider a position x between adjacent panels, as shown in Fig. 13.4(c). The masking angles from the two panels are

$$\theta_1(x) = \tan^{-1}\frac{h}{x} \quad \text{and, } \theta_2(x) = \tan^{-1}\frac{h}{p - x}. \tag{13.22}$$

The average masking angle can be written as

$$\bar{\theta_1} = \frac{1}{p}\int_0^p \theta_1(x)\, dx \tag{13.23}$$

$$= \theta_p + \frac{\ln(\csc\theta_p)}{\cot\theta_p}. \tag{13.24}$$

Due to symmetry, $\bar{\theta_1} = \bar{\theta_2}$. The average diffuse insolation reaching the ground is

$$I_{Gnd:diff} = I_{\text{DHI}} \times \frac{1}{2}(\cos\bar{\theta_1} + \cos\bar{\theta_2}) = I_{\text{DHI}} \times \cos\bar{\theta_1}. \tag{13.25}$$

Note that $\bar{\theta_1}$ is constant throughout the day, and $I_{Gnd:diff}$ is proportional to I_{DHI}. Now, $I_{Gnd:diff}$ can be the diffuse light source for the front (or back) face of the panel. Using Eq. (13.9), we can find the albedo light collection originating from diffuse insolation on the front (or back) face of the panel:

$$P^{(F)}_{PV(Alb:diff)} = P^{(B)}_{PV(Alb:diff)} = hI_{Gnd:diff}R_A \times \frac{1}{2}\left(1 - \tan\frac{\theta_p}{2}\right) \tag{13.26}$$

$$= hI_{\text{DHI}}\cos\bar{\theta_1}R_A \times \frac{1}{2}\left(1 - \tan\frac{\theta_p}{2}\right). \tag{13.27}$$

Next, we can focus on albedo from the direct insolation. In the morning (i.e., $\theta_Z < \pi/2$), the shading on the ground will be configured as shown in Fig. 13.5(a). The shade length s_1 is equal to the period p (i.e., ground fully shaded for the beam component) when the array is turned on at $\theta_e = \theta_p$. The shade goes away ($s_1 = 0$) at noon. At any time of day, we can define the unshaded length $(p - s_1)$ in terms of the angle θ_1^* (see Fig. 13.5(a)):

$$\theta_1^* = \cot^{-1}\left(\frac{p - s_1}{h}\right) = \cot^{-1}(\cot\theta_p - \cot\theta_e). \tag{13.28}$$

Assuming the unshaded region as the reflector, and the front face as the collector, we use Eq. (13.9) to write

$$\textit{morning: } P^{(F)}_{PV(Alb:dir)} = hI_{dir}R_A \times \frac{1}{2}\left(1 - \tan\frac{\theta_1^*}{2}\right). \tag{13.29}$$

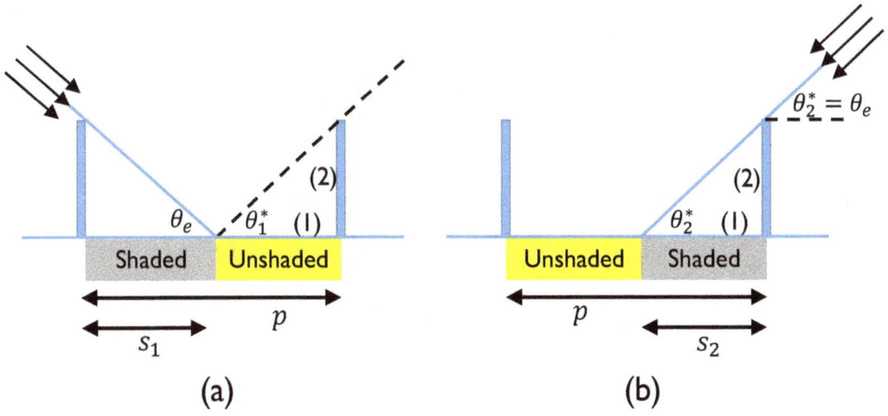

Figure 13.5: Shading on the ground (a) before noon and (b) after noon.

In the afternoon, shading from direct light is adjacent to the front face (Fig. 13.5(b)). We can write

$$P^{(F)}_{noShade} = hI_{dir}R_A \times \frac{1}{2}\left(1 - \tan\frac{\theta_p}{2}\right) \tag{13.30}$$

$$P^{(F)}_{lost} = hI_{dir}R_A \times \frac{1}{2}\left(1 - \tan\frac{\theta_2^*}{2}\right). \tag{13.31}$$

Here, $P^{(F)}_{noShade}$ assumes that the ground is unshaded. $P^{(F)}_{lost}$ corresponds to the albedo light lost due to the shading. Therefore,

$$\textit{afternoon: } P^{(F)}_{PV(Alb:dir)} = P^{(F)}_{noShade} - P^{(F)}_{lost} \tag{13.32}$$

$$= hI_{dir}R_A \times \frac{1}{2}\left(\tan\frac{\theta_e}{2} - \tan\frac{\theta_p}{2}\right). \tag{13.33}$$

Here, we used $\theta_2^* = \theta_e$, as shown in Fig. 13.5(b). As mentioned earlier, the array is turned off for $\theta_e < \theta_p$. We can now summarize the morning and afternoon components of $P^{(F)}_{PV(Alb:dir)}$ as follows:

$$P^{(F)}_{PV(Alb:dir)} = \begin{cases} hI_{dir}R_A \times \frac{1}{2}\left(1 - \tan\frac{\theta_1^*}{2}\right), & \text{for } \theta_Z \le 0 \text{ (i.e., till noon)} \\ hI_{dir}R_A \times \frac{1}{2}\left(\tan\frac{\theta_e}{2} - \tan\frac{\theta_p}{2}\right), & \text{for } \theta_Z \ge 0. \end{cases} \tag{13.34}$$

The net albedo light contribution to the front face ($P^{(F)}_{PV(Alb)} = P^{(F)}_{PV(Alb:dir)} + P^{(F)}_{PV(Alb:diff)}$) is shown by the solid line in Fig. 13.1(ii)(d). For the back face, $P^{(B)}_{PV(Alb:dir)}$ is just the flipped version around noon:

$$P^{(B)}_{PV(Alb:dir)} = \begin{cases} hI_{dir}R_A \times \frac{1}{2}\left(\tan\frac{\theta_e}{2} - \tan\frac{\theta_p}{2}\right), & \text{for } \theta_Z \le 0 \text{ (i.e., till noon)} \\ hI_{dir}R_A \times \frac{1}{2}\left(1 - \tan\frac{\theta_1^*}{2}\right) & \text{for } \theta_Z > 0. \end{cases} \tag{13.35}$$

The net albedo from direct light is

$$P^{(F)}_{PV(Alb:dir)} + P^{(B)}_{PV(Alb:dir)} = \begin{cases} h \times \frac{1}{2}I_{dir}R_A\left(1 - \tan\frac{\theta_1^*}{2} + \tan\frac{\theta_e}{2} - \tan\frac{\theta_p}{2}\right) & \text{for } \theta_Z \le 0 \\ h \times \frac{1}{2}I_{dir}R_A\left(1 - \tan\frac{\theta_1^*}{2} + \tan\frac{\theta_e}{2} - \tan\frac{\theta_p}{2}\right) & \text{for } \theta_Z \ge 0 \end{cases}$$

$$= h \times \frac{1}{2}I_{dir}R_A\left(1 - \tan\frac{\theta_1^*}{2} + \tan\frac{\theta_e}{2} - \tan\frac{\theta_p}{2}\right), \text{for all } \theta_Z. \tag{13.36}$$

The overall albedo light collection by the bifacial panel is

$$P^{(bifacial)}_{PV(Alb)} = \left[P^{(F)}_{PV(Alb:diff)} + P^{(B)}_{PV(Alb:diff)}\right] + \left[P^{(F)}_{PV(Alb:dir)} + P^{(B)}_{PV(Alb:dir)}\right]$$

$$= hI_{\text{DHI}}R_A \cos\bar{\theta}_1\left(1 - \tan\frac{\theta_p}{2}\right)$$

$$+ h \times \frac{1}{2}I_{dir}R_A\left(1 - \tan\frac{\theta_1^*}{2} + \tan\frac{\theta_e}{2} - \tan\frac{\theta_p}{2}\right). \tag{13.37}$$

Remember that, $I_{dir} = I_{\text{DNI}}\sin\theta_Z = I_{\text{DNI}}\sin\theta_e$.

Homework 13.3: Light collection by a vertically elevated horizontal panel

Calculate the view factor of and the light collected by a bifacial panel that is elevated at a height h from the ground.

Solution. Let us first introduce the cross-string method for calculating the view factor. For infinitely long surfaces (along one dimension), we can calculate the view factor from A_i to A_j as follows:

$$F_{A_i - A_j} = \frac{\sum crossed\ string - \sum uncrossed\ string}{2 \times string\ on\ the\ surface\ i}. \tag{13.38}$$

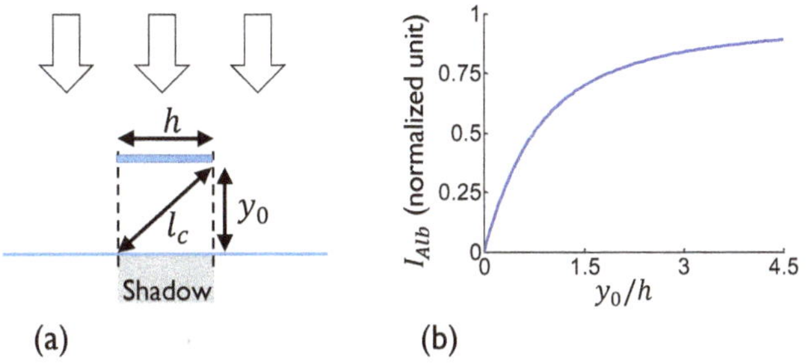

(a) A bifacial panel is placed horizontally at an elevation y_0 from the ground. (b) The amount of albedo light collected depends on the elevation of the solar panel.

Now consider a single horizontal panel elevated at y_0 and illuminated by sunlight at a normal angle. The shadow will be directly below the panel. The back face essentially can see the full ground, and thus the view factor $F_{PV\text{-}Gnd0} = 1$. Of course, we would need to subtract the contribution from the shaded region. The view factor from the back face to the shaded ground can be calculated using the cross-string method:

$$F_{PV\text{-}ShGnd} = \frac{(l_c + l_c) - (y_0 + y_0)}{2h} = \frac{\sqrt{h^2 + y_0^2} - y_0}{h}. \tag{13.39}$$

The view factor from the panel back to the illuminated ground is

$$F_{PV\text{-}LGnd} = F_{PV\text{-}Gnd0} - F_{PV\text{-}ShGnd} = \frac{h + y_0 - \sqrt{h^2 + y_0^2}}{h}. \tag{13.40}$$

Finally, the corresponding albedo collection is $I_{\mathrm{Alb}} = h \times R_A I_{\mathrm{DNI}} \times F_{PV\text{-}LGnd}$. Observe that when the panel touches the ground, no albedo light reaches the back face (set $y_0 = 0$ and check). $F_{PV\text{-}LGnd}$ reaches 0.5 for $y_0/h = 0.75$ and saturates to 1 for very high elevation.

Homework 13.4: Light collection by a vertically elevated periodic array of horizontal panels

Compared to the previous homework about a single panel, the light collected at the back face of the bifacial panel will be reduced in an array setup due to the periodic ground shading. Quantify this intuitive result. Then show that in a practical scenario with $y_0/h < 0.7$, the back face of the panel will only collect 0.3–0.4 of the ground albedo, R_A.

Solution. Assume an array of horizontal bifacial panels with period p and elevation of y_0. Focus on 'Panel #(0)' in the figure below. The illuminated segments on the ground are marked as *LGnd(i)*, where $i = +1, +2, \ldots$ on the right hand of Panel #(0), and $i = -1, -2, \ldots$ on the left. Let us find the view-factor from the panel back to *LGnd(+k)*, i.e., $F_{\textit{PV-LGnd(+k)}}$. This is represented by the shaded region marked 'VF' in the figure.

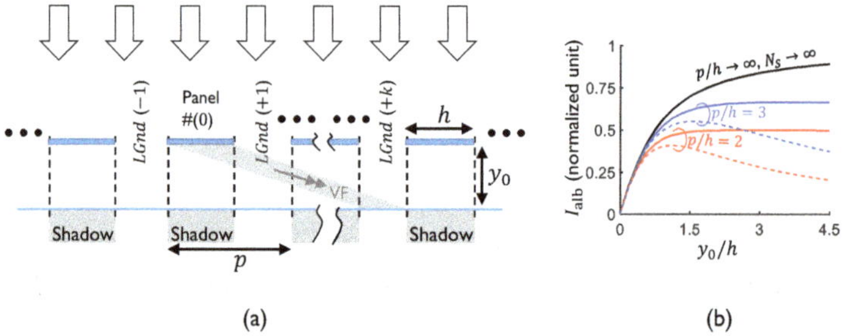

(a) Periodic array of bifacial panels are placed horizontally at an elevation y_0 from the ground. (b) The amount of albedo light collected at the panel back-face depends on both the elevation of the panel and the period of the array (see Eq. (13.42), solid line: exact solution, dashed line: single-term approximation).

For the 'VF' region:
(i) Crossed strings are: $\sqrt{y_0^2 + (kp)^2}$ and $\sqrt{y_0^2 + [(k-1)p]^2}$.
(ii) Uncrossed strings are: $\sqrt{y_0^2 + (kp-h)^2}$ and $\sqrt{y_0^2 + (kp - x_r)^2}$.
Here, $x_r = p - h$ is the length of the unshaded region.

Therefore,

$$F_{\textit{PV-LGnd(+k)}} = \frac{\sum \textit{crossed strings} - \sum \textit{uncrossed strings}}{2 \times \textit{string on surface } i}$$
$$= \frac{\sqrt{y_0^2 + (kp)^2} + \sqrt{y_0^2 + [(k-1)p]^2}}{2h} - \frac{\sqrt{y_0^2 + (kp-h)^2} + \sqrt{y_0^2 + (kp - x_r)^2}}{2h}. \quad (13.41)$$

(*continued on the next page*)

Homework 13.4 (*continued from the previous page*)

The net view-factor from the panel back to the illuminated parts of the ground can finally be calculated using:

$$F_{PV\text{-}LGnd} = 2 \times \sum_{k=1}^{N_S} F_{PV\text{-}LGnd(+k)}. \tag{13.42}$$

Here, the factor '2' at the beginning of the sum is to consider both the $(+k)$ and $(-k)$ segments. N_S is the number of segments considered in the calculations. We would want to set $N_S \to \infty$ for accurate answers. These are shown as the solid lines in figure (b). For a zeroth order estimate, we may want to set $N_S = 1$ — the corresponding answers are shown as dashed lines in figure (b). We see that $N_S = 1$ gives a good estimate when $y_0/h < 0.7$. In a typical solar farm setup, we would see $y_0/h < 0.7$ and $N_S = 1$ would give a very good estimate allowing us to do a back of the envelope estimate of the back face light collection possible by a bifacial panel array. In this practical scenario, we see that $F_{PV\text{-}LGnd}$ only approaches 0.3–0.4. In other words, we will collect $R_A \times F_{PV\text{-}LGnd}$, i.e., 15–20% (for $R_A = 0.5$) of sunlight at the back. These estimates will change slightly if the panels are tilted and the sunlight is not vertical.

Total light collection. Finally, we can write the total light collection by the stand-alone bifacial panel as

$$\begin{aligned} P_{PV}^{(bifacial)} &= P_{PV(dir)}^{(bifacial)} + P_{PV(diff)}^{(bifacial)} + P_{PV(Alb)}^{(bifacial)} \\ &= h \times I_{\text{DNI}} \cos\theta_e + h \times I_{\text{DHI}}(1 - \sin\theta_p) + hI_{\text{DHI}} R_A \cos\bar{\theta_1} \left(1 - \tan\frac{\theta_p}{2}\right) \\ &\quad + h \times \frac{1}{2} I_{dir} R_A \left(1 - \tan\frac{\theta_1^*}{2} + \tan\frac{\theta_e}{2} - \tan\frac{\theta_p}{2}\right). \end{aligned} \tag{13.43}$$

The array is turned off for $\theta_e < \theta_p$. Note that $\theta_e = \pi/2 - |\theta_Z|$. The total bifacial collection is shown in Fig. 13.1(ii)(f).

13.2.3 Energy output of the panel and farm

The daily energy output can be found by integrating $\eta \times P_{PV}^{(bifacial)}$ (or $\eta \times P_{PV}^{(mono)}$) over the hours of the day. Here, η is the conversion efficiency of the panels. We also assumed sunrise and sunset at 6 a.m. and 6 p.m., respectively. The daily energy output per panel from a vertical bifacial panel array as a function of the period p is shown by the blue solid line in Fig. 13.6(a). Here, we set $h = 1\,\text{m}$, $\eta = 20\%$, $R_A = 0.5$. The red dashed line shows the output from a similar-sized monofacial panel (lying flat on the ground). When the period (or p/h) is small, the vertical bifacial panels tend to collect very little due to shading between adjacent panels. For a very large period, the vertical bifacial panel yields more energy compared to a single monofacial panel. This situation, with a very large period, corresponds to the stand-alone case discussed in Sec. 13.2.1.

Homework 13.5: Vertical bifacial solar farm on Mars and Earth

1. The formulation shown here does not specifically mention that the farm is on Earth! If we want to do some calculations for a vertical BPV farm on Mars, should there be any simplifications? [Hint: We should expect very little diffuse light on Mars.]

2. Compare the farm output on Mars versus Earth at noon. Consider only direct sunlight, and revisit Chapter 1 to find the difference in solar irradiance on Mars versus Earth.

3. Earth and Mars have a 24-hour and 25-hour day, respectively. Assume that sunlight is available for 8 hours and 9 hours, respectively. Estimate the farm output per day on these two planets.

Now, in the case of an array of panels (solar farm), we also need to find the energy output per unit farm area — this can be found by integrating $(1/p) \times \eta \times P_{PV}^{(bifacial)}$ over the hours of the day (see Fig. 13.6(b)). As we can see, the vertical bifacial solar farm (blue solid line) has the maximum output per land area with $p/h \approx 1$. The output per land area from the monofacial farm (panels flat on the ground) is shown by the red dashed lines. It appears that the output from a monofacial farm is always higher than the energy output of a vertical bifacial farm. This is not always the case. We will show in the next chapter that from a global

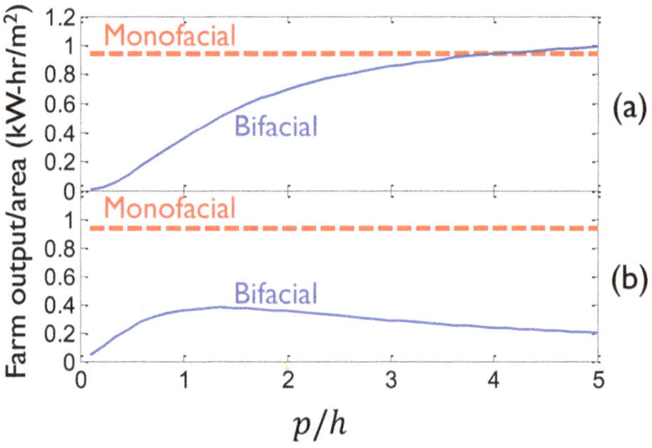

Figure 13.6: Energy output (a) per panel and (b) per farm area for a monofacial (red dashed line) and a vertical bifacial (blue line) panel array as a function of normalized period, p/h.

perspective, the vertical bifacial farm can have a higher yield compared to a monofacial farm in many places on Earth.

13.3 Conclusions

In this chapter, we formulated panel and farm energy outputs for E-W-facing vertical bifacial panels — we assumed that the sun moves directly from east to west. For this sun path, the monofacial panel should be set up horizontally on the ground, thereby collecting the full GHI. A stand-alone vertical bifacial panel can collect more than a monofacial panel with the help of albedo from a large surrounding ground (assuming high R_A). In an array setup, mutual shading among vertical panels significantly degrades the output compared to horizontal monofacial panels.

In a more realistic scenario, the sun path is tilted from the E-W path depending on the latitude. Thus, the monofacial panels need to be tilted toward the sun, and mutual shading starts to contribute to losses. In the next chapter, we will discuss the practical configurations for monofacial and bifacial panel arrays (farms) considering the actual sun path and meteorological variations in insolation for different locations.

Homework 13.6: Energy output of tilted panels

In Homework 13.2, we considered an array of vertical panels. Now consider an array of panels tilted at an angle β with the ground. With a row spacing of r, the period is $p = r + h \cos \beta$.

1. What is the view factor of dl on *front face* at position l along the panel ($l = 0$ is ground position) to sky (F_{dl-sky})? [Hint: As discussed earlier in the text, the view factor from a strip dl onto region A subtended by angles ϕ_1 and ϕ_2 (angles measured normal to dl) is $F_{dl-A} = 1/2 \times (\sin \phi_2 - \sin \phi_1)$.]
2. Find the total view factor $F_{h-sky} = 1/h \times \int_0^h F_{dl-sky} dl$.
3. Use this to find the total diffuse light collected onto the panel.

Repeat the calculations for the back face.

References

[1] Antonio Luque and Steven Hegedus, Editors. *Handbook of Photovoltaic Science and Engineering, Second Edition*. March 2011.

[2] M. F. Modest. *Radiative heat transfer*. Academic Press, New York, *Third Edition*, 2013. OCLC: ocn813855549.

[3] M. Ryyan Khan, Amir Hanna, Xingshu Sun, and Muhammad A. Alam. Vertical bifacial solar farms: Physics, design, and global optimization. *Applied Energy*, 206 (Supplement C):240–248, 2017.

CHAPTER 14

Solar Farms: Practical Perspectives

Chapter Summary

- There are a variety of solar farms being installed worldwide, including fixed-tilt monofacial, fixed-tilt bifacial, sun-tracking E-W and N-S oriented farms, etc.
- The energy output of an optimally-tilted solar farm depends on the geographical location and the season.
- The energy output of vertical solar farms can be increased by landscaping the ground between the rows.
- Land cost is driving multiple innovations in farm design, including agrophotovoltatics, floating solar farms, etc.
- Careful economic analysis determines the farm-type suitable for a given application.

14.1 Introduction

In the previous chapter, we discussed the basic formulation and analysis of the energy yield of a solar farm. For simplicity, we assumed a fixed insolation (GHI), and a sun trajectory that moves directly overhead from east to west. In practice, the insolation depends sensitively on the latitude, season, and weather. In this chapter, we will discuss various farm configurations shown in Fig. 14.1 and their energy yields while considering practical sun paths and meteorological insolation data.

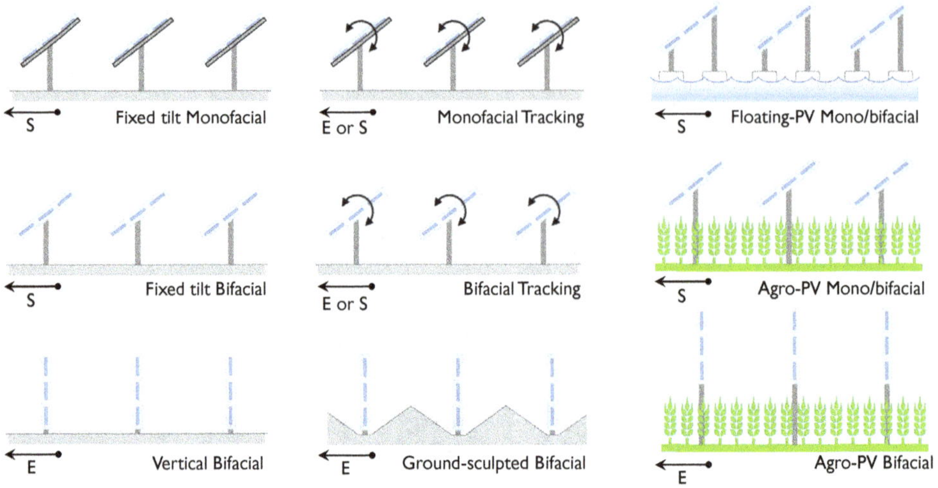

Figure 14.1: Module technology (monofacial, bifacial, tracking) and the land availability (desert, lake, agriculture farmland) dictate the optimum design of a variety of solar farms.

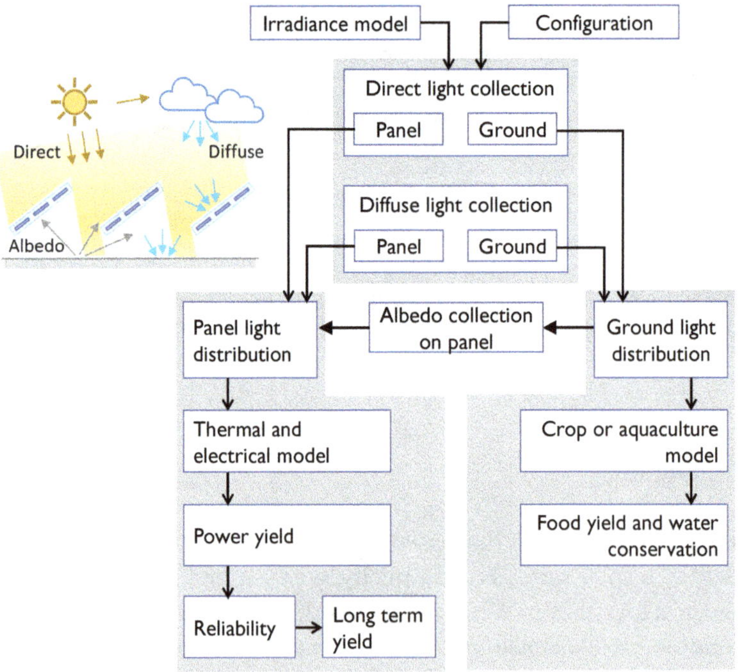

Figure 14.2: A flow diagram describing the various steps for calculating the energy yield of various types of solar farms.

14.2 Global insolation

For standard testing of solar cells and panels, we assume the spectrum AM1.5G, which corresponds to 1 kW/m^2 of radiation power from the sun. However, in practice, the insolation varies from sunrise to sunset. As discussed in Chapter 1, the daily insolation also has seasonal variations. Moreover, the sunlight intensity scales down as we move from the equator toward higher latitudes. The annual integrated insolation (global horizontal illumination, GHI) map is shown in Fig. 14.3(a). The yearly integrated GHI averaged over longitudes is also shown as a function of latitudes in Fig. 14.3(b). As expected, the annual insolation decreases away from the equator. However, there is some change in insolation over the same latitude (see the map in Fig. 14.3(a)) — this is due to variation in average clearness of the sky at different locations. The monthly or seasonal insolation variation can be understood from Fig. 14.3(c). From March to September, the Northern Hemisphere is tilted toward the sun. That is why we expect higher insolation on the Northern Hemisphere (latitudes $> 0°$ in Fig. 14.3(c)) in July. Similarly, the Southern Hemisphere has higher insolation in November. The earth reaches the equinox (both poles are at the same distance from the sun) in March and September. That is why, for example, in March, the insolation versus latitude is approximately symmetric around the equator.

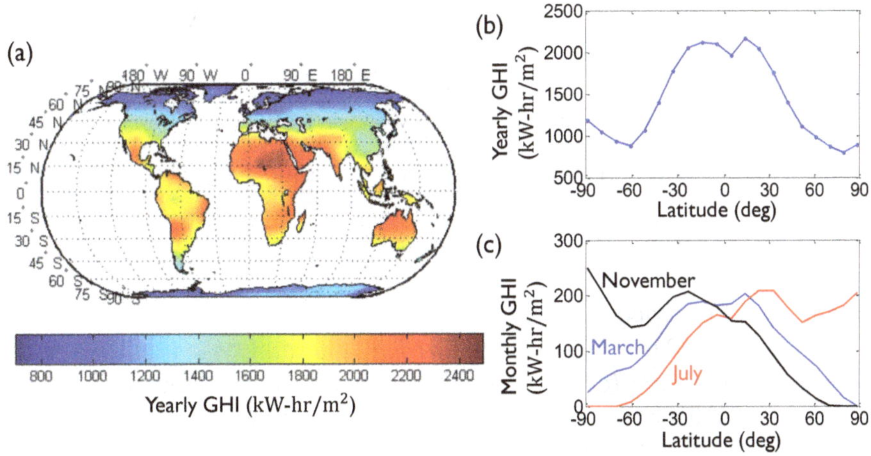

Figure 14.3: (a) Yearly integrated insolation (GHI) map. (b) The yearly integrated GHI in (a) is averaged over longitudes and presented versus latitudes. (c) Monthly variation in insolation as a function of latitude can be understood by the position of the sun in the sky.

To choose among the options shown in Fig. 14.1, we must first consider the location — this will define the tilt of the sun (sun path), and the yearly insolation. For example, the sun paths at latitudes 30° N and 60° N in July are compared in Fig. 14.4(a). The sun path will change through the seasons as shown for 30° N in

Fig. 14.4(b). The sun paths will constrain the optimal tilt of the panel at a given latitude. Then an overall analysis based on the hourly and seasonal change in illumination defines the optimum spacing between rows of panels in the farm.

14.3 Monofacial panel farm: Optimized for minimum shading

Monofacial panels collect light only from the front, transparent face of the panel — this conventional configuration is widely available from different solar panel manufacturers. Ideally, we would want the panel array to face the sun for maximum sunlight collection. At latitude L, the optimum tilt of the monofacial panels β can be estimated as follows:

$$\beta = 0.69|L| + 3.7^\circ. \tag{14.1}$$

The optimum tilt β has been shown versus latitude in Fig. 14.5(a). In the Northern Hemisphere, the sun path is inclined toward the south; therefore, the tilted panels face the south. Conversely, the panels face the north in the Southern Hemisphere.

Next, to define the farm configuration, we need to find the row spacing in the array. In this section we assume the following, conventional constraint for the design: *the annual energy loss due to row-to-row shading should be kept below 5%.* The noontime shadow is the longest in winter when the sun's zenith is $\theta_{Zw} = (|L|+23.5^\circ)$ (see Fig. 14.4). For a panel height h, the apparent height is $h_y = h\sin\beta$, and the corresponding shadow length "beyond panel edge" $g_{\rm s}$ will approximately define the row spacing $R_{\rm s}$. Therefore,

$$R_{\rm s} = g_s = h_y \tan\theta_{Zw}, \tag{14.2}$$

and thus the period is $p = R_{\rm s} + h\cos\beta$. The set-back-ratio (SBR) is,

$$\text{SBR} = \frac{R_{\rm s}}{h_y} = \tan\theta_{Zw}. \tag{14.3}$$

These estimates for period p, row spacing $R_{\rm s}$, and SBR are shown by the dashed lines in Fig. 14.5(a). Here, we assume that $h = 1.5$ m. Obviously, the shadow is longer in the morning (or afternoon) compared to noon. However, sunlight contribution is smaller early or later on a winter's day. That is why the numerically calculated values of p, $R_{\rm s}$, and SBR (dots) are slightly larger than the analytical estimates (dashed lines), as shown in Fig. 14.5(a). The $< 5\%$ shadow loss constraint results in an unreasonably large row spacing beyond 60° latitude — therefore, we will not consider the results for those regions. These high-latitude farms are better designed by maximizing energy output, to be discussed in Sec. 14.4.

Finally, the annual energy production of the designed monofacial solar farm is shown in Fig. 14.5(b). Let us assume that the panel efficiency is 18%. Naturally, more energy is produced closer to the equator due to higher annual insolation. A more detailed inspection of Figs. 14.3(a) and 14.5(b) indicates that the overall farm

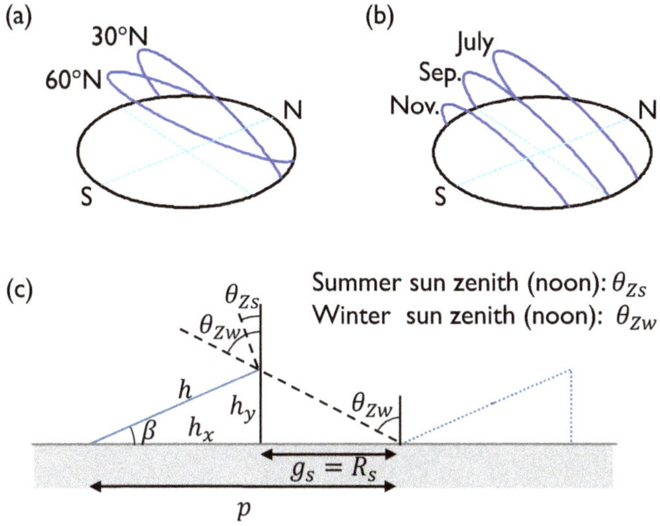

Figure 14.4: (a) Variation in sun path with latitude in July. (b) Seasonal variation in sun path at 30° N. (c) Tilted panel array configuration and relevant design parameters, i.e., tilt-angle β, row-to-row spacing, p, etc.

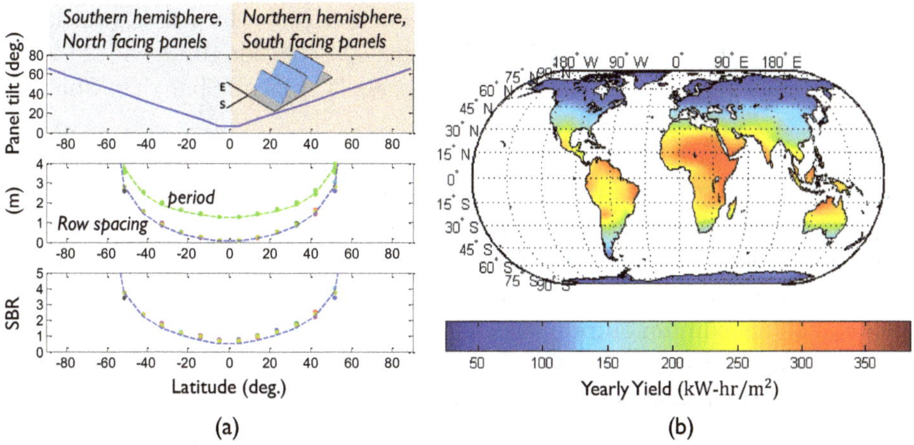

Figure 14.5: (a) Panel tilt, row spacing, period, and SBR shown as a function of latitude. The design assumes a $\leq 5\%$ yearly shading loss constraint. (b) The corresponding yearly yield map of an optimally-tilted monofacial solar farm.

efficiency is ~15–16% close to the equator. Around 45° latitude, the overall farm efficiency reduces to ~10% as longer shadows increase panel spacing.

Homework 14.1: The exact calculation of the setback ratio requires a slightly more complicated calculation

Calculate the *apparent* N-S zenith angle on December 21, 3 hours before the solar noon.

Solution. To calculate the shadow at any location 3 hours before and after the solar noon (e.g., 9 a.m. and 3 p.m.), we must project the actual location of the sun onto the N-S plane, so that

$$\tan \theta_{Zw,a} = \tan \theta_{Zw} \cos \gamma_s \tag{14.4}$$

where γ_s is the azimuth angle and $\theta_{Zw,a}$ is the apparent zenith angle of the sun at 9 a.m. (or, 3 p.m.) for someone in the Northern Hemisphere facing the south. It is easy to show that $\theta_{Zw,a} \leq \theta_{Zw}$. The SBR predicted by the noontime method actually ensures zero row-to-row panel shading. In practice, some installers choose to use SBR based on the 9 a.m. shading condition. This may improve the integrated energy output of a finite size solar farm.

Homework 14.2: Latitude-dependent tilt and spacing of solar modules

Calculate the tilt angle and the spacing of monofacial solar farms for three locations: West Lafayette, Indiana, USA (latitude: 40.27° N; longitude: 86.91° W); Chennai, India (latitude: 12.99° N; longitude: 80.23° E); and Shanghai, China (latitude: 31.23°; longitude: 121.47° E)

Solution. Let us solve the problem in two steps. First, let us consider the tilt angle of the module. The equator is tilted by $\theta_{\text{lat}} = 23.45°$ with respect to the solar plane containing the sun and the earth. Had it not been tilted, the modules must be tilted by an angle equal to the latitude of the solar farm, namely, $\beta = \theta_{\text{lat}}$. For maximum output, one may tilt the angle continuously throughout the year between $\theta_{\text{module}} = \theta_{\text{lat}} \pm 23°$. A fixed-tilt system optimizes for the difference in insolation and length of the day in summer vs. winter, where the tilt angle is given by Eq. (14.1), i.e., $\beta = 0.69\,\theta_{\text{lat}} + 3.7$ degrees. Therefore, the farm at Lafayette, Chennai, and Shanghai should be tilted at 31.79° N, 12.92° N, and 24.83° N, respectively.

(*continued on the next page*)

Homework 14.2 *(continued from the previous page)*

For the row spacing and SBR, we use Eqs. (14.2) and (14.3). The noontime zenith angles (θ_{Zw}) for Lafayette, Chennai, and Shanghai are 63.75°, 36.95°, and 54.68°, respectively. The corresponding SBR are 2.02, 0.75, 1.41. The SBR ratio using the projected zenith angle ($\theta_{Zw,a}$) from Eq. (14.4) would need to be calculated based on a more detailed latitude specific sun path.

Homework 14.3: Estimating the size of a solar farm

What is the size of a solar farm to produce a specific output power, P_{farm}, based on per module output power, P_{module}?

Solution. The array of panels in a farm has a period $p = R_s + h\cos\beta$. Therefore, the land area per period is (assuming a square panel with width and length both equal to h).

$$A_{period} = h \times p = h(R_s + h\cos\beta) = h^2(\text{SBR} \cdot \sin\beta + \cos\beta).$$

Here, we used $R_s = \text{SBR} \times h_y = \text{SBR} \times h\sin\beta$. SBR can be calculated assuming no shading conditions in noon or morning in winter. Therefore, the size of a solar farm equals

$$A_{\text{T}} = A_{period} \times \text{the number of modules} = A_{period} \times \left(\frac{P_{farm}}{P_{module}}\right).$$

14.4 Monofacial panel farm: Maximized farm output

In Sec. 14.3, we calculated the energy yield of a solar farm by requiring that the row-to-row shading is minimized. In this section, we consider an alternative optimization of the monofacial farm by maximizing the yearly energy output *per land area*. First, the panel tilt angle is determined by the average sun inclination at the specific location/latitude — therefore, we use the formula in Eq. (14.1). The corresponding period (or row spacing) maximizes the annual energy output (see Fig. 14.6(a)). Interestingly, up to latitude 35°, the output-optimized array period is approximately the same as that found for the "5% shadow loss constraint" design in the previous section. Therefore, the annual output is the same as before up to latitude 35°.

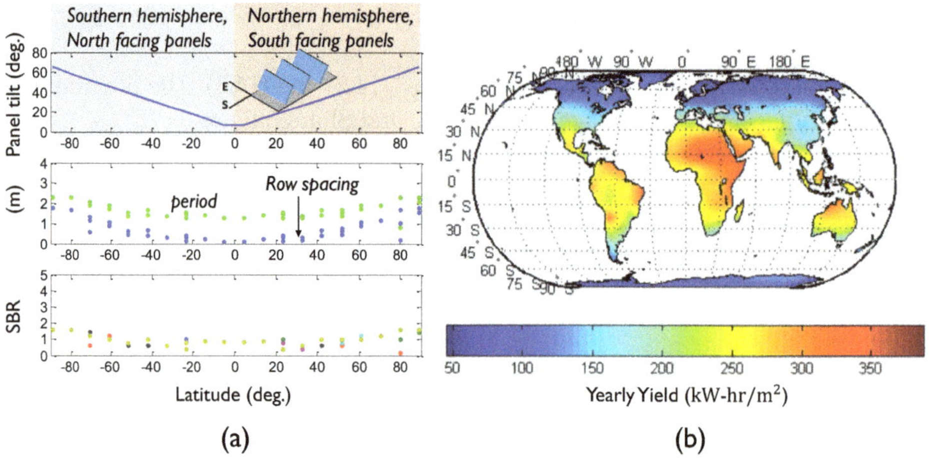

Figure 14.6: (a) Panel tilt, row spacing, period, and SBR shown as a function of latitude. The design maximizes the yearly output for each location. (b) The corresponding yearly yield map of a monofacial solar farm. Compared to minimum-shading design (see Fig. 14.5), the maximum energy based designs have smaller row spacing and SBR, especially at high latitudes.

Homework 14.4: Monofacial farm yearly output

Consider the monofacial panel array (farm) design principles discussed in this section (with no panel-to-panel shadow throughout the year): we know how to find panel tilt β and row spacing R_s as a function of latitude L. We can find the average illumination as a function of L from Fig. 14.3(b). Find the average farm output per land area at latitude $L = 30°$.

Solution. Assume only direct sunlight I_{DNI}. In winter, average sun zenith is θ_{Zw}, and the rays make $(\theta_{Zw} - \beta)$ angle with the normal of the panel resulting in collection $I_{\text{DNI}} \cos(\theta_{Zw} - \beta)$ per panel area. Therefore, the corresponding PV output per land area is,

$$I_{PV}^{(w)} = \eta I_{\text{DNI}} \cos(\theta_{Zw} - \beta) \times \frac{h}{p}.$$

Similarly, for summer:

$$I_{PV}^{(s)} = \eta I_{\text{DNI}} \cos(\theta_{Zs} - \beta) \times \frac{h}{p}.$$

(continued on the next page)

Homework 14.4 (*continued from the previous page*)

Also, remember that, $p = R_s + h\cos\beta$ is related to θ_{Zw} to avoid shading between panels. Thus, with some algebra:

$$\frac{h}{p} = \frac{1}{\sin\beta\,\tan\theta_{Zw} + \cos\beta}. \tag{14.5}$$

For simplicity, let us assume that 50% of the time is winter and 50% is summer in the year:

$$I_{PV}^{(Year)} = \frac{1}{2}\left[I_{PV}^{(s)} + I_{PV}^{(w)}\right].$$

We now have all relevant equations to find the yearly farm output. At $L = 30°$, we have $\beta = 24.5°$, $\theta_{Zw} = 53.5°$, $\theta_{Zs} = 6.5°$, $I_{\text{DNI}} \approx$ Yearly GHI $= 1800$ kWhr/m^2, and $\eta = 18\%$. Thus, $I_{PV}^{(Year)} \approx 257.8$ kWhr/m^2, which is very close to the values estimated by numerical modeling and plotted in Figs. 14.5(b) and 14.6(b).

At higher latitudes, for example in Washington, DC, the optimization results in panel spacing such that the output is maximized per land area although: (i) the reduced row spacing results in more shading losses, and (ii) the farm does not have any power output for 3 months during the winter. It may be undesirable to have no output for several months, even if the annual yield may be maximum. Therefore, at latitudes $> 35°$, there may be a trade-off design between the "shadow loss constraint" and "annual output maximization" depending on the energy demand and storage options in the location of interest.

14.5 Bifacial panel farm: Vertically aligned

Periodic cleaning of the panels in the farm can also contribute to the electricity production cost. (For details, see Chapter 17.) A recent study shows that vertical panels have low dust accumulation while having energy yields similar to conventional tilted panels [2]. That is why vertically aligned panel arrays are of particular interest. However, for vertical panels, we need to collect from both the panel faces. Typically, vertical bifacial panels are aligned to face the east–west direction. In this section, we will discuss the practical design considerations and the global annual energy output of a vertical bifacial farm. A vertical bifacial farm is not optimum for all locations; we will discuss the design of a tilted bifacial design in the next section.

During mornings and afternoons, mutual shading causes non-uniform illumination on the panel, with the lower part of the panel receiving less light than the top. For a panel constructed from a set of series-connected cells, bypass diodes are placed across different subsections of the series-string to avoid reverse breakdown of the shaded cells. (For details, see Chapter 18.)

Due to the array configuration, the front face of a panel is partially illuminated (partially shadowed) in the early part of the day. For example, the bottom part of the panel is shaded when sun elevation is low. In this situation, the bypass diode will turn off the bottom string of the panel, thereby limiting the output from only the top part of the panel. A similar situation occurs for the back face of the panel before sunset.

When the panels are packed close (i.e., small p), the panels on the farm have a bypass-diode-limited operation for a long period of each day — this greatly reduces power generation compared to light collection. Again, at large p, the output of each panel saturates (to the "stand-alone" panel limit), and thus the farm output per unit area decreases with increasing p. Therefore, there is an optimum p for which the power output *per land area* is maximized. The optimum p scales proportionally with h, i.e., the universality of the design holds for the p/h ratio. The optimum p/h as a function of latitude is shown in Fig. 14.7(a) — the spread in the data is due to the small variations along the longitudes. The optimum $p/h \approx 0.8$, close to the equator, and begins to increase above 30° latitude. The optimum p/h is within 0.8–1 for most locations in the world.

We can now look into the global optimization and yield of vertical bifacial solar farms. We assume a constant ground albedo of 0.5. As explained earlier, we expect

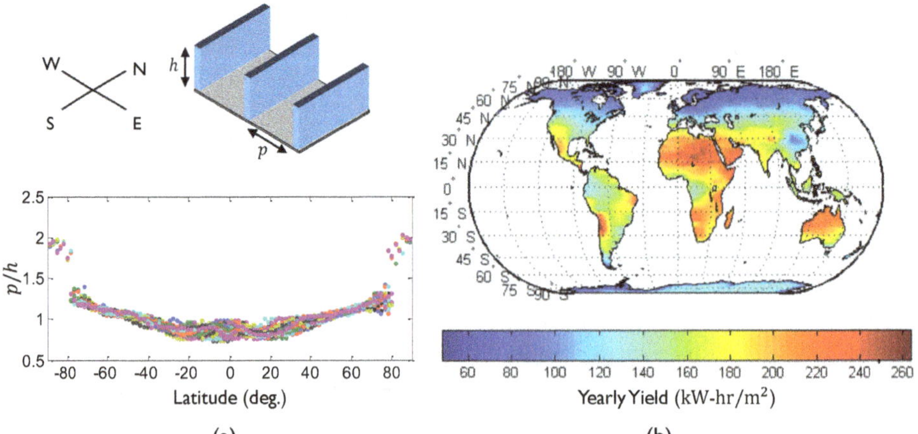

Figure 14.7: (a) Optimum p/h with varying latitude shown for an east-west-facing vertical bifacial solar farm. From Eq. (14.3) and Fig. 14.4, p/h of a vertical farm can also be defined as its SBR, and compared to Figs. 14.5 and 14.6. The scatter in the p/h ratio reflects longitudinal variation of farm design for the same latitude. (b) The yearly maximum yield of the optimized vertical bifacial farm.

a decrease in the GHI and output for increasing latitudes. In addition, there are variations in design and output along the longitudes due to meteorological variations. The global yearly yield and the corresponding optimum p/h are shown in Fig. 14.7. We observe higher output in North Africa and Saudi Arabia compared to India and China due to clearer sky (i.e., higher k_T) and higher GHI.

Close to the equator, the monofacial panels are optimally tilted parallel to the ground, and the optimal row spacing is close to zero. In these locations, the monofacial panels collect the GHI fully, yielding the maximum output for *any* farm configuration. In these locations, in the absence of any soiling considerations, the energy output is twice as large compared to a vertical bifacial farm. The advantage of monofacial farms decreases at higher latitudes. At latitudes $> 60°$, the sun path is highly tilted. This results in prominent shading on adjacent south-facing monofacial panels. The shadows toward the East or West are relatively shorter; therefore, the east-west-facing vertical bifacial panels incur a lower shading loss. The bifacial panels allow the vertical farm to collect more energy both from the sky and from the ground compared to the optimally (and highly) tilted monofacial panel array. In these locations at high latitudes, the bifacial farm produces significantly more energy than monofacial farms. This is observed once we compare the maps in Figs. 14.6(b) and 14.7(b). Finally, even close to the equator, vertical farms could still be attractive if cleaning cost (e.g., water, labor, etc.) is high and an overall reduction in temperature improves farm operating lifetime. Levelized cost of electricity (LCOE) based optimization is essential to accurately quantify the possible gain in utilizing the vertical bifacial farm.

14.6 Bifacial panel farms: South-facing tilted panel

From the discussion in the previous section, we now understand that the east-west-facing, vertical bifacial panel array configuration is not optimum for most latitudes below $60°$. In this section we assume that the bifacial panels are south-north-facing and tilted at the same angle as the monofacial panels. Interestingly, the panel spacing optimizes to values similar to the monofacial case shown in Fig. 14.6(a). This results in the tilted bifacial farm output map shown in Fig. 14.8(b). Closer to the equator, the row spacing is small and there is very little light reflected from the ground — thus, the monofacial and the tilted bifacial farms have similar annual yields. However, with increased row spacing at higher latitudes, there is more opportunity for the bifacial to utilize albedo. For latitudes $> 30°$, the tilted bifacial farm yields $> 10\%$ extra annually compared to a monofacial farm (see Fig. 14.8(a)).

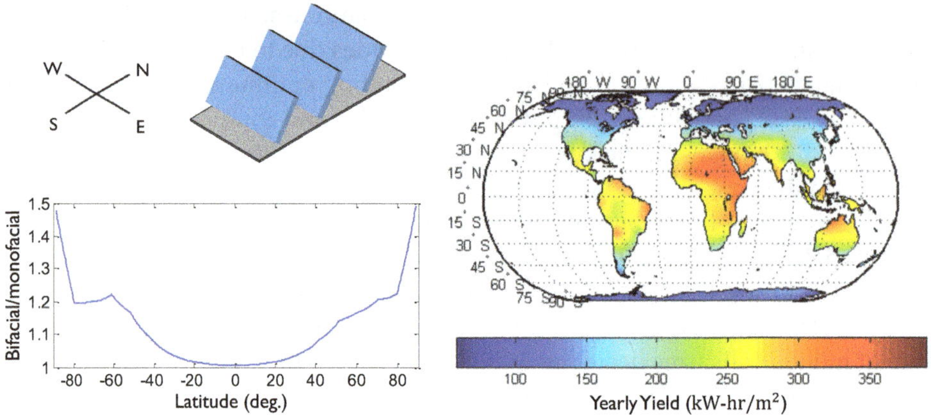

(a) (b)

Figure 14.8: The yearly output of a south-facing, tilted, optimally spaced bifacial panel array is shown in (b). The result is compared to an optimum monofacial farm output in (a).

14.7 Bifacial panel farms: Landscaping

The bifacial panels have the advantage of using ground reflection more effectively than monofacial panels. If a low-cost approach could increase the power output by 20–30%, it is conceivable that bifacial solar cells may outperform monofacial cells over the entire world. One possible approach is to pattern a highly reflective ground to maximize albedo collection on the bifacial panels. To test the hypothesis regarding the bifacial farm, one needs to answer the following questions:

1. What is the optimum orientation of a bifacial panel array?

2. Bifacial panels partially utilize light scattered from the ground. How can we shape the ground for maximum bifacial yield?

3. What is the optimum tilt to minimize soiling while maintaining high output?

These answers will quantify the gain in energy output and lowered LCOE of bifacial panels, assisting in rapid expansion in the market. For example, a ground-sculpted vertical bifacial (GvBF) solar farms can be viewed as an "upgraded" version of the vertical bifacial (vBF) farm for higher albedo collection. In this case, there are additional mutual shading considerations. Consider the example where the ground has been sculpted as a simple upward triangle, as in Fig. 14.9(a)(i). For DHI incident on panels, we must account for diffused light shading among the panels (Fig. 14.9(c)) and the ground (Fig. 14.9(d)). The insolation model must also track the hourly sun path to find the shadow cast on the ground for DNI (direct light). Such shadings are shown for two cases before noon in Figs. 14.9(e, f).

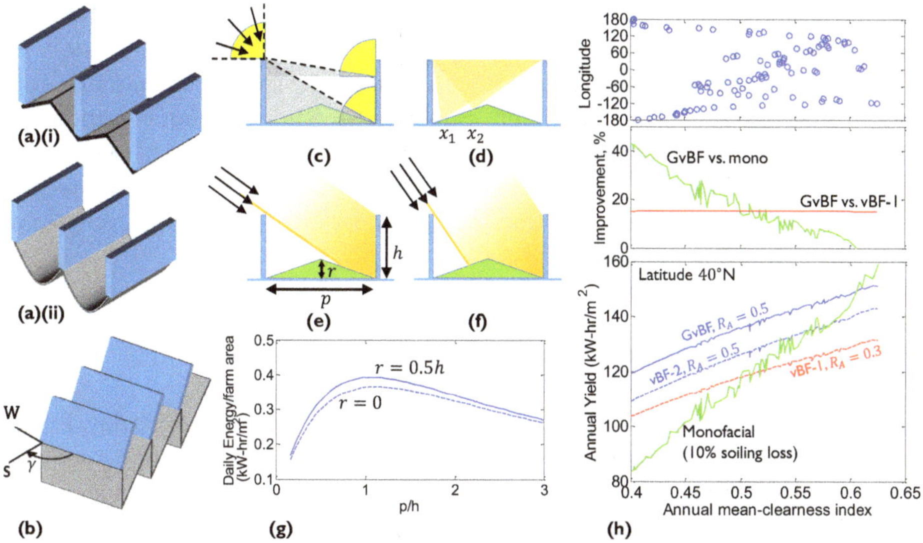

Figure 14.9: (a, b) Various types of ground-sculpting that increase light reflected on the solar panels. (c, d) Diffused sunlight illumination on the panel front face and on the ground, respectively. Partial illuminations from the ground and panel are shown in (e, f). (g) Optimization of farm output in Washington DC on September 22, 2009, with respect to panel gap p and ground shape r. (h) The energy output of monofacial and bifacial farms as a function of the clearness index (at latitude 40° N). Corresponding percentage improvement and longitudes are also shown. The calculation assumes 10% additional soiling loss for the monofacial farm.

Appropriate view-factors are used to calculate the collection of light on both the panel faces reflected from the illuminated portions of the ground (both direct and diffused sunlight).

Consider the energy yield of solar farms located at various longitudes at latitude 40° N. Figure 14.9(h) compares the annual yield of the farms (monofacial, vBF, GvBF) as a function of the annual mean clearness index. High clearness index k_T indicates a clear sky; therefore, GHI is high and DNI has a stronger contribution than DHI. As indicated by the top panel in Fig. 14.9(h), most locations have $k_T < 0.6$. The red dashed line shows the output of vBF with albedo $R_A = 0.3$ (typical for grass). The yield is improved for vBF with $R_A = 0.5$, and further increased for GvBF with $R_A = 0.5$ with patterned/sculptured ground. Such albedo can be achieved artificially, for example, by using white concrete having $R_A = 0.5$–0.6. The output of the GvBF is comparable to or better than the monofacial farm for most locations in the world. Creating a patterned surface with albedo 0.5, e.g., using white concrete, will add to the initial cost. However, the bifacial farm is expected to produce significantly more energy, while minimizing the cleaning cost. Moreover, the reduced infrared heating in bifacial cells will result in a longer lifetime and thus an even higher integrated energy yield over the panel lifetime compared to monofacial panels.

Note that monofacial panel arrays have a certain row spacing to minimize row-to-row shading, especially in winter. Summer has shorter shadows; therefore, there can be a significant amount of sunlight incident on the ground between rows. Optimally tilted bifacial panel arrays (e.g., see Fig. 14.9(b)) will partially recover this sunlight. With additional landscaping, bifacial farms show prospects for always outperforming monofacial farms.

14.8 Bifacial panel farms: Solar tracking

As shown in Figs. 14.4(a, b), the sun's position changes throughout the day. Even if we only trace the sun's position at noon at a given location, there is also a seasonal variation. Intuitively, we expect to collect more sunlight by continuously facing the sun. Such solar tracking system can either directly track the sun or can track the orientation that maximizes the net direct and diffuse light collection.

East–west-facing rows of panels in a tracking system would track the sun throughout the day. This would have a mismatch with the sun path tilted toward south/north. North–south (NS)-facing panel rows, on the other hand, track the seasonal variation of the sun. These panels cannot follow the diurnal positions of the sun from east to west. Figure 14.10(a) shows the global advantage of bifacial EW tracking over fixed-tilt bifacial panels, discussed in Sec. 14.6. The EW tracking underperforms only at high latitudes, where the sun path is highly tilted toward south or north. The bifacial EW tracking in fact yields more than the bifacial NS

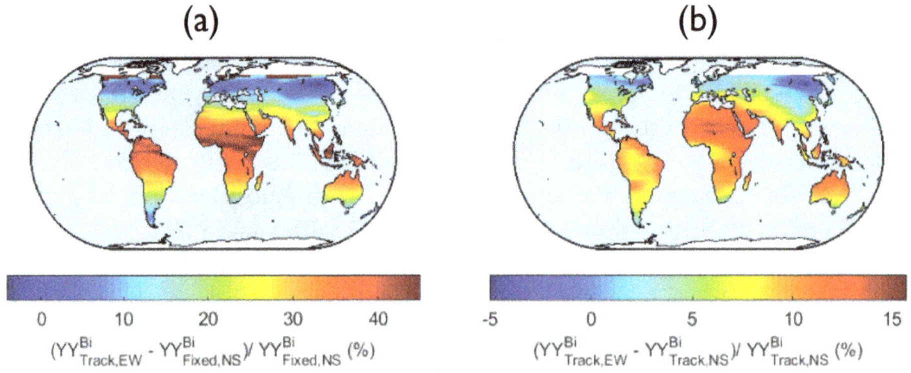

Figure 14.10: (a) Worldwide gain in yearly yield of an EW tracking bifacial PV system vs. a conventional fixed-tilt (north- or south-facing) bifacial PV farm. The gain can be over 40% near the equator. (b) The same plot except the relative advantages of EW tracking over NS bifacial tracking.

tracking systems for most locations over the globe (see Fig. 14.10(b)). While we expect higher output from solar tracking systems, we should remember that it also comes with reliability issues and costs for the additional moving components.

Homework 14.5: How much extra energy can we get by two-axis tracking?

Let us assume that a module is being illuminated only by direct light (I_{DNI}) and the diffuse light is absent. Since the two-axis tracking panel continuously faces the sun, it will receive all the sunlight:

$$I_{2\text{-}axis} = I_{\text{DNI}}.$$

A fixed-tilt NS-facing panel is oriented normal to the sun at the solar noon. As the sun path tilts at different angles across the sky in various seasons, a slow-moving single-axis tracking system continues to orient the panel perpendicular to the sun during the solar noon. Compared to the two-axis tracker, the average collected light by the single-axis tracker is reduced because the angle between the sun and the panel increases before and after noon:

$$I_{1\text{-}axis} = \frac{1}{(t_s - T_r)} \int_{T_r}^{t_s} I_{\text{DNI}} \sin\left(\pi \frac{t - T_\text{r}}{t_\text{s} - T_\text{r}}\right) dt = \frac{1}{\pi} \int_{-\pi/2}^{\pi/2} I_{\text{DNI}} \sin\theta d\theta = \frac{2}{\pi} I_{\text{DNI}}$$

where T_r and t_s are sunrise and sunset times. We have assumed a sinusoidal temporal variation in the angle between the panel and sun. It is clear that $I_{2\text{-}axis}/I_{1\text{-}axis} = \pi/2$, i.e., there can be up to 57% boost in yield if we use a stand-alone two-axis panel. The two-axis tracking is costly and needs more area per module, making them more suitable for concentrator solar cells.

14.9 Emerging solar farm technologies

The declining cost of modules and the scarcity of land area (e.g., in densely populated cities or agricultural farms in rural areas) have led to the design of new types of solar farms with very different cost and reliability considerations. In this section, we discuss two emerging farm topologies called floating solar farms and AgroPV (see Fig. 14.1).

14.9.1 Floating solar farms are being deployed across the world

Floating PV (FPV) is a new type of solar farm with modules floating on relatively calm waters of lakes, ponds, man-made dams, rivers, and coastal areas. The panels are placed on buoyant structures to keep them above the water's surface, and the buoyant structures themselves are either anchored to the ground below or tethered to the shore. These farms are particularly attractive if the land area is limited or expensive, especially in cities or close to agricultural farmlands. We will see in Chapter 16 that since the area is limited (and the module-to-area cost ratio M_L is large), the goal is to maximize the energy yield per unit area by laying the panels flat and staking them end to end. The cooling effect of water is expected to reduce the module temperature T_D (Chapter 5) and increase the efficiency and the energy yield. As additional benefits, FPV reduces water evaporation from dams and the shade may increase the yield of integrated fish farms. There are concerns about moisture-related reliability (e.g., corrosion): while moisture ingress is enhanced due to higher humidity, the reaction rates are reduced at the reduced operating temperature. We will discuss the physics of module corrosion in Chapter 23.

14.10 Agrophotovoltaic solar farms are being tested on a smaller scale

While solar farms are ideally installed in open deserts and open lakes, there is an emerging trend of integrating solar and agricultural farming (APV), which allows crop production under the solar modules. This approach provides dual land use for food energy production and offers the symbiotic benefits of reduced water evaporation and resilience against heatwaves and drought. Experimentally, the overall benefit is described by the concept of land equivalent ratio (LER), originally developed to describe the benefit of intercropping (polyculture) vs. single cropping (monoculture). In the context of APV,

$$\text{LER} = \frac{AY_{\text{APV}}}{AY_{\text{open}}} + \frac{YY_{\text{APV}}}{YY_{\text{PV}}}$$

where AY is the agricultural yield with and without the PV farm, and YY is the solar energy yield with and without the agricultural farm present.

Although LER is easily determined from experiments, it is difficult to predict it by computational modeling. Indeed, the design of solar farms for various crops (that are resilient to different degrees of spatio-temporal shading) involves a complex optimization problem and a predictive modeling of the total yield, as the correlated sum of the output of the solar farm (as discussed in this chapter) and the crop yield (with simulators such as APSIM, CERES, DAISY, EPIC, HERMES, WARM) is essentially impossible. (This is because the crop-yield simulators depend on local soil content and composition data. However, significant progress has recently been made in terms of creating such databases, e.g., Web Soil Survey (SSURGO) and Soil Grids (`https://soilgrids.org/`).) Nonetheless, the benefits of the integration may be described by the light productivity factor (LPF): by adding the relative yields of the PV energy (YY) and the transmitted photosynthetically active radiation (PAR) at the crop level, both of which can be calculated by the solar farm model developed in this chapter. In other words,

$$\text{LPF} = \frac{PAR_{\text{APV}}}{PAR_{\text{open}}} + \frac{YY_{\text{APV}}}{YY_{\text{PV}}}.$$

The first factor describes the loss of (PAR) between 400 and 700 nm due to shading by APV panels installed in an otherwise open field, while the second factor defines the solar energy in APV compared to a fully dedicated solar farm. Despite the losses of individual components, the LPV may be optimized to exceed 1 (subject to the constraint of maximum allowed loss in PAR and food productivity). In general, the APV modules are installed at one-half or one-third density, i.e., the p/h ratio being 2 or 3 times larger than that of a farm exclusively optimized for solar cells. Note that if AY $\propto$ PAR, LPF reduces to LER.

Historically, APV farms have relied on optimally tilted N/S monofacial farms. Bifacial modules have opened new options for farm design. The installation and operating costs are reduced by higher energy yield, lower elevation (especially for vertical modules), reduced soiling, etc. At half-array densities (e.g., $p/h \approx 4$), the optimally tilted N/S farms and the E/W vertical farms produce comparable PAR loss and energy yield, with an overall LPF gain of 30%.

Homework 14.6: Machine learning allows ultra-fast calculation of optimum solar farms

Supercomputers are needed for the parametric sweep calculations to find the optimum spacing, tilt-angle, and energy yield for a given technology (e.g., PERC, HIT, thin-film) deployed in a specific farm topology (e.g., vertical, Fig. 14.7; fixed-tilt, Fig. 14.7; tracking, Fig. 14.10; agrophotovoltaics, etc.). Read the paper "Machine Learning Allows Synthesis and Functional Interpolation of Computational and Field-Data for Worldwide Utility-Scale PV Systems," by T. Patel and M. A. Alam (in Proceedings of Photovoltaic Specialist Conference, 2021) to explain how a relatively simple functional interpolation can reduce the computational time by several orders of magnitude.

Solution. Functional interpolation can be explained as follows. Assume that we wish to fit a set of x-y data by the model equation: $y = a + bx + cx^2$. Here the basis functions are: x^0, x^1, and x^2. Minimizing the root-mean squared error (RMSE) between the data and the model-prediction, we determine the unknown coefficients: a, b, c. Given the parameters, one can predict $y(x)$ for arbitrary x. Indeed, we can use any basis function and determine the corresponding best-fit model coefficients, e.g., $y = w_2 \tanh(w_1 x + b_1) + b_2$ to develop a predictive model of $f(x)$, see Fig. (a) below. This specialized equation for data-fitting, based on sigmoidal-shaped basis functions, is called Machine Learning, where an input bubble containing x is transformed by another bubble representing $\tanh(x; w_1, b_1)$.

(continued on the next page)

Homework 14.6 *(continued from the previous page)*

For a solar farm, the energy yield (YY) is determined by four input variables ($M = 4$), i.e. x_j = latitude, GHI, temperature, and clearness index. If there are N 'nuerons' or 'bubbles', we can calculate, as in Fig. (b), the coefficients w_{2i}, b_2, and $w_{1,ij}$, and b_{1i} for the equation

$$\mathrm{YY} = \sum_{i=1}^{N} \left[w_{2i} \tanh\left(\sum_{j=1}^{M} w_{1,ij} x_j + b_{1i} \right) \right] + b_2$$

by fitting the equation to the super-computer generated energy yield data for a few locations in the world (e.g. Fig. 14.10). As shown in Fig. (c), this fitted equation will then predict, at a fraction of the computational cost, the energy output at any location of the world for arbitrary combination of latitude, GHI, temperature, and clearness-index.

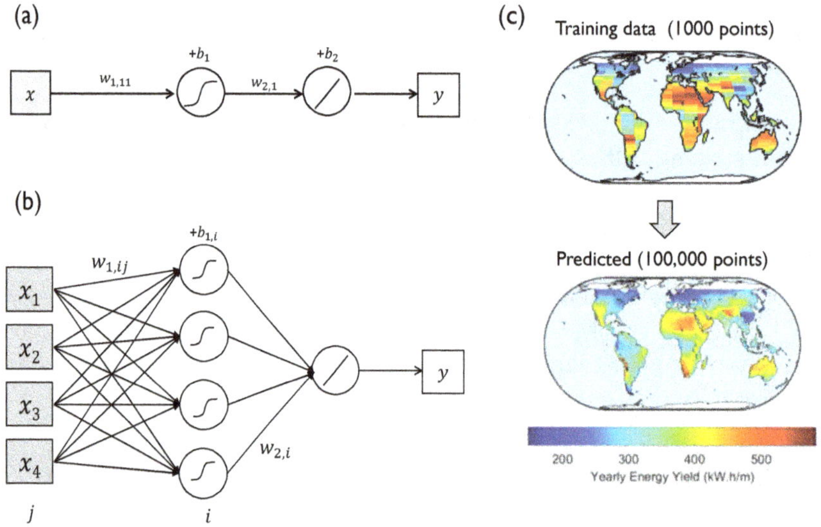

Homework 14.7: Choosing the optimal location of a solar farm involves a multi-criteria optimization

The world maps in Figs. 14.7 and 14.8, for example, can be used to identify locations with high energy yield based on local solar irradiance, ambient temperature, and the module technology. The optimal location of a solar farm,

(continued on the next page)

Homework 14.7 (*continued from the previous page*)

however, involves other factors, such as, geographical information (e.g., the distance from road, forest, residential areas, national boundaries, electrical grid, etc.), topological information (e.g., flatness of the terrain, shadows cast by any neighboring hills, and so on), and weather information (e.g., number of dusty or cloudy days). Different databases, such as, Google Open Street Map (OSM), NASA Digital Elevation Map (DEM), etc. have the relevant information. One assigns relative weights to these factors in a multi-criterion decision system (MCDS) to identify the best possible location for a solar farm. For example, see, Mokarram, Marzieh, *et al.* "Determination of the optimal location for constructing solar photovoltaic farms based on multi-criteria decision system and Dempster-Shafer theory." Scientific Reports 10.1 (2020): 1–17.

14.11 Conclusions

In this chapter, we discussed the annual energy yield of different types of solar farms. The choice and optimum design of a farm depends on several factors: (a) geographical location, which determines the illumination and ambient temperature; (b) module technology, such as monofacial or bifacial; (c) the ratio of module to land cost (M_L^*) included in the levelized cost of electricity (LCOE); (d) type of solar installation (vertical, tilted, tracking); and (e) the target application, e.g., stand-alone vs. utility-scale installation. Once the five factors are specified, one calculates the results presented in this chapter. For example, the results in this chapter demonstrate that the bifacial technology improves energy yield by 5–15%. The daily power output is flatter and has lower temperature sensitivity, and the technology is better suited for integration with agricultural farmland. Similarly, floating solar cells are suitable for locations with low module-to-land cost ratio.

The actual energy output is lower than the idealized energy yield calculated in this chapter due to intrinsic and extrinsic losses. As discussed in Chapter 11, there are extrinsic losses associated with the components of an integrated system, such as inverters. In Part IV of the book, we will discuss the intrinsic losses associated with soiling, shading, solder bond failure, etc.

The most important challenge of a solar farm as a renewable energy source is that the sun does not shine during the night, nor is the power output constant throughout the day. We must store the energy generated so that consumers can best utilize the power when and where they need it. In the following chapter, we will discuss the operation and efficiency of various storage strategies developed for solar cell technology.

References

[1] Antonio Luque and Steven Hegedus, editors. *Handbook of Photovoltaic Science and Engineering, Second Edition*. March 2011.

[2] H. K. Hajjar, F. A. Dubaikel, and I. M. Ballard. Bifacial photovoltaic technology for the oil and gas industry. In *2015 Saudi Arabia Smart Grid (SASG)*, pages 1–4, December 2015.

[3] Greg A. Barron-Gafford, Mitchell A. Pavao-Zuckerman, Rebecca L. Minor, Leland F. Sutter, Isaiah Barnett-Moreno, Daniel T. Blackett, Moses Thompson, Kirk Dimond, Andrea K. Gerlak, Gary P. Nabhan, and Jordan E. Macknick. Agrivoltaics provide mutual benefits across the food–energy–water nexus in drylands. *Nature Sustainability*, 2(9):848–855, September 2019. Number: 9 Publisher: Nature Publishing Group.

[4] H. Marrou, L. Guilioni, L. Dufour, C. Dupraz, and J. Wery. Microclimate under agrivoltaic systems: Is crop growth rate affected in the partial shade of solar panels? *Agricultural and Forest Meteorology*, 177:117–132, August 2013.

[5] Vertical bifacial panels in agrivoltaics by Next2Sun.

[6] Hesan Ziar, Bjorn Prudon, Fen-Yu (Vicky) Lin, Bart Roeffen, Dennis Heijkoop, Tim Stark, Sven Teurlincx, Lisette de Senerpont Domis, Elias Garcia Goma, Julen Garro Extebarria, Ignacio Narvaez Alavez, Daniel van Tilborg, Hein van Laar, Rudi Santbergen, and Olindo Isabella. Innovative floating bifacial photovoltaic solutions for inland water areas. *Progress in Photovoltaics: Research and Applications*, n/a(n/a). _eprint: https://onlinelibrary.wiley.com/doi/pdf/10.1002/pip.3367.

[7] K. Kurokawa, A. Itoh, S. Tatebe, T. Sakurai, and Y. Ueda. Performance Analysis of PV Systems on the Water. *23rd European Photovoltaic Solar Energy Conference and Exhibition, 1–5 September 2008, Valencia, Spain*, pages 2670–2673, November 2008. ISBN: 9783936338249 Publisher: WIP-Munich.

[8] Yanlai Zhou, Fi-John Chang, Li-Chiu Chang, Wei-De Lee, Angela Huang, Chong-Yu Xu, and Shenglian Guo. An advanced complementary scheme of floating photovoltaic and hydropower generation flourishing water-food-energy nexus synergies. *Applied Energy*, 275:115389, October 2020.

[9] Giuseppe Marco Tina, Fausto Bontempo Scavo, Leonardo Merlo, and Fabrizio Bizzarri. Comparative analysis of monofacial and bifacial photovoltaic modules for floating power plants. *Applied Energy*, 281:116084, January 2021.

[10] Jamshad Hussain, Tasneem Khaliq, Ashfaq Ahmad, and Javed Akhtar. Performance of four crop model for simulations of wheat phenology, leaf growth,

biomass and yield across planting dates. *PLOS ONE*, 13(6):e0197546, June 2018. Publisher: Public Library of Science.

[11] Roberto Confalonieri, Marco Acutis, Gianni Bellocchi, and Marcello Donatelli. Multi-metric evaluation of the models WARM, CropSyst, and WOFOST for rice. *Ecological Modelling*, 220(11):1395–1410, June 2009.

[12] M. Ryyan Khan, E. Sakr, X. Sun, P. Bermel, and M. A. Alam. Ground sculpting to enhance energy yield of vertical bifacial solar farms. *Applied Energy*, 241: 592–598, 2019.

[13] Xingshu Sun, Mohammad Ryyan Khan, Chris Deline, and Muhammad Ashraful Alam. Optimization and performance of bifacial solar modules: A global perspective. *Applied energy* 212 (2018): 1601–1610.

[14] M. Tahir Patel, M. Sojib Ahmed, Hassan Imran, Nauman Z. Butt, M. Ryyan Khan, and Muhammad A. Alam. "Global analysis of next-generation utility-scale PV: Tracking bifacial solar farms." *Applied Energy* 290 (2021): 116478.

[15] M. Tahir Patel, M. Ryyan Khan, Xingshu Sun, and Muhammad A. Alam. A worldwide cost-based design and optimization of tilted bifacial solar farms. *Applied Energy* 247 (2019): 467–479.

CHAPTER 15

Storing Energy from Solar Cells

Chapter Summary

- The solar energy must be stored to meet energy needs during the night or on cloudy days.
- The solar energy can be stored as mechanical (e.g. pumped hydro), thermal (e.g. molten salt), electrostatic (e.g. supercapacitors), and electro-chemical (e.g. flow batteries) forms of energy.
- The storage techniques differ in terms of storage capacity and ability to deliver power promptly. Pumped hydro defines very large capacity systems, while supercapacitors can respond quickly.
- Flow-batteries and Li-ion batteries provide large capacity with fast response. These technologies are being developed actively.
- The storage cost must approach \$10 per kW-hr so that solar energy is fully competitive to other sources of energy.

15.1 Introduction

In the preceding chapters, we discussed how to design solar cells, assemble them into modules and panels, and install them in solar farms to generate electrical power directly from the sun (see Fig. 15.1). Electrical power, however, has a major problem — it must be consumed immediately after generation (to drive a motor or to light a bulb, for example); otherwise, we may as well not generate the power to begin with. Indeed, next time you turn on your table lamp, remember that the power being consumed was generated only a fraction of a second ago — the time needed for electricity to move from the power plant to your home through the electrical grid. How does this requirement of immediate consumption affect our strategy to generate solar power? A quick answer would be — quite significantly.

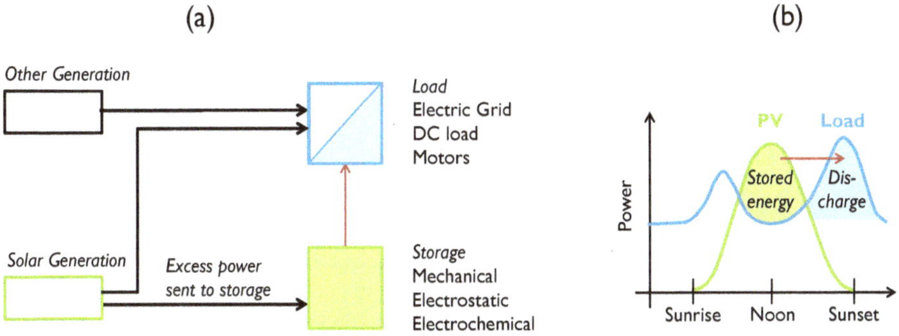

Figure 15.1: The solar energy produced must either be immediately consumed or stored for future use. (a) How PV and other generation sources feed the load. The excess, unused energy from PV is stored and later used by the load. (b) PV generation and the load of a typical day. The green shaded part under the PV curve is unused by the load and is stored. This stored energy is later discharged to support the peak load demand, as shown by the blue shaded region under the load curve.

The main problem arises from the variability of sunlight. The sun does not shine during the night. Even during the day, the power output changes from morning to noon, between seasons, and on a cloudy vs. clear day. If we aim to rely only on solar energy, how can we match the variability between power generation vs. power consumption? Well, if we could bottle the solar power in some "nonelectrical" forms and then reconvert to electricity on demand, we would be able to use the solar energy at arbitrary times, as shown in Fig. 15.1.

Actually, Nature has been doing this type of time-shifted energy conversion all along! Photosynthesis involves converting solar energy to chemical energy. The fossil fuel in our cars is nothing but stored solar energy from millions of years ago. The water cycle converts solar energy to the potential energy of the rain. And river dams allow us to convert the potential energy of water to hydroelectricity days or months afterward. Over the years, researchers have developed analogous systems for solar cells. We will discuss a few of those examples (electrostatic, mechanical, and chemical) in the following sections.

15.2 Electrical energy can be stored mechanically

As an example of "mechanical" storage, solar energy can be converted into potential energy by pumping water up from a lower to a higher reservoir (see Fig. 15.2(a)). We can regenerate electrical power on demand by allowing the water to run down to the bottom reservoir and turning a turbine-connected motor. The process acts as a mini water cycle. Other variants are also being tried: Insert a series of huge hollow spherical reservoirs several hundred feet under water within a river or close to the seashore and store the solar energy by pumping water out of the spheres

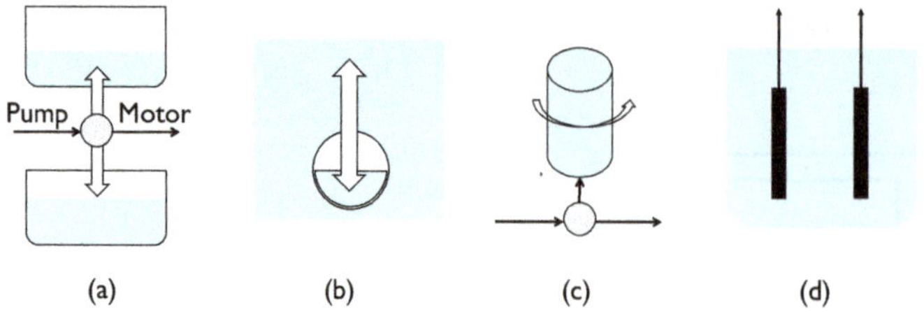

Figure 15.2: Various types of energy storage mechanisms: (a) water pumped to an elevated reservoir, (b) water pumped out of a submerged tank to create pressure difference, (c) heavy flywheel rotated to store angular momentum, (d) electrochemical cell (battery), etc. In general, pumped hydro systems offer larger storage capacity (e.g., GW-scale power for 4–8 hours), but at a lower energy density (0.1–2 W·h/L). On the other hand, electrochemical batteries offer lower storage capacity (e.g., hundreds of megawatts for minutes to hours), but with higher energy density (50–100 W·h/L).

(Fig. 15.2(b)). Electricity is generated as needed by allowing the water to flood back into the spheres. These hydro storage "virtual batteries" could be particularly suitable for floating solar farms, discussed in Chapter 14, Sec. 14.9.

Another approach shown in Fig. 15.2(c) involves turning an ultra-low friction flywheel and storing the solar energy in the angular momentum of large masses. When needed, the energy can be extracted back by connecting an electrical motor — the spinning slows as the energy is extracted out. Overall, many "mechanical" storage schemes are being considered, but the two-reservoir approach is still most popular because these large systems can deliver GW-scale power for a few hours to a full day. Note that in all these approaches, the DC solar energy must first be converted into AC electricity before connecting to the grid. For solar farms, connection to the power grid requires DC-to-AC conversion — therefore, the loss is already accounted for.

Homework 15.1: Hydro power storage with a submerged tank

Hydro power storage is simple and inexpensive, but its low energy density is a challenge. Figure 15.2(b) shows that excess solar energy E_{excess} can be stored by pumping water out of submerged tanks. Calculate the volume V_{tank} of the tank submerged at a depth of h. Also determine the power rating of the pump, P_{pump}.

Solution. For simplicity, let us assume that a floating solar panel (area, A_{PV}) produces E_{excess} uniformly over T seconds when the PV output exceeds the energy limit of the grid (see Fig. 15.1). The energy balance equation (i.e., $E_{\text{excess}} = E_{\text{pump}}$) defines the size of the tank V_{tank}, where

(*continued on the next page*)

Homework 15.1 (*continued from the previous page*)

$$E_{\text{pump}} = \eta_{\text{pump}} \times g\,h \times (\rho V_{\text{tank}})$$

$$E_{\text{excess}} = \eta_{\text{PV}} \times I_{\text{GHI}}(1 - \eta_{\text{load}})A_{\text{PV}}T$$

where η_{load} is the fraction of energy still being supplied to the load during the storage process. Taken together,

$$\frac{V_{\text{tank}}}{A_{\text{PV}}} = \frac{\eta_{\text{PV}}(1-\eta_{\text{load}})}{\eta_{\text{pump}}} \times \frac{I_{\text{GHI}} \cdot T}{\rho g h}. \tag{15.1}$$

Given that $\eta_{\text{PV}} = 0.2$, $\eta_{\text{load}} = 0.75$, $\eta_{\text{pump}} = 0.9$ (see Table 11.1), $I_{\text{GHI}} = 1000\ \text{W/m}^2$, $T = 2.5 \times 3600$ seconds for 2.5 hours of operation, $h = 5$ m, g = 9.81 m/s^2, $\rho = 1000\ \text{kg/m}^3$, we find that $V_{\text{tank}} = 10.19\ \text{m}^3$ for $A_{\text{PV}} = 1\ \text{m}^2$.

If the spheres with diameter d are arranged in a hexagonal configuration (packing factor, $\pi\sqrt{3}/6$), then it is easy to show that $A = (3\sqrt{3}/\pi) \times (V_{\text{tank}}/d)$. With $d = 1$ m, the area needed for the specific example above is 16.5 m^2. This relatively large area needed to store energy generated from 1 m^2 module reflects the low energy density, $\text{W} \cdot \text{h/L} = 2.5 \times 1000/(10.19 \times 1000) = 0.245$, and specific (or gravimetric) energy density, $\text{W} \cdot \text{h/kg} = 2.5 \times 1000/(10.19 \times 1000) = 0.245$, of the system. The numbers are identical because the weight of 1 liter of water is 1 kg. Increasing h will reduce the area and increase the energy density, but will make the tanks more expensive because they must withstand higher pressure. Finally, the power rating of the pump needed to support this small system is given by $P = \eta_{\text{pump}}^{-1}\, g h \rho V_{\text{tank}}/T \sim 60$ W. The power rating of the pump will scale with the size of the system.

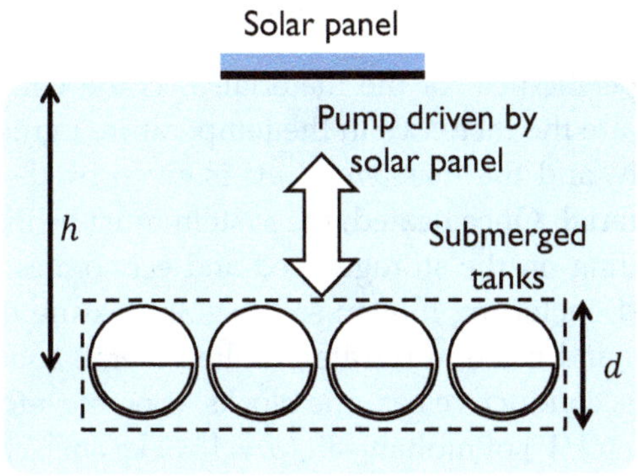

Homework 15.2: Worldwide grid and worldwide pumped hydro storage

Two solutions have been proposed to address the concern that the world cannot run on 100% renewable energy because sunlight is variable. Since the sun is always shining somewhere, a worldwide high-voltage power grid can transfer excess energy produced during the day to other parts of the world where the sun has set. The grid can be stabilized by distributed two-reservoir pumped hydro storage. Despite the low energy density of pumped hydro systems (typically a factor of 100–1000 lower than batteries, see Homework 15.1), these systems are relatively inexpensive and can be scaled to very large ($\sim$ GW) storage capacities. Each reservoir involves a pair of regions in the natural landscape separated by less than a few kilometers and with elevation difference of at least a few hundred meters, with typical height-to-separation ratio is 1–3, with a maximum of 10 (see Lee *et al.* Renewable Energy, 162, 1415–1427, 2020). A global search identifies hundreds of thousands of potential sites: `http://re100.eng.anu.edu.au/global/index.php`. Using the interactive map and Excel database, determine the locations, size, and potential energy storage for the country you were born in.

Homework 15.3: What about thermal storage?

In addition to the pumped hydro storage discussed above and electro-chemical storage to be discussed below, one can use thermal storage to store the solar energy. Thermal energy can be stored by increasing the temperature of a material (Sensible heat storage, SHS) or converting a material from solid to liquid (Latent heat storage, LHS), or a combination thereof, namely,

$$Q = mc_p\,\Delta T_s + mL + mc_p\,\Delta T_l,$$

where c_p is the specific heat of the material, L is the latent heat of fusion, and ΔT_s and ΔT_l are the increases in the temperature in the solid and liquid forms, respectively, and the mass $m = \rho V$ is given by the volume and the density of the material. Once heated, the system must be insulated to reduce heat loss. Depending on the storage need and economics, a variety of materials can be used, including molten salt (e.g., a mixture of sodium nitrate, potassium nitrate, and calcium nitrate), molten metal, solid materials, such as, graphite blocks, conductive ceramic blocks, concrete, etc. Let us calculate the energy density (Q/V) of molten-salt ($\rho = 1549\ \mathrm{kg/m^3}$, $c_p = 1530\ \mathrm{J/kg \cdot K}$)

(*continued on the next page*)

Homework 15.3 (*continued from the previous page*)

heated from 560 K to 840 K vs. conductive ceramic blocks ($\rho = 2000$ kg/m^3, $c_p = 1000$ J/kg · K) heated from 300 K to 2000 K to show that they offer performance comparable to battery technologies.

Solution. For molten salt, the energy density is

$$Q/V = \rho c_p \Delta T = 1549 \times 1530 \times (840 - 560) \times 2.77 \times 10^{-7} = 196 \text{ kWh/m}^3.$$

The last factor converts J to kW · hr. Similarly, for conductive ceramic blocks,

$$Q/V = \rho c_p \Delta T = 2000 \times 1000 \times (2000 - 300) \times 2.77 \times 10^{-7} = 941 \text{ kWh/m}^3.$$

The energy densities are indeed comparable to Li ion batteries at a few hundred kWh/m^3 (compared to 0.25 kWh/m^3 for water pumped to 100 m!). While the energy density of ceramic blocks is higher, molten-salt technology is cheaper. Overall, the storage cost must approach \$10 per kW-hr to make a technology economically viable.

15.3 Electrical energy can be stored in electrostatic capacitors

The relatively low specific energy density of pumped hydro systems (see Homework 15.1) encourages the use of various electrical storage techniques. For example, the electrostatic approach involves using the solar energy to charge a (very large) capacitor. The capacitor is then discharged to generate electricity as needed.

To understand the key considerations, let us consider a simple model of a solar cell connected in series to a storage capacitor, as shown in Fig. 15.3. For simplicity, a solar module is represented by its thermodynamic I-V characteristics of a series-connected module of M cells.

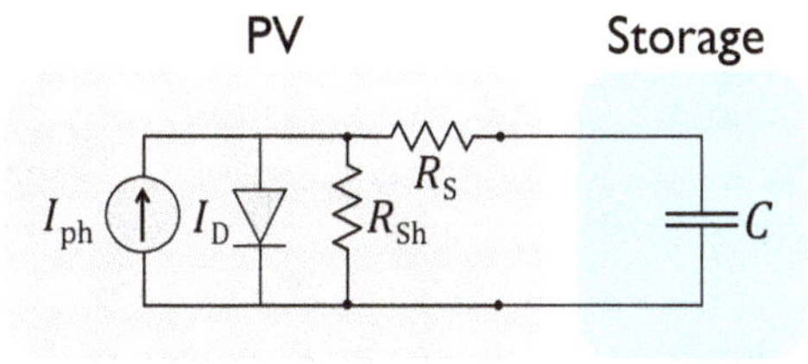

Figure 15.3: A solar cell can charge a capacitor to store energy for future use. In practice, one must use a connecting diode to prevent capacitor discharge when PV is not working, and a voltage regulator to prevent overcharging.

Consider a single solar cell connected to a capacitor. The capacitor will be charged by the current provided by the cell — the charging current relationship can then be written as follows:

$$C\frac{dV}{dt} = I = I_{\mathrm{ph}} - I_0\left(\exp\left[\frac{q(V+IR_{\mathrm{s}})}{k_{\mathrm{B}}T}\right] - 1\right) - \frac{V+IR_{\mathrm{s}}}{R_{\mathrm{sh}}}. \quad (15.2)$$

Here, I_0 and I_{ph} are the dark current and photogenerated current of the solar cell. For now, we will neglect the series and shunt resistances, which simplifies the solar cell I-V to that discussed in Chapter 7. The charging phase is then represented by the differential equation

$$C\frac{dV}{dt} + I_0(\exp(qV/k_{\mathrm{B}}T) - 1) = I_{\mathrm{ph}}. \quad (15.3)$$

If we assume that the capacitor was originally discharged, $V(t=0) = 0$, then (15.3) is solved to obtain

$$\exp(-V(t)/V_0) = b/a + (1 - b/a)\exp(-at) \quad (15.4)$$

where $V_0 = k_{\mathrm{B}}T/q$, $a \equiv (I_{\mathrm{ph}} + I_0)/V_0C$ and $b \equiv I_0/(V_0C)$. The final voltage saturates to $V_{\mathrm{oc}} = (k_{\mathrm{B}}T/q)ln(I_{\mathrm{ph}}/I_0 + 1)$. At this point, when the capacitor is fully charged, all the current is fed to the diode, nothing goes to the capacitor. Also, the stored energy in a capacitor of thickness d and area A_C is ultimately given by

$$U = CV_{\mathrm{oc}}^2/2 \quad (15.5)$$

where $V(t \to \infty) = V_{\mathrm{oc}}$, $C = A_C\kappa\epsilon_0/d$, and $E_{oc} = V_{\mathrm{oc}}/d$. Taken together,

$$U = A_C\kappa\epsilon_0 E_{oc}V_{\mathrm{oc}}/2. \quad (15.6)$$

A large capacitor stores more energy, as expected. An electrolyte-filled supercapacitor is preferred because it has large area (A_C) and high dielectric constant (κ). More importantly, its effective electrode separation (d) is defined by the Debye screening length of the electrolyte that automatically conforms to the shape of the electrode.

Homework 15.4: Charging and discharging time of a capacitor

1. Calculate the time needed to reach 95% maximum storage.
2. Assume that the daily illumination has the form $I(t) = I_{max}\sin(t/t_0)$. Calculate the time t_1 by which the capacitor is 95% charged.

15.4 There are a variety of electrochemical energy (EC) storage schemes

Solar energy can also be stored as electrochemical (EC) energy in traditional and/or flow batteries (see Fig. 15.4). In both batteries, the electrolyte ions shuttle between

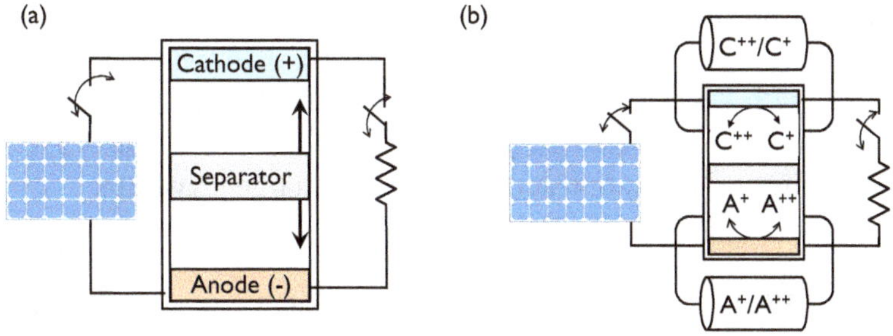

Figure 15.4: (a) A traditional battery stores energy in the electrodes. (b) A flow battery (or a hydrolysis cell) stores energy in separate reservoirs. The charging by solar panels and discharging through the electrical load are also shown.

and "react" with the electrodes. In a traditional battery, the reaction products are stored in the electrodes themselves. In a flow battery (or a hydrolysis cell), the reaction products are stored in separate reservoirs. The electrode reactions are reversible so that the batteries can be charged and discharged every day. For PV applications, these batteries must be low-cost, large, reliable, and respond quickly to power demands.

15.4.1 Battery technologies

Batteries have been around since the 19th century and they start our cars, power cell phones and calculators, and so on. Figure 15.5 shows the configurations and chemical reactions of three commercial battery technologies used in the solar industry.

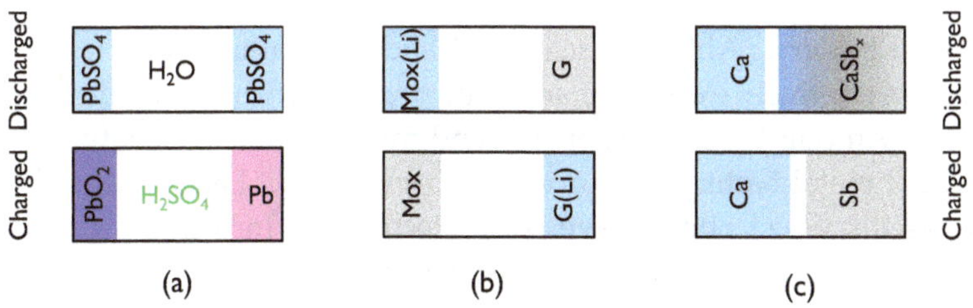

Figure 15.5: The batteries in charged and discharged states: (a) Lead–acid, (b) Li-ion, and (c) liquid metal.

Lead–acid battery. Lead–acid batteries are widely used for smaller standalone systems. The metal electrodes and the liquid solution make the battery heavy, but its low-cost and reliability make it suitable for many applications. A lead–acid battery consists of alternate layers of two electrodes: Pb and PbO_2, separated by an acid, H_2SO_4 (see Fig. 15.4(a)). The discharging (left to right) and charging (right to left) reactions proceed as follows:

$$Pb + PbO_2 + 2H_2SO_4 \underset{\text{charge}}{\overset{\text{discharge}}{\rightleftharpoons}} 2PbSO_4 + 2H_2O.$$

More specifically, the electrode reactions in the discharge phase are:

$$Pb + HSO_4^- \rightarrow PbSO_4 + H^+ + 2e^- \qquad \text{Anode}$$

$$PbO_2 + HSO_4^- + 3H^+ + 2e^- \rightarrow PbSO_4 + 2H_2O \qquad \text{Cathode.}$$

When the battery discharges, the electrodes react with the sulphuric acid to create $PbSO_4$ that deposits on the electrode surfaces, and the acid molecules are replaced by water. The discharged stage of the battery is shown in Fig. 15.5(a) (top). To recharge the battery, the solar cells are connected such that electrons flow from anode to cathode and the $PbSO_4$ ions, already deposited on the anode and the cathode during the discharge phase, react to form the acid. A fully charged battery is shown in Fig. 15.5(a) (bottom).

Lithium-ion battery. Lithium is a light-weight metal. Therefore, lithium-ion batteries are suitable whenever weight is a concern (for integrated storage in rooftop installation). The battery consists of a negative graphite anode and a positive metal oxide (Mox) cathode. In the discharged state, Li ions are inculcated in the cathode among the metal (e.g., Co) and oxygen atoms, see Fig. 15.5(b) (top). When the solar cell is connected, the battery is charged as the Li ions move toward and are eventually embedded within the graphite layers, as in Fig. 15.5(b) (bottom). In a PV system, after sunset, a load connected between the anode and the cathode will receive electrical power as Li ions travel from graphite to metal oxide, and electrons move through the load.

Liquid metal battery. Although lead-acid and lithium-ion batteries are widely used, they are relatively expensive and they degrade with repeated charging and discharging. Among the less-expensive and more reliable alternatives being explored, liquid metal battery is a promising technology. Lead-acid and Lithium-ion batteries discussed above have solid electrodes. In a liquid metal battery, instead, both electrodes are liquid metals, e.g., molten Antimony (Sb) as cathode, and molten Calcium (Ca) for the anode. During the discharge phase, Ca is ionized (i.e. $Ca \rightarrow Ca^{+2} + 2e^-$). The pair of electrons flow through the electrical load, while the Ca^{2+} ions flow through the molten electrolyte to react with and form an alloy of Sb (i.e., $Ca^{+2} + Sb_x + 2e^- \rightarrow CaSb_x$). The discharged state is shown in Fig. 15.5(c) (top). The charging phase involves the reverse process: The solar cell de-alloys $CaSb_x$ by extracting electrons from the Sb electrode, while supplying electrons to neutralize the Ca^{+2} ions returning to the Ca electrode. The charged state is shown in Fig. 15.5(c) (bottom). Calcium and Antimony are commodity products, therefore a liquid metal battery may be less expensive than lead-acid or lithium-ion batteries. Other advantages include high-temperature operation, ability to survive large of number of cycling, etc.

15.4.2 Hydrolysis and flow batteries store energy in separate reservoirs

Figure 15.6 shows that water can be split by electrolysis into H_2 and O_2. Later, as needed, one allows H_2 and O_2 to react in the battery and then uses the heat generated to power a boiler, which will turn a turbine. The turbine can drive a generator to reclaim the stored energy as electricity. If H_2 is difficult to store/transport, one can further convert H_2 through a series of reactions to solid hydrocarbons for future use — this will be discussed later in Sec. 15.9.3 and in Fig. 15.15. Other options include charging batteries (e.g., Li-Ni or salt electrolyte). In the following section, let us consider the energy storage option with hydrolysis and H_2 generation.

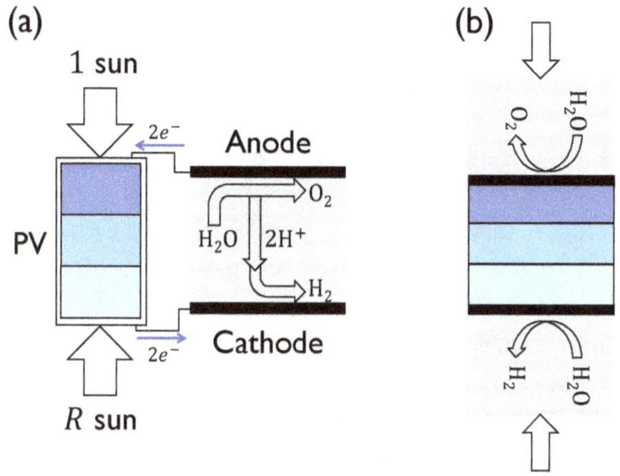

Figure 15.6: (a) An external solar cell, configured either as a multi-cell module or as a tandem cell involving a stack of subcells, drives the hydrolysis of water. (b) The solar cell itself is now immersed in water. The electrodes serve as the anode and cathode for the hydrolysis of water.

15.5 A deeper look into electrochemical storage: water hydrolysis

As discussed in the previous section, an approach to storing solar energy through electrochemical reaction involves water hydrolysis and flow battery. In the rest of this chapter, we will focus on the physics of the hydrolysis of water by solar cells. We will analyze the associated redox reactions to derive current equations and the equivalent circuit of this flow battery. Then we will couple the battery circuit model with the previously discussed solar cell I-V model (Fig. 15.3) to predict the operating conditions of the solar cell-coupled EC charging system. Although the analysis is exemplified through water hydrolysis, the basic concept holds for other electrochemical (EC) battery systems. After all, we can explain any battery charging or discharging through a set of redox reactions.

Now, let us focus on the water hydrolysis-based flow battery. Assume that the positive and negative terminals of a solar cell has been connected to the anode and cathode of the electrochemical cell, respectively, as shown in Fig. 15.6. Sunlight generates electron and hole pairs in the solar cell. The holes from the solar cell flow to the anode contact of the battery and recombine with an electrons. To maintain charge neutrality, the anode will now accept an electron from water, i.e.,

$$\mathrm{H_2O} \leftrightarrow 2\mathrm{H}^+ + \frac{1}{2}\mathrm{O_2} + 2\mathrm{e}^- \tag{15.7}$$

so that the repulsion among the protons breaks the water molecule apart. The O_2 gas is released, and the proton travels to the cathode.

At the cathode, the proton will now accept an electron from the conduction band of the solar cell, as follows.

$$2\mathrm{H}^+ + 2\mathrm{e}^- \leftrightarrow \mathrm{H_2}. \tag{15.8}$$

The hydrogen gas generated will now be saved for future use.

The configuration we discussed involves a solar cell supplying electrons and holes to two electrodes serving as anode and cathode. Other configurations are also possible — for example, Fig. 15.6(b) has the solar cells immersed in water, and the electrodes of the solar cell themselves serve as anode and cathode. The physics and chemistry are essentially the same, but they have different technological challenges. For example, configuration (a) involves an electrical loss in the wire connecting the components, while configuration (b) requires that the cell is encapsulated or otherwise water-stable. In the following discussion, we will focus on the essential physics and chemistry of the problem.

15.6 Redox reactions can be represented by a single diode

In order to understand the fundamental limits of energy storage by water hydrolysis, we need to first understand the chemical processes in the anode and the cathode (see Fig. 15.7(a)). Electrochemists treat each electrode separately by the so-called half-cell reactions. For example, the oxidation/reduction reaction in the anode

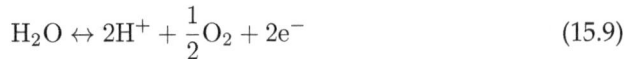

$$\mathrm{H_2O} \leftrightarrow 2\mathrm{H}^+ + \frac{1}{2}\mathrm{O_2} + 2\mathrm{e}^- \tag{15.9}$$

can be described by the following Butler–Volmer equation:

$$\begin{aligned} J_A &= J_{ox,A} - J_{red,A} \\ &= J_{0,A}\left[\exp\left(\frac{V_A - |\mu_A|}{b_{ox,A}}\right) - \exp\left(-\frac{V_A - |\mu_A|}{b_{red,A}}\right)\right]. \end{aligned} \tag{15.10}$$

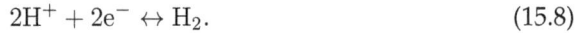

Here, V_A is the anode voltage. The reaction is characterized by the anode half-cell reaction and solution potential μ_A, exchange current density $J_{0,A}$, and the Tafel slopes are $b_{ox,A}$, $b_{red,A}$. The first exponential term describes the dissociation of water into $2H^+$ and O_2, while the second exponential term describes the formation of water from $2H^+$ and O_2. Equation (15.10) can be represented by back-to-back diodes connected via a voltage source μ_A, as in Fig. 15.7(c). The combined half-cell reaction and solution potential μ_A, and hence the anode voltage V_A, vary as a function of the concentrations of the reactants.

To describe the full water-splitting process involving the anode and the cathode, we need to similarly characterize the cathode equation

$$2H^+ + 2e^- \leftrightarrow H_2 \tag{15.11}$$

as follows:

$$\begin{aligned} J_C &= J_{red,C} - J_{ox,C} \\ &= J_{0,C}\left[\exp\left(\frac{V_C - |\mu_C|}{b_{red,C}}\right) - \exp\left(-\frac{V_C - |\mu_C|}{b_{ox,C}}\right)\right]. \end{aligned} \tag{15.12}$$

The voltage across the electrochemical (EC) cell is

$$V_{cell} = V_C + V_A + J_{ec}R_{sol}. \tag{15.13}$$

Here, the solution resistance (R_{sol}) describes the transport of protons through the solution. Note the current direction for $J_{ec} = J_C = J_A$ shown in Fig. 15.7(c).

The general circuit in Fig. 15.7(c) works for an arbitrary voltage impressed on the electrochemical (EC) system. At positive (or negative) bias, only the forward-biased diodes matter, and the reverse-biased diodes can be dropped. When $V_{cell} >$

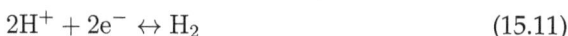

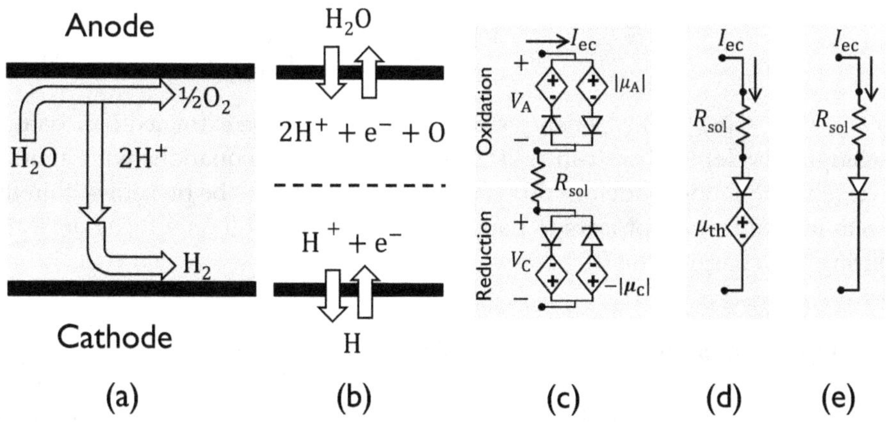

Figure 15.7: (a) The overall reaction splits water into hydrogen and oxygen. (b) Half-cell reactions at the anode and the cathode. (c) Two-diode presentation of the half-cell reaction. (d, e) Single-diode approximation.

$|\mu_A| + |\mu_C|$, oxidation in the cathode and reduction in the anode dominate, and the corresponding diodes ($J_{ox,C}$ and $J_{red,A}$, respectively) are in forward bias. The other two diodes are in reverse bias, and we can neglect $J_{red,C}$ and $J_{ox,A}$. Therefore, we can combine Eqs. (15.10), (15.12) into (15.13) and write (assume $R_{sol} = 0$):

$$\begin{aligned} V_{cell} &= V_C + V_A \\ &= \mu_{th} + b_{red,C} \ln\left(\frac{J_{ec}}{J_{0,C}}\right) + b_{ox,A} \ln\left(\frac{J_{ec}}{J_{0,A}}\right) \quad (15.14) \\ &= \mu_{th} + \ln\left[\left(\frac{J_{ec}}{J_{0,C}}\right)^{b_{red,C}} \left(\frac{J_{ec}}{J_{0,A}}\right)^{b_{ox,A}}\right]. \quad (15.15) \end{aligned}$$

Here, $\mu_{th} = |\mu_A| + |\mu_C|$ is the effective threshold voltage. After some algebra, we get

$$J_{ec} = J_{0,ec} \exp\left(\frac{V_{cell}}{b}\right). \quad (15.16)$$

The effective exchange current density and the effective Tafel slope are given by,

$$J_{0,ec} = J_{0,C}^{(b_{red,C}/b)} J_{0,A}^{(b_{ox,A}/b)} \exp\left(-\frac{\mu_{th}}{b}\right), \quad (15.17)$$

$$b = b_{red,C} + b_{ox,A}. \quad (15.18)$$

The effective exchange current density can be perceived as the weighted average of the exchange current densities of individual electrodes. Clearly, Eq. (15.16) can be represented by a single diode, as in Fig. 15.7(d). The equations (15.16)–(15.18) hold for $V_{cell} > |\mu_A| + |\mu_C|$ when water is split into H_2 and O_2. In this case, energy is being stored into the EC system. If we assume that there are K-EC cells connected in series with a system voltage of $V_{ec} = KV_{cell}$, we can rewrite equation (15.16) as

$$J_{ec} = J_{0,ec} \exp\left(\frac{V_{cell}}{b}\right) = J_{0,ec} \exp\left(\frac{V_{ec}}{Kb}\right). \quad (15.19)$$

For $0 < V_{cell} < |\mu_A| + |\mu_C|$, energy stored in the EC will be extracted (i.e., battery discharging cycle), and we can re-derive a similar set of equations with another (reversed) one-diode model. In this case, H_2 and O_2 need to be present within the system for the reverse process to happen. For now, we will only focus on water splitting, i.e., charging the EC system.

15.7 Optimum energy storage for solar cell

Now that we understand the basic electrochemistry of an EC, we can now optimize its design to maximize storage of solar energy. The solar module may contain either M cells of the same bandgap connected in series or a single tandem cell with a stack of N subcells, or a combination of both.

Let us begin with an N-subcell tandem by recalling the discussion in Chapter 4. Given an albedo of R, we limit the number of cells to $N_{crit} \leq (1+R^{-1})$, so that the bottom cell has the smallest bandgap, E_0. For typical $R = 0.25$ we get $N_{crit} = 5$ and $V_{\rm mp} \sim 3$ volts, which is sufficient for most redox reactions of practical interest.

15.7.1 I-V Characteristics of an N-cell tandem

The I-V characteristics of a tandem cell are given by Homework 4.12:

$$\begin{aligned} J(V) &= J_{\rm ph} - q\theta_{\rm D} \cdot \gamma \cdot \exp(-\langle E_{\rm g}\rangle / k_{\rm B}T_D) \cdot \exp(qV/N\, k_{\rm B}T_D) & (15.20) \\ &= J_{\rm ph} - J_{0,MJ} \cdot \exp(qV/N\, k_{\rm B}T_D) & (15.21) \end{aligned}$$

where $\langle E_{\rm g}\rangle$ is the average bandgap of the tandem cells. Recall that current matching among the subcells defines the following bandgap sequence:

$$E_i = \frac{i}{\beta' N} + \frac{(N-i)\,[(1+R)\beta' E_0 - R]}{\beta' N}. \qquad (15.22)$$

Since $\langle E_{\rm g}\rangle = \sum E_i/N$, we can insert it into Eq. (15.22) to find the minimum bandgap as

$$E_0 = \left(\langle E_{\rm g}\rangle - \frac{(N-1)(1-R)}{2\beta' N}\right) \frac{2N}{N(1+R)+(1-R)} \qquad (15.23)$$

and the maximum bandgap as

$$E_{g,max} = E_{\rm g,p} = \frac{N-1}{\beta' N} + \frac{[(1+R)\beta' E_0 - R]}{\beta' N}. \qquad (15.24)$$

For these bandgaps, the overall output voltage is given by Homework 4.12:

$$\begin{aligned} \frac{qV_{\rm mp}}{N} &= \langle E_{\rm g}\rangle \left(1 - \frac{T_{\rm D}}{\langle E_{\rm g}\rangle}\frac{E_{\rm g,p}}{T_{\rm S}}\right) - k_{\rm B}T_{\rm D} \ln\left(\frac{\theta_{\rm D}}{c\,\theta_{\rm S}}\right) & (15.25) \\ &= qV_{mp,SJ} - \frac{T_{\rm D}}{T_{\rm S}} \times \frac{N-1}{N+1}\left(\frac{\langle E_{\rm g}\rangle}{\beta'}\right). & (15.26) \end{aligned}$$

We have assumed a solar concentration of c. That is why the solar acceptance angle $\theta_{\rm S}$ is replaced by $c\,\theta_{\rm S}$. The last line of the equation can be obtained after a few lines of algebra. (Hint: Relate $\langle E_{\rm g}\rangle$ to E_{SJ} following Homework 4.12. Then, E_{SJ} can be related to $V_{mp,SJ}$.)

Also, the photocurrent at the maximum power point is given by Eq. (4.2):

$$I_{mp} \simeq I_{\rm sc} = cI_{\rm sun}(1 - \beta_{\rm sun} E_{\rm g,p}) = \frac{2I_{mp,\rm SJ}}{N+1}. \qquad (15.27)$$

15.7.2 I-V characteristics of a series-connected M-cell module

In a module with M cells in series, the terminal voltage will be M times that of a single cell:

$$V_{\mathrm{mp}}(N, M, c, R) = MV_{mp,M=1}(N, c, R)$$
$$= \frac{MN}{q}\left[\langle E_{\mathrm{g}}\rangle \left(1 - \frac{T_{\mathrm{D}}}{\langle E_{\mathrm{g}}\rangle}\frac{E_{\mathrm{g,p}}}{T_{\mathrm{S}}}\right) - k_{\mathrm{B}}T_D \ln\left(\frac{\theta_{\mathrm{D}}}{c\theta_{\mathrm{S}}}\right)\right]. \quad (15.28)$$

Here, we have used Eq. (15.25). We assume the series and shunt resistances can be neglected for simplicity.

15.8 Charging an EC system with PV

The general PV-EC system is shown in Fig. 15.8, where a PV panel consisting of M number of N-junction cells in series is connected to K number of EC cells in series. In this section, we will discuss the charging operation of the EC system connected to the PV module. A similar analysis can be done for the discharging cycle by connecting the EC system to a load — we will leave the discharging analysis to the interested readers.

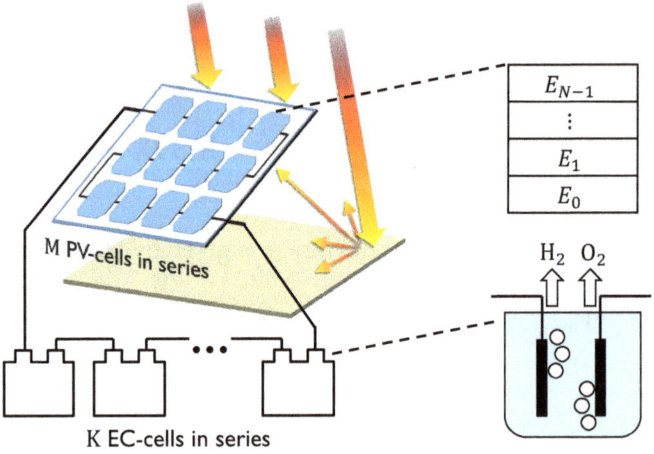

Figure 15.8: A solar module connected to an array of EC cells.

15.8.1 Operating point of the PV-EC system

For a given PV-EC system, we can now find an operating point (V_{op}, J_{op}) by solving for $I_{ec} = I_{pv}$, i.e., $A_{ec}J_{ec}(V_{op}) = A_{pv}J_{pv}(V_{op})$, as shown in Fig. 15.9. Note that the ratio of cell areas ($AF = A_{pv}/A_{ec}$) is another system parameter which will

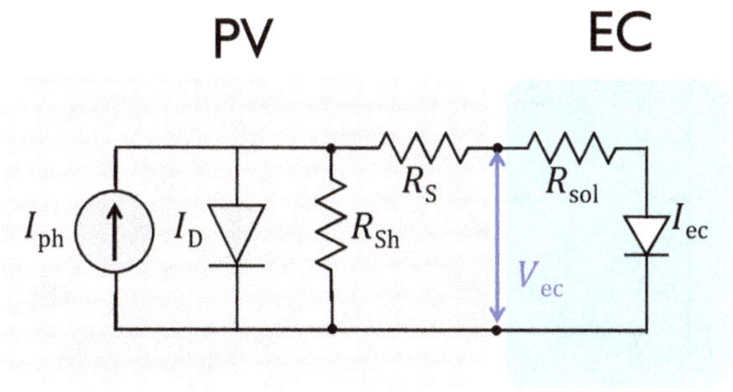

Figure 15.9: A solar cell on the left, configured either as a module or as a tandem cell, is connected to the anode and the cathode of an electrochemical cell that drives the hydrolysis reaction.

appear in the discussion later. A coupling loss described by the difference in maximum power of PV and the operating power should be taken into consideration while analyzing the system.

Intuitively, when the EC is operated at the maximum power point (MPP) of the PV (i.e., $(V, I)_{op} = (V, I)_{mp}$), the coupling efficiency is 100%, and the system is optimized. Therefore, for the global design and optimization of the PV-EC system, we will choose the $(V, I)_{op} = (V, I)_{mp}$ constraint so that the coupling efficiency $\eta_c = 1$. One can numerically find the exact solution for $(V, I)_{op}$.

15.8.2 PV-EC operation: An intuitive picture

The PV system can be configured as a module consisting of series-connected single-junction or multi-junction (tandem) cells. As an illustrative example, we take a water-splitting cell as the electrochemical system (load to the solar module). The parameters that define a water-splitting experiment are $\mu_{th} = 1.23$ V, $J_{0,ec} = 4.06 \times 10^{-36}$ mA/cm^2, $b = 70$ mV/decade .

For an intuitive understanding of the numerical optimization process, consider a single-MJ cell ($M = 1, N = 1, 2, 3$) in a PV system optimized for maximum efficiency (i.e., PV optimized), as shown in Fig. 15.10. We find that this double-junction PV provides the best coupling to the water-splitting EC ($K = 1$) and the highest system efficiency. This is because the point of operation $(V, I)_{op}$ is closest to $(V, I)_{mp}$. Figure 15.10 shows that current is negligible at the point of intersection of the I-V of a single-junction cell and the EC. For a triple-junction ($N = 3, M = 1$) tandem cell, the overall efficiency is also lower than that of a double-junction ($N = 2, M = 1$) cell due to lower coupling efficiency (current matching). This analysis implies that the optimum system efficiency depends on the number of subcells (N) in the tandem PV as well as the number of series-connected cells (M) in the module.

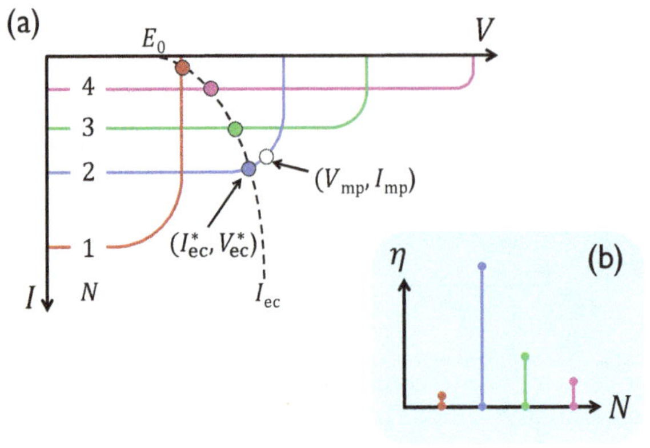

Figure 15.10: (a) Graphical solution for a module connected to an electrochemical cell. Each module has a single-MJ cell ($M = 1$) and the I-V characteristics are labeled with the number of subcells it contains, namely, $N = 1, 2, 3, 4$. The intersection of the $I_{ec}(V)$ with $I_{M,N,R}(V)$ defines the steady-state solution. (b) The conversion efficiency is plotted as a function of N.

Homework 15.5: Optimum number of PV subcells

Referring to Fig. 15.10 for a tandem solar cell driving the hydrolysis reaction, argue that there is an optimum number of subcells (N_{opt}) that maximize energy conversion.

15.9 How efficiently can we store solar energy?

15.9.1 PV-EC system design rule

Using Eq. (15.16) for K electrochemical cells in series, the voltage of the EC cell is given by

$$V_{ec} = bK \ln\left(\frac{J_{ec}}{J_{0,ec}}\right). \tag{15.29}$$

The voltage and current of the EC system should be at the maximum power point of the PV system for optimal storage, i.e.,

$$V_{op} = V_{ec} = V_{\text{mp}}, \tag{15.30}$$

$$\text{and, } I_{op} = A_{ec}J_{ec}(V_{op}) = A_{pv}J_{pv}(V_{op}). \tag{15.31}$$

Next, we substitute V_{mp} from Eq. (15.28) and assume $J_{pv} = J_{mp} \approx J_{sc,N}$ from Eq. (15.27) into Eq. (15.30) to arrive at the very important PV-EC system design formula:

$$bK\ln\left[\frac{cJ_{\rm sun}(1-\beta_{\rm sun}E_{\rm g,p})}{J_{0,ec}}AF\right] = \left(\frac{MN}{q}\right)\left[\langle E_{\rm g}\rangle\left(1-\frac{T_{\rm D}}{T_{\rm S}}\frac{E_{\rm g,p}}{\langle E_{\rm g}\rangle}\right)-k_{\rm B}T_{\rm D}\ln\left(\frac{\theta_{\rm D}}{c\theta_{\rm S}}\right)\right]. \tag{15.32}$$

Here, $AF = A_{pv}/A_{ec}$ is the cell-area ratio. Equation (15.32) determines the optimum parameters (M, N, E_0) for a given EC system $(K, \mu_{th}, J_{0,ec}, b)$ and particular values of c and R. Note that $E_{\rm g,p}$ and $\langle E_{\rm g}\rangle$ are functions of E_0, the smallest bandgap of the tandem cell. Therefore, for a set value of (M, N, c, R) for the PV and a given EC system, one can solve for E_0 from Eq. (15.22) for an optimal design. As we will show later, for a given EC system, a global maximum system efficiency requires: (i) co-optimization of (M, E_0) at a given PV module with(N, c, R), or (ii) co-optimization of (N, E_0) for a given tandem (M, c, R).

Since we find the point of operation, i.e., the intersection of I-V characteristics of PV and EC for the maximum power output, Eq. (15.32) provides the optimum parameters for system design. These parameters can be substituted in the following definition of overall system efficiency to achieve the thermodynamic limit:

$$\begin{aligned}\eta_{sys} &= \eta_{pv}\cdot\eta_c\cdot\eta_{ec}\\ &= \frac{V_{\rm mp}I_{\rm mp}}{cP_{\rm sun}(MA_{pv})}\times\frac{V_{op}I_{op}}{V_{\rm mp}I_{\rm mp}}\times\frac{K\mu_{th}I_{op}}{V_{op}I_{op}}\\ &= \frac{K\mu_{th}I_{op}}{McP_{\rm sun}A_{pv}} = \left(\frac{K\mu_{th}J_{op}^{PV}}{McP_{\rm sun}}\right).\end{aligned} \tag{15.33}$$

Here, $P_{\rm sun}$ is the solar intensity reaching the PV system ($\sim 1\,{\rm kW/m^2}$ for AM1.5G) and $(V, I)_{mp}$ is the maximum power point of the PV module. The power required to initiate the electrochemical process at the thermodynamic equilibrium potential μ_{th} is $\mu_{th}I_{op}$. The factor K accounts for the number of ECs in series. The losses in PV and EC are taken into account with their respective definitions of efficiency. The coupling loss is included in the coupling efficiency, defined as the ratio of the operating power to the maximum power that can be generated by the PV cell.

It is important to note that η_{sys} also comprises Faradaic efficiency, which is assumed to be 100% in this calculation. Moreover, we use the equilibrium potential (lower heating value) of the reaction and not the thermo-neutral potential (higher heating value) because equilibrium potential gives an upper bound to theoretical system efficiency.

15.9.2 PV-EC system efficiency limit

If we revisit Eq. (15.32), we observe that, for an N-junction tandem, we cannot independently set both M and K for an optimized design. In fact, the (M/K) ratio would be another optimization parameter for maximizing η_{sys}.

In most practical cases, for example on rooftops or solar farms, single-junction solar cells are used. Therefore, let us first study the optimum combination of (M, K) for a module of SJ $(N = 1)$ solar cells connected to a K-cell electrochemical system. For any SJ cell (with bandgap E_g) and a known EC, one can readily calculate (M/K) for the optimum design using Eq. (15.32). The corresponding η_{sys} is found from Eq. (15.33). The optimum η_{sys} and the corresponding (M/K) are shown as a function of E_g in Fig. 15.11. For a water-splitting EC system, $\eta_{max} \approx 26.46\%$ for $M/K \approx 1.67 \approx 8/5$, implying that an optimum combination of 8 SJ cells in series with 5 EC cells will yield the best overall system efficiency. Furthermore, this efficiency is achieved at $E_g = 1.33$ eV, which in fact is the optimum SJ PV bandgap. This result can be explained as follows.

Due to the logarithmic change in V_{ec} with current (see Eq. (15.16)), the EC efficiency η_{ec} does not change significantly as long as the change in current is relatively small (i.e., concentration c is essentially a constant). Now, with constant η_{ec} and $\eta_c = 1$, it is obvious that the system η_{sys} will maximize when η_{pv} is maximum. The difference in η_{pv} and η_{sys} arises due to kinetic losses in EC which are incorporated in η_{ec}. Therefore, choosing an (M/K) ratio so as to couple optimum PV to the EC will indeed give the optimum system design. While we have explained the result in the context of SJ-PV and EC coupling, this analysis also holds for tandem PV and EC coupling as well.

Figure 15.11(b) shows optimum η_{pv} and overall optimized η_{sys} for N-junction tandems. The corresponding optimum (M/K) ratio calculated from Eq. (15.32) is also shown in the same plot. The V_{mp} of the optimum tandem increases with N, which is compensated by decreasing the (M/K) ratio to ensure perfect coupling between the PV module and the EC cells. The system efficiency η_{sys} increases from 26.46% to 34.82% for $N = 1$ to 2 and starts to saturate for $N > 4$. From Eq. (15.32), we can predict the ultimate limit of $\eta_{sys} \rightarrow 52.09\%$ as $N \rightarrow \infty$ under 1-sun with no albedo $(c = 1, R = 0)$.

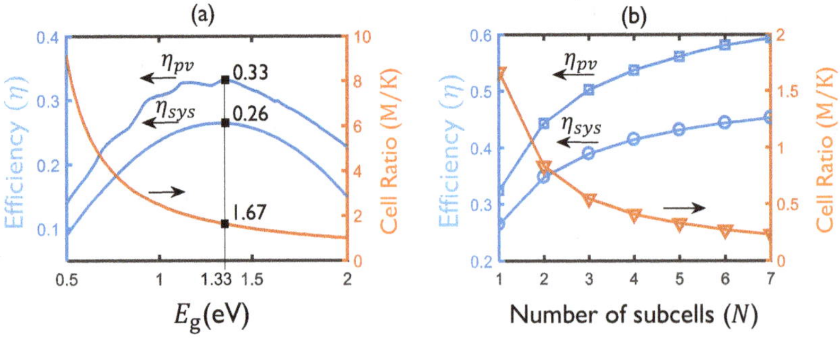

Figure 15.11: Variation in efficiency and the M/K ratio as a function of (a) bandgap for a system with SJ PV connected to EC, and (b) number of subcells (N) for a system with MJ PV connected to EC.

Homework 15.6: It is easy to design an integrated PV-EC electrolyzer system

Let the electrochemical system be a single water-splitting cell with $\mu_{th} = 1.23$ V, $J_0 = 4.06 \times 10^{-36}$ mA/cm^2 and $b = 70$ mV/decade (~19 mV/e). The PV module, on the other hand, can have the following configurations:

1. Module with 2 single-junction solar cells (bandgap: $E_0 = 1.16$ eV)

2. Module with 1 multi-junction solar cell (bandgaps: $E_0 = 0.79$ eV, $E_1 = 1.56$ eV).

Assuming 1-sun illumination (AM 1.5G) and albedo $R = 0$, find the system efficiency for both these configurations and explain why one performs better. For the first system, recalculate the system efficiency with $R = 1$.

Hint. First, calculate the short-circuit current density (J_{sc}) and the system efficiency. The answers are as follows:

For $R = 0$, the single-junction and tandem cell efficiencies are (a) η_{sys} = 25.66% and (b) η_{sys} = 32.53%. The second system is more efficient because the tandem (multi-junction) cells provide a more efficient conversion of solar energy into electrical energy compared to single-junction solar cells.

For $R = 1$, the system based on single-junction solar cells has an efficiency of η_{sys} = 49.69%. Here, bifaciality increases the amount of solar energy absorbed by the PV module and hence improves the overall efficiency. For high enough albedo, the bifacial cells may provide improved energy conversion than tandem solar cells.

15.9.3 Discussion

PV Batteries require careful design. Figure 15.12(a) shows an electrical equivalent circuit of a battery — this is the same as the one discussed earlier in the water-splitting example. The reaction potential μ_{th} in general will change during charging or discharging. We can assume separate constant values for μ_{th} during charging and discharging as long as we do not charge or discharge too much.

If a PV module is connected to the battery for charging, once the voltage exceeds the reaction potential ($V_{\text{mp}} > \mu_{th}$) of a partially discharged battery, it will start to charge. The battery potential rises and falls during the charging (by the solar cell) and discharging (by the load) processes (see Fig. 15.12(b)). The depth of charge indicates the level to which the battery can be charged or discharged. If a lead–acid battery is fully discharged (and $PbSO_4$ covers the electrode surface and the acid is depleted), it is difficult to charge it back. Similarly, during the charging process, one must not allow the battery potential to exceed the threshold for parallel reactions (e.g., hydrolysis for lead–acid battery), which can lead to the formation of hydrogen and oxygen gases and degradation of the battery performance.

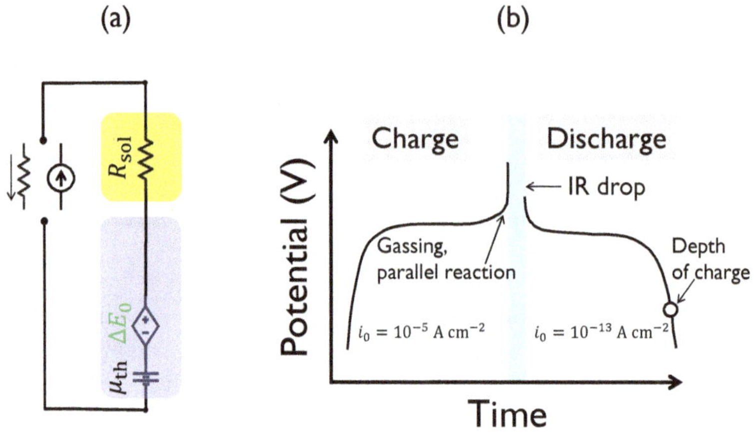

Figure 15.12: (a) Electrical equivalent circuit of a battery. The battery (dependent voltage source on the right arm) is charged by the maximum power point current (indicated by the current source on the left branch) and discharged by energy consumed by the load (resistor). The electrolyte resistance (R_{sol}) and reaction potential of the electrode (μ_{th}) are also shown. (b) The battery charges rapidly to the saturation voltage ($\Delta E_0 \sim k_B T/\eta_c \ln(I_{mp}t/Q_0)$, but discharges slowly through the load. One must neither overcharge nor fully discharge a battery.

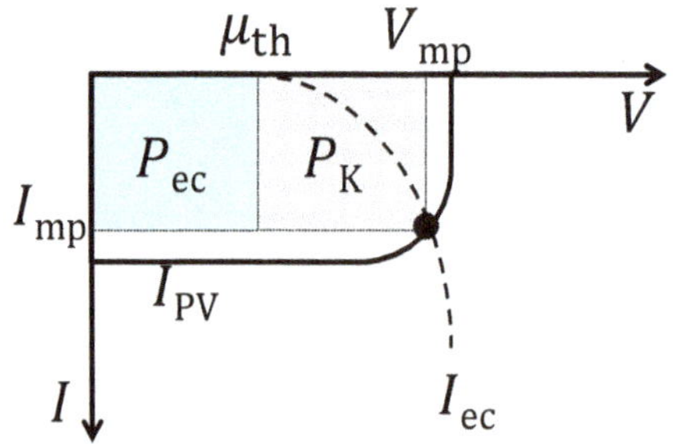

Figure 15.13: Various components of the power during the electrochemical energy storage process. P_{ec} is the power returned through the hydrogen and oxygen reaction. During the discharge phase, the kinetic energy P_K is lost.

Where did the energy go? It is important to compare the energy that can eventually be returned from the stored hydrogen vs. the power provided by the PV system. It is clear from Fig. 15.11(b) that the overall power stored is significantly smaller than the efficiency of the PV system. Figure 15.13 accounts for the lost power. While the system was optimized for V_{ec}^* and I_{ec}^* and the corresponding energy was needed to dissociate water, the energy returned by hydrogen and oxygen reaction is only through the power in the blue box in Fig. 15.13. The kinetic energy associated with the gray box (P_K) is forever lost.

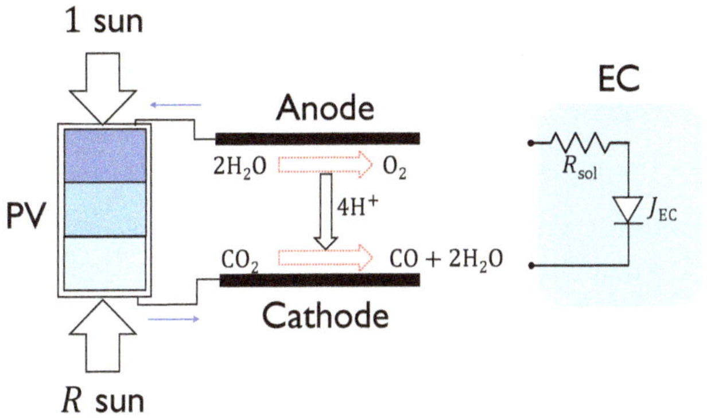

Figure 15.14: (a) The electrochemical reaction involving water and carbon dioxide. (b) The reaction can still be represented by a single diode.

Hydrolysis is but one of the many energy storage reactions. The hydrolysis reaction is easy to explain; however, the gas must be stored at very high pressure to fit in a reasonable volume. Sometimes it is more convenient to use the protons generated at the anode through hydrolysis to drive other reactions whose end product is more conveniently stored. For example, Fig. 15.14 shows how water and carbon dioxide can be used to create CO, which can then be converted to other carbohydrates, similar to what a leaf does during photosynthesis. The analysis proceeds exactly the same way as before, except that we have a new μ_{th} for the diode.

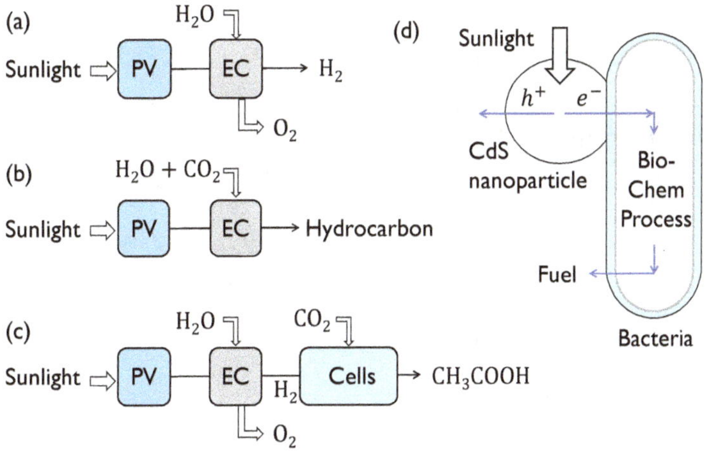

Figure 15.15: (a) Hydrolysis of water. (b) Hydrocarbon can be generated through a reaction involving carbon dioxide and water. (c, d) Bacteria-assisted generation of acetic acid.

Interestingly, researchers have enlisted bacteria to the service of energy storage (see Fig. 15.15(c)). Here, bacteria cells such as *Moorella thermoacetica* use the Wood–Ljungdahl metabolic pathway to convert the protons produced by the hydrolysis of water to acetic acid. We must realize that these pathways themselves have significant losses. Therefore, the conversion has a significant energy penalty.

Homework 15.7: Photosynthesis as a PV to EC energy storage process

Finally, I encourage you to study the photosynthesis reaction shown in Fig. 15.16 and explain the following.

1. Of the two topologies shown in Fig. 15.6, which one corresponds to photosynthetic reaction?

2. A leaf has various light-sensitive molecules, such as chlorophyll a, chlorophyll b, and β-carotene, etc. These cells absorb photons at 400–500 nm and 625–675 nm. Is this a tandem cell or a series-connected cell?

3. Read the paper by Arp *et al.*, "Quieting a noisy antenna reproduces photosynthetic light-harvesting spectra," *Science* 368.6498 (2020): 1490–1495, to see how the specific choice of the two wavelengths is essential to maximize energy yield while reducing the fluctuation of the energy output (and associated plant damage) due to fluctuating ambient light.

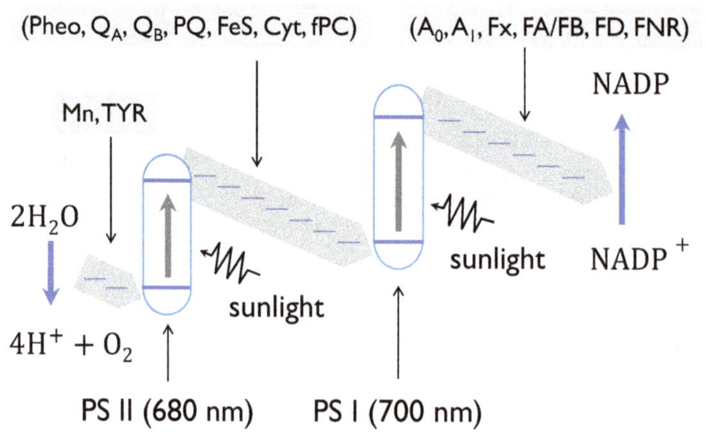

Figure 15.16: A schematic diagram of the photosynthesis process.

15.10 Conclusions

In this chapter, we discussed a variety of energy storage systems, including mechanical, electrostatic, and electrochemical storages. We analyzed the hydrolysis of water to create hydrogen as a potentially important mechanism of energy storage, and how it can be coupled to a PV module for charging. The physics and mathematics are identical for any battery, be it a Li-ion battery or a flow battery.

Among the various energy storage options, pumped hydro and compressed air (i.e., mechanical storage) provide the largest capacity and are most cost-effective in terms of per unit energy stored. The drawbacks include high initial investment cost and finding suitable geographical locations to serve as the reservoirs. The capacity of electrochemical storage options, such as traditional and flow batteries, has increased significantly over the years. In particular, flow batteries are highly reliable and easy to scale to large systems. Some of the storage technologies (e.g., electrostatic supercapacitors and electromechanical flywheels) can store and supply energy quickly, and they are useful in stabilizing grid operation.

The technical design of the solar farm is now complete. We must now discuss how to secure funding for the farm from the bank. The banker is interested in the rate of return on his or her investment and would be keen to know the expected lifetime of the farm. The lifetime of the farm (e.g., 15 years, 25 years, etc.) would depend on the various degradation modes of the module and the weather the module is exposed to. In the next (and final) part of the book, we will prepare to respond to the banker by discussing the reliability physics of solar modules.

References

[1] Mark T. Winkler, Casandra R. Cox, Daniel G. Nocera, and Tonio Buonassisi. Modeling integrated photovoltaic–electrochemical devices using steady-state equivalent circuits. *Proceedings of the National Academy of Sciences*, 110(12):E1076–E1082, March 2013.

[2] William Shockley and Hans J. Queisser. Detailed Balance Limit of Efficiency of p-n Junction Solar Cells. *Journal of Applied Physics*, 32(3):510, 1961.

[3] Louise C. Hirst and Nicholas J. Ekins-Daukes. Fundamental losses in solar cells. *Progress in Photovoltaics: Research and Applications*, 19(3):286–293, 2011.

[4] Javier Farfan and Christian Breyer. Combining Floating Solar Photovoltaic Power Plants and Hydropower Reservoirs: A Virtual Battery of Great Global Potential. *Energy Procedia*, 155:403–411, November 2018.

[5] Marco Rosa-Clot, Giuseppe Marco Tina, and Sandro Nizetic. Floating photovoltaic plants and wastewater basins: An Australian project. *Energy Procedia*, 134:664–674, October 2017.

[6] Mathew Milnes. The Mathematics of Pumping Water.

[7] James R. Bolton, Stewart J. Strickler, and John S. Connolly. Limiting and realizable efficiencies of solar photolysis of water. *Nature*, 316(6028):495–500, August 1985. Number: 6028 Publisher: Nature Publishing Group.

[8] Katherine T. Fountaine, Hans Joachim Lewerenz, and Harry A. Atwater. Efficiency limits for photoelectrochemical water-splitting. *Nature Communications*, 7(1):13706, December 2016. Number: 1 Publisher: Nature Publishing Group.

[9] James R. Bolton and David O. Hall. The Maximum Efficiency of Photosynthesis. *Photochemistry and Photobiology*, 53(4):545–548, 1991. _eprint: https://onlinelibrary.wiley.com/doi/pdf/10.1111/j.1751-1097.1991.tb03668.x.

Part IV

Reliability and Characterization of Solar Cells

We concluded Part III of the book by explaining how the efficiency of a solar farm depends on technology choice, module and farm configurations, and geographical location. A solar farm may be highly efficient, but it may still not be cost-effective: the modules may be too expensive and difficult to clean and/or degrade rapidly when exposed to the heat and humidity of the local environment. Therefore, we need to define a metric that balances the dual requirements of high efficiency and low cost.

The unit cost of electricity (COE) produced by a solar farm is calculated by taking the ratio of the cost of installing and running the solar farm ($C(Y)$) to the total energy produced ($E(Y)$) over its lifetime of Y years. In other words,

$$\mathrm{COE} = \frac{C(Y)}{E(Y)}.$$

The lifetime Y of the solar farm depends on how well the modules are made and the environment it is exposed to. In the deserts of the Middle East, gradual yellowing of the protective polymer encapsulant (as its molecules are broken by UV light) reduces light received by the solar cell, which in turn reduces the power output. In the hot and humid environment of Southeast Asia, the fingers and busbars may corrode over time, and increasing Joule heating of these degraded contacts would reduce the output power. Therefore, understanding the physics of degradation is important, because it helps us design modules with a longer lifetime Y and reduced COE, so that solar energy becomes cheaper than energy produced by burning coal or gasoline.

In addition to the lifetime Y, the total energy output $E(Y)\left(=\int_0^Y P(t)dt\right)$ depends on the instantaneous power output $P(t)$, which varies with multiple timescales. A passing cloud shadow may reduce $P(t)$ only for a few minutes. The daily variation in $P(t)$ due to variable solar intensity (due to overcast sky or smog-filled day) is difficult to predict even a few days in advance, because weather forecasting is imprecise. As the module is soiled and cleaned, $P(t)$ varies in a sawtooth form over the timescale of weeks. And then comes $P(t)$ reduction over the years, as the encapsulants yellow, glass covers and/or backsheets crack, and metal interconnects corrode. Today, even the most advanced commercial solar cell technologies lose 2–2.5% power output during the first year and 0.5–0.7% power output in subsequent years. In this part of the book, we will focus on the slow loss of $P(t)$ over weeks to years due to the climatic conditions at a given location. These processes are not overly sensitive to short spikes in temperature or humidity (weather), and therefore are more predictable.

We will begin the discussion in Chapter 16 with a simple model of the cost of solar energy. The cost model will demonstrate the importance of the reliability and lifetime of solar modules. In subsequent chapters, we will discuss four types of reliability issues that define the lifetime (Y) of solar cells.

1. A *reversible* reliability issue temporarily reduces the energy output of a solar farm, but does not damage it permanently. For example, gradual accumulation of dust, sand, or soil (**soiling**) reduces power output; however, unlike polymer yellowing due to UV exposure, the power of a soiled module can be restored by periodic cleaning. A soft, temporary shadow by a cloud gliding overhead will reduce power, but will not damage the cells irreversibly. We will discuss these topics in Chapters 17 and 18.

2. *Metastable* reliability issues, such as potential-induced degradation (**PID**), light-induced degradation (**LID**), hotspot formation, etc., may recover partially, but not necessarily fully, once the stress is removed or reduced. We will explain these processes in Chapters 20–22.

3. *Irreversible* degradations reduce the power output permanently. Eventually, the power can be restored only by replacing the module. These processes include electrochemical **corrosion** of the metallic lines and stress/electromigration induced failure in solder bonds, which are reflected in increasing series resistance; **stress- and corrosion-induced delamination** and *UV-induced yellowing* of the polymer encapsulant, which reduce light transmission; glass **microcracks** due to seasonal variation in temperature-induced stresses in the module; and *partial shadow* degradation, which is another mechanism reflected in the shunt leakage. These topics are discussed in Chapters 23–24.

4. Other *extrinsic* reliability issues, such as failure of bypass diode or reliability of the DC to AC inverters (which convert the DC energy produced by solar cells to AC energy suitable for the grid), are also important. In this book, we will not discuss these topics in any detail.

Chapter 25 explains how each module must pass through rigorous qualification tests to ensure that the modules are well made and the degradation rates are slow. In this regard, Chapter 26 explains how the degradation mechanisms *collectively* define the intrinsic lifetime of a solar farm. A well-made module, rigorously tested by a sophisticated qualification process, is expected to produce energy for the anticipated intrinsic lifetime of the farm. Nonetheless, there are always surprises when a module is placed for the first time in a new location. We conclude the book in Chapter 27 by highlighting the growing importance of "inverse modeling" to determine the prevailing modes of degradation and the residual lifetime of an existing solar farm. The information gives the farm operator an opportunity to plan ahead and the manufacturers an opportunity to tailor their module design for the local conditions.

CHAPTER 16

Levelized Cost of Electricity Highlights the Importance of Efficiency and Reliability of Solar Modules

Chapter Summary

- Cost-of-energy and levelized-cost-of-energy are two different concepts. One should focus on LCOE.
- The LCOE depends on two learning curves: Swanson law for reduced module price related to volume production and Goetzberger law for improved efficiency due to technology improvement.
- LCOE* is a new concept that decouples local (e.g. land price) from global (e.g. module price) consideration of LCOE, allowing versatile prediction of economic feasibility of solar farms.
- Smart recycling of the modules can reduce LCOE and reduce environmental damage. Making a longer-lived module is more important than making a more recyclable one.
- LCOE is an imperfect measure of the cost-effectiveness of solar energy, because it does not consider the environmental footprint of the manufacturing process.

16.1 Introduction: COE is a simple but an important concept

The cost of electricity (COE), expressed in \$/kWh, allows comparison of the prices of the energy produced from various sources, such as wind, coal, and gas. Specifically, the cost of a unit of solar energy is given by

$$\mathrm{COE} = \frac{C(Y)}{E(Y)} \tag{16.1}$$

where $C(Y)$ is the cost of operating the solar modules for Y years and $E(Y)$ is the energy generated during that period.

16.1.1 A solar farm requires significant investment: An analysis of $C(Y)$

The cost of a solar farm (the numerator of Eq. (16.1)) depends on the initial cost of building the farm, $C_{\rm sys}$, its operation and maintenance over the lifetime, $C_{\rm om}$, and the residual value of the land and the modules, $C_{\rm rv}$, when the farm is decommissioned, i.e.,

$$C(Y) = C_{\rm sys} + C_{\rm om}(Y) - C_{\rm rv}(Y). \tag{16.2}$$

The initial one-time system installation cost is given by

$$C_{\rm sys} = N_{\rm mod}\, C_{\rm mod} + N_{\rm mod}\, C_{\rm BOS,V} + C_{\rm BOS,F}. \tag{16.3}$$

This cost depends on (a) the number of modules installed, $N_{\rm mod}$, and the cost of each module, $C_{\rm mod}$; (b) the variable balance-of-system cost per module, $C_{\rm BOS,V}$, for preparing the land and the hardware and labor needed for wiring, racking, and mounting the system; and (c) the fixed BOS cost, $C_{\rm BOS,F}$, for the permit, monitoring, etc. that are essentially independent of the number of modules. Figure 16.1 shows how various cost components have changed in recent years.

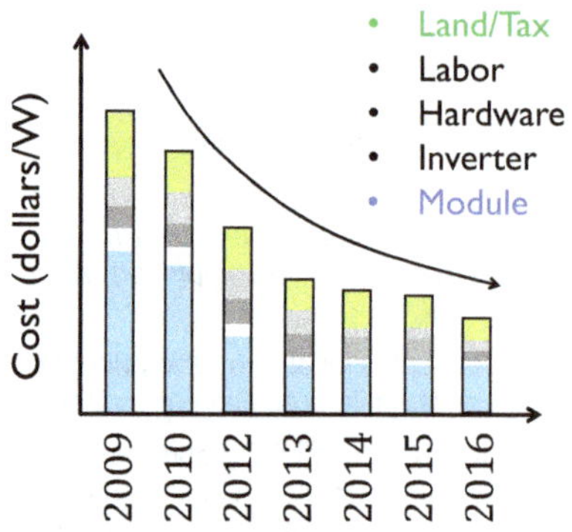

Figure 16.1: The cost components of modules installed from 2009 through 2016. Each column consists of (from bottom to top) module cost, variable BOS costs for inverters that connect the modules to the electrical grid, hardware that connect the modules to each other, labor needed to install the modules, and the cost of land and tax needed to house the farm.

The second term of Eq. (16.2), i.e., ($C_{\rm om}(Y) \equiv \int_0^Y C_{\rm om}(t)\, dt$), includes all the costs incurred for the operation and the maintenance of the farm during its lifetime, based on the yearly rate of $C_{\rm om}$. Third and finally, there is a residual value (rv) of the modules and equipment to be recouped when the farm is decommissioned. As shown in Fig. 16.2, the $C(Y)$ increases (almost linearly) with time due to the recurring cost of maintenance and operation.

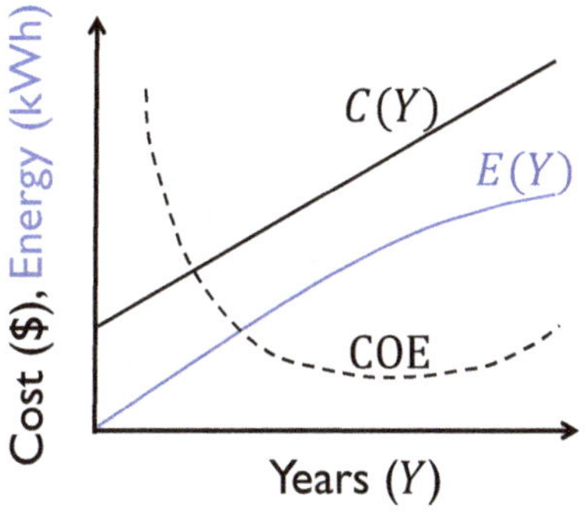

Figure 16.2: The costs of operating a solar farm ($C(Y)$) as well as the energy output from the farm ($E(Y)$) increase with time. However, $E(Y)\left(=\int_0^Y P(t)dt\right)$ saturates due to reduction of $P_0(t)$ due to various types of degradations. Therefore, COE (given by the dashed line) decreases initially, reaches a plateau, and then turns around. The shape is explained in the text.

Homework 16.1: New module installation cost, normalized to the power produced

To understand Fig. 16.1, let us consider a c-Si module with $\eta = 20\%$. It takes 100 dollars to manufacture the module (size 1 m^2) and it is sold at 5% profit. If the BOS cost per module is 100 dollars, calculate the installation cost *per module* and cost per watt for this technology.

Solution. The module cost, including the profit, is 105 dollars. If the variable BOS cost is 100 dollars and fixed BOS cost is negligible, then from Eq. (16.3), $C_{\text{sys}}/N_{\text{mod}} = 205$ \$/module. To calculate the cost per watt, we first calculate the power output of the module, which is $P_0 \equiv \eta I_0 = 0.2 \times 1000\ \text{Wm}^{-2} \times 1\ \text{m}^2 = 200$ W. Therefore, the cost to generate 1 watt of power is $C^*_{\text{sys}} \equiv C_{\text{sys}}/P_0 = 205/200 = 1.025$ \$/W.

16.1.2 A solar farm cannot produce energy forever: The physics of $E(Y)$

The energy output $E(Y)\left(=\int_0^Y P(t)dt\right)$ given by the denominator of Eq. (16.1) initially increases (almost linearly) with time. Eventually, the rate of increase begins to saturate as various degradation modes (e.g., yellowing, corrosion, delamination) erode the efficiency η (and thus the power output, $P(t)$) of the module. For a yearly efficiency degradation rate of d, $E(Y)$ is obtained by summing up the contributions over the years,

$$E(Y) = P_0 + P_0(1-d) + P_0(1-d)^2 + \cdots$$
$$= P_0 \sum_{k=0}^{Y-1} (1-d)^k = P_0 \left[\frac{1-(1-d)^Y}{d}\right].$$

Here, P_0 is the new module power output during the first year, expressed in kWh/year. A plot of $E(Y)$ in Fig. 16.2 shows that for typical degradation ($d \approx$ 2–5%), $E(Y)$ begins to saturate over time because an old farm is less efficient than a new one. Figure 16.2 also shows that the integrated cost of operation and maintenance, C_{om}, increases linearly with time. Initially, $C_{\text{om}}(Y) \ll C_{\text{sys}}$, so that COE $\approx C_{ins}/P_0Y$ decreases rapidly as the installation cost is amortized over the cumulative energy produced. Eventually, when $C_{\text{om}}(Y) \gg C_{\text{sys}}$ and $E(Y)$ begins to saturate, the COE begins to rise as COE $\to (C_{\text{om}}/P_0) \times (Yd)$. Once the COE exceeds a threshold, the farm is dismantled and replaced with more efficient (and hopefully less expensive) next-generation modules.

If the farm did not degrade at all ($d \to 0$), the term $(1-(1-d)^Y)/d \to Y$, so that $E(Y) = P_0Y$ would increase linearly with time. Since the integrated maintenance cost C_{om} also increases linearly as shown in Fig. 16.2, COE would decrease until it reaches a plateau at C_{om}/P_0 when $C_{\text{om}}(Y) \gg C_{\text{sys}}$, and we would forever get low-cost energy just for the price of maintenance. Only if we were so lucky!

16.2 LCOE is a similar but slightly more complicated concept

There is an important limitation of COE: A dollar today is worth more than a dollar tomorrow. Let us assume that the value of a dollar is reduced by a factor $(1+r)$ each year, where r is the discount rate. Therefore, COE must be corrected for r and all the future costs must be collapsed into a single figure (i.e., levelized) so that an apple-to-apple comparison is possible. In other words,

$$\text{LCOE} = \frac{C(Y)}{E(Y)} = \frac{C_{\text{sys}} + \sum_{k=0}^{Y-1} C_{\text{om}}(k)(1+r)^{-k} - C_{\text{rv}}(1+r)^{-Y}}{\sum_{k=1}^{Y} P_0(1-d)^k(1+r)^{-k}} \tag{16.4}$$

where $C_{\text{om}}(k)$, the *yearly* maintenance cost for the k-th year, has been reduced to the present value by accounting for the cumulative discount of $(1+r)^k$.

Less obvious, however, is the scaling of the energy output (in the denominator of Eq. (16.4)) by the same cumulative discount factor, $(1+r)^k$. Does it mean that the a "watt" today is different from a "watt" in the future?! This is indeed the case. If you continue to sell a unit of energy for R\$/watt, the *present* value of the revenue (c) must account for the fact that future earnings are not as valuable as present earnings. Therefore, $c(Y) \neq R\sum_{k=1}^{Y-1} P_0(1-d)^k$, but rather $c(Y) = R\sum_{k=1}^{Y-1} P_0(1-d)^k(1+r)^{-k}$. Therefore, the *present* value of the energy output will have to be scaled by the discount factor as well (as if the farm produced less energy than it actually did). Taken together, LCOE depends sensitively on r. A banker determines r; befriend a banker!

Homework 16.2: Energy output with constant degradation and discount rates

Integrate the denominator of Eq. (16.4) to calculate the total energy output of a solar farm, assuming that the degradation and discount rates are constants. Express the results in terms of location-specific average solar energy, I_0, as well as the efficiency, η, of the module. Find and plot an analytical expression for LCOE(r, d, η, I_0). How does the result compare with the $E(Y)$ curve in Fig. 16.2?

Solution. Integrating the denominator of Eq. (16.4), we find that

$$E(Y) = \eta\, I_0 \left[\frac{1 - \left((1-d)/(1+r) \right)^Y}{1 - (1-d)/(1+r)} \right]. \tag{16.5}$$

The loss of $E(Y)$ due to discount rate r may be viewed as an increase in the effective degradation rate d^* used in the COE calculation, so that $(1 - d^*) \equiv (1 - d)/(1 + r)$.

Homework 16.3: Calculation of LCOE

A city has on average 6 hours of sunlight a day for 200 days a year at an average intensity of 1000 W/m^2. Assume that a 20% efficient module costs \$1/watt to procure and an additional \$0.50/watt to install (including the balance-of-system cost). The module size is 1 m^2 and each module produces 200 W. Assuming $Y = 25$ years, calculate the LCOE for the installed system. For simplicity, neglect the residual value and the maintenance cost, as well as degradation and discount rates.

Solution. The output a single module is $1000 \times 0.2 = 200$ W. Over 25 years, the module will produce

$$E(Y = 25) = (6 \times 200 \times 25) \times (1000 \times 0.2)/1000 = 6000 \text{ kWh of energy.}$$

On the right-hand side, the first term within the bracket is the number of hours the farm would be operated over the 25 years period and the second term is the energy produced in kW.

(continued on the next page)

Homework 16.3 (*continued from the previous page*)

Next, let us calculate the system installation cost of the 200 W module

$$C_{\text{sys}} = 200 \text{ W} \times 1.5 \text{ \$/watt} = 300 \text{ dollars}.$$

Since C_{rv} and C_{om} are negligible, $C(Y) = C_{\text{sys}}$, see Eq. (16.2). Therefore, from Eq. (16.4), LCOE is \$300/6000 kWh = 0.05 \$/kWh. Here LCOE is the same as COE, because we ignored the discount rate. Once the degradation rate, discount rate, and the maintenance cost are included, the LCOE will increase further.

Homework 16.4: LCOE can also be calculated by a web calculator, such as `https://www.nrel.gov/pv/lcoe-calculator/`

In the Baseline column, click **Preset** in the first row to open a dialog box to define the system parameters: mono-Si, glass–polymer backsheet, fixed-tilt utility-scale system for Indianapolis, USA. Using predefined parameters, calculate the LCOE of the city. Can you identify the parameters used in the calculator with the topics discussed in this chapter?

16.3 Two learning curves can be used to project "future" LCOE

The discussion above allows LCOE comparison for a solar farm to be installed right away, because all the parameters, including the module price, are known from the manufacturer's website. Instead, let us assume that we plan to install a solar farm *sometime in the future* and wish to determine the corresponding LCOE. We do not know exactly what the module price or efficiency would be at the time of farm installation, but we can use the historical data judiciously to *predict* these parameters. Based on historical trends, we can safely say that the modules in the future will be more efficient (Goetzberger's law) and less expensive (Swanson's law). The data reflects the empirical fact that the more we do something, the better we get at it. If we could somewhat quantify the rates of improvement (i.e., learning coefficients), we will be able to make better decisions about the economic viability of the solar farm we are thinking about.

Figure 16.1 shows the historical LCOE of farms installed in various years. Interestingly, both the module cost and the BOS cost have declined rapidly, albeit with different rates. We can learn from these trends and project the LCOE of the farms to be installed several years into the future, for example.

16.3.1 Anticipated future growth of the PV industry

As the PV industry expands, N_{mod} installed across the world will increase over time. Additional research and development will improve manufacturing yield and produce more efficient cells. The unit price of the raw materials will fall as manufacturers buy in large quantities. Simply put, learning to produce more modules itself reduces the unit cost of a solar module, that is, $C_{\text{mod}}(N_{\text{mod}})$. Empirically, the idea applies to many industries, e.g., Moore's law for integrated circuits, Wright's law for planes, etc.

Assume that $N_{\text{mod}}(0)$ is the total number of modules installed up to year Y_0 and f_0 is the number of new modules installed during Y_0 *before* our analysis begins. Let us define the yearly growth rate of new installations (g) in a way that $f_0 \times (1+g)$ is the number of *new* modules added during the first year, $f_0 \times (1+g)^2$ is the number of new modules added just in the second year, and so on. Therefore, the total number of modules installed after X years is given by

$$N_{\text{R}} \equiv \frac{N_{\text{mod}}(X)}{N_{\text{mod}}(0)} = 1 + f_0 \sum_{k=1}^{X} (1+g)^k. \tag{16.6}$$

16.3.2 Swanson's law for reduced module price

Figure 16.3(b) defines a quantity called "learning coefficient" (q_{m}), which indicates the percentage reduction in C_{mod} when the N_{R} is doubled, written in the form $1 - q_{\text{m}} \equiv (2)^{\log_2(1-q_{\text{m}})}$. Therefore, the fractional reduction in $C_{\text{mod}}(Y)$ due to increasing N_{R} is given by

$$L_{\text{m}}(Y, q_{\text{m}}, g) \equiv \frac{C_{\text{mod}}(X)}{C_{\text{mod}}(0)} = {N_{\text{R}}}^{\log_2(1-q_{\text{m}})}. \tag{16.7}$$

Figure 16.3(b) also shows that there is a similar reduction in $C_{\text{BOS,V}}$ (e.g., wiring, racking, etc.) as the demand for these items increases with $N_{\text{mod}}(Y)$, but industry-specific learning would be slightly different, q_{BOS}. Taken together, the cost of a new module after X years will be

$$C_{\text{mod}}(X) = C_{\text{mod}}(0) \times L_{\text{m}}(X, q_{\text{m}}, g) \tag{16.8}$$

$$C_{\text{BOS,V}}(X) = C_{\text{BOS,V}}(0) \times L_{BOS,V}(X, q_{\text{BOS}}, g). \tag{16.9}$$

One obtains the q_{m} and q_{BOS} by analyzing the historical data, and then using the parameters to project into the future.

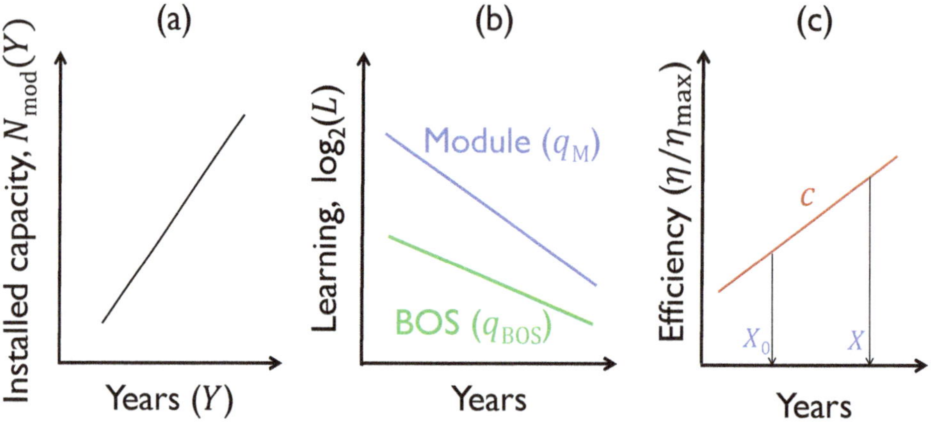

Figure 16.3: (a) Installed capacity increases with time. (b) Module and BOS prices decrease with time due to industry-wide learning through volume manufacturing. Modules and BOS have different learning coefficients, q_M and q_{BOS}. The system cost is obtained as a sum of the two costs. (c) Commercial-grade module efficiency increases with time as well.

16.3.3 Goetzberger's law of improved efficiency

In the previous section, we saw that increasing the production volume reduces the cost to manufacture a solar module. Similarly, the improved process control and ability to manufacture complex (e.g., PERC) solar cells increase the efficiency of the modules produced (see Fig. 16.3(c)). Goetzberger's law describes the historical

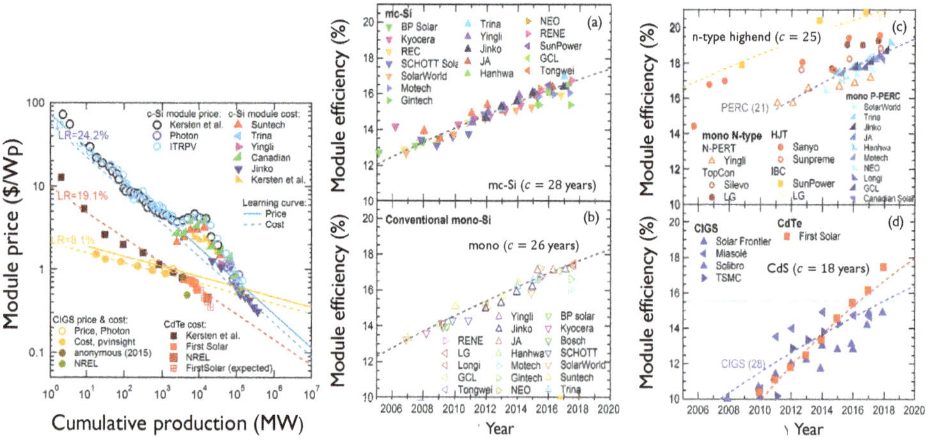

Figure 16.4: (Left) Swanson's law describes the falling module prices with increasing manufacturing volume. (Right) Historical efficiency gain for different technologies can be used to calculate the Goetzberger parameters. (Courtesy: Dr. Pietro Altermatt, Leibniz University of Hanover, Germany)

data by the following equation:

$$\eta(X) = \eta_{\max}\left[1 - \exp\left(X_0 - X\right)/c\right], \tag{16.10}$$

where the fitting parameters $\eta_{\max}$, X_0, and c respectively describe the technology-specific efficiency improvement over the years.

Figure 16.4 summarizes for historical data for various technologies, allowing us to determine the learning coefficients for Goetzberger's law (left) and Swanson's law (right). Given the parameters, we can calculate the future module efficiency ($\eta(X)$). The improved efficiency will increase the energy yield $E(X)$ of the technology and reduce the LCOE of the farm.

One final point: history is not destiny. Even for the same material, significant changes in the module technology (e.g., Al-BSF to PERC cell) can lead to a discontinuous jump in efficiency not reflected in historical curves. Nonetheless, the learning curves do define a conservative estimate of LCOE — a very useful piece of information.

Homework 16.5: Future efficiency of monocrystalline (mc-Si) solar cells

Historical data in Fig. 15.4 show that the efficiency of commercial c-Si solar cells increased from 13% to 17% between 2006 and 2018. If the technology is characterized by an efficiency learning coefficient $c = 28$, use Eq. (16.10) to calculate how long it might take for the efficiency to reach 22%.

Solution. To determine the future efficiency using (16.10), we need to find $\eta_{\max}$ and X_0. Assuming the year 2000 as the reference, the ratio of the efficiencies

$$17/13 = (1 - e^{(X_0-18)/28})/(1 - e^{(X_0-6)/28})$$

gives $X_0 = -15.25$ (or, the year 1984–1985). Using X_0,

$$17 = \eta_{\max}\left(1 - e^{(-15.25-18)/28}\right),$$

we find $\eta_{\max} = 24.45$. Finally, to find the year to reach 22% efficiency, we calculate

$$22 = \eta_{\max}\left(1 - e^{(-15.25-X)/28}\right),$$

to obtain $X = 35$.

In other words, we will have to wait till 2035 to buy a 22% mc-Si module. This is a prediction about industry average efficiency; some companies will achieve it long before others do.

Homework 16.6: Learning curves can be used to predict future LCOE

For a simple analysis of the historical data, recall that the cost per watt of power output is given by $C^*_{\text{sys}} \equiv C(Y)/P_0$. Assume that in 2006, $C^*_{\text{mod}}(0) = 4.2$ \$/watt, $C^*_{\text{BOS,V}}(0) = 0.8$ \$/watt, and $C^*_{\text{BOS,F}} \approx 0$, so that C^*_{sys} is 5 \$/watt. In 2006 alone, 40% new modules were added, i.e., $f_0 = 0.4$. Also assume, $\eta_{\max}$, X_0, and c for commercial mc-Si technology are 24.45%, year 1985, and 28 years, respectively.

1. If the price of a module dropped from 5 \$/watt in 2006 to 2 \$/watt in 2012, use the learning curve to show that the historical data are well represented by $q_{\text{m}} = 0.22$ and $q_{BOS,V} = 0.1$.

2. Use the historical data to project the price at the end of 2017, but with a somewhat lower rate of average growth, $g = 0.4$.

3. How many years will it take for the system cost to reduce below 1 \$/watt? Include the anticipate efficiency gain following the Goetzberger's law.

Solution. For the first part, it is easy to verify that (all costs are in dollars per watt)

$$C^*_{\text{mod}}(X=6) = 4.2 \times \left(1 + 0.4 \times \left(\sum_{k=1}^{6} 1.6^k\right)\right)^{log_2(1-0.22)} = 1.50$$

$$C^*_{\text{BOS,V}}(X=6) = 0.8 \times \left(1 + 0.4 \times \left(\sum_{k=1}^{6} 1.6^k\right)\right)^{log_2(1-0.1)} = 0.50.$$

So, $C^*_{\text{sys}} = 2.00$ \$/watt. For the second part,

$$C^*_{\text{mod}}(X=11) = 4.2 \times \left(1 + 0.4 \times \left(\sum_{k=1}^{11} 1.4^k\right)\right)^{log_2(1-0.22)} = 1.00$$

$$C^*_{\text{BOS,V}}(X=11) = 0.8 \times \left(1 + 0.4 \times \left(\sum_{k=1}^{11} 1.4^k\right)\right)^{log_2(1-0.1)} = 0.42.$$

So that $C^*_{\text{sys}} = 1.42$ \$/watt. These numbers are realistic.

Regarding the third part, verify that the cost is expected to reach 1 dollar per watt by 2020.

Homework 16.7: Future LCOE can also be calculated by a web-calculator, such as `https://www.nrel.gov/pv/lcoe-calculator/`

Given the results of Homework 16.4, we can calculate the LCOE of a farm to be built in the future. Specifically, use the **Proposed** column on the right of the web calculator to determine the LCOE of a farm to be built 5 years from now, for example. The projected costs should be based on the learning coefficients from Homeworks 16.5 and 16.6.

16.4 LCOE* decouples local vs. universal components of LCOE

In the LCOE discussion so far, we have assumed that only module technology matters, but the land for free. That may be true for Sahara Desert, but certainly not true for Tokyo. The mixing of these two issues — local (i.e., land, irradiance, temperature) and global (i.e., module technology) — makes LCOE calculation difficult. The modules produced by a manufacturer with a given technology and efficiency costs about the same (C_{M}) throughout the world (except the shipping cost, obviously). The learning curves discussed in the preceding section provide a good estimate of how the prices will evolve in time. However, the cost of the land the farm will be built on, the incentives a government may provide to build a farm, the interest rates a bank would charge, etc. (C_{L}) depend on the location of the farm (e.g., Sahara vs. Tokyo). Also, total energy yield over the farm lifetime ($E(Y)$) depends on local illumination, temperature, and degradation rates. Historical weather information and field-test data determine these parameters.

A new metric, called LCOE*, resolves this issue by decoupling the local vs. global information. Recall from Eqs. (16.2) and (16.4) that

$$\mathrm{LCOE} = \frac{C_{\mathrm{sys}}(Y=0) + \sum_{k=1}^{Y}(C_{\mathrm{om}}(1+r)^{-k} - C_{\mathrm{rv}}(1+r)^{-k}}{E(Y)}$$

can be rewritten in the form

$$\mathrm{LCOE} \equiv \frac{C_{\mathrm{M}} + C_{\mathrm{L}} + C_{\mathrm{BOS,F}}}{E(Y)}, \tag{16.11}$$

where we have separated the local (e.g., land) and global (e.g., module) costs. These costs include the initial and recurring costs, i.e.,

$$C_{\mathrm{L}} = C_{l,0} + \sum_{k=1}^{Y}(C_{om,l}(k)(1+r)^{-k} - C_{rv,l}(1+r)^{-Y}$$

$$C_{\mathrm{M}} = C_{m,0} + \sum_{k=1}^{Y}(C_{om,m}(k)(1+r)^{-k} - C_{rv,m}(1+r)^{-Y}. \tag{16.12}$$

If we neglect the small balance-of-system fixed cost, $C_{\text{BOS,F}}$ (e.g., cost for obtaining a permit), Eq. (16.11) may be written in terms of per unit area module cost ($\mathbf{C}_{\text{M}}$) and per unit area land cost ($\mathbf{C}_{\text{L}}$)

$$\text{LCOE} = \frac{\mathbf{C}_{\text{M}} \cdot hMZ + \mathbf{C}_{\text{L}} \cdot pMZ + C_{\text{BOS,F}}}{\mathbf{YY} \cdot hMZ \cdot \chi(d, r, Y)},$$

where $\mathbf{YY}$ is the yearly energy yield, h, p, Z, M are the height of a module, the array period, the width of a row, and the number of rows in a farm, and

$$\chi(d, r, Y) = \sum_{k=1}^{Y} (1-d)^k (1+r)^{-k}. \tag{16.13}$$

For constants d and r, this sum is easily integrated; see Eq. (16.5). We can now write a location- and technology-agnostic LCOE* formula in terms of the ratio of the module/land costs and the pitch/height ratio of a solar farm. These parameters also fix the energy yield of a location.

$$\text{LCOE}^* \equiv \frac{\chi \cdot \text{LCOE}}{\mathbf{C}_{\text{L}}} = \frac{M_{\text{L}} + p/h}{\mathbf{YY}(p, h, \beta, ...)} \tag{16.14}$$

where $M_{\text{L}} \equiv \mathbf{C}_{\text{M}}/\mathbf{C}_{\text{L}}$ is an optimization parameter defined by the ratio of the module to land cost.

Once a farm design is specified and optimized purely based on technology and geographical location (longitude and latitude), we can pre-calculate $\text{LCOE}^*(M_{\text{L}}, p/h)$ for all locations of the world as a function of module-to-land (M_{L}) cost ratio. Once the location-specific information is known (e.g., local degradation rate (d) and bank discount rate (r), as well as the specific land cost ($\mathbf{C}_{\text{L}}$)), LCOE* is readily converted into LCOE for the farm by Eq. (16.14).

16.4.1 Solar farm topologies are determined by LCOE and LCOE*

In Chapters 12–14, we optimized solar farms for maximizing $\mathbf{YY}$, with no consideration of its cost. In practice, a farm designed for minimum LCOE may look very different from the one designed for maximum $\mathbf{YY}$. Figure 16.5(a) shows that firm design differs significantly depending on the cost ratio M_{L}. For $M_{\text{L}} \to \infty$ (i.e., $\mathbf{C}_{\text{L}} \ll \mathbf{C}_{\text{M}}$), Eq. (16.14) requires that $\text{LCOE}^* \approx M_{\text{L}}/\mathbf{YY}$, or equivalently, $\text{LCOE} \propto (\mathbf{C}_{\text{M}} + (p/h)\mathbf{C}_{\text{L}})/\mathbf{YY} \to \mathbf{C}_{\text{M}}/\mathbf{YY}$. In this situation, regardless of p/h, LCOE is minimized by maximizing $\mathbf{YY}$ by location-specific optimally tilt of the modules toward the sun, as in the right inset of Fig. 16.5(b). This was exactly the strategy we adopted in Chapters 12–14, where we did not account for module- or land-related costs.

On the other hand, for $M_{\text{L}} \to 0$ (i.e., $\mathbf{C}_{\text{L}} \gg \mathbf{C}_{\text{M}}$), $\text{LCOE}^* \approx (p/h)/\mathbf{YY}$, or equivalently, $\text{LCOE} \propto (\mathbf{C}_{\text{M}} + (p/h)\mathbf{C}_{\text{L}})/\mathbf{YY} \to (p/h)/\mathbf{YY}$. In this situation, LCOE is minimized by maximizing $\mathbf{YY}$ subject to the constraint that p/h (which scales

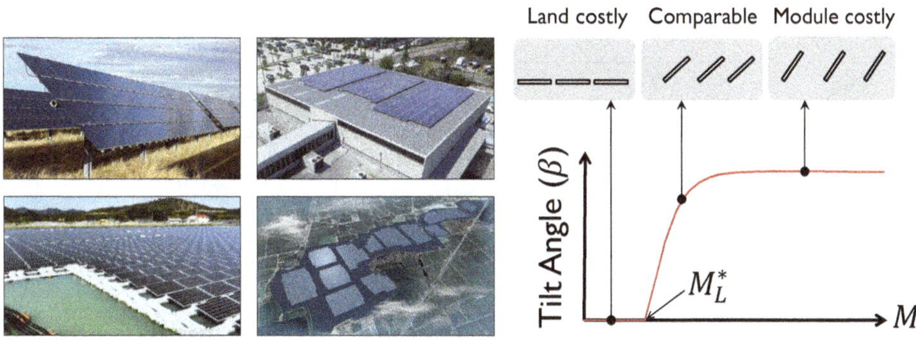

Figure 16.5: (Left) Practical solar farms are configured in various topologies, some optimally tilted toward the sun and others packed flat on the ground. (Right) LCOE* determines the optimum configuration (e.g., orientation, spacing, etc.) of the farm for a specific location.

with area) stays as small as possible. Unlike the strategy adopted in Chapters 12–14, the modules must now be packed flat end-to-end (see Fig. 16.5(b), left inset) as are often seen in area-constrained rooftop or floating solar farms. The transition from flat packing to optimally tilted farm occurs at $M_{\mathrm{L}}^* \approx 4$–6.

16.4.2 An example calculation involving LCOE and LCOE*

Using historical data for the cost, degradation rate, and interest of solar modules, let us explain how technology-specific LCOE* can be converted to LCOE for the following locations with specified longitude and latitude: Phoenix, AZ (33.6° N, $-112.4052°$ W), Kansas City, MO (39.09° N, $-94.8558°$ W), and New York City, NY(40.70° N, $-74.1197°$ W).

Calculation of LCOE* Let us use Eq. (16.14) to pre-calculate $\mathrm{LCOE}^*(M_{\mathrm{L}})$ for MO, NY, and AZ based on technology and irradiance information from public databases, with module-to-land cost ratio ($M_{\mathrm{L}} = \mathbf{C}_{\mathrm{M}}/\mathbf{C}_{\mathrm{L}}$) as an unknown parameter. To convert LCOE* to location-specific LCOE by Eq. (16.13), we will calculate M_{L} and χ based on technology-specific C_{M} and location-specific information ($\mathbf{C}_{\mathrm{L}}, d, r$), respectively.

To determine M_{L} for these three states, we can use a public database (e.g., one that contains cost estimates for residential, commercial, and utility-scale PV systems; see Fig. 16.1). For a given year, we can regroup the costs into C_{M} (i.e., module cost, inverter, hardware BOS, and soft costs to install labor) and C_{L} (i.e., soft costs — others), see Table 16.1. The database presumes that the land cost is the same for the three states, but this may not be the case in practice. The cost also assumes $Y = 25$ years and bank discount rate $r = 6.5\%$, because C_{L} and C_{M} in Eq. (16.12) depend on these parameters.

Table 16.1: 2018 cost estimate for 100 MW fixed-tilt solar farm.

C_{M} ($/W)		C_{L} ($/W)	
Module	0.47	Land acquisition	0.07
Inverter	0.04	Developer overhead	0.03
BOS (electrical)	0.08	Sales tax	0.05
BOS (structural)	0.09	Developer profit	0.05
Labor	0.10	Interconnection fee	0.04
Total	0.78	Total	0.24

Calculating the conversion factors We have a unit problem: $\mathbf{C}_{\mathrm{M}}$ and $\mathbf{C}_{\mathrm{L}}$ must be expressed in $/m, but C_{M} and C_{L} in Table 16.1 are expressed in $/W. In other words, $\mathbf{C}_{\mathrm{M}} = C_{\mathrm{M}} W_{\mathrm{M}}$ and $\mathbf{C}_{\mathrm{L}} = C_{\mathrm{L}} W_{\mathrm{L}}$, where W_{M} is the power produced per unit module height and W_{L} is the power produced per unit land area. Assuming a 1 m wide module, W_{M} is the product of cell efficiency and incident power. If 1000 W/m^2 is incident on a 1 m wide module, 19.1% efficient cell, $W_{\mathrm{M}} = 191$ W/m. Similarly, a typical 1 MW plant requires 16,000 m^2) of land; therefore $W_{\mathrm{L}} = 10^6\ \mathrm{W} \times 1\ \mathrm{m}/16000\ \mathrm{m}^2 = 62.5\ \mathrm{W/m}$. With these conversions, the 2018 data from Table 16.1 give

$$M_L = \frac{\mathbf{C}_{\mathrm{M}}}{\mathbf{C}_{\mathrm{L}}} = \frac{C_{\mathrm{M}} W_{\mathrm{M}}}{C_{\mathrm{L}} W_{\mathrm{L}}} = \frac{0.78 \times 191}{0.24 \times 62.5} = 9.93.$$

Location-specific *p/h* and YY We have discussed in the previous section that for a given M_{L}, the farm must be optimized for location-specific p/h and $\mathbf{YY}$ by changing the tilt angle, β; see Eq. (16.14). To do so, we calculate $\mathbf{YY}(p, h, \beta)$ based on the equations in Chapter 13, and determine the corresponding LCOE* by Eq. (16.14). A farm designed for $M_{\mathrm{L}} = 9.93$ at AZ, MO, and NY, respectively, has pitch-to-height ratio $p/h = 1.18, 1.40, 1.42$; optimum tilt angle $\beta = 22, 19, 17$ degrees; and first-year energy yield $\mathbf{YY} = 322.93, 233.34, 207.32$ kWh/m. Using Eq. (16.14), therefore, LOCE$^* = 0.0346, 0.049, 0.055$.

From LCOE* to LCOE By Eq. (16.14), LCOE* can be converted LCOE if we know χ and $\mathbf{C}_{\mathrm{L}}$. With degradation rate $d = 0.7\%$, bank discount rate $r = 6.5\%$, and farm lifetime $Y = 25$ years, we find that $\chi \approx 12.4$ and $\mathbf{C}_{\mathrm{L}} = C_{\mathrm{L}} \times W_{\mathrm{L}} = 0.24 \times 62.5 = 15$ $/m, so that $\mathbf{C}_{\mathrm{L}}/\chi = 1.21$. The corresponding LCOE for AZ, MO, and NY is LOCE $= 0.042, 0.059, 0.066$ $/kW.h, respectively, for AZ, MO, and NY. It is no surprise that a solar farm is economically viable in sunny Arizona!

Since the costs in Table 16.1 represent the average values across USA, the M_{L}, $\mathbf{C}_{\mathrm{L}}$, and χ values are the same for AZ, MO, and NY. However, the first-year energy yields differ substantially and is reflected in their respective LCOE. Also, based on the available historical data for module and land prices, we can calculate X years into the future the module-to-land cost $M_{\mathrm{L}}(X) \equiv \mathbf{C}_{\mathrm{M}}(X)/\mathbf{C}_{\mathrm{L}}(X)$ and LCOE(X) and design the optimum spacing and orientation of the farm accordingly.

16.5 Smart recycling increases the "residual value" of a solar module

Finally, we saw in Eq. (16.2) that the LCOE can be reduced by recycling the solar modules and increasing the residual value, $C_{\rm rv}$. Without recycling, $C_{\rm rv}$ can turn negative because we would need to pay for the safe disposal and landfills for an extended period of time. To appreciate the potential for recycling, we know that, by mass, a typical crystalline silicon PV panel shown in Fig. 22.3 consists of 75% glass, 10% polymer (encapsulant and backsheet), 8% aluminum (metal frame), 5% silicon, 1% copper and 0.1% silver for interconnected lines. Thin-film solar cells consists of a higher proportion of glass ($> 90\%$). In addition, the metallic racks and cables used to mount the panels in a solar farm contain approximately 55% steel, 20% aluminum, 8% copper, and 6% plastics. Modern electronic waste (e-waste) recycling programs developed by the PV industry can extract both the bulk (e.g., glass, aluminum, and copper) as well as higher-value components (e.g., semiconductor, silver) by first disassembling the metal frame, delaminating the encapsulant and backsheet by various mechanical, thermal, chemical, or optical processes, and then finally recovering glass and other components by mechanical and chemical treatments. Indeed, almost 100% of glass and metal frame can be recovered, and the recovery efficiency of other components are reaching more than 90%. The key remaining challenge is to increase the recovery throughput so that the recycling is cost effective. An efficient recycling program can reduce the module cost by US$ 0.01–0.03 per watt.

16.6 Conclusions: LCOE is an important but imperfect measure of cost-effectiveness of solar cells

Over the years, various institutions have posted a number of LCOE calculators online. Since they use slightly different models for fixed vs. variable costs, include additional details, such as taxes and incentives, and use different databases for weather information and cost estimates, the LCOE estimates may differ considerably. Equally important, one cannot guarantee future discount (r) or degradation (d) rates, for example. Therefore, the projected LCOE must be accompanied by an uncertainty range. In addition, one must assess model assumptions carefully before using the LCOE for decision making.

An LCOE estimate helps decision making in three important ways. First, it allows one to make an economic decision regarding the viability of solar energy at a specific location of the world (compared to other sources of energy such as wind, coal, or nuclear). Second, it allows policy-makers to structure innovative financing, tax credit, etc. so that the initial investment is protected. Finally, the components of LCOE identify the specific costs that must be reduced to substantially reduce the overall cost. In other words, LCOE analysis can support research planning for a specific technology. In particular, a new metric called LCOE* simplifies technology comparison by decoupling local vs. technology-specific costs of a solar farm.

An LCOE estimate does not include several important factors, such as the environmental cost associated with greenhouse gas emission (coal vs. PV, for example), discharge of chemicals during the manufacture of a product (acids released in water/air during the manufacture of solar cells), or the cost of accidents (e.g., failure of a nuclear reactor). Two technologies may have comparable LCOE, but it will be important to choose one that has a more benign environmental impact.

Despite these limitations, LCOE does highlight the importance of a reliable and long-lived solar farm in making solar energy feasible. In the next eight chapters, we will discuss the reliability issues that erode the energy output of a solar cell (with a combined degradation rate, d; see Eq. (16.4)), and thereby increase LCOE.

Homework 16.8: Levelized cost of storage (LCOS) is different from levelized cost of solar storage (LCOSS)

In Chapter 15, we discussed several energy storage options, e.g., pumped-hydro, batteries, etc. Analogous to LCOE, a metric called the levelized cost of storage (LCOS) compares the storage technologies, i.e.,

$$\text{LCOS} = \frac{C_{\text{sys}} + \sum_{k=0}^{Y-1}(C_{store} + C_{\text{om}}(k)(1+r)^{-k} - C_{\text{rv}}(1+r)^{-Y}}{\sum_{k=1}^{Y} E_{store}(1+r)^{-k}} \tag{16.15}$$

where C_{store} is the charging cost, E_{store} is the amount of energy stored and discharged during the farm lifetime, and all other symbols have the same meaning, except that now they apply to a specific storage mechanism. How would the terms compare for pumped hydro vs. batteries? Energy density [$\text{W} \cdot \text{h/L}$] and specific energy density [$\text{W} \cdot \text{h/kg}$] are two important metrics of a storage technology. Which terms in Eq. (16.15) would be affected by the two metrics? (*Hint*: Think of the area cost.) Finally, show that the integrated metric of the levelized cost of solar and storage (LCOSS) is given by LCOSS $=$ LCOE $+ \alpha \cdot$ LCOS, with $\alpha \equiv (f \cdot \eta_c)^{-1}$, where $f \approx 0.25$–0.50 is the fraction of solar energy stored and $\eta_c \approx 0.8$–0.9 is the efficiency of energy storage; see Table 11.1.

References

[1] Seth B. Darling, Fengqi You, Thomas Veselka, and Alfonso Velosa. Assumptions and the levelized cost of energy for photovoltaics. *Energy & Environmental Science*, 4(9):3133–3139, August 2011.

[2] Martin A. Green. Ag requirements for silicon wafer-based solar cells. *Progress in Photovoltaics: Research and Applications*, 19(8):911–916, December 2011.

[3] Nadia Ameli, Mauro Pisu, and Daniel M. Kammen. Can the US keep the PACE? A natural experiment in accelerating the growth of solar electricity. *Applied Energy*, 191:163–169, April 2017.

[4] M. R. Pinto. Silicon Photovoltaics: Accelerating to Grid Parity. In Serge Luryi, Jimmy Xu, and Alex Zaslavsky, editors, *Future Trends in Microelectronics*, pages 194–209. John Wiley & Sons, Inc., 2013.

[5] Nate Blair, Aron P. Dobos, Janine Freeman, Ty Neises, Michael Wagner, Tom Ferguson, Paul Gilman, and Steven Janzou. System advisor model, sam 2014.1. 14: General description. Technical report, National Renewable Energy Laboratory (NREL), Golden, CO., 2014.

[6] Martin A. Green. Silicon photovoltaic modules: a brief history of the first 50 years. *Progress in Photovoltaics: Research and Applications*, 13(5):447–455, August 2005.

[7] ITRPV Working Group. International Technology Roadmap for Photovoltaics Results 2011. *International Technology Roadmap for Photovoltaics Results*, 2011.

[8] Gregory F. Nemet. Beyond the learning curve: Factors influencing cost reductions in photovoltaics. *Energy Policy*, 34(17):3218–3232, November 2006.

[9] National Research Council and others. *Hidden costs of energy: Unpriced consequences of energy production and use*. National Academies Press, 2010.

[10] Financial Forecast Center - Financial Market Forecasts and Economic Outlook, 2017.

[11] M. Tahir Patel, M. Ryyan Khan, Xingshu Sun, and Muhammad A. Alam. A worldwide cost-based design and optimization of tilted bifacial solar farms. *Applied Energy*, 247:467–479, August 2019.

[12] Vasilis M. Fthenakis. End-of-life management and recycling of PV modules. *Energy Policy*, 28(14):1051–1058, November 2000.

[13] Rong Deng, Nathan L. Chang, Zi Ouyang, and Chee Mun Chong. A techno-economic review of silicon photovoltaic module recycling. *Renewable and Sustainable Energy Reviews*, 109:532–550, July 2019.

[14] Yan Xu, Jinhui Li, Quanyin Tan, Anesia Lauren Peters, and Congren Yang. Global status of recycling waste solar panels: A review. *Waste Management*, 75:450–458, May 2018.

[15] Oliver Schmidt, Sylvain Melchior, Adam Hawkes, and Iain Staffell. Projecting the Future Levelized Cost of Electricity Storage Technologies. *Joule*, 3(1):81–100, January 2019.

[16] Verena Jülch. Comparison of electricity storage options using levelized cost of storage (LCOS) method. *Applied Energy*, 183:1594–1606, December 2016.

[17] T. P. Wright. Factors Affecting the Cost of Airplanes. *Journal of the Aeronautical Sciences*. 3 (4): 122–128. doi:10.2514/8.155.

CHAPTER 17

Soiling vs. Cleaning: An Optimization Problem

Chapter Summary

- Soiling by dust and snow reduce the energy output of a solar farm.
- The cleaning cost and the soiling rate determines the optimum cleaning frequency of a farm.
- The soiling rate is determined the soiling parameter, which in turn depends on elevation and tilt-angle of the modules, the geometry and type of the particles, and angle of incidence of the sunlight.
- The soiling-related loss of optical transmission can be calculated by determining the extinction coefficient related to Mie scattering. The photocurrent loss is directly correlated to the area soiled.
- A number of automated cleaning technologies have been developed over the years.

17.1 Introduction: How does soiling affect PV energy output?

The LCOE calculated in the Chapter 16 defines an optimistic lower limit. After all, even if the efficiency remained unchanged over the years (i.e., $d = 0$), soiling will still prevent a module from producing the rated power. A module installed in a hot and humid environment gradually accumulates dust and sand over the glass cover. In cold climates, periodic snow covers the module. In both cases, the amount of sunlight reaching the module is reduced, with the corresponding reduction in the power output $P(t)$, as shown in Fig. 17.1(a). Globally, soiling-related power loss is estimated to be 3%–4%, resulting in billions of dollars in lost revenue. To restore $P(t) \to P_0$, the module will need to be cleaned periodically, either by sending a person with a bucket and a washcloth or by mounting a cleaning robot. Neither is cheap. Moreover, transporting water to the desert is difficult and

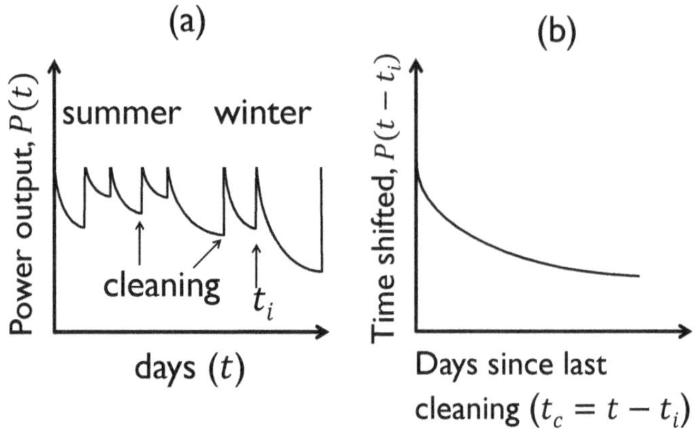

Figure 17.1: (a) Soiling gradually reduces the photocurrent and thus the power output, until the original power is restored either by rain or cleaning. The soiling depends on the season, among other things. (b) The soiling rates can be calculated by folding back the individual curves onto a same universal curve based on the days since the last cleaning.

expensive. Under these circumstances, how frequently should you clean so that cost per watt of power produced (LCOE) is minimized? In the next section, we will answer this question based on the analysis by J.K. Jones, *IEEE JPV*, 6(3), 2016.

17.2 What is the cleaning cost to produce an extra watt of power?

Assume that the power output of a pristine, dust-free module is P_0. If the soiling function $L_s(t)$ defines the fractional power loss due to soiling, the total power lost is given by $\Delta P(t) \equiv L_s \cdot P_0$. If power sells at R dollars per watt, the revenue lost until the module is cleaned at time t_c is

$$V_L = \int_0^{t_c} RP_0 L_s(t)dt. \tag{17.1}$$

It is easy to see from Fig. 17.1(b) that $t_c = t - t_i$, where t is the time since installation, and t_i is the last time the module was cleaned. Figure 17.2 shows the revenue earned (V_S, white area) and revenue lost (V_L, dark area) must sum up to the potential earning capacity of the farm, i.e., $RP_0 t_c$. Therefore,

$$V_S = \int_0^{t_c} RP_0 \left(1 - L_s(t)\right) dt. \tag{17.2}$$

One must clean the module at a cost of C_s to restore power. Therefore, the *marginal* cost to produce one unit of energy is

$$H(t_c) \equiv \frac{V_L + C_s}{V_S}. \tag{17.3}$$

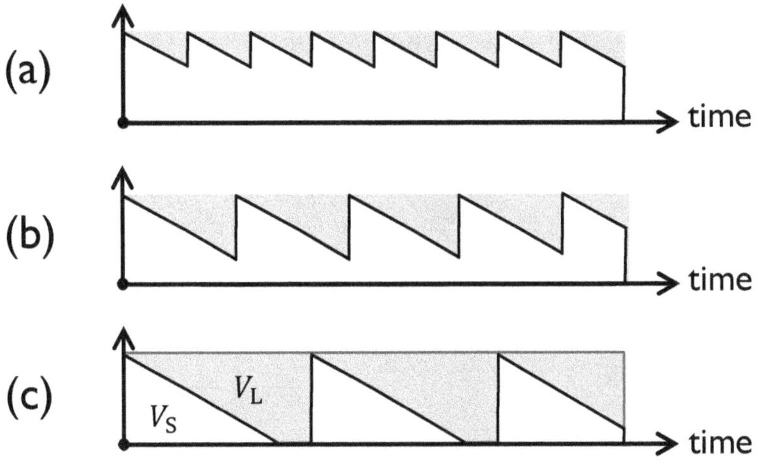

Figure 17.2: Plots of power output as a function of time and cleaning frequency. The area under the curve is the amount of energy produced, and the gray area is the amount of energy wasted because the module was not cleaned. The periodic jumps of the curve indicate that cleaning has restored the power of the original module. (a) Frequent cleaning produces more energy, but total cleaning cost will add up quickly. (b) Less frequent cleaning will reduce total energy output, but the cleaning cost is reduced as well. (c) If the cleaning frequency is too low, the solar cells will not produce any power for long periods of time, which is not desirable.

17.3 Optimized cleaning is defined by a cost–benefit analysis

Equation (17.3) suggests that if the module is cleaned too frequently (i.e., t_c is small), then the module would produce very little energy between successive cleaning (i.e., $V_S \to 0$), but the recurring cost of cleaning will make $H(t_c)$ very large and the cost of energy produced very high. Yet, if we wait too long ($t_c \to \infty$), the module will cease to produce power, the revenue lost (V_L) will mount, and H will become unacceptably large as well. There must be an optimum somewhere in between, obtained by setting $dH/dt_c = 0$. With a few lines of algebra to be discussed below, the minimization of H produces the following condition:

$$\frac{V_L + C_c}{V_S} = \frac{V_L'}{V_S'} = \frac{L_s}{1 - L_s}. \tag{17.4}$$

Once the form of L_s is specified, the implicit equation can be solved for t_c.

Homework 17.1: Calculus review, just in case!

Derive Eq. (17.4).

Solution. Recall that if $F(x) \equiv G(x)/H(x)$, then the quotient rule says

$$F'(x) = \frac{G'(x)H(x) - G(x)H'(x)}{H(x)^2}.$$

Also recall that $F(x) = \int_0^x g(t)dt$, then $F'(x) = g(x)$. Using these equations, show that

$$\frac{d}{dt_c}H(t_c) = \frac{d}{dt_c}\left(\frac{V_L + C_c}{V_S}\right) = \frac{V_L'V_S - (V_L + C_c)V_S'}{V_S^2} = 0 \quad (17.5)$$

where $V_L' \equiv dV_L/dt_c = R\,P_0L(t_c)$ and $V_S' \equiv dV_S/dt_c = RP_0(1 - L(t_c))$. Inserting these relationships in Eq. (17.5) produces the desired result.

Homework 17.2: Optimized cleaning for "exponential" soiling

The empirical soiling function is given by $L_s(t) = 1 - e^{-at}$, where the location-specific soiling constant (a) is obtained by fitting the field data shown in Fig. 17.1(b). Show that the optimum cleaning time t_c is defined by the following relationship:

$$\frac{C_c}{RP_0} = t_c + \frac{e^{at_c} - 1}{a}. \quad (17.6)$$

Solution. Since $L_s(t) = 1 - e^{-at}$, therefore from Eq. (17.4) we find that

$$\frac{V_L + C_c}{V_S} = \frac{L_s}{1 - L_s} = e^{at_c} - 1,$$

so that

$$C_c = V_S\,(e^{at} - 1) - V_L \quad (17.7)$$

$$V_L(t_c) = \int_0^{t_c} R(t)P(t)L_s(t)dt = RP_0\left(t_c + \frac{e^{-at_c} - 1}{a}\right)$$

and

$$V_S(t_c) = RP_0\left(\frac{1 - e^{-at_c}}{a}\right).$$

Inserting these expressions for V_S and V_L in Eq. (17.7), we can determine the only unknown parameter, t_c.

Homework 17.3: Free cleaning by rain!

We did not include cleaning by rain in the preceding analysis. If periodic rainfall reduces the number of cleanings per year from $N(\equiv 1/t_c)$ to M, argue that one can still use Eq. (17.7) by simply scaling the cost, C_c by M/N.

Homework 17.4: Minimizing the revenue lost: Optimum cleaning cycle with linear soiling

Instead of marginal cost, the cleaning frequency can also be determined by focusing exclusively on the revenue lost. Given the constant soiling rate a, use the linear and exponential soiling rate models to calculate the optimum cleaning interval t_c.

Solution. For the linear soiling model, the cumulative power loss is given by $L_s(t) = a\,t$. Therefore, the revenue lost due to soiling is: $V_L = \int_0^{t_c} RP_0 L_s(t)dt = RP_0 a\, t_c^2/2$. Including the cleaning cost C_c, the loss per cleaning cycle is $V_L + C_c$. The profit is maximized by minimizing the average of revenue loss, i.e., $(V_L + C_c)/t_c$. Minimizing the function, we find that

$$t_c = \sqrt{\frac{2C_c}{aRP_0}}.$$

Had the soiling rate been nonlinear, i.e., $L_s = 1 - e^{-at}$, the same analysis leads to the implicit equation for the optimum cleaning interval

$$t_c = \sqrt{\frac{C_c}{RP_0(1 - e^{-at_c})}}.$$

At short time scales, (i.e., $t_c < a^{-1}$), we can expand the exponential function to the second order in time to show that both linear and nonlinear models lead to identical cleaning intervals.

17.4 The soiling parameter a depends on a number of variables

The parameter a in the empirical soiling function $L_s(t)$ defines the rate at which soiling erodes the power output of a solar cell. The ability of sand/silt/clay particles to stick to the surface depend primarily on two factors: (1) the geometry and type of sand or snow that have accumulated over the module (and its interaction with wind, moisture, temperature), and (2) the configuration of the modules and the farm (including the tilt angle β, ground clearance, row spacing, etc.).

17.4.1 Soiling depends on the tilt angle

To understand the tilt-dependence of soiling, it is obvious that a horizontal panel ($\beta = 0$) has the highest soiling rate $a_{(\beta=0^\circ)}$. With increasing tilt, its horizontal area projection changes by the factor $\cos\beta$. Had the dust particles travelled vertically and landed on the panel surface, the dust accumulation would be proportional to the projected area, i.e.,

$$a_\beta = a_{(\beta=0^\circ)} \times \cos\beta.$$

Given a certain amount of accumulated dust, the loss in transmittance is additionally dependent on the angle of incidence (see Al-Hasan, 1994).

Homework 17.5: The weight of a monolayer of sand

If the average size of a sand particle is $10\,\mu$m, and its specific density is $\rho = 1500$ kg/m^3, then show that $w_{\text{mono}} = 20$ g/m^2 sand provides a monolayer coverage over the module. If particles of sizes 0–4, 4–8, 8–16, 16–32, and 32–64 μm occur in equal proportion, find the weight of a monolayer of sand that will cover the module.

Solution. The number of particles with radius r_{i} that occupies a fractional surface area $f_{\text{i}}(r_{\text{i}})$ is given by

$$n_{\text{S}}(r_{\text{i}}) = \frac{f_{\text{i}}}{\pi r_{\text{i}}^2}$$

and the mass of a single particle is

$$w(r_{\text{i}}) = \frac{4}{3}\pi r_{\text{i}}^3 \times \rho.$$

Therefore, the weight of a single monolayer of sand is

$$w_{\text{mono}} = \sum_{r_{\text{i}}} w(r_{\text{i}}) \times N(r_{\text{i}}) = \sum_{r_{\text{i}}} \frac{4}{3} f(r_{\text{i}}) r_{\text{i}} \rho.$$

For the first part of the problem, $r_{\text{i}} = 10 \times 10^{-6}$ m, and $f(r_{\text{i}}) = 1$, so that $w_{\text{mono}} = 19.95$ g/m^2. For the second problem, $i = 5$, $f(r_{\text{i}}) = 0.2$, and $r_1 = 2, r_2 = 6, \cdots$. The monolayer coverage density is comparable to the first part. These densities are comparable to the experimentally measured critical density of 10–20 g/mm^2, beyond which reflection/absorption increases dramatically.

In practice, the pattern of the wind-flow defined to the farm configuration (i.e., row spacing, ground clearance) weakens the tilt-angle dependence of the soiling rate. Moreover, once the module is tilted beyond a critical angle ($\beta > \beta_c$), gravity may overcome the friction between the particles and the glass cover, and the dust cannot accumulate. Various factors, such as particle shape, moisture, etc. increase stiction, and the corresponding critical angle.

For $\beta < \beta_c$, soiling is increased closer to the equator where the modules must be placed parallel to the ground to maximize collection of sunlight. At a higher latitude, accumulation of snow offers a similar challenge. Since the module tilt angle is high, snow accumulates at the bottom of the module. We will see in Chapter 18 that this partial shadowing reduces the short-circuit current. It may be possible to find a location-specific tilt angle to maximize the energy output by requiring less frequent cleaning (increasing t_c).

Homework 17.6: Origin of the exponential soiling model: Nonlinear light blocking by accumulated dust

If w is the weight of randomly placed dust particles on a module, then show that the area *not* covered by dust is given by $A_{\mathrm{f}} = e^{-w(t)/w_{\mathrm{mono}}}$. This is why the light transmission and the short-circuit current decrease nonlinearly with accumulated dust coverage.

Solution. It is important to realize that dust accumulation is a random sequential absorption process. Therefore, the rate of decrease in A_{f} as new particles land in an unoccupied area is given by

$$\frac{dA_{\mathrm{f}}}{dN} = -\pi r^2 A_{\mathrm{f}}, \text{ therefore } A_{\mathrm{f}} = e^{-\pi r^2 N(t)} = e^{-N(t)/N_0} = e^{-w(t)/w_{\mathrm{mono}}}$$

where N is the number of dust particles (per m^2) that have accumulated since the last cleaning and $N_0 = \left(\pi r^2\right)^{-1}$ is the number of dust particles needed for monolayer coverage. If the soiled region suppresses light transmission completely (for additional discussion see Fig. 17.3), we find that the fractional light lost is given by

$$L_{\mathrm{s}}(t) \propto 1 - A_{\mathrm{f}} = 1 - e^{-\pi r^2 N(t)}.$$

If $N(t) = b\,t$, where b is the location, season, and tilt-specific constant, then $a = \pi r^2 b$ interprets the origin of exponential soiling model discussed above.

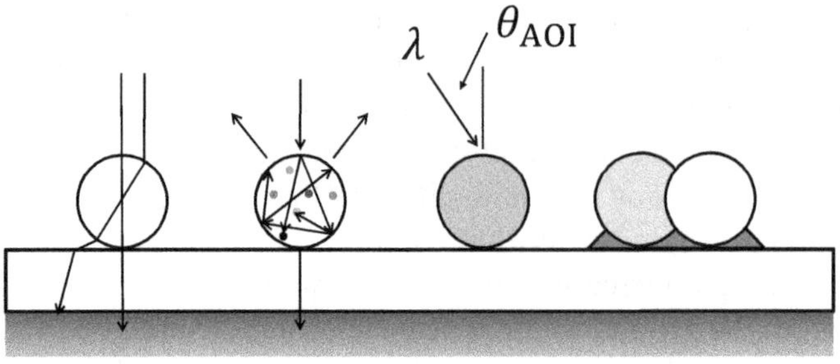

Figure 17.3: (a) A pure silica particle is a weak scatterer of light. (b) Impurities included within sand dramatically increase the reflection and absorption. (c) The light transmission depends on the wavelength, radius, and angle of incidence. (d) Moisture partially dissolves the particles, which are then cemented together as water evaporates. These particles completely block sunlight from reaching the solar cell.

17.4.2 Photocurrent reduction also depends on the geometry and type of soiling particles

In the previous section, we implicitly assumed that light transmission locally reduces to zero ($T = 0$) as soon as a particle lands on a module. This is not quite right! Simply because a particle has attached to the module tilted below the critical angle does not imply that it will block the incident sunlight completely (see Fig. 17.3). After all, sand, silt, clay, all share a refractive index of $n \approx 1.5$, similar to the glass cover of the module. Also, the particle size (s) of the sand, silt, and clay particles ranges from 100 μm to 1–2 mm, 4–100 μm, and 1–4 μm, respectively. These particles are either comparable to or much larger than the solar wavelengths of interest, namely 0.3–1 μm. Therefore, the light scattering is relatively weak and is described either by Mie scattering ($s \approx \lambda$) or simply by geometric scattering ($s \gg \lambda$). Moreover, calcium-, iron-, aluminum-, and sodium-related impurities with ($s \ll \lambda$) are embedded within the sand, silt, or clay. These particles absorb/scatter light depending on the wavelength and give sand its characteristic color. Let us now calculate T for particles of different shapes and sizes.

Homework 17.5 showed that if $W(r_\mathrm{i})$ is the weight of particles accumulated on a module, the areal density $n_\mathrm{S}(r_\mathrm{i}) = W/\rho v$ is given by the specific density ρ and the volume $v = 4/3\pi r_\mathrm{i}^3$ of a particle. Assuming that the particles are uniformly distributed over the surface, the spectrum-weighted light transmission through a soil thickness of $2r_\mathrm{i}$ can be calculated by the Beer–Lambert law (i.e., $e^{-\alpha_\mathrm{s} \times 2r_\mathrm{i}}$, see Eq. (6.6)):

$$T_\mathrm{i} \equiv \frac{I_\mathrm{sc}(W)}{I_\mathrm{sc}(W=0)} = \frac{q\int_{E_\mathrm{g}}^{\infty} n_\mathrm{S}(E)\, e^{-\alpha_{\mathrm{s,i}}(W, r_\mathrm{i}, \lambda)\times 2r_\mathrm{i}}\, dE}{q\int_{E_\mathrm{g}}^{\infty} n_\mathrm{S}(E)\, dE}. \tag{17.8}$$

Homework 17.7: Weight of sands and loss of current

Assuming a relatively flat spectrum over the energy range of interest, express Eq. (17.8) in terms of $W(t)$, the accumulated weight of sand.

Solution. At low soiling densities, the extinction coefficient is calculated as the product of: (a) extinction efficiency per particle Q_1^*, (b) its cross section $A(r_i) = \pi r_i^2$, and (c) its volume density $N_V = n_S/(2r_i)$, with $n_S = W(t)/\rho v$ and $v \equiv (4/3)\pi r_i^3$. Taken together,

$$\alpha_s(W, r_i, \lambda) = Q_1^*(r_i, \lambda)\, A(r_i)\, N_V(r_i) = \frac{3Q_1^*}{8r_i^2}\, W(t).$$

For relatively flat spectrum, Eq. (17.8) can be rewritten as

$$I_{sc}(W) = q \int_{E_g}^{\infty} n_S(E) e^{-2r_i\alpha_s} = I_{sc}(W=0)e^{-\frac{3Q_1^*}{4r_i}W(t)}.$$

For larger particles (i.e., $r_i \gg \lambda$), the extinction efficiency $Q_i^* \approx 2$ from geometric optics or Mie theory. Therefore, the exponent reduces with particle size and so does the photocurrent loss due to soiling. For a given W, the photocurrent loss is related primarily to intermediate-size particles with $r_i \approx \lambda$ and $Q_i^* > 2$.

Did you notice that there are a few paradoxes in the discussion above? First, $Q_i^* \approx 2$ implies that the a particle with geometric cross section $A(r_i)$ scatters light as if it is twice as big! This extinction paradox has been explained by the fact that a particle can diffract waves beyond its geometric cross section. Second, why do larger particles transmit more light? Given W, the particle number reduces as r_i^3. Despite having a larger decay length, $2r_i$, the fewer larger particles scatter a smaller fraction of the total sunlight. Finally, using $n_S = f_i(t)/(\pi r_i^2)$ from Homework 17.5, instead of $n_S = W/\rho v$ as we did in the discussion above, we find that

$$I_{sc}(W)/I_{sc}(W=0) = e^{-Q_1^* f_i(t)}$$

suggesting that the photocurrent loss depends on the fractional area soiled f_i regardless of the particle size. This is an amazing and a bit anticlimactic result: you can measure the short-circuit loss simply by visually measuring the fractional soiled area and doubling the Beer–Lambert exponent!

Here $n_S(E)$ is the solar spectrum, and the soiling-related extinction (or decay) coefficient α_s is calculated by Mie theory or geometric optics (generalized to include the flat glass-substrate) and depends on the volume density i.e., $N_V = n_S/(2r_i)$

and the size $s = 2r_\mathrm{i}$ of the particles. We will see in Homework 17.7 that the coefficient α_s is relatively constant for particles comparable to the wavelength of light ($s = 2r_\mathrm{i} \approx \lambda$), but then falls off with N_v for larger r_i. As a result, T_i is mostly affected by particles with sizes comparable to the wavelength (few μm) because these particles maximize $\alpha_\mathrm{s} r_\mathrm{i}$. Unfortunately, these are precisely the type of particles that are easily airborne at typical windspeeds.

Finally, it is important to distinguish between extinction and absorption. The classical lossless Mie scattering (without absorption) randomizes the photon direction so that the direct incident beam of light appears attenuated or extinguished at the *far-field*. The total photon number, integrated over the solid angles, is actually conserved. These randomly directed photons would be easily absorbed by the (near-field) solar absorber (see Fig. 17.3(a)) and unlike Eq. (17.8), the short-circuit current should remain unaffected! In practice, Mie scattering leads to bouncing of photons within each particle, as in Fig. 17.3(b), leading to absorption by the impurities. The scattered photons are indeed extinguished/absorbed and the short-circuit current reduced, as in Eq. (17.8). Additionally, the reflection by the dust particles are higher at steeper incidence of the sunlight on the panels. For a conventional south-facing panel, the fractional soiling loss L_s would therefore deem more pronounced in the morning and the afternoon.

Homework 17.8: The scattering efficiency Q_i^* can be determined by a web calculator

In Homework 17.7, the soiling photocurrent loss was expressed in terms of Q_i^*. Use Prahl's web calculator `https://omlc.org/calc/mie_calc.html` to calculate the radius dependence of Q_i^* for mono-disperse sand particles for $0.5 < r_\mathrm{i} < 20\,\mu\mathrm{m}$. Assume that $n = 1.5$ and average sunlight $\lambda = 1\,\mu\mathrm{m}$. In these weakly absorbing media, the absorption coefficient (given by the imaginary refractive index) can be neglected. Compare scattering efficiencies for particles with $r_\mathrm{i} \approx \lambda$ and $r_\mathrm{i} \gg \lambda$.

17.5 A number of technologies have been developed to clean solar farms

Figure 17.4 shows a number of ways a module can be cleaned. For a specified C_c, Eq. (17.7) suggests that t_c can be increased by increasing the tilt angle to reduce the dust accumulation rate a. Cleaning robots that span the module height and are mounted on the far end of a row of modules can slide across the row to clean the modules by water or pressurized air. The cost of the robots must be included in C_sys of the LCOE calculation (see Chapter 16). Self-cleaning arrays of electrodes, with alternating polarities, can sweep away the particles by dielectrophoretic force. The approach could be useful for space missions. Finally, anti-soiling coatings (with alternate stripes of hydrophobic and hydrophilic regions) could reduce the amount of water needed for cleaning. In general, the economic viability of more advanced cleaning techniques for large solar farms remains an open question and an interesting research problem.

Homework 17.9: Dust particles comes in various shapes and sizes

For poly-disperse particles, show that $T = \prod_i T_{\mathrm{i}}$, where T_{i} is determined by the light transmission through particles of radius r_{i}, given by Eq. (17.8).

Solution. At low densities, the Beer–Lambert law for poly-disperse particles is given by sum of their individual extinction coefficients:

$$I_{\mathrm{sc}}(W) = \int_{E_{\mathrm{g}}}^{\infty} qn_{\mathrm{S}}(E)\, e^{-\sum_i \alpha_{\mathrm{s,i}}(N_{\mathrm{V,i}},\lambda)\times 2r_{\mathrm{i}}} \tag{17.9}$$

where i represents particles of a specific radius r_{i} and $\alpha_{\mathrm{s,i}}$ is the extinction ratio associated with those particles. The volume density $N_{\mathrm{V,i}}$ depends on the probability density of poly-disperse particles. For example, if the particle size distribution is log–normal, with average size μ and standard deviation σ, then

$$N_{\mathrm{V}}(r_{\mathrm{i}}) = \frac{k}{2r_{\mathrm{i}}} \frac{1}{\sigma\sqrt{2\pi}} \cdot \exp\left[-\left(\frac{\ln(2r_{\mathrm{i}}) - \ln(2\mu)}{\sqrt{2}\sigma}\right)^2\right]. \tag{17.10}$$

The constant k ensures that the mass integrates to the weight of the dust accumulated.

The exponential of a sum can be written as the product of the individual exponentials. Exchanging the product and the integral in Eq. (17.9), we find that

$$\begin{aligned} I_{\mathrm{sc}}(W) &= \prod_i \left[\int_{E_{\mathrm{g}}}^{\infty} dE\, qn_{\mathrm{S}}(E)\, e^{-2r_{\mathrm{i}}\alpha_{\mathrm{s,i}}(N_{\mathrm{V,i}},E)}\right] \\ &= \prod_i \left[\int_{E_{\mathrm{g}}}^{\infty} dE\, qn_{\mathrm{S}}(E)\, e^{-\frac{3Q_{\mathrm{i}}^*(E)W_i(t)}{8r_{\mathrm{i}}^2}}\right] \\ &= \prod_i \left[\int_{E_{\mathrm{g}}}^{\infty} dE\, qn_{\mathrm{S}}(E)\, e^{-Q_{\mathrm{i}}^*(E)f_{\mathrm{i}}(t)}\right]. \end{aligned} \tag{17.11}$$

Here, f_{i} and W_i are the area fraction and the weight covered by a particle of size r_{i}. Dividing both sides by the constant $I_{\mathrm{sc}}(W = 0)$, we find that $T = \prod_i T_{\mathrm{i}}$. In this product, T_{i} is smallest when the particle sizes are comparable to the wavelength, so that the overall transmission is determined by these particles.

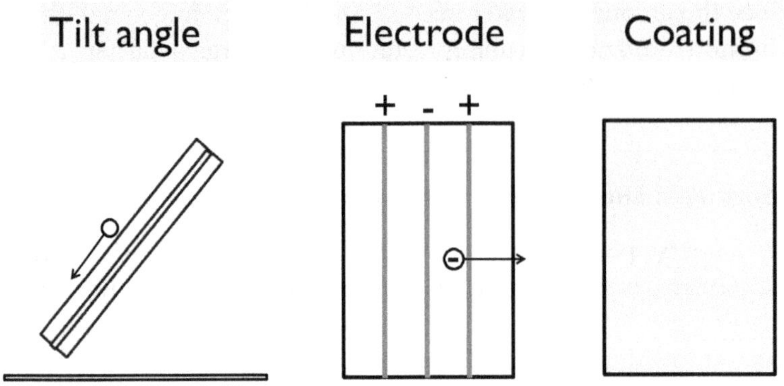

Figure 17.4: Soiling reduction strategies include increased tilt angle, dry cleaning, and soil-resistant coating.

17.6 Conclusions: Optimized cleaning maximizes cost-effective energy output of a solar cell

In this chapter, we discussed how soiling erodes the power output of a solar module, and how one must optimize the cleaning frequency to maximize the energy output of such a system. Fortunately, the effect is reversible — that is, the initial rated power can be restored once the modules have been cleaned.

Soiling is one of the several ways in which the short-circuit current is reduced and the farm may not produce the rated output. In the next chapter, we will discuss the reliability issue of shadow degradation: the reduction of short-circuit current when a shadow blocks the sunlight. In principle, the power loss due to shadowing and soiling is reversible, because the original power can be restored when the panel is cleaned or the shadow is removed. Unfortunately, under extreme conditions, a shadow may permanently/irreversibly damage a module, just as cleaning by hard water or excessive pressure may leave behind scratch marks to permanently reduce the power output.

References

[1] Travis Sarver, Ali Al-Qaraghuli, and Lawrence L. Kazmerski. A comprehensive review of the impact of dust on the use of solar energy: History, investigations, results, literature, and mitigation approaches. *Renewable and Sustainable Energy Reviews*, 22:698–733, June 2013.

[2] J. J. John, V. Rajasekar, S. Boppana, S. Chattopadhyay, A. Kottantharayil, and G. TamizhMani. Quantification and Modeling of Spectral and Angular Losses of Naturally Soiled PV Modules. *IEEE Journal of Photovoltaics*, 5(6):1727–1734, November 2015.

[3] Hassan Qasem, Thomas R. Betts, Harald Müllejans, Hassan AlBusairi, and Ralph Gottschalg. Dust-induced shading on photovoltaic modules. *Progress in Photovoltaics: Research and Applications*, 22(2):218–226, 2014.

[4] Neil S. Beattie, Robert S. Moir, Charlslee Chacko, Giorgio Buffoni, Simon H. Roberts, and Nicola M. Pearsall. Understanding the effects of sand and dust accumulation on photovoltaic modules. *Renewable Energy*, 48:448–452, December 2012.

[5] R. K. Jones, A. Baras, A. A. Saeeri, A. Al Qahtani, A. O. Al Amoudi, Y. Al Shaya, M. Alodan, and S. A. Al-Hsaien. Optimized Cleaning Cost and Schedule Based on Observed Soiling Conditions for Photovoltaic Plants in Central Saudi Arabia. *IEEE Journal of Photovoltaics*, 6(3):730–738, May 2016.

[6] Klemens Ilse, Leonardo Micheli, Benjamin W. Figgis, Katja Lange, David Daßler, Hamed Hanifi, Fabian Wolfertstetter, Volker Naumann, Christian Hagendorf, Ralph Gottschalg, and Jörg Bagdahn. Techno-Economic Assessment of Soiling Losses and Mitigation Strategies for Solar Power Generation. *Joule*, 3(10):2303–2321, October 2019.

[7] Md. Mahamudul Hasan Mithhu, Tahmina Ahmed Rima, and M. Ryyan Khan. Global analysis of optimal cleaning cycle and profit of soiling affected solar panels. *Applied Energy*, 285:116436, March 2021. tex.ids: mithhu_global_nodate.

[8] Craig F. Bohren and Donald R. Huffman. *Absorption and scattering of light by small particles*. Wiley-VCH, Weinheim, 2004. OCLC: 254937169.

[9] H. Moosmüller and C. M. Sorensen. Small and large particle limits of single scattering albedo for homogeneous, spherical particles. *Journal of Quantitative Spectroscopy and Radiative Transfer*, 204:250–255, January 2018.

[10] P. A. Bobbert and J. Vlieger. Light scattering by a sphere on a substrate. *Physica A: Statistical Mechanics and its Applications*, 137(1):209–242, July 1986.

[11] Jasper F. Kok, Eric J. R. Parteli, Timothy I. Michaels, and Diana Bou Karam. The physics of wind-blown sand and dust. *Reports on Progress in Physics*, 75(10):106901, September 2012. Publisher: IOP Publishing.

[12] Benjamin Figgis, Ahmed Ennaoui, Bing Guo, Wasim Javed, and Eugene Chen. Outdoor soiling microscope for measuring particle deposition and resuspension. *Solar Energy*, 137:158–164, November 2016.

[13] H. Qasem, A. Mnatsakanyan, and P. Banda. Assessing dust on PV modules using image processing techniques. In *2016 IEEE 43rd Photovoltaic Specialists Conference (PVSC)*, pages 2066–2070, June 2016.

[14] Wenjie Zhang, Shunqi Liu, Oktoviano Gandhi, Carlos David Rodriguez-Gallegos, Hao Quan, and Dipti Srinivasan. Deep-Learning-Based Probabilistic Estimation of Solar PV Soiling Loss. *IEEE Transactions on Sustainable Energy*, pages 1–1, 2021. Conference Name: IEEE Transactions on Sustainable Energy.

[15] A. Y. Al-Hasan, "A new correlation for direct beam solar radiation received by photovoltaic panel with sand dust accumulated on its surface," Solar Energy, vol. 63, no. 5, pp. 323–333, Nov. 1998, doi: 10.1016/S0038-092X(98)00060-7.

[16] A. A. Hegazy, "Effect of dust accumulation on solar transmittance through glass covers of plate-type collectors," Renewable Energy, vol. 22, no. 4, pp. 525–540, Apr. 2001, doi: 10.1016/S0960-1481(00)00093-8.

CHAPTER 18

A Transient Partial Shadow May Cause Permanent Damage

Chapter Summary

- A partially shadowed cell loses photo-current. In a series-connected module, the cell may have to go to reverse breakdown to supply the lost current.
- The power-loss is substantially more than the fraction of cell shadowed. A few shadowed cell may lead to nearly complete power-loss of the module.
- The reverse-breakdown leads to substantial heat dissipation: the module output is reduced and the cell may be damaged irreversibly.
- Bypass diodes protect Si solar modules from the detrimental effects of partial shadowing.
- Innovative module design can also reduce the effect of partial shadowing for thin-film solar cells.

18.1 Introduction: The danger of a partial shadow

The power output of a solar module depends on the intensity of sunlight incident on it. If a soft, semitransparent shadow covers the module completely, it causes a temporary (reversible) dip in the power output. Once the shadow is removed, the power is restored and the module operates as if nothing had happened.

Sometimes, however, a shadow may block illumination to *parts* of a module, as shown in Fig. 18.1. These *partial shadows* may arise from clouds overhead, a fallen leaf, a nearby building, or a neighboring solar panel. Naively, one may think that shadowing one of the N subcells in a module would reduce the total power to $(N-1)P_{cell}$. For typical $N \approx$ 50–100, the power loss would be negligible.

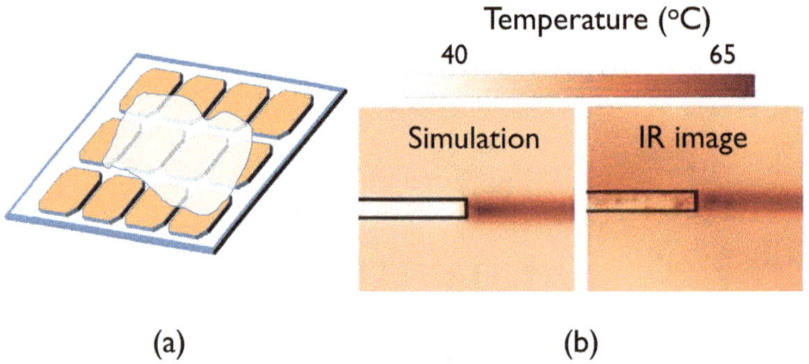

Figure 18.1: (a) A partial shadow, marked by the white region, blocks illumination to parts of a solar module. (b) Both the numerical simulation result and the experimental infrared image show that the regions next to the shadow become extremely hot, which may severely damage the solar cell.

In practice, even a small partial shadow may cause dramatic power loss in this series-connected system. More importantly, power dissipation within the shadowed subcell may become so high and the region so hot that the cell may be damaged permanently. In the worst case, a fire may start. Interestingly, this reliability issue was first understood not in a solar farm on the ground, but in NASA satellites orbiting the earth. As the narrow shadow of the satellite boom bisected the solar wings, the power dropped precipitously. Indeed, the satellite could be lost if the situation remained unresolved. In this chapter, we will discuss the physics of partial shadowing and the ways to mitigate its effects.

18.2 A module is optimized for shadow-free operation

In Chapter 10, we determined the optimum number of series-connected subcells (N) that minimizes the series resistance loss of the module, see Fig. 18.2. The optimization assumed that the module is uniformly illuminated — a partial shadow was not anticipated.

Under uniform illumination, a subcell is characterized by its operating point, $V_{\mathrm{mp},1}$ and $I_{mp,1} \equiv I_{\mathrm{mp}}$, see Fig. 18.3(a). Recall from Homework 4.12 that $V_{\mathrm{mp}}(N, M)$ of an idealized M-junction, N-subcell module (without series resistance) can be written in terms of $V_{\mathrm{mp}}(N=1, M=1) = V_{\mathrm{mp},1}$ as follows:

$$V_{\mathrm{mp}}(N, M) = N\left(V_{\mathrm{mp},1} - \Delta(M)\right) \tag{18.1}$$

where $\Delta(M) \equiv \frac{T_{\mathrm{D}}}{T_{\mathrm{S}}}(E_{\mathrm{g,p}} - E_{\mathrm{g,a}}) = \frac{T_{\mathrm{D}}}{T_{\mathrm{S}}}\frac{M-1}{M+1}(\beta_{\mathrm{sun}}^{-1} - E_{\mathrm{g,a}})$. Here, $E_{\mathrm{g,p}}$ and $E_{\mathrm{g,a}}$ are respectively the largest and the average bandgap of the M-junction tandem cell. For a single-junction solar cell ($M = 1$ or $E_{\mathrm{g,p}} = E_{g,a}$), $\Delta = 0$ and $V_{\mathrm{mp}}(N) = NV_{\mathrm{mp},1}$, as expected. The actual V_{mp} is obtained by adding the extra voltage drop due to series resistance. Also, note that the same photocurrent, I_{mp}, flows though the N

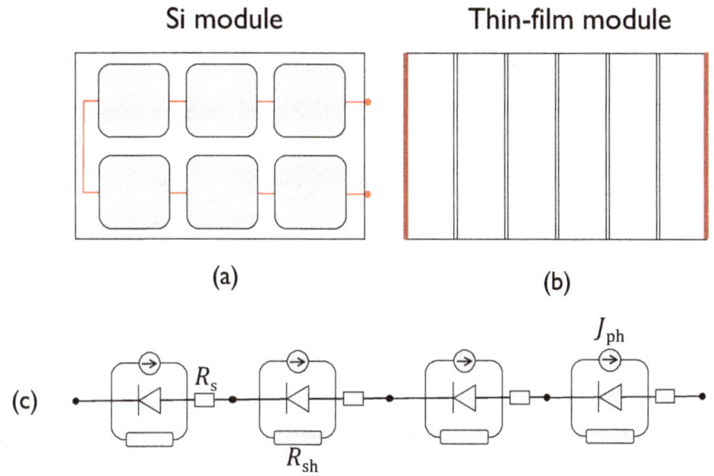

Figure 18.2: (a) Series-connected c-Si solar module with six subcells ($N = 6$). (b) Series-connected thin-film solar module, also with 6 subcells. (c) Each subcell may be represented by a current source (J_{ph}), a diode, and a shunt resistance (R_{sh}). A series resistance, R_s, connects a pair of subcells. The figure shows a string of four subcells.

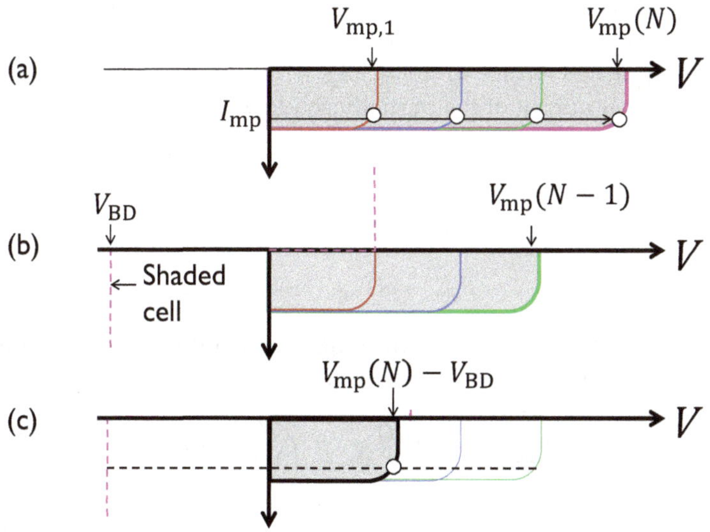

Figure 18.3: (a) The composite I-V characteristics of a series-connected module ($N = 4, p = 0$) is obtained by adding — at each current level to ensure current continuity — the voltages of the four individual subcells. (b) The dark I-V characteristic of a fully shaded subcell (dashed line) is defined by its breakdown voltage, V_{BD}. The I-V characteristics of the remaining three, fully illuminated, subcells are also shown. (c) The integrated response of shadowed module ($N = 4$, $p = 1$) — obtained by adding up voltages at each current level — explains the reduction in the operating voltage.

series-connected cells of a module. From Eq. (4.2), the photocurrent is approximately given by

$$I_{\mathrm{mp}} \approx I_{\mathrm{ph}} = A_{subcell} J_0 \left[1 - \beta_{\mathrm{sun}} E_{\mathrm{g}}\right]. \tag{18.2}$$

With Eqs. (18.1) and (18.2) for V_{mp} and I_{mp}, the module power output is

$$P_{\mathrm{out}}^{(i)} = V_{\mathrm{mp}}(N) \times I_{\mathrm{mp}} \simeq V_{\mathrm{mp}}(N) \times I_{\mathrm{ph}}. \tag{18.3}$$

The superscript i represents the initial power output of the module, before the arrival of the shadow.

18.3 A shadow decreases the power output dramatically

When a subcell is fully shadowed, it loses its photocurrent. Since the other subcells continue to produce a photocurrent, the shadowed subcell is forced into reverse breakdown (V_{BD}) to ensure current continuity. Assuming p out of N subcells are shadowed, the voltage

$$V_{\mathrm{mp}}(N,p) = (N-p)V_{\mathrm{mp},1} - pV_{\mathrm{BD}}. \tag{18.4}$$

In other words, the module not only loses the voltages of the cells shadowed (first term on the right), but also the voltage needed to put the shadowed cells into breakdown (second term on the right). Since $V_{\mathrm{BD}} \gg V_{\mathrm{mp},1}$, it is clear the module voltage output will reduce dramatically, which in turn will be reflected in the reduced *final* power output of the partially shadowed cell:

$$\begin{aligned} P_{\mathrm{out}}^{(f)} &= I_{\mathrm{mp}} \times V_{\mathrm{mp}}(N,p) \\ &= I_{\mathrm{ph}} \times \left[(N-p)V_{\mathrm{mp},1} - pV_{\mathrm{BD}}\right] \\ &\sim P_{\mathrm{out}}^{(i)} - p(V_{\mathrm{mp},1} + V_{\mathrm{BD}})I_{\mathrm{ph}}. \end{aligned} \tag{18.5}$$

The term proportional to p indicates the magnitude of the power loss due to shadowed cells. The corresponding loss of efficiency is given by

$$\frac{\delta\eta}{\eta_0} \equiv \frac{P_{\mathrm{out}}^{(i)} - P_{\mathrm{out}}^{(f)}}{P_{\mathrm{out}}^{(i)}} \approx \frac{p}{N}\left(1 + \frac{V_{\mathrm{BD}}}{V_{\mathrm{mp},1}}\right). \tag{18.6}$$

One expects the p/N factor intuitively, because p of the N cells are shadowed and have stopped producing power. The second factor, involving the ratio of the breakdown voltage to the maximum power point voltage (with $V_{\mathrm{BD}} \gg V_{\mathrm{mp},1}$), reflects the series connection of the module, which amplifies the effect of the shadowed cells. The formula suggests that the efficiency loss due to partial shadowing could be greatly reduced by reducing V_{BD}, a topic we will return to at the end of the chapter.

Homework 18.1: Efficiency loss due to partial shadowing: A numerical example

A 20% efficient c-Si module consists of 72 cells. The cells are asymmetrically doped with donor doping of 10^{18} cm^{-3} and acceptor doping of 5×10^{16} cm^{-3}. Use Eq. (18.6) to calculate the efficiency loss if 3 cells are shaded. Assume that $V_{\text{mp},1} = 0.6V$.

Solution. Here, $N = 72$, $p = 5$, and $V_{\text{mp},1} = 0.6$ V. To calculate V_{BD}, we recall that a reverse-biased diode breaks either by Zener tunneling or by avalanche multiplication. The exact expressions for the breakdown field (E_{BD}) are complicated. For a one-sided p-n junction (as is the case here), a simpler empirical equation suffices:

$$E_{BD} = \frac{4 \times 10^5}{1 - 0.33 \log(N_{\text{A}}/10^{16})} \text{ V/cm}$$

where $N_{\text{A}} = 5 \times 10^{16}$ cm^{-3} is the doping of the lower-doped side. The corresponding breakdown voltage is

$$V_{\text{BD}} = \frac{\kappa_{\text{s}}\epsilon_0 E_{BD}^2}{2qN_{\text{A}}} \approx 10 \text{ V}.$$

Therefore, the efficiency loss is $\delta\eta = 20 \times (3/72)(1 + 10/0.6) = 14.7\%$. In other words, the module has lost two-thirds of the output power with only 3 cells (out of 72) shaded! No wonder, partial shading is such an important consideration.

Homework 18.2: The power output of a partially shaded tandem module is easily derived

Re-derive Eq. (18.6) where each of the N series-connected cells is actually an M-junction tandem cell. Use the information from Homework 4.12 to complete the derivation.

Solution. For an M-junction optimized tandem cell with maximum bandgap $E_{\text{g,p}}$, and average bandgap $E_{\text{g.a}}$, the maximum power point voltage is given by

$$\frac{V_{\text{mp}}}{M} = E_{\text{g,a}} \left(1 - \frac{T_{\text{D}}}{T_{\text{S}}} \frac{E_{\text{g,p}}}{E_{\text{g,a}}}\right) - k_{\text{B}} T_{\text{D}} \ln\left(\frac{\theta_{\text{D}}}{\theta_{\text{S}}}\right) \tag{18.7}$$

(continued on the next page)

Homework 18.2 (*continued from the previous page*)

and from Eq. (4.4)

$$I_{\mathrm{mp}}(M) = \frac{I_0}{M+1} \approx \frac{2I_{mp,\mathrm{SJ}}}{M+1} \tag{18.8}$$

because $I_0 = 2I_{\mathrm{mp}}(M=1) \equiv 2I_{mp,\mathrm{SJ}}$, with optimum single-junction bandgap, $E_{g,\mathrm{SJ}} = 1.34$ V. Similarly, the dark current of a tandem cell is given by

$$J_{\mathrm{dark}}(V) = 2\pi q \langle \gamma_g \rangle \, e^{-\langle E_{\mathrm{g}} \rangle / k_{\mathrm{B}} T_{\mathrm{D}}} \, e^{qV/N k_{\mathrm{B}} T_{\mathrm{D}}} \tag{18.9}$$

where $\langle \gamma_g \rangle$ is the geometrical average of the density of states factors associated with each bandgap, i.e., $\gamma_{\mathrm{i}}(E_{\mathrm{g}}) \equiv \frac{2k_{\mathrm{B}}T_{\mathrm{D}}}{c^2 h^3}\left(E_{\mathrm{g}}^2 + 2k_{\mathrm{B}}T_{\mathrm{D}}E_{\mathrm{g}} + 2k_{\mathrm{B}}^2 T_{\mathrm{D}}^2\right)$, see Sec. 3.5. When N of these cells are connected in series to form a module, then

$$\frac{V_{\mathrm{mp}}(M,N)}{N} \to V_{\mathrm{mp}}(M). \tag{18.10}$$

Now replace $V_{\mathrm{mp},1}$ in Eq. (18.6) with the new expression for $V_{\mathrm{mp}}(M)$ to complete the derivation.

18.4 Semitransparent shadows produce complex I-V characteristics

The shadows are seldom so opaque that they block the sunlight completely. Direct sunlight may be blocked, but scattered light will still illuminate (at a reduced intensity α) the subcells under shadow. In this case, Eq. (18.4) is modified slightly, as follows.

Figure 18.4(a) shows the I-V characteristics of the fully illuminated subcell. The I-V characteristics are obtained by adding the voltages at each current level. The dashed line in Fig. 18.4(b) is the modified I-V characteristic for a partially illuminated subcell with semitransparent shadow. Here, the short-circuit current has been suppressed by α. Figure 18.4(c) shows that the full I-V characteristics of the module are obtained by summing the voltages of the shaded and the unshaded cells.

The composite I-V characteristic has two plateaus, αI_{ph} and I_{ph}, with the transition voltage, V_{t}. For $I < \alpha I_{\mathrm{ph}}$, the partially shaded cell can support the current and the output voltage is high (operating point A). For $I_{\mathrm{ph}} < I < \alpha I_{\mathrm{ph}}$, the shaded cells must go into reverse breakdown to support the current needed. The voltage is now reduced and the operating point moves to B. It is easy to see that the two-plateau curve transitions between the I-V characteristics of a fully illuminated module ($\alpha = 1$, Fig. 18.3(a)) and that of a module with a few fully shaded subcells ($\alpha = 0$; Fig. 18.3(c)).

Which of the two operating points — A or B — would produce more power? If the number of partially shaded cells p is less than a critical number p_c, then the

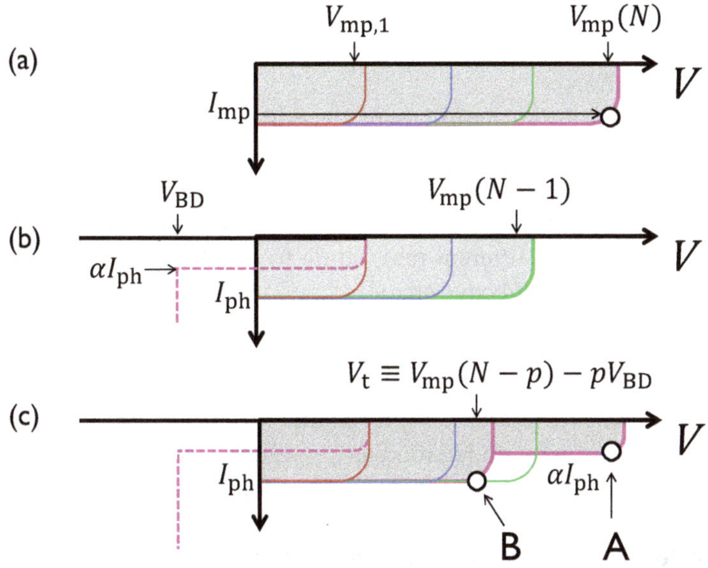

Figure 18.4: (a) The composite I-V characteristics of the fully illuminated module. (b) The dashed line is the I-V characteristic of a partially illuminated subcell — the other cells are fully illuminated. (c) The composite I-V is obtained by adding the subcell voltages at each current. At low currents, the I-V characteristic resembles that of a fully illuminated module, and the module operates at point A on the right. At higher currents, the partially shaded cell must go into reverse breakdown and the I-V characteristics resemble a module with fully shadowed subcells. The module now operates at point B. The maximum power point (A or B) depends on the number of cells shaded and the breakdown voltages.

module produces maximum power if the shadowed cells are allowed to break, and the power output (associated with B) is given by

$$P_{out}^{(f)} = I_{ph} \times [V_{mp}(N-p) - pV_{BD}], \quad (p < p_c). \tag{18.11}$$

In this case, the efficiency loss is similar to that of a fully shaded subcell discussed in the preceding section (see Eq. (18.6)).

On the other hand, when $p > p_c$, the output power may increase if all the subcells operate at forward bias, limited by the current produced by the shadowed subcell (point A), as follows:

$$P_{out}^{(f)} = \alpha I_{ph} \times V_{mp}(N). \quad (p \geq p_c). \tag{18.12}$$

The module would now operate at point A to maximize power output, and the efficiency loss is given by

$$\frac{\delta\eta}{\eta_0} = 1 - \alpha. \tag{18.13}$$

Although the power loss is significant in either case, we will see in the next section that point A operation may be preferred to avoid self-heating and long-term reliability degradation.

We just discussed the scenario where a semitransparent shadow ($0 < \alpha < 1$) completely covers a cell area A_{cell}. A different scenario is also possible: a part of the cell (A_1) may be covered by a semitransparent shadow of one magnitude (α_1), while the remaining part ($A_2 \equiv A_{\text{cell}} - A_1$) may have been covered with a semitransparent shadow of a different magnitude (α_2). The loss of the photocurrent can be represented by an effective shadow

$$\alpha \equiv \frac{\sum_i \alpha_i A_i}{A_{\text{cell}}}.$$

Interestingly, we will see later in Chapter 23 that the effects of metal grid corrosion on cell performance can be understood as a semitransparent shadowing problem, with $\alpha_1 = 0$ for the corroded section, and $\alpha_2 = 1$ for the uncorroded section. Thus, the I-V characteristics of a module with a few corroded cells can be interpreted by the multi-plateaued characteristics shown in Fig. 18.4.

> **Homework 18.3: The expression for p_c is obtained by comparing the two efficiency formulae**
>
> Equate the efficiency loss formula for $p < p_c$ (Eq. (18.6)) and $p \geq p_c$ (Eq. (18.13)) to derive an expression for p_c.

18.5 Partial shadows cause irreversible damage

Did you realize that the "lost" power, $P_{\text{lost}} = I_{\text{ph}} \times pV_{\text{BD}}$, is dissipated within the p fully shadowed subcells and adds to the self-heating of a shadow-free solar module discussed in Chapter 5, Eq. (5.15)? The excess temperature of the cell is given by

$$\Delta T_{\text{D,sh}} \approx \frac{P_{\text{lost}}}{2h},$$

where h is the convection heat transfer coefficient. The increased self-heating will accelerate the bond dissociation and shunt formation rates (see Eq. (22.2)):

$$R_{\text{D}} = A_{\text{s}}\, e^{-E_{\text{A}}/(k_{\text{B}}(T_{\text{D}}+\Delta T_{\text{D,sh}}))},$$

where E_{A} is the activation energy for creating a defect and A_{s} is an empirical prefactor. These recombination centers, due to broken bonds, will irreversibly increase diode recombination current and shunt leakage, i.e.,

$$R_{\text{sh}} = R_{\text{sh}}(t=0)\left[1 + \int_0^{t_{\text{s}}} R_{\text{D}}\, dt\right]^{-1},$$

where t_{s} is the total time a module experiences shadowing.

This increase in the dark current (due to junction leakage and shunt formation) will permanently reduce the efficiency of the solar cell. This is exactly what happened in Fig. 18.1(b). This discussion has important implications for partial shadowing discussed in the preceding section: it may be preferable to operate at point A to avoid self-heating in reverse-biased diodes, even if point B produces more output power from the module (see Fig. 18.4).

18.6 Strategies to mitigate the effect of partial shadows

There are two ways to reduce the detrimental effects of a partial shadow: using a bypass diode to reduce the "effective" V_{BD} of a crystalline silicon solar cell (see Eq. (18.6)) or adopting the specialized module geometry to reduce the possibility of creating an $\alpha = 0$ shadow over the entire subcell.

18.6.1 Bypass diodes reduce partial shadow degradation in a c-Si module

The detrimental effect of partial shadowing can be reduced significantly if we can reduce V_{BD} without sacrificing the forward-bias performance of the solar cells. Various solar cell designs with specialized doping profiles to reduce V_{BD} have been proposed, but they have not been commercially adopted. Instead, commercial systems use a module-based approach shown in Fig. 18.5(a). A bypass diode is connected to a subcell with reverse polarity so that the diode is reverse-biased and turned off during normal operation, but it is turned on once the corresponding

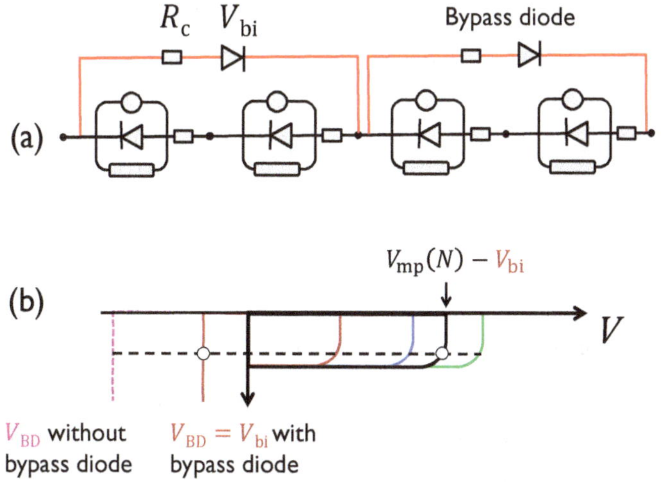

Figure 18.5: (a) A string of c-Si cells is being protected by a bypass diode (red line). The diode contact resistance R_c is also shown. (b) Such protection improves the power output of the solar cells. The shadowed subcells are protected.

subcell is shadowed and a reverse-bias develops across it (forward-biasing the bypass diode in the process). As shown in Fig. 18.5(b), the magnitude of the power loss is reduced from

$$P_{\text{out}}^{(f)} = P_{\text{out}}^{(i)} - pI_{\text{ph}}\left(V_{\text{mp},1} + V_{\text{BD}}\right)$$

to

$$P_{\text{out}}^{(f)} = P_{\text{out}}^{(i)} - pI_{\text{ph}}\left(V_{\text{mp},1} + V_{\text{bi}}\right)$$

where $V_{\text{bi}}(\ll V_{\text{BD}})$ is the built-in voltage of the bypass diode. The expression for the efficiency loss is now rewritten as

$$\frac{\delta\eta}{\eta_0} = \frac{p}{N}\left(1 + \frac{V_{\text{bi}}}{V_{\text{mp},1}}\right). \tag{18.14}$$

The subcell is now protected and the power loss is greatly minimized, but since the entire current of the cell $I_{\text{ph}} = A_{\text{cell}}J_{\text{ph}}$ is being dissipated into a small-area bypass diode, frequent shadowing may lead to failure of the bypass diode itself. In practice, instead of placing bypass diodes for each cell, a string of cells is protected by a single diode. The number of cells protected by a bypass diode is dictated by economic and cell layout considerations.

Homework 18.4: A bypass diode reduces the efficiency loss significantly

Use Eq. (18.14) to calculate the efficiency loss of the solar cell discussed in Homework 18.1. We also assume $n_{\text{i}} = 8.3 \times 10^9\ \text{cm}^{-3}$ at 25°C.

Solution. From Homework 18.1, we find that $p = 3$, $N = 72$, $\eta_0 = 20\%$, $N_{\text{D}} = 10^{18}\ \text{cm}^{-3}$, $N_{\text{A}} = 5 \times 10^{16}\ \text{cm}^{-3}$. The turn-on voltage of the bypass diode is given by

$$V_{\text{bi}} = k_{\text{B}}T_{\text{D}}\ \ln\left(\frac{N_{\text{A}}N_{\text{D}}}{n_{\text{i}}^2}\right) = 0.026\ \ln\frac{10^{18} \times 5 \times 10^{16}}{(8.3 \times 10^9)^2} = 0.89\,\text{V}.$$

Therefore, $\delta\eta = 20 \times (3/72)(1 + 0.89/0.6) = 2.06\%$.

Homework 18.5: Is diode self-heating a reliability concern?

A module consists of 72 Silicon subcells, each with an area of 100 cm^2. If one of the cells is shaded and its current is bypassed by a protection diode (area 0.1 cm^2), calculate the power dissipated within the diode and the degree of self-heating associated with this shading event.

(continued on the next page)

Homework 18.5 (*continued from the previous page*)

Solution. Using Eq. (4.2), we find that $I_{sc} \approx I_{mp} = A_{cell} \cdot J_{sun}(1 - 0.428E_g) = 100 \times 83.75(1 - 0.428 \times 1.1) = 4.432$ A. Commercial silicon bypass diodes can handle a 3–8 A current. With breakdown voltage slightly larger than $V_{bi} = 0.89V$ (see Homework 18.4), the power lost is $P \approx 4.432 \times 0.89 = 3.94$ W, or equivalently, $P_{lost} = 3.94/0.1 = 39.4$ W/cm^2. Since the temperature of a convection-cooled semiconductor device is given by $\Delta T_D = P_{lost}/2h$, where $h = 5$ W/cm$^2\cdot$ K, therefore $\Delta T_D \approx 3.9$ K. This temperature rise itself is unlikely to accelerate degradation. However, the diodes are encapsulated in a small junction box attached to the back of a solar module. Therefore, the actual diode operating temperature is the sum of the module temperature (Homework 5.2) and the self-heating temperature, ΔT_D. The increased temperature accelerates moisture corrosion of the contact, a topic we will discuss in Chapter 23. Once the diode contacts are corroded (indicated by R_c in Fig. 18.5(a)), the resistive self-heating accelerates contact failure, leading to the resumption of module degradation due to partial shadowing.

18.6.2 Bypass diodes cannot protect thin-film solar modules

We recall from Chapter 10 that a thin-film module are laser-scribed and monolithically reconnected; i.e., the subcells are not individually accessible (see Fig. 18.2(b)). Therefore, unlike a c-Si solar module, a bypass diode cannot protect individual subcells of a thin-film module. Protection against partial shadowing is still possible if we observe that "a shadow is typically rectangular, but a laser-scribed thin-film subcell need not be." In other words, the subcell geometry can be modified to reduce the effects of partial shading.

First, let us consider the power loss in a W by L thin-film module, composed of N rectangular subcells of height (L/N) and width W (see Fig. 18.6). The subcells are connected in series, with electrodes placed at the top and bottom contacts. We wish to understand how the power output of the fully illuminated thin-film module (Fig. 18.6(a)) changes when a semitransparent ($\alpha = 0.5$) shadow of size $W_{sh} \times L_{sh}$ is overlaid on the module. A quantitative analysis requires careful numerical simulation (see the references at the end of the chapter), but the key results are easily understood.

Let us consider four cases. First, if the module is shadow-free ($W_{sh} = L_{sh} = 0$; Fig. 18.6(a)), the module produces its rated power of 800 W, indicated by the point on the bottom left corner (deep red) of Fig. 18.6(e). Second, if the module is fully shaded ($W_{sh} = W$; $L_{sh} = L$; Fig. 18.6(b)), the light intensity scales by transparency factor ($\alpha = 0.5$), and the output power reduces to 400 W, indicated by the point on the top right corner of Fig. 18.6(e) (deep blue). Third, for a vertical shadow

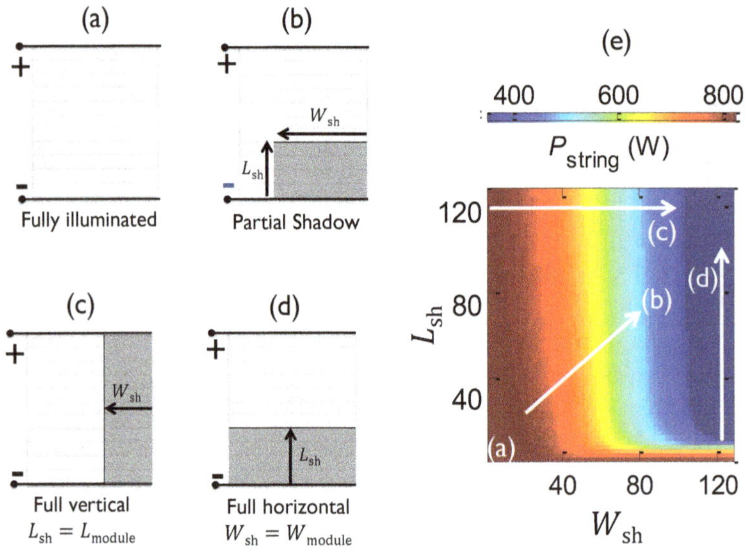

Figure 18.6: (a) A fully illuminated module. (b) A partially shadowed module. (c) The narrow vertical shadow covers the full length of the module. (d) A narrow horizontal shadow covers the full width of the module. (e) The power output obtained from numerical simulation. The arrows correspond to the subfigures (a), (b), (c), and (d). See the text for details.

(Fig. 18.6(c)), the current of *each* subcell is reduced, but by an equal amount. As a result, the power reduces linearly from 800 W to 400 W with the increasing width of the shadow, as indicated by the horizontal arrow in Fig. 18.6(e) that goes from the top left corner (deep red) to the top right corner (deep blue). Finally, for the full horizontal shadow in Fig. 18.6(d), we already know from the analytical results in Sec. 18.4 and Fig. 18.4 that the current in the partially transparent shadowed cells (αI_{ph}) controls the module output. The module output reduces precipitously even if $p \ll N$, as discussed in Homework 18.1. Indeed, as indicated by the vertical arrow on the right edge of Fig. 18.6(e), the power output is immediately reduced to the lowest level (400 W) even if only a few subcells are shadowed ($p \approx$ 4–5, corresponding to $L_{sh} \approx$ 4–5 cm). Other shadows interpolate among these four extreme limits; see the diagonal arrow in Fig. 18.6(e).

It is clear from the discussion above that even a very thin shadow ($L_{sh} \to 0$, $W_{sh} = 0.7$–0.8 m) will significantly reduce the power output of the module. What can we do to improve the shadow tolerance of thin-film solar modules? We will discuss an interesting geometry-based strategy in the next section.

18.6.3 Spiral-shaped subcells improve shadow performance

We recall from Chapter 10 (Sec. 10.5) that a module composed of spiral-shaped subcells (Fig. 18.7(a)) improves power output by decreasing the series resistance loss. Since the shadows are typically rectangular, the spiral shapes ensure that the worst-case shading scenario, namely, the shadow covering the width of the module (see Fig. 18.6(d)) can only occur for very large shadows. Compared to Fig. 18.6(e), Fig. 18.7(b) shows that the module retains its maximum power for a broader range of shadows. Only for very large shadows covering nearly the entire module ($L_{sh} \rightarrow L_{module}$, and $W_{sh} \rightarrow W_{module}$), we see a significant loss of output power (top right corner of Fig. 18.7(b)). The idea of module geometry solving a partial shadowing problem is intriguing, but the manufacturability of such modules remains an open question.

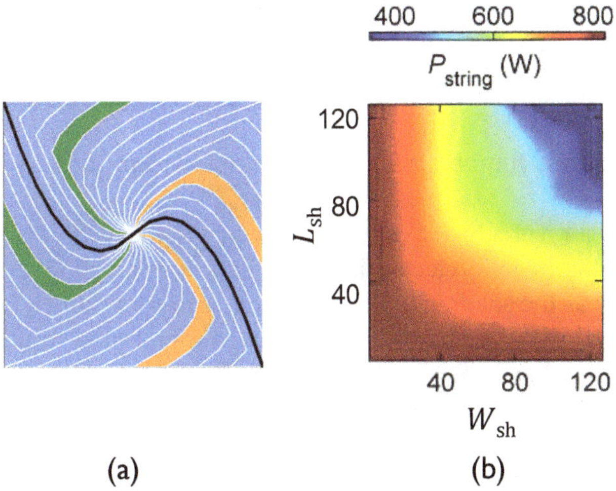

Figure 18.7: (a) A module composed of spiral subcells. (b) Plot of the power output in response to shadows of size $W_{sh} \times L_{sh}$.

Homework 18.6: Understanding the role of vertical shadows in a traditional thin-film solar cell

Assume that a CdTe *subcell* has the dimension of $W = 1$ m, and $L = 1$ cm. A module consists of 120 series-connected subcells, with a total power output of 800 W. Calculate the loss of power for a vertical shadow of $L_{sh} = 1.20$ m and $W_{sh} = 3$ cm. Does any of the subcells go to reverse breakdown? How does the output voltage change as a function of W_{sh}?

Solution. None of the subcells will go into reverse breakdown, and unless the series resistance is significant, the maximum power-point voltage is insensitive to the loss of the photocurrent due to the shadow of width W_{sh}.

Homework 18.7: A numerical simulation tool provides additional insights

Use the simulator **PVPanelSim**
`https://nanohub.org/resources/pvpanelsim`
to simulate the shading performance of a 10 cm by 10 cm CdTe thin-film module subdivided into 25 elements, each represented by its own compact model consisting of a photocurrent source, a diode, a shunt resistance, and a series resistance, as in Fig. 18.2(c). Starting from the bottom left, shade an increasingly larger fraction of the module to create a plot similar to that of Fig. 18.6. How can you simulate the shading response of a tandem cell using PVPanelSim?

18.7 Conclusions: A module is designed/installed for shadow tolerance

Shadow degradation emerged as a reliability concern for early satellites that relied on solar energy to power its electronics. Engineers worried about permanent damage to solar cells as the shadow of the boom cuts across the long wings of the solar module. One had to be careful; otherwise, the entire mission could be destroyed. In this chapter, we saw that the shadow degradation arises as a consequence of the series connection of subcells in a solar module. The series connection reduces series resistance, but increases the susceptibility to shadow degradation, with dramatic loss of power output. Similar to soiling, frequent shadowing reduces power output and increases the LCOE of the specific installation. The self-heating related to shadowing increases junction and shunt recombination. We will see in the next chapter that the hotspots associated with these weaker diodes and stronger shunts not only reduce overall power output permanently, but also accelerate other degradation modes.

For c-Si solar cells, bypass diodes provide an effective strategy to mitigate the effects of partial shadowing. One must, however, ensure that the diodes themselves are reliable and would not fail due to current crowding, self-heating, or corrosion. Mitigating shadow degradation in thin-film solar cells is more difficult. Several groups have explored the options of monolithically integrating bypass diodes, engineering devices with very low breakdown voltage, and modifying the geometrical shapes of the subcells. In addition, during field installation, the subcells of the modules are oriented perpendicular to the ground. Shadow degradation is minimized because horizontal shadows from the neighboring rows affect all subcells equally, see Fig. 18.6(c). Indeed, row spacing are designed to ensure "no-shadow" condition even for the shortest day of the year. Creating shadow-robust modules as well as the methodology to identify the worst-case shading scenario remain interesting topics of ongoing research.

In the next chapter, we will discuss photodegradation of solar cells. It appears that even if a shadow never touches a module, the module will still degrade due to energetic photons incident on it.

References

[1] Ralph M. Sullivan. Shadow effects on a series-parallel array of solar cells. Technical report, NASA Goddard Space Flight Center, Greenbelt, MD, 1965.

[2] Hajime Kawamura, Kazuhito Naka, Norihiro Yonekura, Sanshiro Yamanaka, Hideaki Kawamura, Hideyuki Ohno, and Katsuhiko Naito. Simulation of I–V characteristics of a PV module with shaded PV cells. *Solar Energy Materials and Solar Cells*, 75(3):613–621, February 2003.

[3] S. Dongaonkar, M. A. Alam, Y. Karthik, S. Mahapatra, D. Wang, and M. Frei. Identification, characterization, and implications of shadow degradation in thin film solar cells. In *2011 International Reliability Physics Symposium*, pages 5E.4.1–5E.4.5, April 2011. ISSN: 1938-1891.

[4] S. Dongaonkar, C. Deline, and M. A. Alam. Performance and Reliability Implications of Two-Dimensional Shading in Monolithic Thin-Film Photovoltaic Modules. *IEEE Journal of Photovoltaics*, 3(4):1367–1375, October 2013. Conference Name: IEEE Journal of Photovoltaics.

[5] Sourabh Dongaonkar and Muhammad A. Alam. Geometrical design of thin film photovoltaic modules for improved shade tolerance and performance. *Progress in Photovoltaics: Research and Applications*, 23(2):170–181, 2015. _eprint: https://onlinelibrary.wiley.com/doi/pdf/10.1002/pip.2410.

[6] T.J. Silverman, M.G. Deceglie, Xingshu Sun, R.L. Garris, M.A. Alam, C. Deline, and S. Kurtz. Thermal and electrical effects of partial shade in monolithic thin-film photovoltaic modules. *IEEE Journal of Photovoltaics*, 5(6):1742–1747, November 2015.

[7] X. Sun, J. Raguse, R. Garris, C. Deline, T. Silverman, and M. A. Alam. A physics-based compact model for CIGS and CdTe solar cells: From voltage-dependent carrier collection to light-enhanced reverse breakdown. In *2015 IEEE 42nd Photovoltaic Specialist Conference (PVSC)*, pages 1–6, June 2015.

[8] Mehdi Hosseinzadeh and Farzad Rajaei Salmasi. Determination of maximum solar power under shading and converter faults — A prerequisite for failure-tolerant power management systems. *Simulation Modelling Practice and Theory*, 62:14–30, March 2016.

CHAPTER 19

Dangerous Hotspots are Caused by Weak Diodes and Strong Shunts

Chapter Summary

- Local hotspots arise from processing imperfections.
- The hotspots suppress current collection from the neighboring regions and reduce the efficiency of solar cells.
- Hotspots are acerbated by the proximity to the electrical contacts because contacts extend the effect over a larger region.
- Various post-processing techniques (e.g. striping, etching) have been proposed to reduce the effects of hotspots on cell performance.

19.1 Introduction: The origin of hotspots in solar modules

In the last chapter, we saw how the non-uniform illumination due to partial shadowing leads to localized power dissipation (hotspots). The output is reduced and various degradation modes are accelerated. We will see later in Chapter 23 that hotspots may also form when c-Si grids are delaminated and photocurrent collection from the affected region is suppressed. It is important to reduce hotspot formation due to partial shadowing and localized corrosion.

Unfortunately, hotspots can form even when the illumination is uniform and grid corrosion is absent. Natural manufacturing variation across a large-area module makes the diode built-in voltage (and thus the dark current and photoluminescence efficiency) spatially inhomogeneous. We have also seen in Chapter 9 that the shunt resistances are highly localized and their magnitudes are broadly (Weibull) distributed. This spatial non-uniformity of diodes and resistances serve as an additional source of hotspot formation.

In this chapter, we will explain the danger posed by hotspots around the most conductive shunts (strong shunt) and/or diodes with smallest low turn-on voltage (weak diodes). Our analysis applies to both intrinsic hotspots (due to processing non-uniformity) or extrinsic hotspots (created as a consequence of partial shading, corrosion, or potential-induced degradation). Let us begin with weak diodes and then we will discuss strong shunts.

19.2 Process non-uniformity creates weak diodes

Each cell in a thin-film solar cell is typically 300–400 nm thick, 1 cm long, and 1 m wide. The films are deposited in a big deposition chamber where various reactant gases swirl, find nucleation points on the surface, and gradually build-up the layer. A small variation in deposition temperature or non-uniformity in reactant flow will lead to variability in the film thickness. Local variation in composition could create atomic configurations that act as recombination centers for the electrons and holes. The thinner films reduce photocurrent, J_{ph}, while the recombination increases the dark current, J_{dark}. Taken together, these regions are defined by lower $V_{\mathrm{oc}} = (k_{\mathrm{B}}T/q)\ln(1 + J_{\mathrm{ph}}/J_{\mathrm{dark}})$. In a well-controlled manufacturing process, there will be relatively few spots with very low V_{oc}. In this section, we will follow V. G. Karpov *et al.*, *Physical Review B*, 2004 to show that even a few "weak spots" can significantly reduce the power output of the solar cell.

19.3 Light I-V characteristics can be expressed in a diode-like form

We discussed in Chapter 7 and that the net current density *into* the solar cell is given by the superposition of dark and light currents, namely,

$$I/A = J_{\mathrm{total}} = J_0(e^{qV/k_{\mathrm{B}}T_{\mathrm{D}}} - 1) - J_{\mathrm{ph}}(V) \tag{19.1}$$

where A is the subcell area. At V_{oc}, $I = 0$, so that $J_{\mathrm{ph}}(V_{\mathrm{oc}}) = J_0(e^{qV_{\mathrm{oc}}/k_{\mathrm{B}}T_{\mathrm{D}}} - 1) \approx J_0\, e^{qV_{\mathrm{oc}}/k_{\mathrm{B}}T_{\mathrm{D}}}$. Therefore, the dark current prefactor is related to the photocurrent: $J_0 = J_{\mathrm{ph}}(V_{\mathrm{oc}})e^{-qV_{\mathrm{oc}}/k_{\mathrm{B}}T_{\mathrm{D}}}$. Inserting it back into Eq. (19.1), we find that

$$\begin{aligned} I/A &\sim J_{\mathrm{ph}}(V_{\mathrm{oc}})(e^{q(V-V_{\mathrm{oc}})/k_{\mathrm{B}}T}) - J_{\mathrm{ph}}(V) \\ &\sim J_{\mathrm{ph}}(V_{\mathrm{oc}})(e^{q(V-V_{\mathrm{oc}})/k_{\mathrm{B}}T_{\mathrm{D}}} - 1). \end{aligned} \tag{19.2}$$

Equation (19.2) is derived by assuming that $J_{\mathrm{ph}}(V) \approx J_{\mathrm{ph}}(V_{\mathrm{oc}})$, that is, the photocurrent is voltage independent. We know from Chapter 7, Eq. (7.20), that the photocurrent in thin-film solar cells is actually voltage-dependent. Nonetheless, the approximate equation reproduces the correct limits: $I = 0$ at V_{oc}, and $I \approx -A \times J_{ph}$ at $V = 0$. Did you notice that the dark current is apparently gone, but its memory is actually hidden in V_{oc}?

> **Homework 19.1: The exact expression for the total current is easily derived**
>
> Derive an expression for the total current in terms of photocurrent (similar to Eq. (19.2)), but without assuming that the photocurrent is voltage independent; see Eq. (7.20). Explain the difference between the two expressions at $V \approx V_{oc}$.

19.4 Low-V_{oc} diodes sink the photocurrent generated in the neighboring region

Let us (conceptually) divide a solar cell into small segments of cross-sectional area πl^2 (see Fig. 19.1). Except for a few "weak-diode" spots with $V_{oc,w}$, the solar cells are presented by a network of diodes with normal V_{oc}, so that $V_{oc,w} < V < V_{oc}$, where V is the operating point of the integrated system. Figure 19.1(c) shows that with $V_{oc,w} < V$, the weak diode sinks the current produced by the normal cells with $V < V_{oc}$ with area πL^2. Like bad company, not only does the weak diode fail to contribute, but it steals the photocurrent produced by the neighboring regions and wastes it as heat. These hotspots accelerate various degradation modes.

To calculate the radius of influence, L, we need to balance the current produced by the normal region with the current lost in the weak diode (see Fig. 19.1(c)):

$$\pi l^2 \times J_{ph}(V_{oc,w})(e^{q(V-V_{oc,w})/k_BT} - 1) + \pi L^2 \times J_{ph}(V_{oc})(e^{q(V-V_{oc})/k_BT} - 1) = 0. \quad (19.3)$$

If the density of photocurrent generated in the weak vs. normal areas is the same (i.e., for now we ignore film thickness variation), then $J_{ph}(V_{oc,w}) = J_{ph}(V_{oc})$. Also, we can drop the term $e^{q(V-V_{oc})/k_BT_D}$ on the right, because $V_{oc,w} < V < V_{oc}$. Therefore,

$$L^2 \approx l^2 \times e^{q(V-V_{oc,w})/k_BT_D}. \quad (19.4)$$

The exponential term in Eq. (19.4) suggests that a weak diode sinks current from a much larger area compared to its size (i.e., $L \gg l$) (see Figs. 19.1(b, c)). The size cannot be arbitrarily large, however. Rather, it must be defined by the need to support the current through a resistive voltage drop associated with the lateral sheet resistance, R_s. Therefore, as shown in Fig. 19.2, $V - V_{oc,w} = \alpha(J_{ph}\pi L^2) \times R_s$, where α accounts for the distributed current collection toward the weak diode, as discussed in Chapter 10. Typically, $\alpha \approx 1$ at the middle of a cell. Close to a grid or a finger, $\alpha \gg 1$ — this will be discussed in Sec. 19.6. Inserting the relationship in

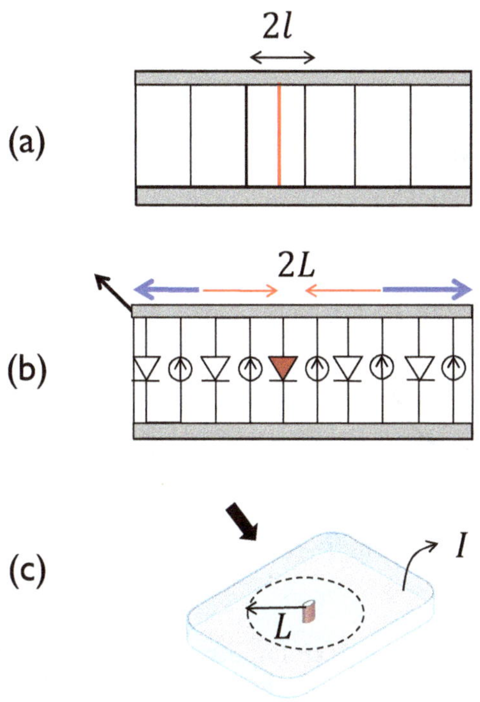

Figure 19.1: (a) Side view of a thin-film solar cell, hypothetically divided into segments with radius l. The segment marked by the red line has higher than average defect concentration. (b) The representation of each region with an equivalent circuit consisting of a photocurrent source and a diode. For simplicity, we have omitted the shunt and series resistances. The defective region is marked by a "weak" diode, a diode that turns on at a lower $V_{\mathrm{oc,w}}$. The weak diode sinks the current generated by neighboring "normal" diodes. (c) A perspective view showing that a weak diode (of size l) sinks the photocurrent from a much larger area (of radius L). The dissipated power through the diode will make it appear hotter than the neighboring region.

Eq. (19.4), we find an implicit equation for L, i.e.,

$$L \approx \sqrt{\frac{1}{\pi\alpha R_{\mathrm{s}}}\frac{k_{\mathrm{B}}T_{\mathrm{D}}}{qJ_{\mathrm{ph}}}\ln\left(\frac{L}{l}\right)}. \tag{19.5}$$

The affected area $A_{\mathrm{dark}} = \pi L^2$ is inversely proportional to the photocurrent J_{ph} and the series resistance R_{s}. Larger resistance insulates the weak spots from the neighboring regions by incurring a voltage-drop penalty. Finally, defining the diode resistance $R_{\mathrm{D}} \equiv k_{\mathrm{B}}T_{\mathrm{D}}/qJ_{\mathrm{ph}}$, we can rewrite Eq. (19.5) in an equivalent form:

$$L \equiv \sqrt{\frac{R_{\mathrm{D}}}{\pi\alpha R_{\mathrm{s}}}\ln\left(\frac{L}{l}\right)}.$$

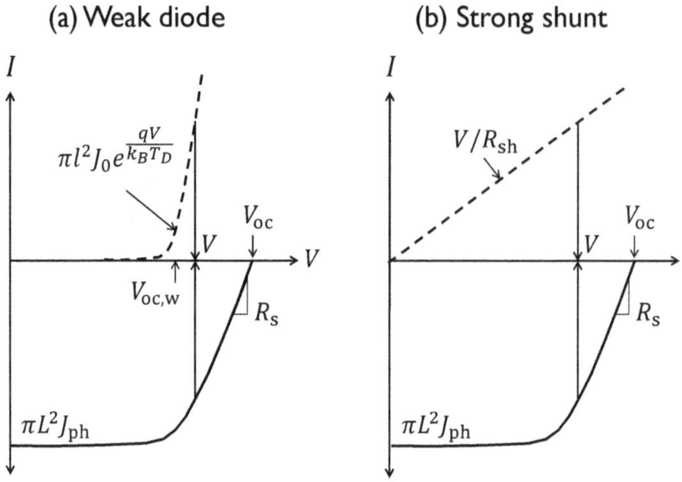

Figure 19.2: (a) The current of the weak diode is supported by the integrated photogenerated current of the affected area. (b) The corresponding curves for a strong shunt.

We conclude this section with an important observation. Equation (19.5) suggests that the affected region is independent of the "weakness" of the diode. In other words, since $V_{\mathrm{oc,w}}$ does not enter explicitly in Eq. (19.5), it does not affect L directly. This unphysical conclusion arises from the approximation that relates voltage drop to the current flow. In practice, L will increase weakly as $V_{\mathrm{oc,w}}$ is reduced, because even a small increase of ΔL increases the current collected significantly by $2\pi L \Delta L J_{\mathrm{ph}}$. This increase supports the extra current sunk by a weaker diode.

Homework 19.2: A simpler, physically transparent derivation of Eq. (19.4)

Assume that at the operating voltage V, only the weakest diode within the hotspot has been turned on. The other diodes are weakly conducting and can be neglected. Balance the photo current from area πL^2 by the current sunk by the diode (area πl^2) to obtain the desired expression.

Solution. The flux balance gives us the expression

$$\pi L^2 J_{\mathrm{ph}} = \pi l^2 J_0 e^{qV/k_{\mathrm{B}} T_{\mathrm{D}}}.$$

Since $J_{\mathrm{ph}} = J_0 e^{qV_{\mathrm{oc,w}}/k_{\mathrm{B}} T_{\mathrm{D}}}$, we can use the relationship in the previous expression to complete the derivation.

Homework 19.3: Equation (19.5) **must be solved iteratively.**

Calculate the affected size, power dissipation, and temperature rise of a hotspot in a CdTe solar cell. Assume that the absorber is 300 nm thick, its subcell size is 1 cm, and the series resistance is 10 ohms/square.

Solution.

1. Begin by assuming l = 300–500 nm, equal to the film thickness. Since Eq. (19.5) is implicit, let us begin by assuming the worst case, i.e., $L_{max} = 0.5$ cm, half the typical subcell size. Estimate $\ln(L_{max}/l)$.
2. Use the bandgap of CdTe ($E_g = 1.5$ eV) to calculate by Eq. (4.2) the thermodynamic limiting current $J_{ph} = J_{sun}(1-\beta_{sun}E_g) = 83.75(1-0.425E_g)$. With $R_s = 10$ ohms/square, use Eq. (19.5) to re-estimate L and $A_{dark} = \pi L^2$. Iterate Steps 1 and 2 a few times to convergence.
3. Calculate by Eq. (4.1) the thermodynamic limit for $V_{mp} \approx V_{oc}$ for a CdTe cell. The power dissipated within the hotspot is approximately given by $P = \pi L^2 J_{ph} V_{mp}$.
4. The hotspot temperature is given by Eq. (5.11), i.e., $\Delta T = P_{conv}/2h$, where $h \approx 5W/K$ is the convection coefficient of still air. The temperature rise adds to the already heated module by above-bandgap and sub-bandgap photons, as discussed in Chapter 5.

19.5 Strong shunts and weak diodes have similar effects on module performance

A weak diode sinks photogenerated current from the neighboring regions. A highly conductive (strong) shunt does the same. As discussed in Chapter 9 (Fig. 9.2), when metal ions diffuse and destroy a junction, the shunt resistance ($R_{sh,w}$) is reduced, and the leakage current increases significantly. This current is provided by the neighboring regions, as follows. We assume that in the region affected, the voltage is too low to turn on the diodes and leakage current through the normal shunts are negligibly small. Thus, the photogenerated current from the neighboring area equals the leakage current through the strong shunt.

$$\frac{V_{oc} - V}{R_s} = \frac{V}{R_{sh,w}} = \alpha J_{ph}(\pi L^2)$$

so that

$$V = V_{oc}\left(\frac{R_s + R_{sh,w}}{R_{sh,w}}\right) = \alpha J_{ph}(\pi L^2) R_{sh,w}.$$

The radius of influence is

$$L = \sqrt{\frac{1}{\alpha\pi}\frac{V_{\text{oc}}}{J_{\text{ph}}}\left[\frac{1}{R_{\text{s}}} + \frac{1}{R_{sh,w}}\right]}. \tag{19.6}$$

The similarity to Eq. (19.5) is obvious and expected. Figure 19.2 explains the connection geometrically.

19.6 Hotspots are acerbated by the proximity to the electrodes

Equations (19.5) and (19.6) show that the area affected by a hotspot depends inversely on the series resistance, R_{s}. If the hotspot occurs in between the metal grid fingers of a c-Si solar cell, or close to the subcell boundary of a thin-film solar cell, the low-resistivity metal grid dramatically increases the area affected, with a corresponding increase in power dissipation (see Fig. 19.3). Moreover, the excess heating so close to the metal lines accelerates corrosion and delamination.

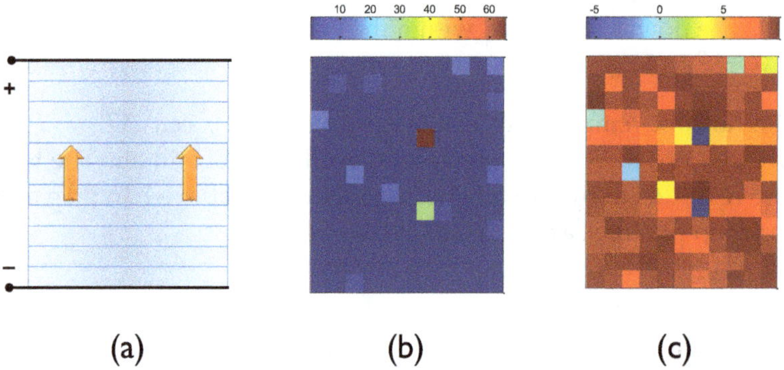

Figure 19.3: (a) A 11 × 11 cm^{-2} thin-film submodule. There are 11 cells connected in series. (b) The plot of the shunt current density (in mA/cm^2). The red square marks the strongest shunt. The green square marks a somewhat weaker shunt. (c) Corresponding power generation (in mW/cm^2) plot for the submodule. The shunted subcells not only sink power locally (blue), but the proximity to the metal contacts suppresses power generation of the neighboring regions within the same cell.

19.7 Solution strategies: There are different ways to reduce hotspots

Hotspots may be removed by improving the uniformity of the deposited films or controlling the uniformity of the doping profile. However, there may be limits to process improvement for very large modules. In this case, post-fabrication design strategies that can mitigate the effect of hotspots can be helpful. In this section, we will discuss two such approaches for hotspot control.

19.7.1 Process improvement solves the root cause of hotspot formation

Several strategies have been reported to improve the uniformity of the films deposited (see Fig. 19.4). First, *process improvements*, such as improved temperature uniformity across the wafer, or chemical annealing to prune weaker bonds during the film growth by He/Ar ions, etc., improve film uniformity and reduce the number of weak diodes and strong shunts. Second, even after the processing is complete, the film can be homogenized by immersing in an *electrochemical solvent*. The region containing the weak diodes reacts faster with the solvent and are etched away. Finally, following contact metal deposition, an electroluminescence map can identify the hotspots. *Laser ablation* can locally remove the regions containing strong shunts or weak diodes.

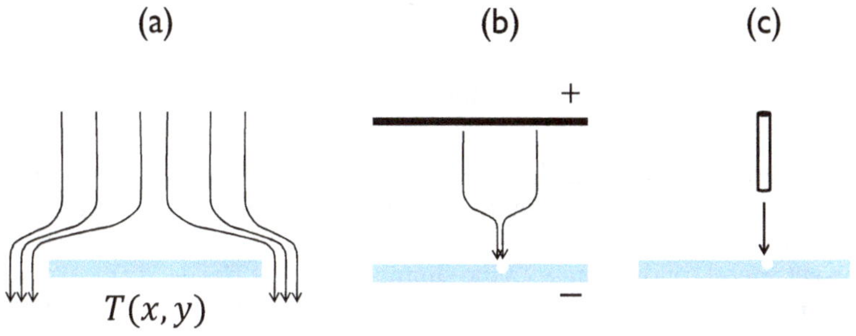

Figure 19.4: (a) Spatial control over the process temperature, (b) effective use of electrochemical etching, and (c) laser ablation of defective regions can improve the uniformity of the films deposited.

19.7.2 Striping suppresses the effects of hotspots formed

We know that the proximity to electrodes increases the size of the hotspots, because photocurrent from a wider region can use the low-resistance electrodes to reach the weak spot. Similar to a controlled burn that serves as a forest firewall, a vertical patterning (striping) offers a way to disrupt this lateral flow of photocurrent to the weak spot. The top row of Fig. 19.5 shows three different ways the module may be patterned. *Full vertical strips* isolate neighboring subcells regardless of the locations of the weak spots. Increasing the number of stripes localizes the effects of weak spots more effectively, but with the corresponding penalty for dead area. *Partial vertical stripes* and *localized stripes* reduce the dead area penalty significantly, provided the shunt locations are known accurately by photoluminescence imaging, for example. Figure 19.5(e) shows that striping offers significant improvement, especially for thin-film solar cells, where the subcells cannot be sorted for current matching.

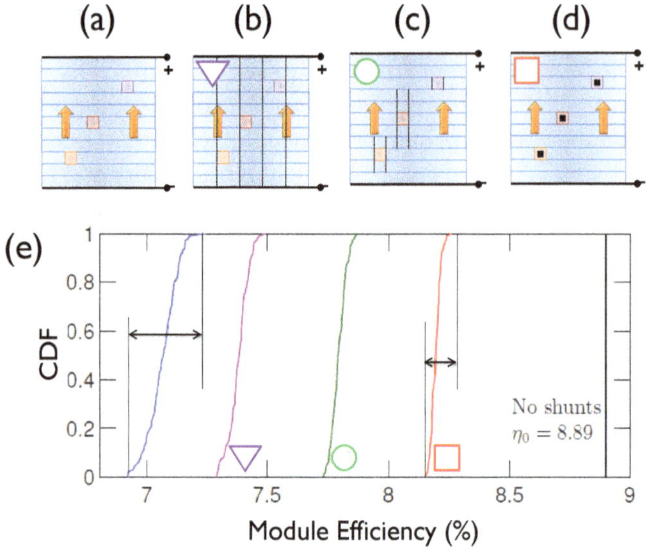

Figure 19.5: (a) An 11 × 11 cm^2 a-Si thin-film submodule. There are 11 subcells in series. Only three of the 121 subcells are shunted. (b) Vertical stripes prevent lateral conduction among the subcells of a given cell. The stripes increase the dead area loss. (c) Partial vertical stripes prevent lateral conduction from the shunted subcells. The dead area is reduced, but one must know the location of the shunts. (d) Localized striping isolates the shunts, if their positions are known in advance. (e) The efficiency distribution associated with various striping strategies. Without striping, an average efficiency is only 7%. Various striping strategies improve the efficiency toward the shunt-free perfect module (8.9%).

19.8 Conclusions: Hotspots and partial shadowing must be reduced

In this chapter, we explained why even a relatively few defective regions reduce the power output of a solar module significantly. More importantly, hotspots resulting from non-uniform current distribution accelerate delamination, corrosion, and electromigration. Obviously, as discussed in Chapter 18, hotspots could also be generated due to partial shadowing. This is why improving process uniformity and reducing partial shadowing are important considerations for modern solar cell technology. Even if the cells are perfect and uniformly illuminated, they will degrade through other intrinsic degradation modes. We will discuss these intrinsic degradations in the next chapter.

References

[1] V. G. Karpov, A. D. Compaan, and Diana Shvydka. Random diode arrays and mesoscale physics of large-area semiconductor devices. *Physical Review B*, 69(4):045325, January 2004. Publisher: American Physical Society.

[2] V. G. Karpov, A. D. Compaan, and Diana Shvydka. Effects of nonuniformity in thin-film photovoltaics. *Applied Physics Letters*, 80(22):4256–4258, May 2002. Publisher: American Institute of Physics.

[3] Y. Roussillon, D. M. Giolando, Diana Shvydka, A. D. Compaan, and V. G. Karpov. Blocking thin-film nonuniformities: Photovoltaic self-healing. *Applied Physics Letters*, 84(4):616–618, January 2004. Publisher: American Institute of Physics.

[4] S. Dongaonkar, J. D. Servaites, G. M. Ford, S. Loser, J. Moore, R. M. Gelfand, H. Mohseni, H. W. Hillhouse, R. Agrawal, M. A. Ratner, T. J. Marks, M. S. Lundstrom, and M. A. Alam. Universality of non-Ohmic shunt leakage in thin-film solar cells. *Journal of Applied Physics*, 108(12):124509, December 2010. Publisher: American Institute of Physics.

[5] V. G. Karpov, G. Rich, A. V. Subashiev, and G. Dorer. Shunt screening, size effects and I/V analysis in thin-film photovoltaics. *Journal of Applied Physics*, 89(9):4975–4985, April 2001. Publisher: American Institute of Physics.

[6] Sourabh Dongaonkar, Stephen Loser, Erik J. Sheets, Katherine Zaunbrecher, Rakesh Agrawal, Tobin J. Marks, and Muhammad A. Alam. Universal statistics of parasitic shunt formation in solar cells, and its implications for cell to module efficiency gap. *Energy & Environmental Science*, 6(3):782–787, February 2013. Publisher: The Royal Society of Chemistry.

[7] S. Dongaonkar and M. A. Alam. In-Line Post-Process Scribing for Reducing Cell to Module Efficiency Gap in Monolithic Thin-Film Photovoltaics. *IEEE Journal of Photovoltaics*, 4(1):324–332, January 2014.

CHAPTER 20

Photodegradation of Solar Cells Due to UV Exposure

Chapter Summary

- UV exposure degrades the transparent encapsulant, back-sheet, and top-interface of the solar cell.
- The UV-assisted yellowing of the encapsulant reduces the photocurrent, and interface degradation increases recombination current.
- The photo-degradation is accelerated by UV intensity, but not the module temperature.
- The UV-assisted backsheet degradation accelerates moisture ingress and corrosion.
- A number of techniques have been developed to improve UV robustness, including new polymers, addition of photo-stabilizers, filtering of high-energy photons, etc.

20.1 Introduction: Photodegradation of solar absorber, polymer encapsulant, and backsheets are intrinsic reliability concerns

In Chapters 17 and 18, we discussed how soiling and partial shadowing reduce power output of a solar module. We may think that the power loss, however significant, is temporary. Once the shadow is removed or the module cleaned, the original efficiency will be restored. Unfortunately, this is not true. Recall that shadowing increases self-heating and hotspot formation, leading to increasing vibration of the molecules and occasional bond dissociation. These broken bonds (defects) increase the probability of the electron–hole recombination. The corresponding increase in the dark current permanently reduces the open-circuit voltage, and the power output of the solar cell. The shadow is long gone, but its memory persists in broken bonds.

The bonds are broken not only by shadows, but also by light, especially if an atom within a module absorbs a high-energy photon of the solar spectrum. As shown in Fig. 20.1, photodegradation may occur in the absorber layer of the solar cell, the polymer encapsulant that protects the cells from moisture, and the glass cover/polymer backsheet that make the cell mechanically stable. To initiate photodegradation, the incident photon energy E_{ph} must be comparable to the bond energy E_0 of these materials. As shown in Fig. 20.1, the light-induced degradation (LID) of the absorber layer increases the dark (diode) current, the photo-induced yellowing of the polymer encapsulants reduces the short-circuit (photocurrent) current, and cracking of the backsheet leads to moisture ingress and corrosion-induced increase in the series resistance. Taken together, these irreversible processes reduce the cell efficiency and farm output permanently. The bonds are strong and high-energy photons relatively few; therefore, photodegradation proceeds slowly. Over the module lifetime of 25–40 years, even a slow degradation significantly reduces power output.

20.2 Yellowing of polymer encapsulant

20.2.1 Why do we need an encapsulant

Figure 20.1(a) shows that for environmental protection a solar cell array is first encapsulated by a thin polymer layer (EVA) and then sandwiched between a glass front cover and the polymer backsheet. All these pieces are then held together by a metal frame. The polymer layer protects against moisture diffusion that may otherwise seep through any gaps between the frame and the glass, or through any cracks in the glass or backsheet. Also, the glass and cells are not separated conformally over a large module. Therefore, if the glass was assembled directly on top of the solar cell, isolated contact points will localize stress and fracture the cell. The polymer encapsulant redistributes the stress point to improve the mechanical performance of the system. In this regard, a polymer encapsulant is similar to a shoe sole, which protects our feet from sharp stones on the road by redistributing the localized stress over the larger region.

20.2.2 Chemical composition of encapsulants

The polymer encapsulant must be transparent to sunlight, with bandgap $E_g > 3$–4 eV. Commercially, poly(ethylene-vinyl acetate) (PEVA or EVA) has been an encapsulant of choice for many years. As shown in Fig. 20.1(c), EVA is a copolymer of ethylene and vinyl acetate. The last H of the acetic acid (CH_3COOH) combines with the first H of vinyl (H–CH–CH–) to make vinyl acetate (CH_3COO–CH–CH– ... –CH–CH– ...), and so on. Once the ethylene comes along (C_2H_4), the C double bond opens up, and the new chain (CH_3COO–(CH–CH)–CH_2–CH_2– ... CH_2–CH_2) keeps repeating. Any of the C–H, C–O, C–C bonds may break through

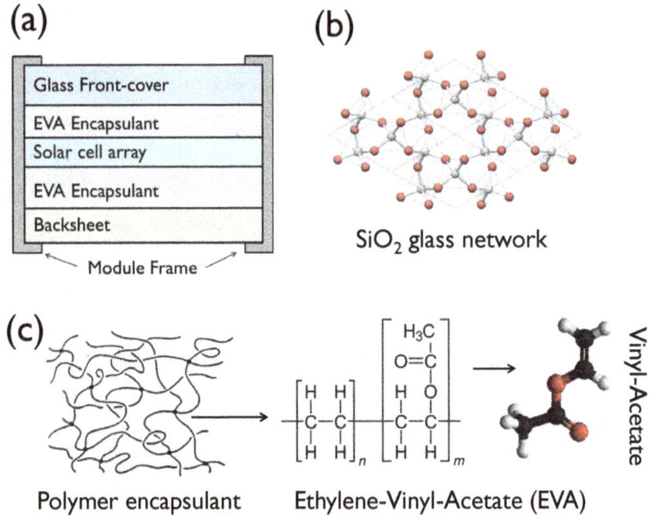

Figure 20.1: (a) A module consists of a glass cover, an encapsulant (typically EVA), a solar absorber, and a backsheet, all held together by a metal frame. (b) Glass is composed of a random network of SiO_2 bonds. Each Si atom defines a tetrahedra with the oxygen atoms in the corner. Each oxygen atom is shared between a pair of tetrahedra. (c) Each strand of the EVA polymer consists of alternate segments of ethylene (n molecules) and vinyl acetate (m molecules). High-energy photons can break the bond between vinyl and acetate. The broken bonds absorb visible light and give the degraded polymer its characteristic yellow or brown color. The polymer backsheet is also susceptible to degradation by high-energy photons that pass between the gaps of the cells.

photoexcitation. The bonds broken by UV light will now absorb light in the visible range, and the film will turn yellow or brown over time. The loss of light transmission (with the corresponding loss of short-circuit current) is proportional to the number of bonds broken, expressed in terms of yellowing index (YI).

Table 20.1: Approximate bond energies of materials in a solar module

Bond	C-C	C=C	H_2	O=O	C-H	H_3C-Vinyl	Si-Si	Si-H	O-H
E_0 (eV)	3.65	6.32	4.52	5.15	4.25	1.25	3.40	3.10	4.75

Homework 20.1: Photodegradation is also called UV degradation

Recall that the least energetic ultraviolet photons have a wavelength of 400 nm (3.1 eV). Refer to Fig. 20.1(c) and Table 20.1 to explain why UV photons are necessary for photodegradation of solar cells. Which material do you think is most susceptible to UV degradation? Can UV degradation damage a silicon solar cell?

20.3 A phenomenological model for UV degradation

Let us assume that ($I(t)$, in $\mathrm{kW/m^2}$) is the time-dependent sunlight intensity incident on a module. Only a fraction of the incident photons (f_{uv}) is energetic enough to break chemical bonds. If the average bond dissociation efficiency of a photon is $\sigma(E)$, the time-dependent defect density (D_{uv} per m^3) is given by

$$D_{\mathrm{uv}}(t) = \int_{t_i}^{t} \sigma\, I(t)\, f_{\mathrm{uv}}\, dt. \tag{20.1}$$

Here, t is expressed in hours and t_i is "incubation period," to be discussed later.

Homework 20.2: High-energy photons in a solar spectrum

Calculate the number of photons in the UV spectrum. Assume that approximately 5.5% of the AM1.5 light has UV frequency (280–400 nm, 3.2–4.8 eV range), i.e., $f_{\mathrm{uv}} \approx 0.055$.

Solution. If all the electron–hole pairs generated by photons could be collected, then $I(t) = 1000\ \mathrm{W/m^2}$ of sunlight would produce 70 $\mathrm{mA/cm^2}$ of current. Therefore, $1000 \times 0.055 = 55\ \mathrm{W/m^2}$ of UV flux would produce $I_{\mathrm{uv}}(t) = 70 \times 0.055 = 3.85\ \mathrm{mA/cm^2}$ or 38.5 $\mathrm{A/m^2}$ current. The number of UV photon is therefore given by

$$N_{\mathrm{uv}} = I_{\mathrm{uv}}/q = 38.5/1.6 \times 10^{-19} = 2.40 \times 10^{20}\ \mathrm{m^{-2}\,s^{-1}}.$$

Therefore, the maximum number of bonds broken per second cannot exceed this number. In practice, far fewer bonds will be broken, because only a fraction (σ) of these high-energy photons are absorbed by the relatively thin polymer encapsulant.

When the UV light breaks encapsulant polymer bonds, the film gradually turns brownish-yellow and the light transmission (and therefore the photocurrent J_{ph}) is reduced. As discussed in Secs. 17.4.1 and 17.4.2, the Beer–Lambert law says that light transmission through a material (e.g., EVA) containing spatially uniform defect density D_{uv} (per $\mathrm{m^{-3}}$) decays exponentially with film thickness t_{EVA}, expressed in m, and the absorption cross-section (γ), expressed in $\mathrm{m^2}$. In other words,

$$\frac{\Delta J_{\mathrm{ph}}(t)}{\Delta J_{ph,\infty}} = 1 - e^{-\gamma\, t_{\mathrm{EVA}}\, D_{\mathrm{uv}}(t)} \tag{20.2}$$

where ΔJ_{ph} is the photocurrent loss at time t, while $\Delta J_{ph,\infty}$ is maximum loss after prolonged exposure (i.e., $t \to \infty$). On the right-hand side. The UV-created defects are smaller than incident wavelength ($s \ll \lambda$); therefore, we cannot use Homework 17.7 (for sand particles with ($s \gg \lambda$) to calculate the extinction efficiency γ.

Instead, γ can be calculated theoretically by the Raleigh scattering or determined empirically through optical measurements.

Homework 20.3: Time dependence of UV degradation

Using Eqs. (20.1) and (20.2), show that the power loss due to time-independent UV exposure is given by

$$\frac{\Delta P(t)}{\Delta P_\infty} \simeq 1 - e^{-k_p(t-t_i)} \tag{20.3}$$

where $k_p \equiv \sigma\, \gamma\, t_{\mathrm{EVA}}\, I_0 f_{\mathrm{uv}}$ is the time constant for UV degradation.

In addition to breaking polymer bonds and reducing the photocurrent, UV photons are energetic enough to break Si–Si and Si–H bonds (see Table 20.1) and degrade the top surface of the solar cell. These defects increase the dark current through increased surface recombination (especially for passivated contact solar cells, PERC). One finds empirically that

$$\Delta J_{02}(t) \equiv J_{02}(t) - J_{02}(t=0) = (mt)^n \tag{20.4}$$

with $m \approx 2$ and $n = 0.25$–0.5, see Fig. 20.2. This power-law degradation implies that the initial degradation proceeds quickly, but then it saturates at a later time. The reaction–diffusion theory discussed in Chapter 21 (see Eq. (21.8)) explains the power-law dynamics in terms of photo-induced dissociation and diffusion-limited self-passivation of the interfacial Si–H bonds. The theory also explains why n

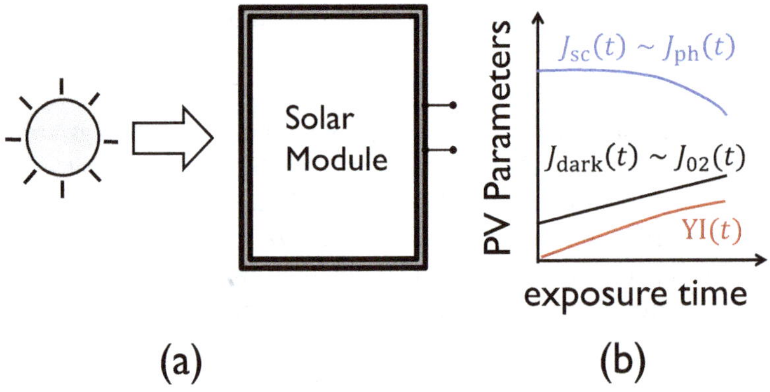

Figure 20.2: (a) With sustained exposure to sunlight, (b) the photo-induced degradations are reflected in, for example, the increasing dark current (J_{dark}) associated with bond dissociation in the active layer and/or in the gradual reduction in the short-circuit current (J_{sc}) associated with the yellowing of the encapsulant. The degree of UV damage is quantified by the yellowing index, YI.

does not depend on light intensity or rate constants for dissociation, diffusion, or passivation.

Homework 20.4: UV damage increases surface recombination

Use the energy-dependent photon absorption coefficient plotted in Fig. 6.2(c) to explain why the UV damage is confined within a few nanometers of the top surface of a solar cell.

The phenomenological theory does not explain a set of puzzles

The phenomenological theory described above is often used to characterize the UV degradation of a solar module and the parameters are derived from measurement. The simple theory, however, cannot answer a number of important questions. For example:

- What fraction of solar spectrum is responsible for photodissociation of polymer molecules? In other words, how should one calculate f_{uv}?
- Why does UV degradation depend only weakly on temperature? Do other variables, such as RH, play a role in defining UV degradation?
- What is the physical origin of the strong wavelength dependence of UV degradation?
- In Eq. (20.1), how should we explain the existence of the incubation period (t_i) in some encapsulant, but not in others?
- Why does the photocurrent-loss saturate?

We will discuss a simple physical theory that can answer these questions better. Incidentally, UV exposure also increases the brittleness of the glass cover and the polymer backsheet, although the processes have not been described by simple empirical equations. The theory described below applies to these degradation modes as well.

20.4 A physical theory of UV degradation

20.4.1 Number of high-energy photons in a blackbody spectra

To understand the rates of the photodegradation process (and calculate f_{uv}), let us calculate the number of high-energy solar photons above a critical energy E_*, that is, with $E > E_*$.

$$n_{\text{ph}} = \int_{E_*/h}^{\infty} n_S(f, T) df \tag{20.5}$$

where $E(\equiv hf$, h is Planck's constant) and $n_S(f,T)$ is the number of photons of frequency f contained within the solar spectrum, namely

$$n_S(f,T) \equiv \frac{2f^2}{c^2}\frac{1}{e^{hf/k_BT}-1}.$$

Therefore,

$$n_{ph} = \sigma_1 \int_{x_0}^{\infty} \frac{x^2}{e^x - 1} dx \tag{20.6}$$

where $\sigma_1 \equiv 2c\left(\frac{k_BT}{hc}\right)^3$ and $x_0 = \frac{E_*}{k_BT}$. For $x_0 \gg 0$, the integrant simplifies to x^2e^{-x}, so that

$$n_{ph} = \sigma_1 \exp(-x_0)(x_0^2 + 2x_0 + 2) \approx \sigma_1 x_0^2 \exp(-x_0).$$

Therefore,

$$n_{ph}(E > E_*) \approx \frac{2k_BT_S}{h^3c^2}E_*^2 e^{-E_*/k_BT_S}. \tag{20.7}$$

Note that the temperature T_S reminds us that the photons are arriving from the sun. Equation (20.7) shows that the number of high-energy photons decreases rapidly with increasing photon energy. Equation (20.7) suggests that the number of high-energy photons with $E > E_0$, where E_0 is the polymer bond energy, decreases exponentially with E_0. Therefore, stronger bonds are exponentially harder to dissociate than weaker bonds.

20.4.2 Rate of polymer degradation

To calculate the number of broken bonds (N_{BB}) in a solid exposed to sunlight, assume that the solid contains N_0 bonds with bond energy E_0. The probability that a bond is already in an excited state with energy E is given by the Boltzmann distribution, namely,

$$N_{BB}(E) = \frac{N_0}{k_BT_D}e^{-E/k_BT_D} \tag{20.8}$$

where T_D is the cell temperature. To break such a bond by photoexcitation, the photons must contribute the additional energy E^* needed to overcome the energy barrier, i.e., $E_* + E \geq E_0$. There are $n_{ph}(E > E_*)$ such photons in the solar spectrum, given by Eq. (20.7).

Therefore, the bond dissociation rate has two components: thermal self-dissociation rate R_D and photon-assisted dissociation rate R_{uv}, i.e.,

$$\frac{dN_{BB}}{dt} = R_D(E > E_0) + R_{uv}(E \leq E_0). \tag{20.9}$$

With time-independent rates, the number of broken bonds increases linearly with time.

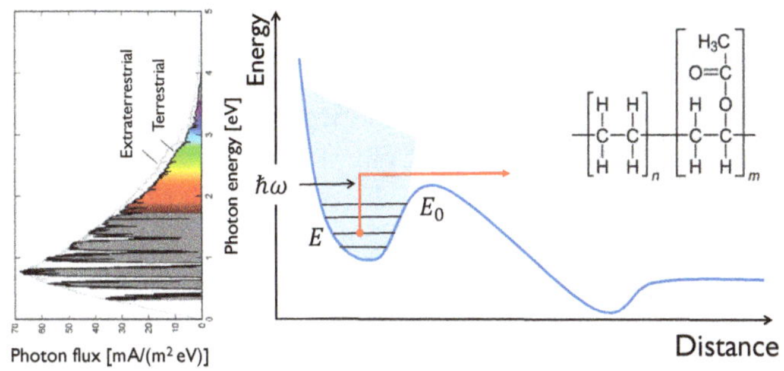

Figure 20.3: The bonds of EVA can be described by a parabolic potential well, with binding energy E_0. In equilibrium with the device temperature T_D, the bonds have different vibration energy E. If a bond at energy E absorbs a photon of energy $\hbar\omega$ (from the AM1.5 solar spectrum shown on the left; recall Fig. 1.6) that pushes it over the energy barrier, the atoms may be displaced far from its equilibrium position, and the bond may be permanently broken.

The first term involves self-dissociation by thermal excitation, so that any bond that has acquired an energy $E > E_0$ has the near unity probability (k_D) of being dissociated. The number can be calculated as follows:

$$R_D = \int_{E_0}^{\infty} k_D N_B(E) dE = k_D \frac{N_0}{k_B T_D} e^{-E_0/k_B T_D}. \tag{20.10}$$

At $T_D \approx 350$ K and $E_0 \approx 2$–4 eV, only a few bonds will self-dissociate over the lifetime of the solar module. The activation energy of such bond dissociation, obtained as a slope of N_{BB} vs. $(k_B T_D)^{-1}$, is E_0, as expected.

The second term of Eq. (20.9) describes photon-assisted dissociation rate and is given by the following integral:

$$R_{uv}(E < E_0) = \int_0^{E_0} k_{uv}\, N_B(E) n_{ph}(E_0 - E) \cdot f(E_0 - E) dE. \tag{20.11}$$

Since $N_B(E)$ decreases with E, but $n_{ph}(E_0 - E)$ increases with E, the temperature dependence is suppressed. The $f(E_0 - E)$ is the location-specific atmospheric transmission coefficient of photons with energy $E_* \equiv E_0 - E$. For example, only a very small fraction ($< 5\%$) of the UV photons with ($E_* > 4.43$ eV) actually reaches a rooftop solar module, but a much larger fraction is incident on the solar wings of a satellite/spacecraft. For now, let us set $f = 1$. Inserting Eqs. (20.8) and (20.7), the integral evaluates to

$$R_{uv} = k_{uv} \frac{N_0}{k_B T_D} \frac{2 k_B T_S}{h^3 c^2} e^{-E_0/k_B T_S} \left[\frac{E_0^2}{\beta} - \frac{2E_0}{\beta^2} + \frac{2}{\beta^3} \right] \tag{20.12}$$

with $\beta \equiv \left[(k_B T_D)^{-1} - (k_B T_S)^{-1} \right]$. The explicit expression connects all the fundamental variables related to photodissociation of bonds into a single expression.

Homework 20.5: Equation (20.12) has a simple physical interpretation

Equation (20.12) can be interpreted as follows. Each of the N_0 bonds has binding energy E_0 and an small amount of vibration energy $k_B T_D$. All these bonds will be broken by photons with energy $E_* > E_0$, Eq. (20.7). Therefore, the dissociation rate could be simply written as the product of the number of bonds and the number of photons energetic enough to break the bonds:

$$R_{uv} \approx K_{uv} \times N_0 \times \left[\frac{2k_B T_S}{h^3 c^2} e^{-E_0/k_B T_S} E_0^2 \right].$$

You should convince yourself that this is the first term of Eq. (20.12), provided $E_0 \gg k_B T_S$ and $T_S \gg T_D$, so that $\beta \approx (k_B T_D)^{-1}$. Can you explain the origin of the second and third terms based on the fact that $\beta^{-1} \to 0$ when $T_D \to 0$?

Homework 20.6: Equation (20.12) can be used to predict the temperature dependence of UV degradation

A solar cell is tested for UV degradation at 50, 75, 100, and 125°C. Assuming the $E_0 = 3.5$ eV, $T_S = 5777$ K, determine the degradation rates for the test temperatures. For simplicity, collect all the unknown parameters as a prefactor A_{eff}. Plot the $\log - R_{uv}$ vs. $1/(k_B T_D)$ curve and show that the activation energy E_0 could be determined approximately by the slope of the line.

20.4.3 Photodegradation is accelerated by UV intensity, but not by module temperature

It is clear from Fig. 20.1 that direct photodissociation of bonds with energy $E_0 > 3$–5 eV requires high-energy photons (as in the ultraviolet part of the solar spectrum with a wavelength range of 10–400 nm, or 3–12 eV). The number of high-energy photons are so small (especially after it has been filtered by the ozone layer) that it takes years of exposure to normal sunlight to turn a polymer yellow. To test the UV degradation of the polymer within a short time, one must accelerate the degradation process. Special UV sources such as xenon arc lamps and metal halide lamps are used to increase the intensity (by 3-fold and 30-fold, respectively) to accelerate the degradation. After 1 to 2 weeks of exposure, the samples are taken out and their yellowing index (YI) measured. The corresponding loss of light transmission is given by ΔL.

Solar cells are expected to operate at various regional temperatures, T_D. To predict the temperature-dependent degradation, YI is measured (see Fig. 20.4) for a fixed duration t_0 of UV exposure and at a fixed temperature, T_D. The experiment

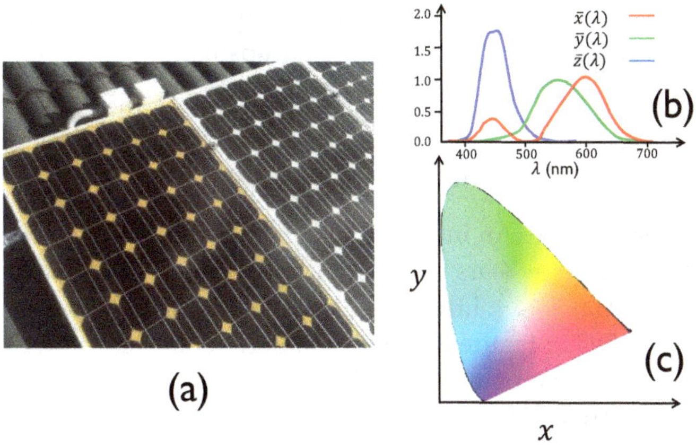

Figure 20.4: (a) The broken bonds give a polymer its characteristics yellow/brown color. (b) The degree of "yellowing" can be quantified my measuring the wavelength spectrum of the emitted light, then decomposing them into red, green, and blue by respectively multiplying it by the three-color spectrum: $x(\lambda)$ for red, $y(\lambda)$ for green, and $z(\lambda)$ for blue. The X, Y, and Z values so obtained are normalized to 1, i.e., $x \equiv X/(X+Y+Z)$. (c) Typical examples of the x, y, and z values associated with a color. (d) Yellowing index is calculated based on an agreed-upon formula (e.g., $\mathrm{YI} = (100/y)(1.274641506\,x - 1.057434092\,z)$); see **ASTM E313**.

is then repeated for multiple T_D. Assuming that

$$\Delta L \approx A_\mathrm{eff}\, e^{-E_\mathrm{eff}/k_\mathrm{B}T_\mathrm{D}},$$

is a simplified version of Eq. (20.12), we can respectively calculate E_eff and A_eff from the slope and the intercept of the plot $\ln(\Delta L)$ *vs.* $(k_\mathrm{B}T_\mathrm{D})^{-1}$. With these polymer-specific empirical parameters, UV degradation can be predicted at all locations of the world.

Of the two processes responsible for UV degradation, the activation for *self-dissociation* (see Eq. (20.10)) is $E_\mathrm{eff} \approx E_0$. A strong temperature dependence is expected, but the prefactor is so small that self-dissociation does not contribute much to yellowing. On the other hand, Eq. (20.5) suggests that the *photodissociation* is essentially independent of T_D, and therefore $E_\mathrm{eff} \approx 0$, and $A_\mathrm{eff} \propto R_\mathrm{s}$. In fact, given the distribution of bond energies, Eq. (20.12) does depend on temperature, but the dependence is weak, a conclusion supported by experiments which find that $E_\mathrm{eff} \approx 0.1$–$0.25$ eV.

Thus, the strong location dependence of UV degradation and EVA yellowing arises not from the local temperature, but by the location-specific, wavelength dependent filtering of the high-energy photons by the local atmosphere. For example, the annual UV dose in Florida is 80 kWh/m^2, while that of Arizona is 93 kWh/m^2. However, since the degradation is (essentially) temperature independent, one cannot increase the sample temperature to significantly shorten the test time of a newly developed polymer encapsulant.

Homework 20.7: Interpretation of experimental data

Read the paper "PVLife: An Integrated Model for Predicting PV Performance Degradation over 25+ Years" by M. A. Mikofski *et al.* (DOI: 10.1109/PVSC.2012.6317932) http://ieeexplore.ieee.org/stamp/stamp.jsp?tp=&arnumber=6317932. Figure 6 contains experimental results regarding UV degradation at two temperatures and various wavelengths. Show that the data can be interpreted by $E_{\mathrm{eff}} \approx 0.12$–$0.25$ eV. How does the results compare to the theoretical results from Homework 20.6?

20.5 Solution strategies: Techniques to reduce encapsulant yellowing

There are three ways to improve UV robustness: remove the UV spectrum before they reach the UV-susceptible bonds, remove the bonds susceptible to UV, or add UV-quencher molecules that relax photo-excited polymer molecules.

20.5.1 Filter high-energy photons before it reaches the encapsulant

By mid-1990s, a number of groups have demonstrated that a glass containing Cerium Oxide blocks sunlight above 3.2 eV (380 nm). Unfortunately, as shown in Fig. 20.5, when the Ce glass is first exposed to sunlight, its optical transmission degrades significantly. Therefore, although Ce glass suppresses EVA yellowing, the steady-state light transmission degrades considerably. Hence, the technology has not been adopted commercially.

20.5.2 Use stronger polymers

The second strategy involves replacing EVA by silicone, a more UV-stable polymer. Silicones are constructed from inorganic-organic monomers and is described by the chemical formula $[R_2\text{-SiO}]_n$, where R is an organic group such as an alkyl (methyl, ethyl) or phenyl. Another way to understand the structure of silicone is to compare with silica glass. In a silica glass, each silicon atom is bonded to four oxygen atoms in a tetrahedral configuration (see Fig. 20.1(c)). In silicone, two of the Si–O bonds are replaced by Si–R bonds, where R is an organic compound, such as CH_3. The pair of R molecules do not connect to others; therefore $[R_2SiO]$ becomes a unit of one-dimensional polymer chain, similar to EVA.

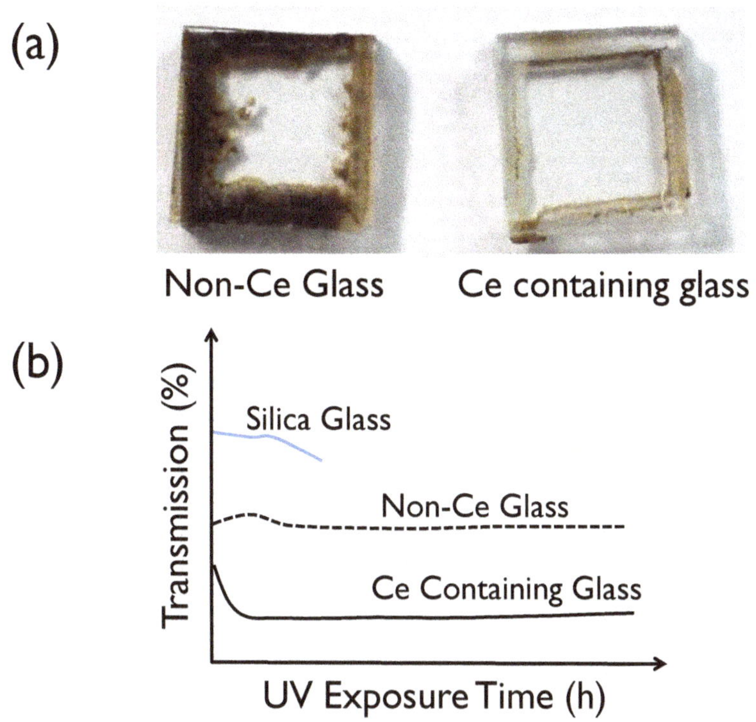

Figure 20.5: (a) Ce-doped glass absorbs the high-energy photons and protects the polymer underneath. (b) Unfortunately, the light transmission through the Ce-containing glass is poorer and the dissociation of weaker bonds leads to fast initial degradation of light transmission.

20.5.3 Add photo-stabilizer molecules

Not all polymer molecules absorb UV light; only the molecules with the right absorption gap (known as chromophores) do. When UV quenchers are added, the photo-excited chromophores can transfer the excess energy to these molecules. The chromophores are therefore not broken as frequently. As long as the additives themselves do not discolor the polymer, the relative simplicity of the approach makes it an attractive strategy. The technique is not a panacea, because the quenchers themselves degrade following repeated absorption of excess energy. Once the "guardian" molecules are gone, the chromophores are no longer protected and the natural degradation of the polymer resumes.

One final comment before we conclude this section. In this chapter, we explained how UV photons degrade the polymers through photodissociation of molecules. However, this is not the only way the polymers discolor. The molecules can also dissociate and polymers can brown by chemical reaction with the moisture — a topic we will return to in Chapter 23. Overall, it is not easy to develop cost-effective, UV-robust, and moisture-resistant alternatives to standard and field-tested glass-EVA combination, to justify the increased cost by increased energy yield over the years.

Homework 20.8: The time dynamics of quencher stabilized polymers can be described by a time delay

If the rate of quencher molecule degradation is k_q and that of polymer degradation is k_p, then the differential equations that describe their time-dependent concentrations are given by

$$dQ/dt = -k_q Q \tag{20.13}$$

$$dN_{\mathrm{BB}}/dt = -k_p(Q_0 - Q)N_{\mathrm{BB}} \tag{20.14}$$

where Q_0 is the initial quencher concentration. If none of the polymer bonds are initially broken (i.e., $N_{\mathrm{BB}}(t = 0)$), then calculate the time evolution of $N_{\mathrm{BB}}(t)$ and $\Delta P(t)$.

20.6 The backsheet polymer is also damaged by UV radiation

We know that the backsheet provides the structural rigidity for the module, just as the glass cover does for the front side. Unless it is a bifacial solar module, the backsheet need not be transparent. Therefore, strong polymers such as PET (polyethylene terephthalate), PVF (polyvinyl fluoride), or PVDF (polyvinylidene fluoride) are common choices. Although the bond energies are somewhat different, the physics and mathematics of photodissociation of bonds are essentially the same. For example, backsheet degradation often begins with photodissociation of the ester bonds. This step initiates a chain reaction called the Norrish reaction, the detailed physics of which is still being investigated. As the bonds break, one measures the degradation by the same YI, as we did for EVA. The UV robustness of the backsheet can be improved by incorporating additives that suppress UV absorption by the polymer. A laminated PVF-PET-PVF stack also appears to be relatively more resistant to UV degradation.

Backsheet degradation affects the cell performance indirectly. Obviously, the yellowing or cracking of the backsheet do not affect the short-circuit current. However, backsheet cracking reduces the mechanical strength and increases moisture penetration, leading to hydrolysis of the polymer and faster corrosion of the back contacts. We will discuss the topics of mechanical integrity and moisture penetration in Chapters 24 and 23, respectively.

Homework 20.9: Additional reliability concerns for solar cells used to power satellites, space-stations, and interplanetary missions

The UV intensity increases with altitude ($\sim 1\%/100$ m) because the thinner atmosphere absorbs/scatters fewer UV photons. At even higher altitude, earth's magnetic field traps electrons and protons. A satellite-mounted solar module exposed to these harsh conditions would suffer from "Displacement dose damage (DDD)", where atoms are knocked off by the high energy particles, and the resultant defects increase diode and shunt leakage. Use the software SPENVIS (The Space Environment Information System) to calculate the proton and electrons fluxes (cm^{-2}) associated with an orbit. Then, read the paper by S.R. Messenger *et al.* titled "SCREAM: A new code for solar cell degradation prediction using the displacement damage dose approach," Proc. of 35th IEEE PVSC, p. 1106, 2010, to calculate the DDD (D_d, MeV/g) by integrating over the fluxes from SPENVIS and the Non-ionizing energy loss (NIEL(E), $\mathrm{MeV} \cdot \mathrm{cm}^{-2} \cdot \mathrm{g}^{-1}$) of the absorber material to see why the power-degradation of a solar module ($\Delta P(t) \equiv P(t=0) - P(t)$), due to altitude-dependent, time-integrated radiation dose (D_d), is given by the semi-empirical formula:

$$\frac{\Delta P(t)}{P(t=0)} = B \log\left(1 + \frac{D_d}{D_c}\right)$$

where B and D_c are fitting parameters.

20.7 Conclusions: UV degradation affects absorber materials as well

In this chapter, we focused on the UV degradation of encapsulant and backsheet. We explained why the degradation depends linearly on UV intensity, but is relatively insensitive to device temperature. A number of strategies have been developed to improve UV robustness of these materials, but the cost-effectiveness of the proposed approaches remains an open question. Our focus on the encapsulant and the backsheet does not imply that the absorber itself does not suffer from UV radiation damage. In fact, UV-related degradation has been reported for many absorbers, including c-Si modules. The c-Si bonds are strong, but a continuous UV radiation does lead to defect generation and increased dark current. The encapsulation by UV-resistant EVA reduces UV degradation of the absorber material. Now, the absorber material is more susceptible to degradation from visible light for which the encapsulants offer no protection. We will discuss this (visible) light-induced degradation of solar absorbers in the next chapter.

References

[1] W. H. Holley, S. C. Agro, J. P. Galica, L. A. Thoma, R. S. Yorgensen, M. Ezrin, P. Klemchuk, and G. Lavigne. Investigation into the causes of browning in EVA encapsulated flat plate PV modules. In *Proceedings of 1994 IEEE 1st World Conference on Photovoltaic Energy Conversion - WCPEC (A Joint Conference of PVSC, PVSEC and PSEC)*, volume 1, pages 893–896 vol. 1, December 1994.

[2] Michael Kempe, M. Reese, A. Dameron, and T. Moricone. Types of encapsulant materials and physical differences between them. In *NREL PV Module Reliability Workshop, Denver West Marriott, Golden, Colorado*, 2010.

[3] Michael D. Kempe. Ultraviolet light test and evaluation methods for encapsulants of photovoltaic modules. *Solar Energy Materials and Solar Cells*, 94(2):246–253, February 2010.

[4] M. D. Kempe, M. Kilkenny, T. J. Moricone, and J. Z. Zhang. Accelerated stress testing of hydrocarbon-based encapsulants for medium-concentration CPV applications. In *2009 34th IEEE Photovoltaic Specialists Conference (PVSC)*, pages 001826–001831, June 2009. ISSN: 0160-8371.

[5] C. R. Osterwald, J. Pruett, and T. Moriarty. Crystalline silicon short-circuit current degradation study: Initial results. In *Conference Record of the Thirty-first IEEE Photovoltaic Specialists Conference, 2005*, pages 1335–1338, January 2005. ISSN: 0160-8371.

[6] Polymer Solutions Incorporated. Let There Be Light Stabilizers!, November 2015. Section: Materials Science.

CHAPTER 21

Light-induced Degradation in Solar Cells

Chapter Summary

- Light induced degradation is important reliability problem for all solar cells, especially immediately after the light-exposure.
- Classical LID in a:Si-H solar cells is a consequence of a depassivation of Si-H bonds and diffusion of the hydrogen within the film.
- LID in n^+-p Silicon involves displacement of Boron atoms and formation of Boron-oxide midgap recombination states.
- Temperature enhanced LID (LeTID) has been an important concern for modern solar cells with back-side passivation.

21.1 Light-induced degradation has been known since the 1970s

There is no point in fighting the second law of thermodynamics — even in isolation, everything eventually falls apart. Exposure to sunlight accelerates the degradation. In fact, the photons need not have high energies; even visible photons can do considerable damage. This was illustrated most dramatically when Staebler and Wronski reported in the 1970s that thin-film a-Si solar cells lose 20–30% of their output power just after a few hours of light socking, see Fig. 21.1. This light-induced degradation received widespread attention because thin-film a-Si solar cells appeared promising as a low-cost alternative to c-Si solar cells. Eventually, researchers realized that even c-Si is not immune to light-induced degradation (LID). In fact, LeTID (Light and elevated Temperature Induced Degradation), alias Carrier Induced Degradation (CID), is an important reliability concern for PERC c-Si solar cells. In this chapter, we will discuss the physics of light-induced degradation in a-Si and c-Si solar cells. The degradation in a-Si layers continues to

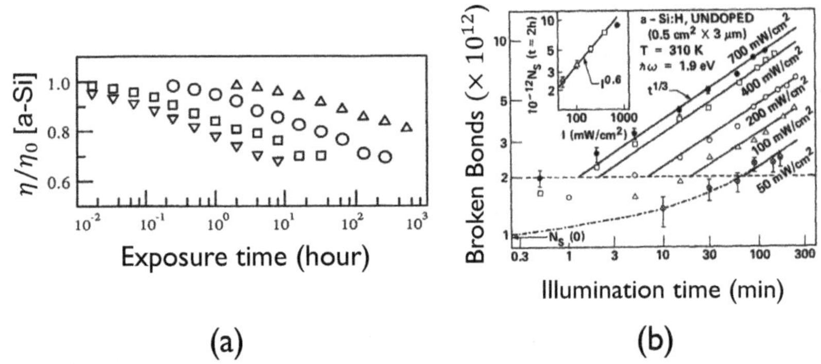

Figure 21.1: (a) Experimental evidence of the rapid degradation of a-Si solar when exposed to sunlight. (b) The degradation depends on time as a sub-linear power-law, i.e., $t^{1/3}$. The cells are not encapsulated by EVA or covered by glass. (Taken from Y. Nakata *et al.*, JJAP, 1992)

be relevant, because they are used as passivating layers for highly efficient silicon-heterojunction solar cell technology.

21.2 Physics and mathematics of LID in a-Si solar cells

21.2.1 a-Si:H is passivated by Si–H bonds

To understand the LID, we need to understand the atomic structures of amorphous vs. crystalline silicon. In c-Si, each Si atom is connected to four of its equidistant neighbors in a tetrahedral structure. These tetrahedra fit within a diamond unit cell (with 8 atoms). The motif is repeated to make a crystalline solid. The atomic structure is stable and relaxed because all the atoms are stabilized/separated by the same distance.

Each Si atom in a-Si is also connected to four of its neighbors. The tetrahedra, however, are connected with random orientation. Despite the randomness, Polk in 1971 showed that a 3D a-Si can in principle have all the bonds satisfied (albeit with large twisted angles). The strain of the system can be relaxed if some bonds are allowed to break and the broken bonds are passivated by atomic hydrogen (see Fig. 21.2(a)). With all the bonds satisfied (either as Si–Si or Si–H), a-Si can be used as a high-quality, direct-bandgap, thin-film solar absorber.

Once the LID was reported by Staebler and Wronski, many researchers immediately implicated the photodissociation of relatively weak Si–H bonds (see Table 20.1) as the root cause of the degradation. Subsequently, Stutzmann *et al.* used the electron spin resonance experiment to demonstrate that the number of Si–H bonds broken at time t is given by

$$N_{\mathrm{BB}} \propto n_{\mathrm{ph}}^{2n}\, t^{n} \tag{21.1}$$

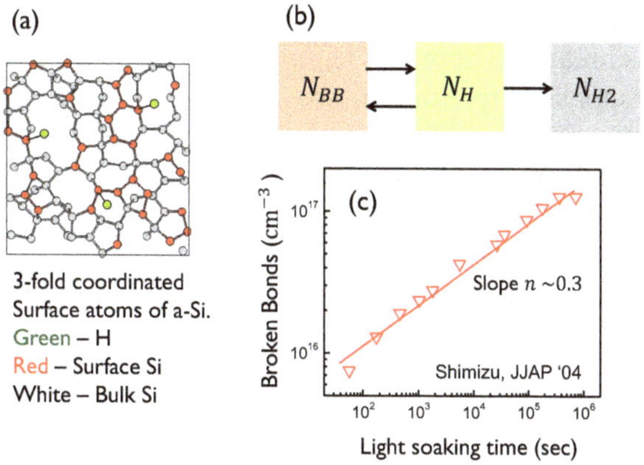

Figure 21.2: (a) A schematic diagram of a-Si:H (viewed from the top) with randomly arranged Si–Si and Si–H bonds. Some of the broken Si–Si bonds have been passivated by H (green). The Si–H bonds are weaker than Si–Si bonds. (b) The H atoms are released during the photodissociation of Si–H bonds. They may either return to passivate the broken bonds (BB) or may be permanently lost in the form of H_2. (c) The numerical solution of Eqs. (21.2) and (21.3), plotted as a function of the light soaking (LS) time, produces the same time exponent as is observed in experiments.

with the power-law exponent $n = 1/3$. The increase in the defect density is reflected in the corresponding increase in the dark current and the decrease in the efficiency, as shown in Fig. 21.1. Unlike Eq. (20.9), the N_{BB} increases sub-linearly with time and light intensity as shown in the inset of Fig. 21.1(b). In the next section, we will develop a physical model to explain these dependencies quantitatively.

> **Homework 21.1: Euler's rule and Maxwell's constraints characterize the geometry of solids and defects**
>
> Review Lectures 4 and 5 of the "Reliability Physics of Nanotransistors" course posted at `https://nanohub.org/resources/16560.` to explain that an amorphous material can be defect-free, and how the addition of hydrogen reduces the strain in an otherwise defect-free a-Si system.

21.2.2 A model for light-induced degradation

Several models have been proposed to explain LID, also known as Staebler-Wronski effect [T. Shimizu, JJAP, 2004; R. Biswas, PRL, 2000, 2002]. We will use a simple model based on hydrogen kinetics that explains the key experimental observations. Briefly, the H-model suggests that the rate of free hydrogen generation (dN_{H}/dt) due to photodissociation of Si-H bonds is determined by three processes, see the middle box in Fig. 21.2(b) and Eq. (21.2). The first term on the right of Eq. (21.2) describes the number of hydrogen atoms released when energetic photons break Si–H bonds (N_{BB}). This term is proportional to the number of photons responsible for bond dissociation n_{ph} and the initial concentration of unbroken Si–H bonds (N_0). Once released from Si–H bonds, the mobile hydrogen atoms (density, N_H) can diffuse within the absorber, until they either restore some other broken Si–H bond (second term, Eq. (21.2)), or combine with each other to form a stable H_2 molecule (the third term, Eq. (21.2)). One assumes that the H_2 molecules will never again dissociate into atomic H and participate in bond passivation.

$$\frac{dN_H}{dt} = k_F n_{ph}(N_0 - N_{BB}) - k_R N_{BB} N_H - 2k_c N_H^2. \tag{21.2}$$

Here k_F, k_R, and k_c are the rate constants for forward dissociation, repassivation, and dimerization, respectively. As we will see later, the characteristic time exponent of LID does not depend on these rate constants!

Photons break Si–H bonds in two different ways. First, sufficiently energetic photons from the sun can directly break Si–H bonds. Second, photons may initially excite electron–hole pairs. The energy released from the recombination of these photogenerated carriers may break Si–H bonds. Obviously, the two different rate constants k_F characterize the direct and the recombination-assisted processes.

Let us calculate the rate of broken bonds by accounting for the forward photodissociation and reverse bond restoration processes (see the left box in Fig. 21.2(b)):

$$\frac{dN_{BB}}{dt} = k_F n_{ph}(N_0 - N_{BB}) - k_R N_{BB} N_H. \tag{21.3}$$

We can solve the coupled Eqs. (21.2) and (21.3) numerically to find $N_{BB}(t)$, the number of dangling/broken bonds generated as a function of time. In practice, numerical simulation provides insights that allow an analytical solution of the coupled equations.

First, the solar cell efficiency degrades significantly even with $N_{BB} \ll N_0$; therefore the first term of Eq. (21.2) can be simplified to $k_F n_{ph} N_0$. Second, three fluxes on the right of Eq. (21.2) nearly balance each other, so that $dN_H/dt \approx 0$, i.e.,

$$k_F n_{ph} N_0 - k_R N_{BB} N_H - 2k_c N_H^2 = 0. \tag{21.4}$$

Inserting Eq. (21.4) into Eq. (21.3), we find that

$$\frac{dN_{BB}}{dt} = 2k_c N_H^2. \tag{21.5}$$

Also, the third term in Eq. (21.4) is much smaller than the first two terms, so that $k_F n_{ph} N_0 \approx k_R N_{BB} N_H$, or

$$N_H = (k_F/k_R) N_0 n_{ph} N_{BB}^{-1}. \tag{21.6}$$

Inserting Eq. (21.6) into Eq. (21.5), we find that

$$\frac{dN_{BB}}{dt} = 2k_c \left(\frac{k_F N_0 n_{ph}}{k_R}\right)^2 N_{BB}^{-2}. \tag{21.7}$$

Integrating the differential equation results in the following expression,

$$N_{\mathrm{BB}} = 6k_{\mathrm{c}}^{\frac{1}{3}} \left(\frac{k_{\mathrm{F}} N_0 n_{\mathrm{ph}}}{k_R} \right)^{\frac{2}{3}} t^{\frac{1}{3}}. \tag{21.8}$$

The equation suggests that N_{BB} increases rapidly as soon as an a-Si module is exposed to sunlight, but the rate of growth slows markedly once the initial degradation is over (see Fig. 21.2). The rate is initially high because N_{H} and N_{BB} are so low that the repassivation of broken bonds (given by the $k_{\mathrm{R}} N_{\mathrm{BB}} N_{\mathrm{H}}$ term) is small. With time, the densities increase, and so does the repassivation. These counterbalancing effects of generation and repassivation dramatically reduce the overall rate of $N_{\mathrm{BB}}(t)$. Eventually, $N_{\mathrm{BB}} \to N_0$ after a prolonged period of light soaking.

Homework 21.2: There is an interesting correlation among densities of H_2 molecules, mobile atomic H and broken Si–H bonds

Show that after an initial rise in the N_{H} concentration due to bond dissociation, Eq. (21.6) suggests that $N_{\mathrm{H}} \propto t^{-1/3}$. Can you explain the rise and fall of N_{H} concentration by calculating the time-dependent increase of H_2 molecules?

Homework 21.3: The correlation between the density of broken bonds and efficiency loss of an a-Si solar cell

Explain how the power-law increase in the density of broken Si–H bonds (Eq. (21.8)) is directly reflected into power-law loss of efficiency observed in the experiments (Fig. 21.1 and Eq. (21.2)). Do you expect the relationship to hold for other types of solar cells as well?

Solution. Recall from Sec. 7.4.2 that the photocurrent in an a-Si p-i-n solar cell is given by $J_{\mathrm{ph}} \propto \mu \mathcal{E} \tau$, where μ is the mobility, $\mathcal{E}$ is the electric field, and τ is the recombination lifetime of the intrinsic a-Si layer. In other words, only the electrons generated close to the contact within $L \approx v\tau = \mu \mathcal{E} \tau$ can escape recombination to contribute to photocurrent. Since the reduction in carrier lifetime is proportional to N_{BB}, therefore the loss of photocurrent $\Delta J_{\mathrm{ph}} \propto N_{\mathrm{BB}}^{-1} \approx t^{1/3}$. The performance of p-n and heterojunction solar cells is characterized by a more complicated functional relationship between the defect density and cell efficiency. Thus, the simple relationship is not expected for other cells.

Many groups have explored processing improvement to suppress LID in a-Si solar cells. One interesting approach involves *chemical annealing* of the a-Si film grown in successive layers. Here, chemical annealing allows pruning of the weaker Si–H bonds (and thus change the chemical microstructure) by using He ions. For 2–3 nm of film deposition, the layer is annealed for a minute by a plasma flux of He ions (or other inert ions, such as Ar). The deposition/annealing cycle is repeated until the desired thickness is achieved. The improvement in LID has been attributed to the improvement of the a-Si microstructure and ion-assisted removal in weaker Si–H bonds.

21.3 Crystalline silicon cells suffer from LID too

21.3.1 LID of bulk silicon

Light-induced degradation is most pronounced in a-Si cells, but n^+-p c-Si solar cells also suffer from light-induced degradation. Obviously, Si–H bonds do not exist in the bulk of the c-Si; therefore LID in c-Si originates from a different physical phenomenon.

Recall that boron is used as the dopant for the p-base of the cell, red dots in Fig. 21.3(a). During the crystal growth by the Czochralski (Cz) method, oxygen is incorporated into the substrate as well (white dots). Sunlight releases these weakly bonded atoms, which can then diffuse through the crystal and eventually bind to form a boron oxide defect site (blue dot), which increase recombination of photogenerated carriers, as shown in Fig. 21.3(b). Therefore, although the kinetics of defect generation is different compared to a-Si LID (see Fig. 21.2), the result is the same: following prolonged light exposure, the performance of c-Si cells decrease due to increasing bulk recombination assisted by boron oxide defects.

Once the basic mechanism is understood, it is easy to suggest various solutions. For example, LID is reduced in a p^+-n solar cell. Other methods of crystal growth (e.g., float zone) reduces the oxygen content, the precursor for LID. Or, one can reduce boron doping (high-resistivity substrate) or even replace boron by

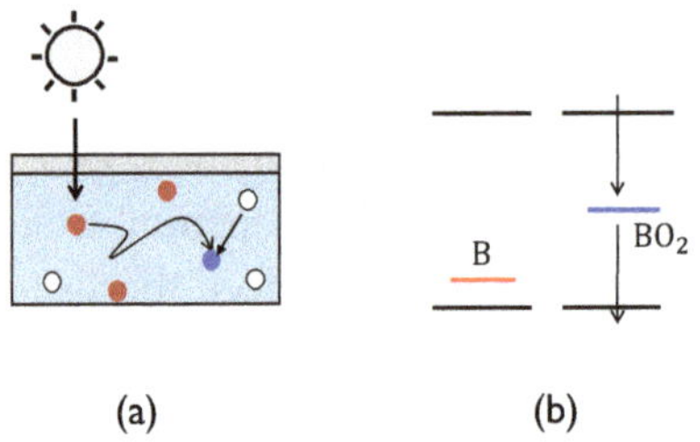

Figure 21.3: (a) An n^+-p cell: photons pass through the thin n-layer and generate electron–hole pairs in the p-layer. Also shown are boron (red), oxygen (white), and boron oxide (blue) atoms. The energy released from the recombination of electron-hole pairs dislocates the B dopants. (b) The conversion of boron to boron oxide creates mid-gap states. Bulk recombination associated with these defects increases the dark current and reduces the efficiency of the cell.

gallium as the p-dopant to obviate boron oxide formation. Finally, light-induced regeneration (LIR), either by forward-biasing the solar cell in dark for several minutes or by high-intensity illumination using halogen lamp, LED, or laser for 5–20 seconds, appears to restore (regenerate) and sustain the efficiency, presumably by dissociating the boron–oxygen complexes.

Homework 21.4: A mathematical model for LID for c-Si

Write three rate equations, respectively, for mobile boron concentration (N_{B}), mobile oxygen concentration (N_{O}), and the concentration of the boron–oxygen complex (N_{BO}). Solve the kinetic equations either numerically or analytically to explain how the time kinetics of c-Si LID differs from that of a-Si LID.

21.3.2 LID of the rear-side passivation

You may recall from Sec. 7.5.3 and Fig. 7.5 that modern c-Si solar cells use thin rear passivation layers (e.g., Al_2O_3/SiN_x) to "geometrically" prevent wrong-contact recombination and improve cell efficiency. The broken bonds at the Si/oxide interface are passivated by H, forming Si-H covalent bonds involving (and held together by) a pair of electrons. One must optimize the relatively high temperature needed for metal contact formation to avoid breaking the Si–H bonds and degrading the cell efficiency prior to field installment. Once fielded and exposed to sunlight, the photogenerated holes are captured by Si-H covalent bonds. The removal of one of the two electrons holding a Si–H bond together weakens it. The light-induced self-heating of the module (see Chapter 5) as well as the direct photodissociation break the weakened bonds. The broken bonds increase (wrong-contact) surface recombination and degrade cell efficiency. This light and elevated temperature induced degradation, abbreviated L(eT)ID, occurs in a variety of passivated contact technologies and must be carefully managed to avoid significant power loss during field operation. LeTID is also known as CID (carrier-induced degradation) or HID (hydrogen-induced degradation) because photogenerated holes and Si–H bonds are involved in the degradation process. Similar to other LID mechanisms discussed above, LeTID can be described by a reaction–diffusion theory, except now the reaction coefficients depend on the carrier concentration and self-heating of the module.

21.4 Conclusions: Photodegradation affects all the components of a solar module

Light-induced degradation is a fact of life. One must account for this degradation to calculate the total energy output from a solar farm over its lifetime. There are technical solutions to reduce the degradation rate, but the commercially viable solutions must be inexpensive.

In this chapter, we have focused on LID in a-Si and c-Si solar cells. Other types of cells are also not immune, for example perovskite and bulk heterojunction solar cells (Chapter 8) also show significant instability under light exposure. Despite the superficial similarities, however, the physics of LID differs significantly from one solar cell technology to the next. The solution strategies, therefore, must be specifically tailored for the cell technology.

Sunlight damages the cells directly by photodissociation, but the indirect effects are equally important. First, the increase in the temperature accelerates various thermally activated degradation mechanisms exponentially. Second, the photodissociation of EVA produces acetic acid, an important precursor to corrosion. Finally, photodegradation may itself be catalyzed by moisture ingress. In Chapter 23, we will discuss the reliability challenges associated with corrosion of metal contacts, hydrolysis of EVA, etc. Before we do so, we wish to discuss an important reliability issue involving potential-induced degradation.

References

[1] Robert A. Street. *Hydrogenated amorphous silicon*. Cambridge solid state science series. Cambridge Univ. Press, Cambridge u.a, new edition, 2005. OCLC: 845372037.

[2] D. L. Staebler and C. R. Wronski. Reversible conductivity changes in discharge-produced amorphous Si. *Applied Physics Letters*, 31(4):292–294, August 1977. Publisher: American Institute of Physics.

[3] Martin Stutzmann. Microscopic Aspects of the Staebler-Wronski Effect. *MRS Online Proceedings Library Archive*, 467, January 1997.

[4] H. Fritzsche and Tucson Az. Search for Explaining the Staebler-Wronski Effect. *MRS Online Proceedings Library (OPL)*, 467, 1997. Publisher: Cambridge University Press.

[5] Tatsuo Shimizu. Staebler-Wronski Effect in Hydrogenated Amorphous Silicon and Related Alloy Films. *Japanese Journal of Applied Physics*, 43(6R):3257, June 2004. Publisher: IOP Publishing.

[6] R. Biswas and Y.-P. Li. Hydrogen Flip Model for Light-Induced Changes of Amorphous Silicon. *Physical Review Letters*, 82(12):2512–2515, March 1999. Publisher: American Physical Society.

[7] Daniel Chen, Phillip G. Hamer, Moonyong Kim, Tsun H. Fung, Gabrielle Bourret-Sicotte, Shaoyang Liu, Catherine E. Chan, Alison Ciesla, Ran Chen, Malcolm D. Abbott, Brett J. Hallam, and Stuart R. Wenham. Hydrogen induced degradation: A possible mechanism for light- and elevated temperature- induced degradation in n-type silicon. *Solar Energy Materials and Solar Cells*, 185:174–182, October 2018.

[8] Nanlin Wang and Vikram L. Dalal. Improving stability of amorphous silicon using chemical annealing with helium. *Journal of Non-Crystalline Solids*, 352(9):1937–1940, June 2006.

[9] Radovan Kopecek, Joris Libal, and Lejo J. Koduvelikulathu. Is LeTID degradation in PERC cells another degradation crisis even worse than PID?, November 2018.

[10] Meng Xiajie. The complexity of LID, LeTID and HID, January 2019. Jensen, Mallory A., Ashley E. Morishige, Jasmin Hofstetter, David Berney Needleman, and Tonio Buonassisi. Evolution of LeTID defects in p-type multicrystalline silicon during degradation and regeneration. *IEEE Journal of Photovoltaics 7*, no. 4 (2017): 980–987.

[11] M. A. Alam and S. Mahapatra. A comprehensive model of PMOS NBTI degradation. *Microelectronics Reliability*, 45(1):71–81, January 2005.

CHAPTER 22

Potential-induced Degradation is a Serious Reliability Issue

Chapter Summary

- Potential induced degradation (PID) depends on the maximum voltage of series-connected solar modules as well as the sodium content within the glass.
- The PID leakage involves complex pathways through the encapsulant, glass-cover, and the frame.
- PID damages a solar cell though shunt-formation and creation of localized defects that enhances trap-assisted recombination current.
- A semi-quantitative PID model explains many features of PID degradation, including, delayed onset, linear voltage dependence, and recovery during the off-state.
- A number of methods have been developed to suppress PID, including the use of low-sodium glass cover.

22.1 Introduction: PID is a reliability problem with a long history

A solar farm produces DC electricity, but a typical electrical grid carries AC electricity, see Fig. 22.1. You may recall from Chapter 11 that an electronic device called an "inverter" converts DC electricity to AC electricity. The inverter works best at high voltages; therefore, a set of solar modules are connected in series to increase the output voltage, $V_{\max}$, to several hundred volts, before the DC voltage is inverted to grid-compatible AC electricity.

In the 1980s, when the first solar farms were being installed in various parts of the world, engineers were alarmed to find that the last module in the series (e.g., M_3 in Fig. 22.1) mysteriously lost 50%–90% of the rated power within a few months of installation. The phenomenon was named potential-induced degradation (PID)

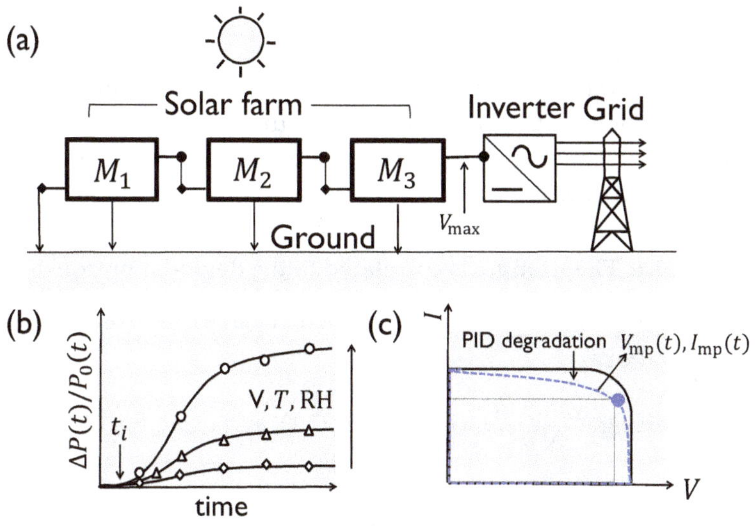

Figure 22.1: (a) A solar farm is connected to the grid by an inverter. The frame of each module are grounded. (b) The module with the highest voltage experiences a significant power loss. The power loss depends on voltage, temperature, and relative humidity. (c) PID degradation is reflected in the loss of FF associated with the increase in the shunt resistance, indicated by the increasing slope of the I-V characteristics at the short-circuit condition, i.e., $R_{sh} = (dV/dI)_{V=0}$ (see Homework 23.5).

because the power loss depends on the magnitude of the highest voltage of the system, V_{max}. PID was also affected by operating temperature, T, and percent relative humidity [RH], and the time duration t the module is exposed to high voltage.

22.2 An empirical formula summarizes the experimental observations

Empirically, the time-dependent power loss $\Delta P(t > t_i)$ is characterized by an onset delay (t_i), a rapid rise, an asymptotic saturation at ΔP_∞, as shown in Fig. 22.1(b). In other words, the following logistic function describes the power loss as a function of time:

$$\frac{\Delta P(t)}{\Delta P_\infty} = \left[1 + \left(\frac{\Delta P_\infty}{\Delta P(t_i)} - 1\right) e^{-(t-t_i)R_D}\right]^{-1}. \tag{22.1}$$

Here, $\Delta P(t_i)$ and ΔP_∞ are the initial and saturation power losses, respectively. The degradation constant (R_D)

$$R_D = A\,V^n e^{-E_A/k_B T}[\text{RH}]^B, \tag{22.2}$$

depends on the moisture content (RH), ambient temperature (T), and the maximum voltage (V) a module is exposed to. Here, $n = 1 - 2$, E_A is the activation energy, B is the humidity constant. Over the years, PID experiments have shown that the constant A depends on the type of solar cell (n^+-p vs. p^+-n), the resistances of various layers that make up a module, namely, the glass cover, encapsulation, and the anti-reflection coating, etc. The goal of this chapter is to explain the essential physics of PID (namely, how it depends on voltage, temperature, time, cell type, etc.) and techniques to control it.

Homework 22.1: The lifetime is correlated to the degradation constant, R_D.

If a module fails when $\frac{\Delta P(t_{\mathrm{MTF}})}{\Delta P_\infty} = 0.5$. Calculate an expression for the mean time to failure, t_{MTF}.

Solution. Inserting $\frac{\Delta P(t_{\mathrm{MTF}})}{\Delta P_\infty} = 0.5$ in Eq. (22.1), and using Eq. (22.2), we find that

$$\begin{aligned} t_{\mathrm{MTF}} &= t_i + R_{\mathrm{D}}^{-1} \ln\left(\frac{\Delta P_\infty}{\Delta P(T_{\mathrm{i}})} - 1\right) \\ &\propto V^{-n} e^{E_{\mathrm{A}}/k_{\mathrm{B}}T} [\mathrm{RH}]^{-B}. \end{aligned} \tag{22.3}$$

In other words, the mean time to failure depends inversely on the voltage, temperature, and relative humidity, as expected.

Homework 22.2: Empirical PID equation can be used to compare relative PID degradations in various climatic conditions

A research group reports that when a Si module is stressed at 1000 V, its initial PID power loss is given by a variant of Eq. (22.1):

$$\Delta P(t)/\Delta P_\infty = C \cdot V^{2n} e^{-2E_{\mathrm{A}}/k_{\mathrm{B}}T} [\mathrm{RH}]^{2B} t^2 \tag{22.4}$$

with the following parameters: $C \cdot V^{2n} = 2 \times 10^{-8}$, $E_{\mathrm{A}} = 0.88$ eV, $B = 7.12$, and t expressed in hours.

Explain that the quadratic time dependence in Eq. (22.4) as an approximate Taylor series expansion of Eq. (22.1). Using these parameters, calculate the relative PID power loss of the same module and at the same voltage stress following 10 years of operation in a hot and humid climate (e.g., $T = 40°$C, and RH $= 70\%$) vs. a cold/dry environment (e.g., $T = 30°$C, and RH $= 20\%$).

In this chapter, we will answer the following questions related to Eq. (22.1).

1. Why does PID depend on voltage, temperature, and relative humidity? What are the physical interpretations of n, E_{A}, and B?
2. Why is the PID power loss primarily correlated to shunt formation?

3. What determines the onset delay (t_i) and the maximum power degradation (P_∞)?

4. PID occurs close to the metal frames. What effect does it have in terms of localized self-heating (hotspot formation) and increased degradation of a solar module?

5. How should we reduce or suppress PID degradation?

If we understand the physics of PID, we will be able to answer these questions.

22.3 PID occurs when the modules are connected in series and the frame is grounded

Let us focus on the cells within the module M_3 in Fig. 22.1(a), which has been redrawn for clarity in Fig. 22.2. We recall from Chapter 10 that a c-Si module is composed of a number of series-connected c-Si cells, with $V_{\rm mp} = 0.6 - 0.7\,\rm V$, as in Fig. 22.2(a). Similarly, a thin-film module is divided into smaller cells (~ 0.5–1 cm wide) and reconnected in series to complete the module, as in Fig. 22.2(b). In both cases, the series connection reduces the resistive loss during current collection, and the output voltage is given by $V_{\rm m} \equiv V_{\rm f} - V_{\rm i} \approx M \times V_{\rm mp}$, as shown in Fig. 22.2(c). If K such modules are connected in series, the output voltage of the string is

$$V_{\rm max} = K\,V_{\rm m} = K\,M\,V_{\rm mp}. \tag{22.5}$$

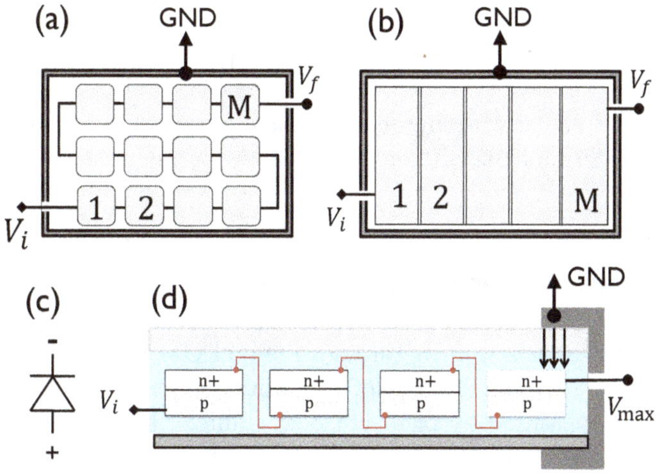

Figure 22.2: Series-connected cells in a (a) c-Si module ($M = 12$) and (b) thin-film module ($M = 5$). (c) Each p-n$^+$ cell of the module can be represented by the diode of the specified polarity. (d) The series connection allows the (negative) voltage to build-up with respect to the frame. The voltage difference between grounded frame and the last cell of the last module is $V_{\rm max}$.

To see if $V_{\max}$ is positive or negative, consider a typical c-Si solar cell with p-type substrate and n^+ emitter, shown in Fig. 22.2(c). If the substrate is grounded, the n side of the forward-biased cell (operating at V_{mp}, for example) will develop a negative voltage. Once these negative voltages are added up, $V_{\max}$ becomes a large negative quantity, say -500 to -600 volts. Obviously, the opposite is true for a cell with an n substrate and p^+ emitter: $V_{\max}$ is now positive, $+500$ to $+600$ V, for example.

An important feature of a solar farm is that the metallic frame of a module is always grounded, as shown in Fig. 22.2. Therefore, the cells in the last module develop very high voltages ($V_{\max}$) relative to the ground. As a result, a significant voltage drops across the dielectric stack (glass, EVA, and possibly SiN anti-reflection coating, etc.) that separates the cells from the module frame (see Figs. 22.2(c) and 22.3). Obviously, the perimeter cells close to the metal frame are most susceptible to PID degradation. PID worsens on a rainy day. As the rainwater covers the glass top of the module, the conductivity of the water is high enough to make the entire glass-top the same potential as the metal frame (i.e., zero). Now, *every* cell, even the ones far away from the perimeter, experiences a significant voltage stress.

Homework 22.3: Number of cells and modules to be connected in series

Consider a module with 96 p-n^+ c-Si solar cells, i.e., $M = 96$. The maximum power point for each cell is given by $V_{\text{mp}} = 0.57$ V and $I_{\text{mp}} = 5.98$ A.

- Calculate the output voltage of each module.
- What is the minimum number of modules needed (K) to make $V_{\max} \geq 600\ V$?
- The 96 subcells of a rectangular module is arranged in a 12 by 8 configuration. Determine the voltages of cells at the perimeter of the 7th module.

22.4 PID leakage involves complex pathways

A leakage current begins to flow through the dielectric stack in response to the voltage developed between the cell and the grounded metal frame. This dielectric leakage is easily measured by connecting an ammeter between the metal frame and the ground. And this leakage current is a direct precursor to PID damage (No leakage, no PID!).

Let us carefully consider the leakage current in a module where the maximum cell voltage is negative with respect to the frame, see Fig. 22.3. The dielectric leakage may occur through several parallel pathways, such as 1-2-3-4 or 5-1-2′-2″-3-4, etc. For example, (1) may involve flow of positive ions (mostly Na^+) from

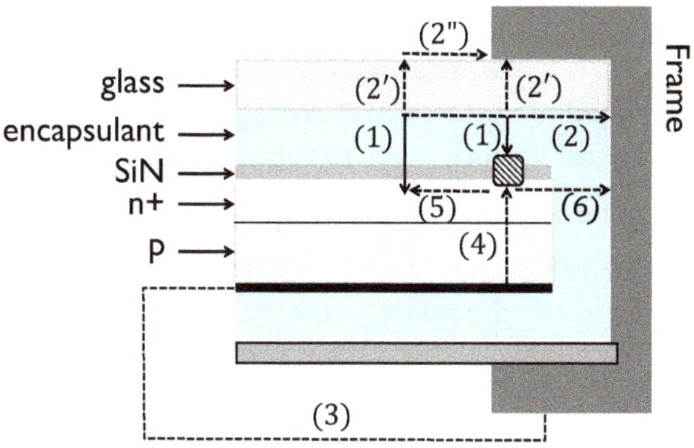

Figure 22.3: A schematic sideview of a c-Si solar cell, in close proximity to the metal frame that holds the cell together. A number of leakage pathways involving the cell and the metal frame contribute to PID. The solid arrows define ion transport, while the dashed arrows define electron–hole transport.

the polymer/glass interface to the cell electrode where it is neutralized by electrons. If the ions terminate on the SiN coating, it can be neutralized by electrons from the top electrode, as indicated by (5). For regions close to the metallic frame, (2) involves electron current to the frame through the glass/encapsulant interface. If the glass is not excessively resistive, then (2′) involves current flow through the glass to the metal frame. The electrons continue from the frame to the substrate of the first cell of the module (which is also grounded), as indicated by the dashed line (3). An excess leakage through the diode (4) completes the circuit. On a wet day, an additional path (1-2′-2″-3-4-5) may open up. Since, (2″) involves conduction through the thin water layer accumulated over the cell surface, locations far away from the frame can now be affected. Finally, other leakage paths (e.g., 6-3-4) may also exist, however, (1-2′) is typically less resistive than the bulk conduction through (6) (the figure is not drawn to scale); therefore these additional pathways may be neglected.

Homework 22.4: PID leakage can be described by an equivalent circuit diagram

To characterize PID leakage each current segment in Fig. 22.3 may be represented by a (possibly nonlinear) resistance.

- Draw the circuit diagram that captures the various series-parallel pathways for current conduction.

(continued on the next page)

Homework 22.4 (*continued from the previous page*)

- If PID is correlated to the leakage current (to be discussed below), use the diagram to suggest various ways to suppress PID.
- If $R_{2'} \gg R_2$, determine the voltage drop across the glass, V_g.
- Redraw the current pathways if the solar cell electrodes are positively biased with respect to the metal frame.

22.5 The voltage at glass–polymer interface is sufficient to pull Na out of the glass and push them toward the negative electrode of the solar cell

The leakage pathway (1) in Fig. 22.3(c) may either involve electron flow from the electrode to the glass/encapsulant interface or ion flow in the opposite direction. Chemical analysis has established that Na^+ ions are deposited close to the electrodes, indicating that ionic conduction along the pathway (1) dominates and dictates PID. But where do these Na atoms come from?

Recall from Chapter 20 that pure silica glass (SiO_2) is composed of long chains of tetrahedral units (Si at the center are covalently bonded to four oxygen atoms at the corners) connected by the oxygen atoms. The strong bonds make the melting temperature high. The incorporation of Na reduces the melting temperature and simplifies glass processing.

A fraction of the PID voltage V_g drops across the glass and ionizes the Na atoms. The positive ions travel through the glass and the polymer encapsulant toward the negative electrode of the solar cell. The electrons are pulled through the glass (2′) or glass/encapsulant interface (2) to complete the current path.

22.6 Multiple processes occur once Na^+ ions reach the cell electrode

Once Na^+ ions reach the top solar cell electrode and build up at the polymer/SiN interface, they may affect performance of a solar cell in two ways, see Fig. 22.4: increased dark current (PID-s) and enhanced surface polarization (PID-p).

22.6.1 As a mid-gap state, Na^+ increases shunt and diode leakage

Once the Na^+ ions are neutralized (by the electrons from the top electrode) and their concentration builds up over time, they will diffuse through the SiN layer into the solar cell, as shown in Figs. 22.4(a, b). Na atoms form mid-gap states in silicon; these mid-gap states increase junction and bulk recombination, as shown

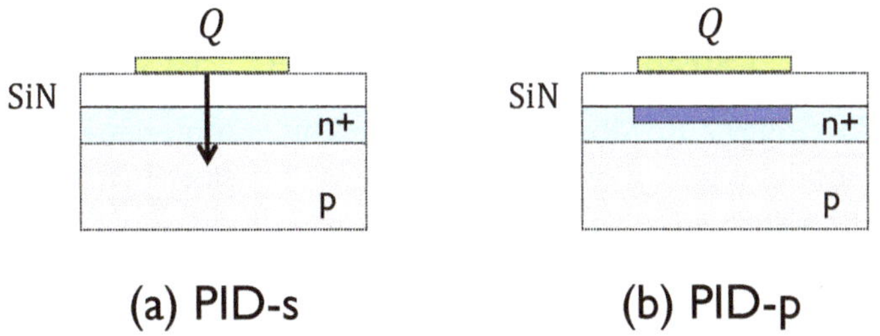

Figure 22.4: Accumulation of Na+ atoms (indicated by Q, green box) leads to two types of potential-induced degradation: (a) PID caused by shunts, and (b) PID caused by surface polarization.

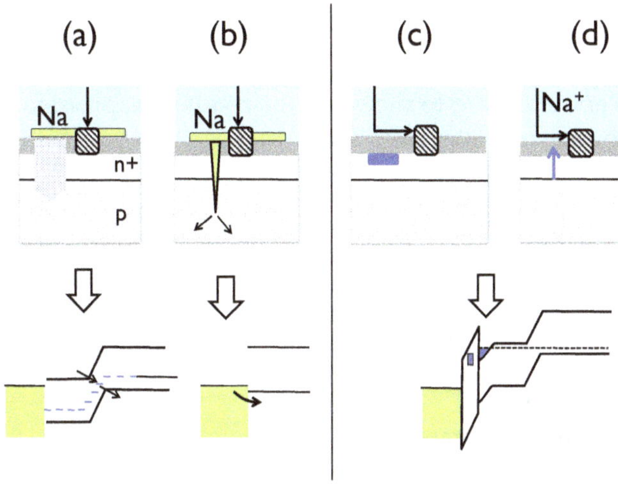

Figure 22.5: PID perturbs the junction either by introducing recombination centers (a, b) or through electrostatic interactions (c, d). The bulk and junction recombinations are shown in (a), while the shunt formation is shown in (b). The trapping of the charges in SiN passivation layer shown in (c) or electrostatic accumulation of the charges in the emitter (d) have been attributed to PID degradation as well.

in Fig. 22.5(a). In addition, Na may diffuse into the stacking faults. Once Na concentration becomes sufficiently high, it destroys the junction locally and behaves as a metallic contact to the base, as shown in Fig. 22.5(b). We recall from Chapter 9 that this is precisely how shunts are formed. The junction recombination and the shunt formation increase dark current of a solar cell, which in turn reduces the fill factor and the open-circuit voltage of the affected cell. We will call this phenomenon PID shunt (or PID-s for short).

Homework 22.5: Cell efficiency decreases with reduced shunt resistance

We recall from Fig. 7.1(d) that an ideal solar cell is represented by a current source in parallel with a diode, with zero series resistance ($R_s = 0$) and very large shunt resistance ($R_{sh} \to \infty$). For this ideal diode, show that the loss of output power due to PID-s is given by

$$\frac{\Delta P(t)}{P_0} = \frac{R_L(t=0)}{R_{sh}(t)}, \tag{22.6}$$

where R_L is a load that allows the cell to operate at the maximum power point. Moreover, if the time-dependent evolution of shunt resistance is written in the form

$$R_{sh}(t) = R_{sh}(t=0) / \left(1 + \int_0^t R_D(u)du\right). \tag{22.7}$$

Use Eq. (22.2) and Eq. (22.6) to show that the resulting expression has a similar asymptotic form to that of Eq. (22.1).

Solution. If the series resistance R_s is negligible, the maximum power point of the current source and the ideal diode combination must be satisfied at all times by the parallel combination of the load and the shunt resistances, i.e.,

$$\frac{V_{mp}}{I_{mp}} = R_L(t=0) = \frac{R_L(t) \cdot R_{sh}(t)}{R_L(t) + R_{sh}(t)}.$$

Moreover, since shunt resistance is the only source of power loss, therefore

$$\frac{\Delta P(t)}{P_0} = \frac{V_{mp}^2 / R_{sh}(t)}{V_{mp}^2 / R_L(t=0)} = \frac{R_L(t=0)}{R_{sh}(t)}.$$

Next, let Eq. (22.7) be a linear approximation of the more general equation:

$$R_{sh}(t) = R_{sh}(t=0) \times e^{-\int_0^t R_D(u)du} \approx R_{sh}(t=0) \left[1 + \int_0^t R_D(u)du\right]^{-1}.$$

Since $\Delta P(t) = P_0 R_L(t=0)/R_{sh}(t)$ by Eq. (22.6), the ratio of shunt resistances at two different times is given by

$$\frac{\Delta P(t)}{\Delta P_\infty} = \frac{R_{sh}(\infty)}{R_{sh}(t)} = \left[\frac{R_{sh}(t=0)}{R_{sh}(\infty)} \times e^{-\int_0^t R_D(u)du}\right]^{-1}.$$

(continued on the next page)

Homework 23.10 (*continued from the previous page*)

If the degradation begins with a delay t_i, the expression is rewritten in a slightly general form:

$$\frac{R_\mathrm{sh}(t)}{R_\mathrm{sh}(\infty)} = \left[1 + \left(\frac{R_\mathrm{sh}(t=t_\mathrm{i})}{R_\mathrm{sh}(\infty)} - 1\right) e^{-\int_{t_\mathrm{i}}^{t} R_\mathrm{D}(u)du}\right]^{-1}. \qquad (22.8)$$

With $\Delta P(t) \propto R_\mathrm{sh}^{-1}$ and time-independent R_D, we recover Eq. (22.1).

22.6.2 Positive Na^+ ions attract negative charges leading to surface polarization

Until being discharged by the electrode, the positively charged Na^+ ions accumulate close to the electrode, as shown in Figs. 22.5(c, d). The positive ions $\rho(x, y)$ induce negative charges within the n^+ layer of the diode (Fig. 22.5(d)) and a perpendicular electric field ($\mathcal{E}(z) = \kappa_\mathrm{SiN}\epsilon_0\rho(x, y)$ develops across the SiN dielectric layer. If $\mathcal{E}(z)$ is sufficiently high, charges may be injected and trapped into the SiN layer (see Fig. 22.5(c)). For an n^+-p diode, the increase in majority carrier close to the interface increases surface scattering and reduces mobility, but it does not affect the performance of the solar cell significantly.

Homework 22.6: Surface polarization, emitter depletion, and increased recombination in a p^+-n cell

PID degradation and its mitigation strategies depend on the solar cell structure. The importance of PID degradation was first highlighted in an "interdigitated back-contact" (IBC) solar cell, see R. Swanson *et al.*, "Preventing Harmful Polarization of Solar Cells," US Patent 7,554,031 B22009. Read the patent carefully to explain how PID in traditional $\mathrm{n}^+ - \mathrm{p}$ solar cells discussed above compares to the PID $\mathrm{p}^+ - \mathrm{n}$ IBC solar cell discussed in the patent. Which of the mitigation strategies can still be used in traditional front-contacted solar cells?

22.7 Solution strategies: How to reduce potential-induced degradation

The discussion above suggests that PID can be suppressed either by reducing the leakage current, or by reducing the effects of leakage that do occur, or a combination thereof. Several methodologies have been developed, although their cost-effectiveness is not fully understood.

1. Quartz, borosilicate glass, and aluminosilicate glass have only 3–6% Na compared to conventional soda–lime glass with 16% Na. These unconventional glasses have high resistance to electron flow. If the soda–lime glass is replaced by these alternate glasses, leakage is reduced and so does the PID.

2. One can use ultra-high resistance encapsulants, such as silicone, polyolefins, or ionomers. Since μ_{EVA} is small, ions and electron transport are suppressed and PID is reduced.

3. Another option is to insert a high conductivity transparent layer at the glass/encapsulant interface. The layer will neutralize that ions as soon as they leave glass and the current will be shunted laterally away from the solar cells.

4. Modification of the SiN layer can also be effective. An Si-rich SiN layer neutralizes the ions and makes (neutral) ion diffusion through SiN layer difficult. A two-layer stack, composed of Si-rich SiN close to the cell and the standard SiN layer on the encapsulant side, ensures that PID resistance is obtained without compromising the anti-reflection properties of the SiN layer.

22.8 A semi-quantitative model for PID

The physics of PID is well understood, and we can deterministically suppress PID based on the understanding, as discussed in the previous section. Given the complexity of the various processes involved, however, a quantitative model that can precisely predict the PID degradation following outdoor environmental exposure is still lacking. Instead, researchers use empirical models (e.g., Eq. (22.2)) with technology-specific calibrated parameters to predict PID degradation of various modules installed in different climatic conditions. The goal of the section is to phenomenologically justify the appearance of various terms in the empirical equation.

22.8.1 Na transport through EVA controls PID-s

In order to understand the appearance of voltage exponent n, activation energy E_{A}, relative humidity [RH], and onset delay t_{i}, let us reasonably assume that Na transport through the polymer is the rate-limiting process. Once the Na reaches the stacking fault, it quickly diffuses through the fault to form the shunt.

22.8.2 Physics of t_i: Delayed onset of PID-s

If the mobility of the Na ions in the polymer is μ_{p} and the electric field across the polymer is $\mathcal{E} = V_{\mathrm{p}}/L_{\mathrm{p}}$, then ions drift at a velocity $v = \mu_{\mathrm{p}}\mathcal{E}$. The time taken for these ions to cross the polymer layer (and eventually damage the p-n junction) is

$$t_i = \frac{L_p}{\mu_{\mathrm{p}}\mathcal{E}} = \frac{L_p^2}{\mu_{\mathrm{p}}V_{\mathrm{p}}}. \tag{22.9}$$

This explains the delayed onset of PID for $t < t_i$. Once the Na ions reach the cell electrode, PID voltage exponent n is determined by the voltage and temperature dependence of the leakage current through the EVA, as follows.

22.8.3 Case 1: Linear voltage dependence of PID-s

The number of Na atoms ionized within glass, p_0, is determined by the electric field across it, namely, $\mathcal{E}_g = V_g/L_g$, where V_g is the voltage drop across the glass, and L_g is the glass thickness. If p_0 is low or if Na mobility in polymer encapsulant (μ_p) is high, then the PID leakage is given by the drift current, associated with transport of stored charge $Q_p (= q p_0 L_p)$ every t_p seconds (see Fig. 22.6), i.e.,

$$J_{Na} = Q_p/t_p = q p_0 \mu_p V_p / L_p. \tag{22.10}$$

Unlike a semiconductor or a metal, ion transport in a polymer involves repeated trapping into and release from localized energy states (with potential depth E_A). Since the release rate from the localized states increases exponentially with temperature, therefore

$$\mu_p = \mu_0^* e^{-E_A/k_B T}.$$

The mobility increases significantly in the presence of moisture which ingresses within the polymer during periods of high relative humidity, RH. The actual physics involves a complex process involving moisture-induced suppression of the localized trapping states. Phenomenologically, the process can be described by an RH-dependent increase of the prefactor, μ_0^*, i.e.,

$$\mu_0^* = \mu_0 [\mathrm{RH}]^B.$$

Combining with Eq. (22.10), the degradation rate, R_D, scales with the Na flux:

$$R_D \propto J_{Na} = A\, e^{-E_A/k_B T} [\mathrm{RH}]^B V_p \tag{22.11}$$

where $A \equiv q p_0 \mu_0 / L_p$ is a technology-dependent prefactor. The prefactor can be reduced by using low-Na glass (p_0) or increasing the polymer thickness (L_p), and using a new encapsulant with lower intrinsic mobility (μ_0). If $V_p \propto V_{max}$, the overall PID voltage dependence would also be linear (i.e., $n = 1$), see Eq. (22.2).

22.8.4 Case 2: Nonlinear voltage dependence of PID-s

If Na ionization in the glass p_0 is high and/or if the polymer mobility μ_p is low, then the voltage drop across the polymer is no longer linear, as shown in Fig. 22.6(b). Even in this case, the leakage current through the polymer is easily calculated by the same charge transport formula $J_{Na} \approx Q_p/t_p$, where Q_p is related not to the amount of Na^+ ionized, but to the amount that can be transported by the low-mobility polymer, $Q_p = C_p V_p$. If the cell and the module frame define a parallel

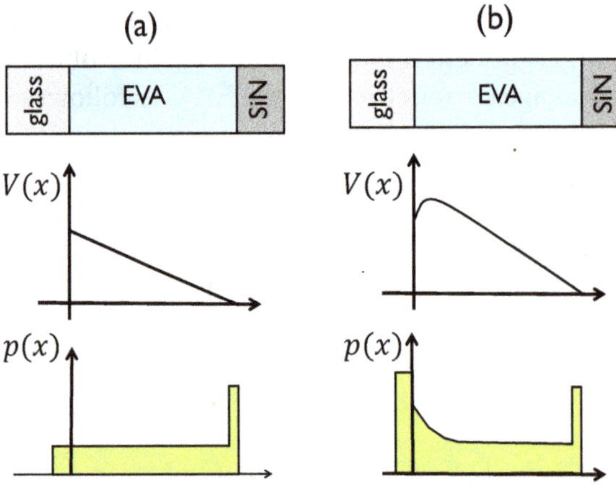

Figure 22.6: Na^+ ion-drift through the encapsulant from (a) low-Na glass vs. (b) high Na glass. The corresponding profiles of potential $V(x)$ and Na concentration $p(x)$ are shown. In both cases, charges build up next to the SiN layer (on the right).

plate configuration, then $C_p = \kappa_s \epsilon_0 \mathcal{E}_p = \kappa_s \epsilon_0 V_p / L_p$, therefore

$$Q_p = C_p V_p = \kappa_s \epsilon_0 V_p^2 / L_p.$$

Since $t_p = L_p^2 / \mu_p V_p$, therefore, the space charge limited ion current is

$$J_{Na} = Q_p / t_p = c\, \kappa \epsilon_0 \mu_p \frac{V_p^2}{L_p^3}.$$

You may realize that this is the Mott–Gurney law, originally derived in Sec. 9.3 the context of space charge limited shunt conduction. Although our simple derivation gives $c = 1$, a more precise calculation would have given $c = 9/8$. Inserting back the dependence of μ_p on T and RH (i.e., $\mu_p = \mu_0 e^{-E_A/k_B T_D} [\mathrm{RH}]^B$), we find that

$$J_{Na} = A\, e^{-E_A/k_B T} [\mathrm{RH}]^B\, V_p^2 \tag{22.12}$$

where $A \equiv \kappa \epsilon_0 \mu_0 / L_p^3$ is a technology-dependent prefactor, which can be reduced by thicker (and/or lower mobility or lower dielectric constant) encapsulant. If $V_p \propto V_{max}$, the overall PID will be dependent quadratically on voltage ($n = 2$). In fact, $n > 2$ is possible if the trapping states that have multiple distribution of activation energies, E_A, see Eq. (9.33). Equation (22.12) explains the nonlinear voltage dependence of PID observed in some experiments, consistent with Eq. (22.2).

Finally, note that Eq. (22.12) does not depend on p_0, the Na concentration in glass. When the voltage is turned on, the number of ions injected (F_{inj}) is too large to be transported by a steady-state Mott–Gurney flux, (F_{out}). The flux imbalance

$$\frac{d\rho}{dt} = F_{inj} - F_{out}$$

creates an ion build-up at the injection point, shown in the plot for $p(x)$ in Fig. 22.6(b) at $x = 0$. The corresponding potential barrier in $V(x)$ throttles the ion flow, to allow only a small fraction of Na ions to escape over the virtual electrode. The excess field reduces the transit time slightly, as follows:

$$t_{\mathrm{i}} = \frac{L_{\mathrm{p}}^2}{\mu_{\mathrm{p}}(V_{\mathrm{p}} + \Delta V)}. \tag{22.13}$$

This shortening of transit time is important at low voltages, but not necessarily for voltages relevant for PID.

Homework 22.7: PID leakage current in practical module geometry

The discussion above introduces a powerful way to calculate space charge limited leakage current in complex structures. In practice, the module frame can be viewed as an infinite strip placed at a distance L_{p} above an infinite two-dimensional plane defined by the electrodes of the solar cell. You may want to look up the expression for C_p for such a structure and recalculate the leakage current. How does the new formula differ from that derived using the parallel plate assumption? Would an experimentalist notice the difference based on the leakage current measurement made?

22.8.5 The physics of PID saturation

Once the atoms traverse the SiN interface and are neutralized, they will diffuse through the stacking faults with diffusion coefficients, D_{S}. It takes approximately, $T_{\mathrm{D}} \equiv W^2/2D_{\mathrm{S}}$ for the atoms to reach the junction, where W is the thickness of the emitter. The junction concentration $p(W,t)$ is related to SiN concentration $p(0)\left(= \int_{t_{\mathrm{i}}}^{t} dt\, J_{\mathrm{Na}}\right)$ by

$$\frac{p(0)}{\sqrt{2D_{\mathrm{S}}t}} = \frac{p(W,t)}{\sqrt{2D_{\mathrm{S}}t} - W}. \tag{22.14}$$

Therefore, the ion build-up within the stacking fault is given by the

$$p(W,t) = \int_{t_{\mathrm{i}}+T_{\mathrm{D}}}^{t} dt\, J_{\mathrm{Na}} \times \left[1 - \sqrt{\frac{T_{\mathrm{D}}}{t}}\right] = p(0)\left[(t - t_*) - \frac{\sqrt{2}W}{\sqrt{D_S}}\left(\sqrt{t} - \sqrt{t_*}\right)\right] \tag{22.15}$$

with $t_* \equiv t_{\mathrm{i}} + T_{\mathrm{D}}$. Once the $p(W,t) \approx p_{sat}$, the saturation density of Na atoms that can be accommodated in the stacking fault, no further increase in concentration is possible. The junction is shorted and the shunt formation is complete. The time-dependent formation of shunt (with onset delay, with a superlinear increase in current, followed by a saturation) is reflected in the time dependence of the power loss.

22.8.6 Na^+ ions also increase the diode dark current

We have already noted that Na is a fast diffuser in n^+ silicon. Therefore, in addition to reducing the shunt resistance discussed above, we recall from Fig. 22.5(a) that an isolated Na atom introduces a donor level at approximately 0.77 eV below the conduction band. As a result, Na adds to the bulk and junction recombination and reduces the overall efficiency. The bulk recombination current is given by the diode equation with ideality factor $n = 1$, namely,

$$J_{\text{dark}} = J_0(e^{qV/k_\text{B}T} - 1) \tag{22.16}$$

where $J_0 \equiv qn_\text{i}^2\left(\sqrt{D_\text{n}/\tau_n} + \sqrt{D_p/\tau_p}\right)$, with minority carrier lifetime $\tau_n \equiv (\sigma n_\text{t} v)^{-1}$. Here, n_t proportional to Na concentration, $p(W, t)$, σ is the capture cross-section, and v is the thermal velocity of electrons.

Similarly, the diffusion of Na into the junction increases *junction recombination.* The junction recombination is given by the diode equation with ideality factor $n = 2$, i.e.,

$$J_R = q\left(\frac{k_\text{B}T\, W_J}{V_\text{bi} - V}\right)\frac{n_\text{i}}{\tau}\left(e^{qV/2k_\text{B}T_\text{D}} - 1\right) \tag{22.17}$$

where $\tau = (\sigma n_\text{t} v)^{-1}$ proportional to Na concentration at the junction and W_J is the p-n junction depletion width. The bulk and junction recombination currents increase slowly over time.

22.8.7 Recovery of PID-s due to Na out-diffusion

The Na atoms that create the shunts and increase junction/bulk leakage current are not permanently bonded to the lattice. Therefore, PID-s can recover during the night when the farm is turned off and PID voltage supporting the Na flux is removed. The ions within the shunt ($p(W, t)$) given by Eq. (22.15) decreases as Na atoms diffuse back into the polymer. An intentionally applied reverse-PID voltage can accelerate the recovery. This recovery phenomenon increases the PID lifetime of the solar modules.

Homework 22.8: Deep donors, such as Na in silicon, cannot affect device doping

There is a common misconception that Na ions can increase the n^+ emitter doping or compensate the p base doping of an n^+-p solar cell. Explain why this cannot occur. Repeat the argument for a p^+-n solar cell.

22.9 Conclusions: PID is a system-level reliability issue

In this section, we have discussed PID as an important electrical reliability concern for large-scale, high-voltage PV systems. Experiments have correlated leakage current related to Na ions as being the most important contributor to PID. Therefore, PID is suppressed by suppressing the ion-leakage current. For example, PID is reduced by reducing $V_{\max}$, or increasing of encapsulant/glass resistance, etc. Due to its exponential dependence on temperature, PID accelerates when the module temperature rises during the noon, and it recovers during the night when the temperature is cooler and system is turned off. It is important to realize that PID not only reduces the power output, but the hotspots formed (centered on the shunts) can accelerate other degradation modes. One of such degradation modes is moisture-induced corrosion of the electrical grids. We will discuss the physics of corrosion in the next chapter.

References

[1] R. Swanson, M. Cudzinovic, D. DeCeuster, V. Desai, Jörn Jürgens, N. Kaminar, W. Mulligan, L. Rodrigues-Barbarosa, Doug Rose, D. Smith, A. Terao, and K. Wilson. The surface polarization effect in high-efficiency silicon solar cells, page 410, January 2005.

[2] Wei Luo, Yong Sheng Khoo, Peter Hacke, Volker Naumann, Dominik Lausch, Steven P. Harvey, Jai Prakash Singh, Jing Chai, Yan Wang, Armin G. Aberle, and Seeram Ramakrishna. Potential-induced degradation in photovoltaic modules: A critical review. *Energy & Environmental Science*, 10(1):43–68, January 2017. Publisher: The Royal Society of Chemistry.

[3] Volker Naumann, Dominik Lausch, Angelika Hähnel, Jan Bauer, Otwin Breitenstein, Andreas Graff, Martina Werner, Sina Swatek, Stephan Großer, Jörg Bagdahn, and Christian Hagendorf. Explanation of potential-induced degradation of the shunting type by Na decoration of stacking faults in Si solar cells. *Solar Energy Materials and Solar Cells*, 120:383–389, January 2014.

[4] G. J. M. Janssen, Maciej Stodolny, B.B. van Aken, J. Loffler, Ma Hongna, Zhang Dongsheng, and Shi Jinchao. *Cell modifications for preventing potential-induced degradation in c-Si PV systems*. Petten: ECN, 2017.

[5] Murray A. Lampert. Simplified Theory of Space-Charge-Limited Currents in an Insulator with Traps. *Physical Review*, 103(6):1648–1656, September 1956. Publisher: American Physical Society.

CHAPTER 23

Humid Environment Causes Electrode Corrosion

Chapter Summary

- Moisture-assisted electrode corrosion is an important source power-loss in a solar module.
- There are two types of corrosion: dark and light.
- Dark corrosion is accelerated by the acetic acid generated by EVA-moisture reaction.
- Light corrosion involves hydrolysis of water and erosion of the electrodes.
- In c-Si cells, corrosion does not increase series resistance, but it reduces power by reducing the collection of photo-current.

23.1 Introduction: The corrosion of electrodes reduces power output

When a module is installed in a hot and humid environment or stressed in a chamber with high temperature (T) and high relative humidity ($[\mathrm{RH}]\%$), the module power loss (i.e., $\Delta P(t) \equiv P_0 - P(t)$) increases as shown in Fig. 23.1. Similar to PID (Eq. (22.1)), a simple empirical formula describes the time-, temperature-, and $[\mathrm{RH}]$-dependent power loss of a module:

$$\frac{\Delta P(t)}{\Delta P_\infty} = \left[1 + \left(\frac{\Delta P_\infty}{\Delta P(t_i)} - 1\right) e^{-(t-t_i)R_\mathrm{D}}\right]^{-1}. \qquad (23.1)$$

Here, $\Delta P(t_i)$ and ΔP_∞ are the initial and saturation power losses, respectively. The degradation constant (R_D) is empirically given by Peck's equation:

$$R_\mathrm{D} = K e^{-E_\mathrm{A}/k_\mathrm{B}T} [\mathrm{RH}]^B, \qquad (23.2)$$

where K is a module-specific fitting parameter, $E_\mathrm{A} \approx$ 0.8–1 eV is the activation energy, and $B \approx 2$ is the relative humidity exponent, and stress time t is expressed

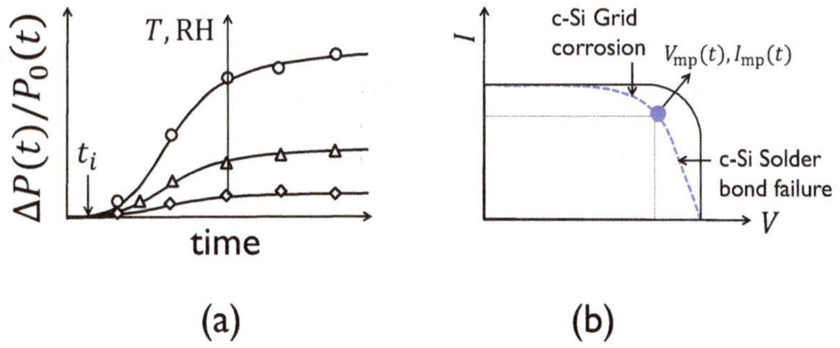

Figure 23.1: (a) Module power loss in a hot and humid weather depends on RH and temperature. (b) The loss of fill factor indicates that the power loss is associated with (a) the increase in series resistance due to solder bond failure in c-Si solar cells and/or *apparent* increase in shunt resistance due to corrosion of front metal grids of c-Si cells. For thin-film solar cells, the interconnect corrosion increases only the series resistance, R_s. The series and shunt resistances can be extracted from the slopes of the module I-V characteristics, see Homework 23.5.

in hours. These activation energy and humidity exponent are different from those of PID. Given this power loss, the module lifetime (t_f) would be

$$t_f = Ae^{E_A/k_B T}[\text{RH}]^{-B} \tag{23.3}$$

where A is a constant that depends on the percentage of power loss (e.g., 10% or 20%) that can be tolerated before the module must be replaced.

In this chapter, we will answer the following questions related to moisture-assisted corrosion of metal grids:

1. Why does corrosion depend on temperature and relative humidity? What are the physical interpretations of the activation energy E_A and humidity power exponent B? How do these parameters compare to those of PID?

2. The formula for PID and corrosion are apparently similar. Is this accidental, or is there a physical reasoning?

3. Corrosion increases the local specific resistance of the front metal grids of a c-Si cell. Why doesn't the increased local resistance increase the module series resistance, indicated by the slope of the I-V curves at V_{oc}, i.e., $R_s = (dV/dI)_{V_{oc}}$?

4. How does the non-uniform corrosion lead to localized self-heating (hotspot formation) and increased degradation of a solar module?

5. How can we reduce or suppress corrosion?

Homework 23.1: Similar empirical equations describe power loss in a hot/humid environment vs. power loss under PID stress conditions

Equation (23.2) is apparently similar to Eq. (22.2). Highlight the similarities and differences. Also, can we distinguish between PID and corrosion by extracting the series and shunt resistances from the time-dependent I-V characteristics?

Homework 23.2: An example problem explains the calculation of corrosion-limited lifetime

In the 2016 paper "Global Acceleration Factors for Damp Heat Tests of PV Modules," G. Kimball *et al.* report results from accelerated test of SunEdison modules. The modules consist of p-type wafers, packaged with glass in the front, and polymer sheet in the back. They find that the module lifetime can be described by the following formula:

$$t_{\mathrm{f}} = A\, e^{E_{\mathrm{A}}/k_{\mathrm{B}}T} [\mathrm{RH}]^{-B} \tag{23.4}$$

where t_{f} is calculated in hours and RH is the relative humidity percent. Other symbols have their usual meaning. Analyzing the data, they find that $E_{\mathrm{A}} = 0.89 \pm 0.11$ eV and $B = 2.2 \pm 0.8$. The prefactor A depends on the magnitude of power loss. For example, $A = 3.7 \times 10^{-10}$ for 5% P_{max} loss; 5×10^{-10} for 10% P_{max} loss; and 6.4×10^{-10} for 20% P_{max} loss. Based on the information provided, calculate the corrosion-limited lifetime for any four cities in the world. You can find the information regarding the temperature and relative humidity from the internet.

Solution. For RH $= 50\%$, $E_{\mathrm{A}} = 0.89$ eV, $T = 300$ K, $B = 2.2$ and $A = 6.4 \times 10^{-10}$, we find that $t_{\mathrm{f}} = 4.15 \times 10^{6}$ hours, or equivalently, greater than 400 years (8760 hours make a year). Before you begin to boast that your module is never going to fail, remember that even a modest self-heating (e.g., 30 K, see Chapter 5) could reduce the corrosion-limited module lifetime to just 20 years! In practice, the lifetime reduction is not so dramatic. Self-heating reduces local RH around the module to suppress moisture ingress. For the moisture that do seep in, self-heating decreases the moisture concentration close to the metal grid by spreading it within the module.

23.2 Corrosion involves a sequential diffusion–reaction process

The loss of power in a hot and humid environment given by Eq. (23.1) is generally attributed to moisture-induced corrosion of the metal electrodes. Moisture may enter the module in several ways, see Fig. 23.2: (1) microcracks in the glass or imperfect edge sealing between the frame and the glass, (2) the plug that seals the module opening needed to bring out the electrical wiring, and/or (3) diffusion through the (possibly UV-degraded) backsheet. Subsequently, moisture would diffuse through the polymer, and eventually corrode the metal electrodes.

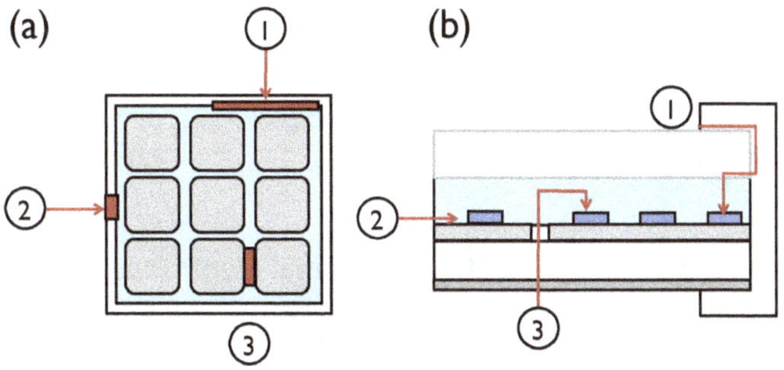

Figure 23.2: Various pathways for moisture ingress that lead to corrosion and delamination of metal grids. (a) Top view. (b) Side view.

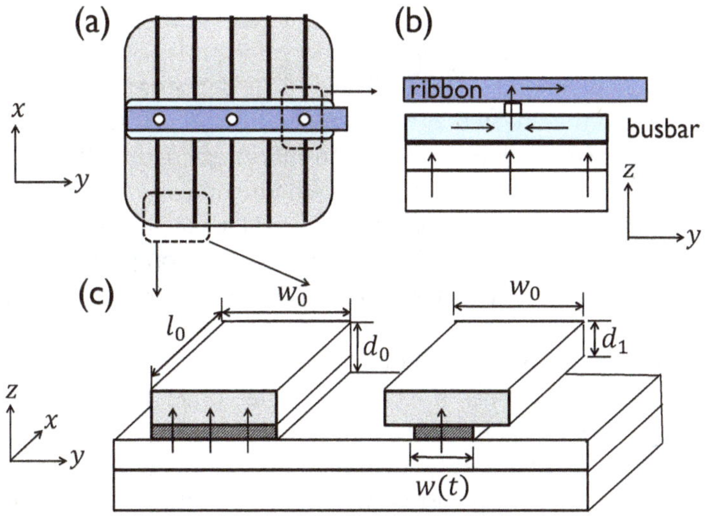

Figure 23.3: Increase in series resistance due to metal corrosion and solder bond failure. (a) Top view of a module with metal fingers, busbar, and metal ribbons. (b) The busbars are connected to ribbons by solder bonds. (c) Sideview of an uncorroded finger (left) vs. corroded finger (right). Corrosion reduces the cross-section of the finger and/or delaminates the finger by dissolving the metal–semiconductor interface.

23.2.1 Diffusion barriers and breakthrough time

Low diffusivity glass, backsheet, edge seals, and plug act as the first line of defense to prevent moisture ingress into the module and to prolong the breakthrough/incubation time, i.e.

$$t_i = \xi L^2/KD$$

where L and D are respectively the thickness and the diffusion coefficient of the barrier region being considered. The factor K is determined by the limit of moisture ingress (i.e. K = 2, for 50%; K = 18.3 for 5% limit). The factor ξ depends on the ambient temperature, daytime operating temperature, duration of night and day at a given geographical locations. Naturally, developing new materials for these diffusion barriers is a topic of broad interest.

23.2.2 Solar cells need special types of metal electrodes

To understand how metal grids corrode, it is important to understand that the front grid of a solar module requires special low-temperature processing. Instead of using a solid Ag, solar cells use Ag paste that can be screen-printed and processed at a lower temperature. The paste is composed of a mixture of inorganic particles, such as Ag powder for adequate line conductivity, glass frit for electrical contact to the Silicon surface and for optical index control, and other additives to control adhesion to the substrate. The inorganic particles are dispersed into a mixture of organic resin and solvents. The paste is deposited on the solar cells by screen printing and then heated at high temperature ($\sim$ 600°C) for a few seconds to allow the metal to diffuse through the anti-reflection coating. In the process, the organic resin evaporates, leaving behind a conducting Ag metal film bound together by various additives. Corrosion often involves moisture reacting not with the metal, but with the additives. As a result, the corrosion process leaves behind a highly resistive 3D perforated structure that makes photocurrent collection difficult.

In the following section, we will discuss the physics of two types of corrosion that leads to the four types of corrosion-related phenomena discussed above. Dark corrosion is a chemical process that occurs whether the module is producing power or not (although the rate of degradation does depend on power output). Light-induced electrochemical corrosion occurs only when the module is illuminated and the voltage induced in the front contact electrodes can hydrolyze water. In the following discussion, we will assume vertical moisture-ingress through cracked-glass or damaged backsheet. The lateral water ingress through the edge seal or plugs have slightly different diffusion geometry, but essentially identical functional dependence.

23.2.3 Metal corrosion and charge collection

The metal corrosion affects photogenerated charge collection in four ways.

1. **Reduced cross-section.** The electrode cross-section reduces as metal ions are dissolved by moisture-related reactions. The increased resistance prevents photocurrent collection from the semiconductor.
2. **Metal/semiconductor delamination.** Moisture-assisted reactions dissolve the interface between metal/semiconductor. The delaminated metal contacts will not collect current and will reduce power output.
3. **Metal/EVA delamination.** As the electrode cross-section is reduced, EVA may delaminate from the metal. This delaminated section will reflect sunlight and reduce photocurrent.
4. **Solder bond failure.** The solder bonds connect the ribbons to the busbar. A temperature-induced failure of the solder bond makes charge collection difficult. Solder bond failure is typically caused by mechanical stress induced by thermal cycling (see Chapter 24), but corrosion can play a secondary role in weakening the bonds.

23.3 Physics of dark corrosion that persists during daytime

23.3.1 Dark corrosion involves acids produced by moisture-EVA reaction

Figure 23.4(a) shows one of the ways the metal electrode (finger or busbar) may corrode due to moisture ingress. We recall from Chapter 20 that commercial modules are encapsulated by a polymer called EVA (ethylene-vinyl acetate, $(C_2H_4)_n(C_4H_6O_2)_m$), where n and m represent the relative fractions of the chemical species in the polymer. As moisture (H_2O) diffuses through the encapsulant and reacts with a molecule of ($C_4H_6O_2$) within the EVA polymer chain, it produces a molecule of acetic acid (CH_3COOH), and leaves behind a molecule of ethanol (C_2H_3OH). The acetic acid diffuses through the polymer to reach the metal electrode. The contact resistance ($R_{S,c}$) increases as the acid dissolves the glass frit between metal and the semiconductor. The reaction proceeds even when a module is turned off, at night for example.

23.3.2 A phenomenological model for dark corrosion

As shown in Fig. 23.4, the dark corrosion process involves four steps: diffusion of water (W), reaction with EVA to produce acetic acid (A), diffusion of acetic acid, and reaction with the metal electrode to reduce the contact thickness by ΔW. The coupled system of equations is easy to solve; however, a simpler analytical model offers considerable insight.

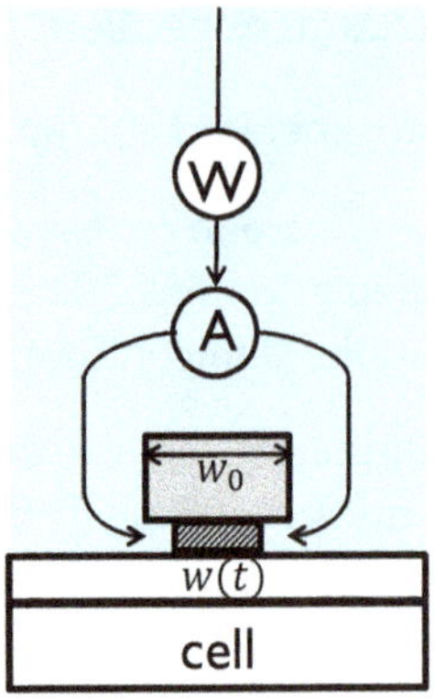

Figure 23.4: Side view of a finger/busbar subjected to dark corrosion. Dark corrosion depends on water (W) reacting with EVA to produce acetic acid (A). The acetic acid corrodes (dissolves) the metal–semiconductor interface leading to electrode thinning as well as grid delamination.

The width of glass frit layer corroded ($\Delta w(t) \equiv w_0 - w_t$, see Figs. 23.3 and 23.4) is proportional to the concentration of acetic acid produced at a time t, $a(t)$. Since $\Delta P(t) \propto \Delta w(t)$ (due to fake shunting, see below), therefore, the power loss is directly proportional to $a(t)$. Also, as EVA offers large number of reaction sites, acetic acid production is controlled by the amount of water present within the EVA, $w(x,t)$ (not to be confused with the loss of finger width $\Delta w(t)$). This is easily obtained by integrating the amount of water that will have diffused a distance $\sqrt{D_{\mathrm{w}} t}$ within the module given the moisture ingress duration, t. In other words,

$$a(t) = k_a \int w(x,t)dx = k_a \, [\mathrm{RH}] \sqrt{D_{\mathrm{w}} t}$$

where $D_{\mathrm{w}} = D_0 e^{-E_{\mathrm{A,w}}/k_{\mathrm{B}}T}$ is the diffusion coefficient of moisture in EVA. Once $a(t_{\mathrm{f}})$ reaches a critical concentration, the interface corrosion would increase the series resistance and reduce the power output. Rewriting the expression, we find that

$$t_{\mathrm{f}} = A[\mathrm{RH}]^{-B} e^{E_{\mathrm{A,w}}/k_{\mathrm{B}}T}.$$

Here, the constant $A \propto D_0^{-1}$. This analysis provides a phenomenological justification of Eq. (23.3). In particular, it explains that the moisture exponent ($B = 2$) arises as a consequence of the diffusion physics of water within the polymer. In practice, the water diffusion in a polymer does not follow Fick's law of diffusion, because a water molecule repeatedly reacts/binds to the polymer matrix as it is diffusing through the EVA. The anomalous diffusion leads to $B > 2$. The more general numerical solution of the coupled equations identifies the functional dependence of the prefactor A in terms of the diffusion coefficient of the acetic acid, corrosion/dissolution rate of the interfacial layers, etc.

Homework 23.3: The exact formula for moisture ingress is easily obtained

In Sec. 23.3.2, we used an approximate expression for the moisture ingress within the polymer to derive an expression for the corrosion-limited lifetime of a solar module. Show that the exact expression of moisture-ingress for a EVA of thickness L is given by the following analytical formula involving complementary error-functions:

$$w(x,t) = \frac{[\mathrm{RH}]}{1+\mathrm{erfc}(2L/2\sqrt{D_{\mathrm{w}}t})} \cdot \left(\mathrm{erfc}\left(\frac{x}{2\sqrt{D_{\mathrm{w}}t}}\right) + \mathrm{erfc}\left(\frac{2L-x}{2\sqrt{D_{\mathrm{w}}t}}\right)\right).$$

23.3.3 Strategies to suppress dark corrosion

Dark corrosion can be reduced by suppressing any of the four steps involved: ingress/diffusion of water, reaction to produce acetic acid, diffusion of acetic acid, and reaction with the metal electrode. For example, a well-made frame and a high-quality backsheet reduce moisture ingress. When EVA is replaced by silicone, acetic acid will not be produced. These predicted improvements have indeed been confirmed experimentally.

23.4 Physics of light-induced corrosion

23.4.1 Light corrosion involves moisture hydrolysis by the electrodes

We know from our discussion on potential-induced degradation (Chapter 22) that a series-connected module reaches a very high voltage (with respect to the frame). Once moisture diffuses through the encapsulant and arrives at the electrode surface, it picks up an electron, see Fig. 23.5. In other words, the electrode hydrolyzes the water molecule by supplying an electron.

The pH change associated with (OH^-) ions dissolves metal additives (MOH^-), with gradual thinning of the metal fingers and corresponding increase of finger/busbar resistance. In addition, the H_2 gas may accumulate at the metal/semiconductor interface and the pressure build-up can delaminate the grid, with the corresponding increase in the contact resistance.

Light corrosion proceeds only if the electrochemical circuit is complete. Once water is hydrolyzed after picking up an electron from the metal, MOH^- is subsequently neutralized by PID-induced Na^+ ions arriving on the electrode surface, see Chapter 22 for details. The electrons left behind by Na^+ atoms then complete the circuit by sequentially flowing through the frame, the ground contact, the absorber, and finally the top electrode. Simply put, the electrochemical circuit involves two fluxes: moisture diffusion/hydrolysis and drift flux of Na^+ ions. Let us first develop a simple model for a system where the water diffusion/hydrolysis is the rate-limiting process.

23.4.2 A phenomenological model for light corrosion: Water hydrolysis

The hydrolysis begins once moisture reaches the metal surface, so that the diffusion flux through the polymer of thickness L_p is described by the diffusion equation, namely,

$$J_1 = D_\mathrm{w}(\mathrm{RH} - C)/L_\mathrm{p} \tag{23.5}$$

where RH is the ambient moisture concentration and C is the moisture concentration at the electrode. The water hydrolysis involves a complex reduction-oxidation process, approximately given by

$$J_2 = k_\mathrm{s} C. \tag{23.6}$$

For quasi-steady-state hydrolysis, $J_1 = J_2 \equiv J_{ss}$, therefore,

$$J_{ss} = \frac{D_\mathrm{w} k_\mathrm{s}}{D_\mathrm{w} + k_\mathrm{s} L_\mathrm{p}} \mathrm{RH}. \tag{23.7}$$

The time to failure (t_f) is determined by the integrated-flux needed to corrode the metal fingers by ΔW_max and increase the finger resistance by $\Delta \rho_\mathrm{c}$. In other words,

$$\Delta \rho_c \propto \Delta w_\mathrm{max} \propto J_{ss} t_\mathrm{f} = \frac{D_\mathrm{w} k_\mathrm{s}}{D_\mathrm{w} + k_\mathrm{s} L_\mathrm{p}} \cdot [\mathrm{RH}] \cdot t_\mathrm{f}. \tag{23.8}$$

Note that the diffusion coefficient is given by $D_\mathrm{w} = D_0 e^{-E_{\mathrm{A,w}}/k_\mathrm{B}T}$ and the reaction rate is given by $k_\mathrm{s} = k_0 e^{q(V-E_0)/k_\mathrm{B}T}$, where E_0 is the reaction potential and V is

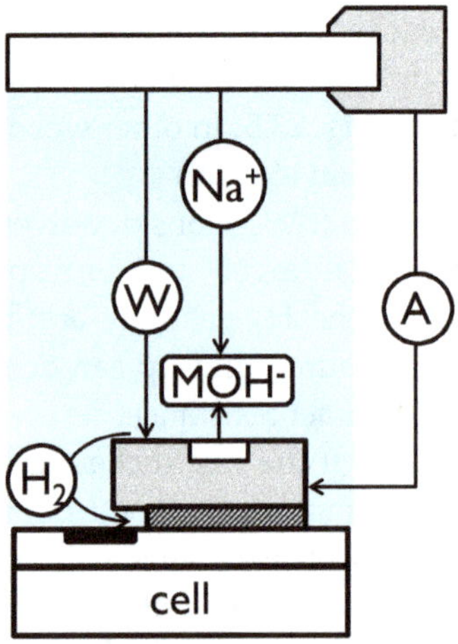

Figure 23.5: Light-induced corrosion involves a sequence of steps including hydrolysis of water.

the electrode voltage. Collecting these expressions, we find that

$$t_{\mathrm{f}} \approx A\, e^{E_{\mathrm{A}}/KT}\,[\mathrm{RH}]^{-B}, \tag{23.9}$$

where $B = 1$ and E_{A} is determined by the rate limiting process between the water diffusion vs. hydrolysis reaction at the metal surface.

23.4.3 A phenomenological model for light corrosion: PID current

At high humidity, hydrolysis can generate sufficient OH^- ions to support the corrosion process. Based on the discussion regarding PID in Chapter 22, we know that the ion flux is given by

$$J_i = \frac{9}{8}\kappa\epsilon_0\mu_{\mathrm{p}}\frac{V_{\mathrm{p}}^2}{L_p^3}, \tag{23.10}$$

where V_{p} and L_p are voltage drop and thickness of the polymer, respectively. Here, $\mu_{\mathrm{p}} = \mu_0 e^{-E_{\mathrm{A}}/k_{\mathrm{B}}T}$. The prefactor μ_0 and activation energy E_{A} depend on the relative humidity [RH]. Once again, t_{f} is determined by the integrated flux needed to corrode the metal fingers by ΔW_{max} and increase the finger resistance by $\Delta\rho_{\mathrm{c}}$. In other words,

$$\Delta\rho_{\mathrm{c}} \propto \Delta w(t) \propto J_i \cdot t_{\mathrm{f}}. \tag{23.11}$$

Therefore, the failure time (t_{f}) depends on the activation energy (E_{A}), device temperature (T_{D}), EVA thickness (L_p), and applied voltage (V_{p}) as follows:

$$t_{\mathrm{f}} \approx A e^{E_{\mathrm{A}}/k_{\mathrm{B}}T_{\mathrm{D}}}\frac{L_p^3}{V_{\mathrm{p}}^2}. \tag{23.12}$$

Here, the activation energy is associated with the trapping level of the ions with the encapsulant. The RH dependence is implicit through E_{A}. This nonlinear voltage dependence of daytime corrosion has been confirmed through multiple experiments.

Homework 23.4: Moisture build-up initiates light corrosion

In the discussion above we have calculated an expression for the steady-state light corrosion. The corrosion cannot initiate until the moisture-front reaches the electrode. Use Homework 23.3 to show that the breakthrough/incubation time, t_i, is related to increase of the moisture at the EVA-metal interface and is given by the following expression:

$$w(L,t) = 2\frac{[\mathrm{RH}]}{1+\mathrm{erfc}(2L/2\sqrt{D_{\mathrm{w}}t})}\cdot \mathrm{erfc}\left(\frac{L}{2\sqrt{D_{\mathrm{w}}t}}\right).$$

23.4.4 Strategies to suppress light corrosion

Light corrosion is suppressed by self-heating of a solar cell exposed to sunlight, see Chapter 5. A heated module reduces RH of the surrounding air, thereby suppressing moisture intake. Moreover, the heated electrodes force the moisture away from the metal grid. In addition, light corrosion may be suppressed if the electrochemical circuit is disrupted. Initially, water hydrolysis limits the rate of corrosion. The strategies that suppress dark corrosion (by reducing water diffusion) are also effective in reducing light corrosion. Note however that replacing EVA by silicone to suppress acetic acid production does not help in this context. Once the hydrolysis rate exceeds the Na^+ flux, light corrosion is dictated by PID current. Any technique that suppresses Na^+ flux (e.g., low-Na glass) or diverts Na^+ flux away from the contact by inserting a conducting plane would improve light corrosion.

23.5 Corrosion does not increase the series resistance

23.5.1 The puzzle defined

You may expect that corrosion-related finger thinning and/or delamination will increase the overall the series resistance of the module, R_s. Recall that R_s is easily measured as the slope of the cell or module I-V characteristics at V_{oc} (see Homework 23.5). We also know that

$$\Delta P(t)/P_0 = \Delta R_s(t)/R_L,$$

where R_L is the optimum load resistance for an ideal solar cell with $R_s = 0$. You could conclude that the increase of $R_s(t)$ due to corrosion would reduce power $P(t)$ and define the lifetime of a module. The argument would be perfectly logical and yet completely wrong! If you planned to catch time-dependent corrosion in the act by measuring the $R_s(t)$, paradoxically you will return empty-handed. Before we explain why, let us first do a quick Homework 23.5 to define R_s and R_{sh} from the module I-V characteristics.

Dark and light corrosion increase the metal resistance associated with the corroded/delaminated metal fingers. Remarkably, the metal grid is so overdesigned and picks up the current in such a way that grid thinning or delamination is reflected in the module I-V characteristics not as $R_s(t)$, but either as (fake) shunt resistance ΔR_{sh} (Fig. 23.6(a)), or as a loss of photocurrent ΔJ_{ph} (Fig. 23.6(b)). Let us explain the surprising statement.

Homework 23.5: Series and shunt resistances from I-V characteristics

Time-dependent degradation of series and shunt resistances can be determined from the cell and module I-V characteristics. Recall that the I-V characteristics of an ideal solar cell is given by a current source in parallel with a diode, without series or shunt resistances. The power output of the diode is maximized with the load $R_L \equiv V_{mp}/I_{mp}$.

1. Show that the I-V characteristics of a practical diode with series and shunt resistance can be written as

$$I = I_{ph} - I_0\, e^{\frac{q(V+IR_s(t))}{k_B T}} - \frac{V + IR_s(t)}{R_{sh}(t)}.$$

2. At $V = V_{oc}$, the diode current is much larger than the shunt current. Neglecting the shunt conduction, calculate the slope of the I-V characteristics at $V = V_{oc}$ to show that

$$\left|\frac{dV}{dI}\right| = \left[R_s(t) + \frac{k_B T}{qI_{ph}}\right] \approx R_s(t).$$

Since $I = 0$ at $V = V_{oc}$, therefore $I_{ph} = I_0 \exp\left(qV_{oc}/k_B T\right)$.

3. At $V = 0$, the diode current is much smaller than the shunt current. Neglecting the diode conduction, calculate the slope of the I-V characteristics at $V = 0$ to show that

$$\left|\frac{dV}{dI}\right| = R_{sh}(t).$$

23.5.2 Finger thinning reflected in an increasing (fake) shunt resistance

Figure 23.6(a) shows the top view of a partially corroded electrode. The initial electrode cross-section is ($W \approx 70$ μm; $H \approx 30$ μm). We recall from Eqs. (10.11) and (10.15) and Fig. 10.8 that as the finger cross-section $f(x)$ is reduced by corrosion, the resistance increases by ΔR_s. Initially, ΔR_s is small enough that even with high current, the voltage drops over the resistance across points A, B, C, and D in Fig. 23.7 are negligible. Therefore, the I-V characteristics shown by the solid line in Fig. 23.6(b) remains unchanged even after significant loss of cross-section and corrosion of metal/semiconductor interface.

As the finger thinning continues, however, the I-V begins to change noticeably, see Fig. 23.6(b). In particular, the slope of the I-V characteristics changes significantly for $0 < V < V_{mp}$, but remains essentially unchanged for $V \approx V_{oc}$. The result can be understood by referring to Fig. 23.7. As ΔR_s increases due to corrosion, so

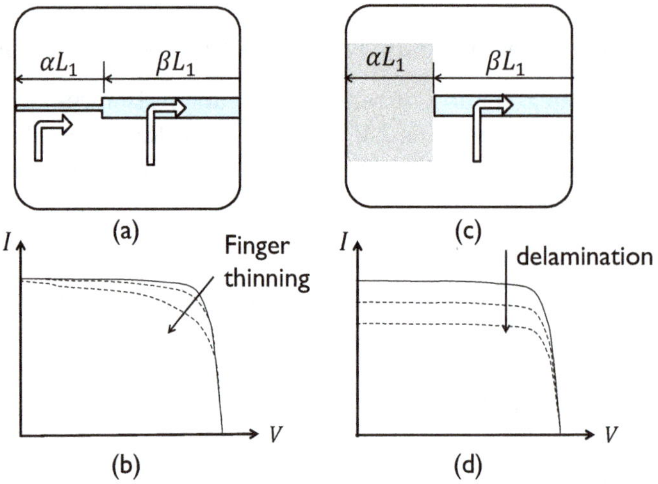

Figure 23.6: (a) Partially corroded finger makes current collection difficult. (b) The corresponding I-V characteristics showing apparent increase in shunt resistance. (c) Fully corroded or delaminated finger cannot collect current at all. (d) Loss of photocurrent depends on the extent of delamination.

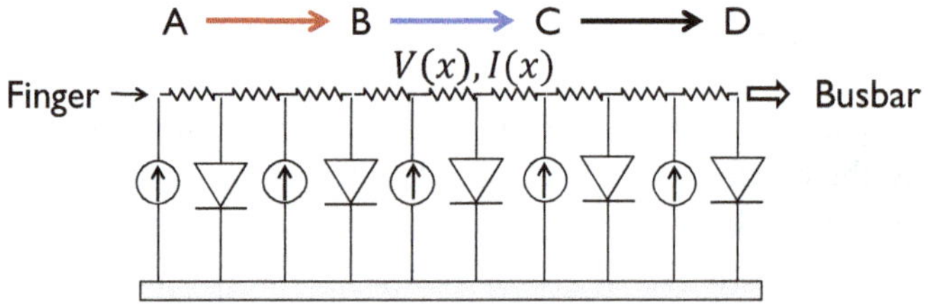

Figure 23.7: Current collection by a finger described by a distributed equivalent circuit. Points A through D show the direction of the collected current, such that $V_A > V_B > V_C > V_D$. Point D is connected to the busbar. The magnitude of the series resistance determines the bias of different points. A high local voltage is detrimental to photocurrent collection, because then the photocurrent is dissipated through the local diode.

does $\Delta V \equiv V_A - V_D = \int_A^D I(x)\, dR_s$. Initially, at $V_D \approx V \to 0$, the voltage drop across the finger ΔV (or, equivalently V_A) is significant, but not yet large enough to turn the diodes on, therefore J_{sc} can be viewed simply as sum of the photocurrent sources, somewhat independent of the level of corrosion. With increasing V, however, V_A begins to exceed V_{bi}, the built-in voltage of the local diode. The left-most diode at point A now begins to siphon off photocurrent generated at A, so that the external photocurrent is reduced. The degree of siphoning increases with V, as successive diodes turn on from locations A to D, and the photocurrent collection becomes increasingly inefficient. The change in the I-V slope at $0 < V < V_{mp}$ can be interpreted as an effective reduction in shunt resistance. Unlike Chapter 19, this shunt is not real: it is just that the nonlinear diode turn-on due to the finger resistance appears as a fake shunt-resistance in the output characteristics. Finally, this dramatic effect of corrosion disappears as $V \to V_{\rm oc}$. Now all the diodes are all on sinking the locally generated photocurrent. The current on the finger is greatly reduced so that $V_A \approx V_D \to V_{\rm oc}$. The finger resistance no longer plays a role in selectively turning on the diodes; therefore the slope of output I-V characteristic remains unaffected by the changes in $\Delta R_{\rm s}$, suggesting as if the series resistance remains unaffected by finger corrosion!

Homework 23.6: Analytical formulation of $V(x)$ for a distributed network

Derive an expression for current distribution along the finger based on the distributed network shown in Fig. 23.7.

Solution. The voltage drop between two points x and $x + dx$ associated with the local finger resistance per unit length, $\rho(x)$ [ohm/meter^2], is given by $V(x) - V(x + dx) = I(x)\rho dx$, which implies that $-dV/dx = I(x)\rho(x)$. Integrating the voltage equation, we find that

$$V(x) = V_{\rm m} - \int_0^x I(x)\rho dx$$

where $V_{\rm m}$ is the maximum voltage along the finger that occurs on the left-most side of the finger. The current along the finger resistor network involves a source associated with photocurrent and a sink associated with the local diode. If the finger separation is L_0, then $I(x+dx) - I(x) = J_0 L_0 dx - J_D L_0 dx$. In other words,

$$dI/dx = J_0 L_0 - J_D L_0 = J_0 L_0 - J_{01}(e^{qV(x)/k_{\rm B}T_{\rm D}} - 1).$$

Inserting the expression for $V(x)$, we find a self-consistent expression for current $i(x)$

$$I(x) = J_0 L_0 x - J_D L_0\, e^{qV_{\rm m}/k_{\rm B}T_{\rm D}} \int_0^x e^{-\frac{q\rho}{k_{\rm B}T_{\rm D}} \int_0^{x'} I(x')dx'} dx.$$

The intrinsic expression for $I(x)$ can be calculated by iteration to obtain an expression for $I_m \equiv I(x = L)$ and $V_{\rm L} \equiv V(x = L)$. Initially, $\rho(x) \approx$ const.; after corrosion, $\rho(x)$ depends on local corroded thickness. The voltage and current distribution at any point x shows that $V(x)$ drops across the finger from the left to right, while $I(x)$ increases as it picks up current along the finger.

23.5.3 Corrosion-induced delamination is reflected in the loss of photocurrent

The situation changes dramatically once a section of the finger is fully corroded/delaminated, as in Figs. 23.6(c, d). Here, the increasing delamination is reflected in the increasing loss of photocurrent. Once again, the effect can be understood with the reference to Fig. 23.7. Now, the resistance between A to C (for example) is so high that all the diodes are on in the delaminated section: the photogenerated current is locally dissipated through the diodes. The dead area is reflected in reduced short-circuit current. The side view of the delaminated finger shown in Fig. 23.8 and its equivalent circuit provide an intuitive explanation of the photocurrent loss.

In short, the increasing finger series resistance is reflected in the reduction in effective shunt resistance or effective loss of photocurrent. The power loss shows up as a hotspot surrounding the delaminated finger.

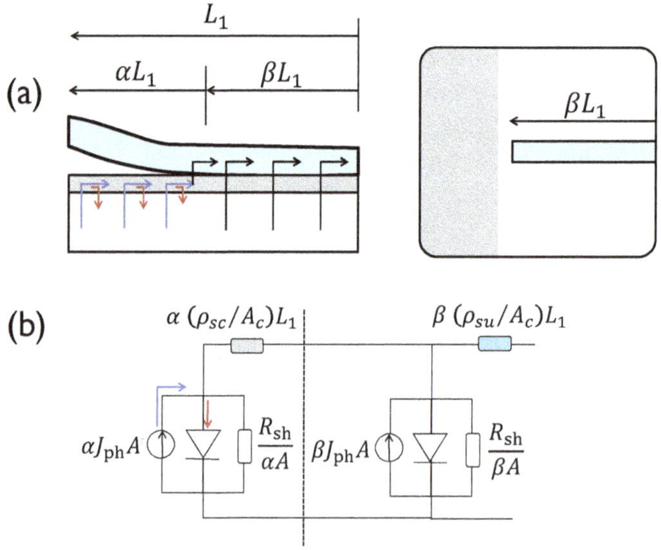

Figure 23.8: (a) Side view of a partially delaminated finger, after the interface has been corroded, or the (αL_1) section of the finger has been dissolved, the βL_1 section remains. (b) Top view of the same finger. The delaminated section is not shown. (c) The delaminated section and the uncorroded sections are represented by their respective equivalent circuit. Here, ρ_{sc} and ρ_{su} represent the series resistances associated with the corroded and uncorroded sections, respectively.

23.5.4 Solder bond failure is reflected in the module series resistance

The degradation mechanisms discussed so far in the book either degrades R_{sh} (e.g., grid thinning, PID, hotspot formation) or I_{sc} (e.g., UV degradation, partial shading, corrosion-induced grid delamination). We will now show that the commonly observed R_s degradation in the fielded modules is related to solder bond failure. The causes of solder bond failure, namely, glass cracking or stress build-up due to seasonal temperature fluctuation will be discussed in Chapter 24.

Homework 23.7: A simplified representation for a delaminated finger

The distributed photocurrent collection by a finger as shown in Fig. 23.7 can be simplified by a two-section network: one for the delaminated section, the other which is still unaffected. Convince yourself that the two-element network should capture the essential physics of charge collection, including the loss of photocurrent due to delamination.

As shown in Figs. 23.9(a, b), a solder bond failure increases the module series resistance, indicated by the I-V slope at V_{oc} in Fig. 23.9(c). Unlike finger thinning or delamination, the short-circuit current and shunt resistance remain unchanged because by the time electrons reach the busbar/ribbon, the fingers have already collected the photocurrent. Moreover, solder bond failure is a discrete event; therefore in a characteristic signature, the series resistance increases by quantized jumps. Finally, a number of solder bonds connect the ribbon to the busbar. Therefore, the failure of one or two (non-neighboring) solder bonds do not affect the module I-V significantly because a number of parallel current paths are involved (see Fig. 23.9(b)). When the neighboring solder bonds begin to fail,

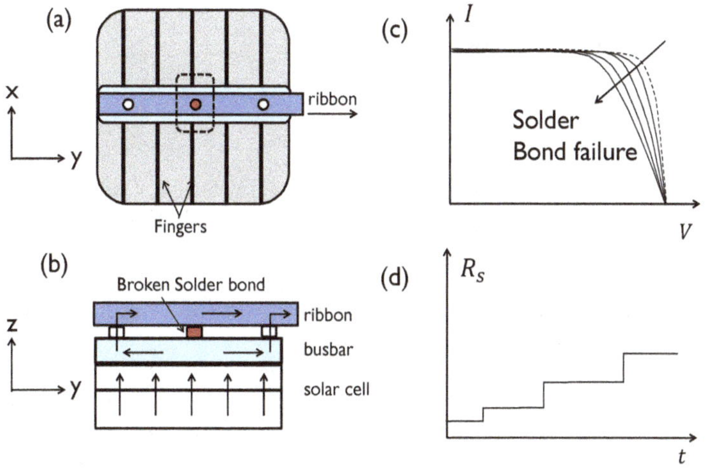

Figure 23.9: (a) Top view of a c-Si cell within a module. One of three solder bonds connecting the ribbon and the busbar is broken. The broken solder bond in the middle of the ribbon is shown in red. (b) Side view of region containing the broken solder bond (red) shows rerouting of the collected photocurrent to the unbroken solder bonds (white). (c) The successive loss of solder bonds increases the series resistance, reflected in the I-V characteristics. (d) The resistance increase is expected to rise in discrete steps.

however, the differential increase in series resistance is significant and its effect is obvious on the module I-V characteristics.

> **Homework 23.8: Probability of consecutive solder bond failure**
>
> A module contains K cells and each cell consists of S solder bonds. If the bond failure is statistically independent, what is the probability that k neighboring bond failures. Plot the probability of series resistance increase as a function of number of neighboring bond failures.

23.5.5 The combination of corroded and uncorroded cells determine module performance

A c-Si module consists of a series-connected set of cells selected so that they have similar short-circuit current. Even for a thin-film module, where the cells cannot be sorted for current matching, the non-uniformity of the current is typically small enough that we may still view the module as a collection of essentially similar cells. Then, the module output voltage is just the sum of individual cell voltages, and the module output current is given by the cell current. Similar to partial shading discussed in Chapter 18, as corrosion affects a subset of the cells, the current through the cells are no longer matched. For simplicity, assume that M cells are corroded, while N cells are unaffected, as shown in Fig. 23.10. Therefore, the module voltage at a specific current I is given by

$$V(I) = MV_{\mathrm{c}}(I) + NV_{\mathrm{nc}}(I)$$

where $V_{\mathrm{c}}(I)$ and $V_{\mathrm{nc}}(I)$ are, respectively, the voltage drops across corroded and non-corroded cells. If cells are corroded at different levels, $V(I) = \sum_{i=1}^{M+N} V_{i,cell}(I)$.

23.5.6 Is it possible to determine the degradation mechanisms from the terminal I-V characteristics alone?

We can calculate the module I-V from the cell I-V (forward problem), but can we infer the cell characteristics from the module I-V (inverse problem)? If we could, then we could determine if the cells are corroding simply based on the terminal characteristics.

In general, we know the total number of cells in a module, their geometrical arrangements, and the light and dark I-V characteristics of the pristine cells. For c-Si cell, we can represent the module by a series of connected cells, each described by a 6-parameter model shown in Fig. 23.8(c), for example.

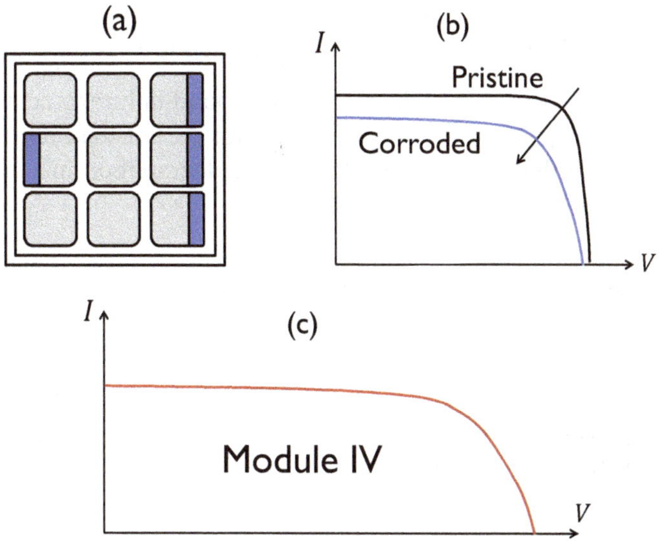

Figure 23.10: (a) A subset of cells ($M = 4$, in this example) is corroded, while the rest ($N = 5$) are unaffected. (b) The I-V characteristics of the corroded vs. nominally unaffected cells. (c) One constructs the module I-V by adding the voltage for each current; see text for details.

Next, the module can be stressed in a stress chamber with high humidity (85%) and high temperature (85°C), in a so-called damp heat qualification test (see Chapter 25). As the module degrades, its time-dependent I-V characteristics can be continually fitted by a 6-parameter equivalent circuit (plus the solder bond resistance) shown in Fig. 23.8(c), and the parameters are plotted as a function of time. The plot of the parameters would allow us to determine the degradation modes as follows.

If the series resistance of the module begins to degrade but the short-circuit current and shunt resistance remain essentially unchanged, we can attribute it primarily to solder bond failure. The number of bond failures can be estimated from the magnitude of the series resistance change. On the other hand, any loss of photocurrent or reduction in the shunt resistance in a damp heat test, however, must be attributed to corrosion. Initially, only the edge cells are degraded. The I-V characteristics of the damaged cells can be obtained by subtracting from the module I-V from the pristine I-V characteristics of uncorroded cells. The remaining I-V characteristics can be fitted by the two-section model described in Fig. 23.8. A plot of time-dependent α demonstrates how delamination is progressing with time.

The inverse modeling of damp heat test is simplified because other degradation modes (i.e., UV yellowing, PID, partial shadings, etc.) are absent. Nonetheless, the modeling provides an electrical probe into the real impact of corrosion on the module performance.

Homework 23.9: Power loss due to various degradation mechanism

Calculate $P(t)$ as a function of $R_s(t)$ for solder bond failure/cracking, $R_{sh}(t)$ for PID or contact grid corrosion, and $I_{mp}(t)$ for partial shading or grid delamination. Note that one cannot identify the degradation mechanisms by analyzing the power (efficiency) degradation.

Solution. The power output of the diode is maximized with the load $R_L \equiv V_{mp}/I_{mp}$. As $R_s(t)$ or $R_{sh}(t)$ change with time, the load resistance must be adjusted to track the maximum power point.

- We have seen that $R_s(t)$ increases due to solder bond failure. Since I_{mp} does not change significantly with $R_s(t)$, we can write

$$\frac{P(t)}{P(0)} = \frac{I_{mp}^2 R_L'}{I_{mp}^2 R_L}.$$

  Since the power delivered by the ideal cell must be dissipated by the series resistance and the load resistance, therefore $R_L'(t) + R_s(t) = R_L$. Inserting the expression, we find that

$$P(t) = P(0)\,[1 - R_s(t)/R_L]$$

  where $R_L = V_{mp}/I_{mp}$. With I_{mp} a constant, $\mathrm{FF(t)} = \mathrm{FF}(0)\,[1 - R_s/R_L]$ as well.

- We also know that $R_{sh}(t)$ decreases due to PID or grid thinning. Since V_{mp} does not change significantly with $R_{sh}(t)$, we can write

$$\frac{P(t)}{P(0)} = \left(\frac{V_{mp}^2/R_L'}{V_{mp}^2/R_L}\right) = \frac{R_L}{R_L'}.$$

  Since the power delivered by the ideal cell must be dissipated by R_{sh} and R_L', therefore $R_L'(t)^{-1} + R_{sh}(t)^{-1} = R_L^{-1}$. Inserting the expression, we find that

$$P(t) = P(0)\,[1 - R_L/R_{sh}(t)]$$

  where $R_L = V_{mp}/I_{mp}$. With I_{mp} a constant, $\mathrm{FF(t)} = \mathrm{FF}(0)\,[1 - R_L/R_{sh}(t)]$ as well.

(*continued on the next page*)

Homework 23.9 (*continued from the previous page*)

- Finally, photocurrent may be lost to UV degradation (Chapter 20), partial shading (Chapter 18), or corrosion- or stress-induced finger delamination. Since $V_{\rm mp}$ does not change with $I_{\rm mp}$, we can write:

$$\frac{P(t)}{P(0)} = \frac{V_{\rm mp} I_{\rm mp}(t)}{V_{\rm mp} I_{\rm mp}(0)} = \frac{I_{\rm mp}(t)}{I_{\rm mp}(0)}.$$

Equivalently, $\mathrm{FF(t)} = \mathrm{FF}(0)\,[1 - I_{\rm mp}(t)/I_{\rm mp}]$.

If we know $R_{\rm sh}$ or $R_{\rm sh}$, we can predict efficiency loss, $P(t)/P(0)$, and FF degradation, i.e.,

$$\mathrm{FF} = \mathrm{FF}(0)\left(1 - \frac{R_{\rm s}}{R_{\rm L}}\right)\left(1 - \frac{R_{\rm L}}{R_{\rm sh}}\right).$$

As an aside, since $R_{\rm L} = V_{\rm mp}/I_{\rm mp}$, therefore $R_{\rm s}/R_{\rm L} \propto c$ and $R_{\rm L}/R_{\rm sh} \propto 1/c$, where c is the relative concentration normalized to 1-sun. We will see in Chapter 27 (Sec. 27.4.2) that the c-dependent variability of FF suggests an opportunity to determine $R_{\rm s}(t)$ and $R_{\rm sh}(t)$ without disconnecting the module.

Homework 23.10: Cell efficiency decreases with series resistance

Sometimes the time-dependent increase in series resistance due to solder bond failure is written in the form

$$R_{\rm s}(t) = R_{\rm s}(t=0)\left(1 + \int_0^t R_{\rm D}(u)du\right). \tag{23.13}$$

Show that the resulting expression has a similar asymptotic form to that of Eq. (23.1).

Solution. Let Eq. (23.13) be a linear approximation of the more general equation:

$$R_{\rm s}(t) = R_{\rm s}(t=0) \times e^{\int_0^t R_{\rm D}(u)du} \approx R_{\rm s}(t=0)\left[1 + \int_0^t R_{\rm D}(u)du\right].$$

(*continued on the next page*)

Homework 23.10 (*continued from the previous page*)

Since, $P(t)/P_0 = (1 - R_s(t)/R_L)$ or equivalently, $\Delta P(t) = P_0 R_s(t)/R_L$. Therefore,

$$\frac{\Delta P(t)}{\Delta P_\infty} = \frac{R_s(t)}{R_s(\infty)} = \left[\frac{R_s(\infty)}{R_s(t=0)} \times e^{-\int_0^t R_D(u)du}\right]^{-1}.$$

If the degradation begins with a delay t_i, the expression is rewritten in a slightly general form:

$$\frac{R_s(t)}{R_s(\infty)} = \left[1 + \left(\frac{R_s(\infty)}{R_s(t=t_i)} - 1\right) e^{-\int_{t_i}^t R_D(u)du}\right]^{-1}. \tag{23.14}$$

With $\Delta P(t) \propto R_s$ and time-independent R_D, We recover Eq. (23.1).

23.6 Conclusions: corrosion is an important PV degradation mechanism

In this chapter, we discussed intrinsic dark and photo (light) corrosion of metal grid lines. Dark corrosion involves moisture-assisted generation of acetic acid in the EVA encapsulant. The acid corrodes the metal fingers, leading to thinning and delamination. Photo-corrosion involves voltage-assisted hydrolysis of moisture and corresponding erosion of the metal grid and accumulation of bubbles that delaminates the grid. Dark corrosion is suppressed by improved sealing of the module, encapsulants that do not produce acetic acid, etc. In addition to intrinsic self-heating, photo-corrosion is also suppressed by reducing PID ion flux and by using low-Na glass, low-mobility encapsulants, etc. As an aside, the equations describing PID (Eq. (22.3)) and corrosion (Eq. (23.3)) look similar, but the physics is different. For example, RH is directly implicated in corrosion but indirectly assists ion transport in PID. The temperature activation (E_A) is defined by the corrosion kinetics, while it is defined by the ion hopping in PID. Another important difference is that ion mobility makes PID strongly field dependent. But corrosion is diffusion limited and therefore field independent.

In this chapter, we also established the surprising fact that the power loss associated with corrosion of metal grid appears as a fake shunt (for finger thinning) and photocurrent loss (for finger delamination). Since a fake shunt involves nonlinear diode turn-on, it has a very different intensity and voltage dependence compared to real shunt, allowing us to easily identify its characteristics features and attribute the degradation to corrosion. The corroded subcells may be confined to the module edge (moisture ingress through the frame) or all the cells in a module (ingress through the backsheet), leading to different terminal characteristics of the moisture-corroded module.

Finally, we saw that series resistance increase is exclusively attributed to solder bond failure (or degradation of module connectors). Intrinsically, solder bond failure arises primarily from the daily and seasonal temperature cycling and stress build-up. We will discuss this topic of stress-induced reliability of a solar module in the next chapter.

References

[1] M. Koehl. Moisture as stress factor for PV-modules. In *2013 IEEE 39th Photovoltaic Specialists Conference (PVSC)*, pages 1566–1570, June 2013. ISSN: 0160-8371.

[2] G. M. Kimball, S. Yang, and A. Saproo. Global acceleration factors for damp heat tests of PV modules. In *2016 IEEE 43rd Photovoltaic Specialists Conference (PVSC)*, pages 0101–0105, June 2016.

[3] D. S. Peck. Comprehensive Model for Humidity Testing Correlation. In *24th International Reliability Physics Symposium*, pages 44–50, April 1986. ISSN: 0735-0791.

[4] Barry Ketola and Ann Norris. Degradation mechanism investigation of extended damp heat aged PV modules. *26th EUPVSEC*, 2011.

[5] Michael D. Kempe, Gary J. Jorgensen, Kent M. Terwilliger, Tom J. McMahon, Cheryl E. Kennedy, and Theodore T. Borek. Acetic acid production and glass transition concerns with ethylene-vinyl acetate used in photovoltaic devices. *Solar Energy Materials and Solar Cells*, 91(4):315–329, February 2007.

[6] K. Whitfield, A. Salomon, S. Yang, and I. Suez. Damp heat versus field reliability for crystalline silicon. In *2012 38th IEEE Photovoltaic Specialists Conference*, pages 001864–001870, June 2012. ISSN: 0160-8371.

[7] Benoît Braisaz, Chloé Duchayne, Mike Van Iseghem, and Khalid Radouane. PV aging model applied to several meteorological conditions. In *Proceedings of the 29th European Photovoltaic Solar Energy Conference (EU PVSEC), Amsterdam, the Netherlands*, pages 22–26, 2014.

[8] Jichao Li, Yu-Chen Shen, Peter Hacke, and Michael Kempe. Electrochemical mechanisms of leakage-current-enhanced delamination and corrosion in Si photovoltaic modules. *Solar Energy Materials and Solar Cells*, 188:273–279, December 2018.

[9] Jeremy D. Fields, Md Imteyaz Ahmad, Vanessa L. Pool, Jiafan Yu, Douglas G. Van Campen, Philip A. Parilla, Michael F. Toney, and Maikel F. A. M. van Hest. The formation mechanism for printed silver-contacts for silicon solar cells. *Nature Communications*, 7(1):11143, April 2016. Number: 1 Publisher: Nature Publishing Group.

[10] R. Asadpour, X. Sun, and M. A. Alam. Electrical Signatures of Corrosion and Solder Bond Failure in c-Si Solar Cells and Modules. *IEEE Journal of Photovoltaics*, 9(3):759–767, May 2019. Conference Name: IEEE Journal of Photovoltaics.

[11] Reza Asadpour, Muhammed Tahir Patel, Steven Clark, Nick Bosco, Timothy J. Silverman, and Muhammad A. Alam. Worldwide Physics-Based Lifetime Prediction of c-Si Modules Due to Solder-Bond Failure. *IEEE Journal of Photovoltaics*, (2022).

CHAPTER 24

Physics of Glass, Cell, and Backsheet Cracking: Mechanical Reliability of Solar Modules

Chapter Summary

- Glass and backsheet cracking as well as solder bond failure is related to the stress created by temperature cycling between day and night.
- Stress concentration around even small micro-cracks can dramatically accelerate the large-scale crack formation and interfacial delamination.
- Fatigue failure is related to the accumulated stress under repeated temperature cycling. Stress-induced delamination leads to solder bond failure and increase in the series resistance.
- Numerical modeling can be used to predict the stress-induced delamination and solder bond failure.

24.1 Mechanical integrity is essential for module operation

So far, we have discussed how photons (UV yellowing, LID, partial shadowing), moisture (corrosion, delamination), and high voltage (potential-induced degradation) slowly erode the performance of solar cells. Implicit in the analysis is the assumption that the module has not been compromised mechanically, i.e., front glass cover and backsheet have remained intact, the solder bonds are not broken, and Glass/EVA (or EVA/Si) delamination has not occurred. When they eventually crack or delaminate, moisture will rush in to corrode the grid and the module will fail soon thereafter. We need to understand the physics of cracking and delamination to design a reliable module.

Continuous exposure to moisture makes glasses fragile, and exposure to UV light makes backsheets brittle and breaks the interfacial bonds that hold the layers together. As these "chemically" damaged cells are exposed to mechanical impact stresses due to wind/snow/hail or differential thermal stress due to day vs. night (or summer vs. winter) temperature cycles, the module becomes increasingly susceptible to mechanical damage, see Fig. 24.1(a). Given that a module is exposed to "baking sun, bitter cold, heavy snow, pelting hail, buffeting wind, and falling pinecones," it is a miracle that the modules survive as long as they do.

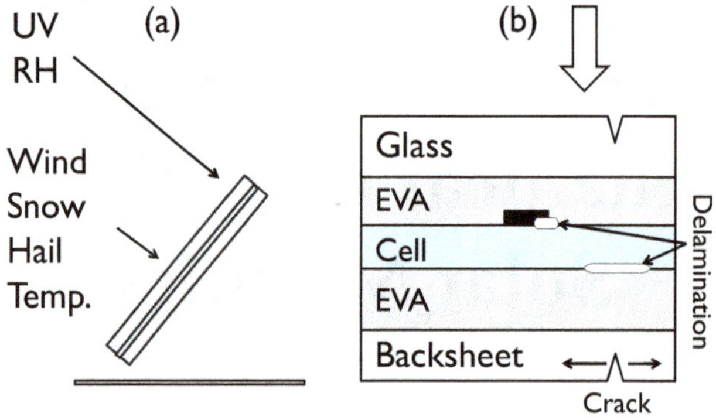

Figure 24.1: (a) Side view of a module exposed to various types of thermal and mechanical stresses from the environment. In addition, exposure to humidity and UV further degrade the mechanical integrity of the cell. (b) The stresses may lead to cracking of the backsheet or the glass cover, as well as delamination of the interfaces and grids.

Table 24.1: Material properties of various layers of a solar module.

Material	Thickness [μm]	Modulus [GPa]	CTE 10^{-6} K^{-1}
Glass	4000	73	9
EVA	500	0.015–0.080	270
Silicon	200	98	2.6
Backsheet	350	3.5	50.4
Aluminum frame	1000	–	25

24.2 Cracking and delamination must be avoided

24.2.1 Interfacial delamination

Interfacial delamination is characterized by the breaking of the interfacial bonds that hold the two layers of a module together, see Fig. 24.1(b). For example, once the glass/EVA interface delaminate, the uneven, non-conformal delaminated surface increases reflection and reduces short-circuit current. Also, any moisture that seeps through the edges of the metal frame and the glass cover will now move through the delaminated interfaces very rapidly and corrode the contacts and hydrolyze the encapsulant. The module edges will begin to show the signs of gradually spreading corrosion, with the corresponding decrease in the power output.

24.2.2 Cracking of glass and backsheet

Stress-induced cracking of the glass and the backsheet is another important challenge, as shown in Fig. 24.1(b). The glass cover experiences enormous stress at various stages of its lifecycle: when it is first framed by the metal during manufacture, when the modules are stacked, transported, and installed at the site, and as it experiences daily/seasonal temperature cycling, and exposed to pressure from wind and snow. The stress may eventually crack the glass. Both large scale and microscopic cracking have been observed. Large cracks formed during manufacture and installation break the solar cells into multiple pieces, the modules must be discarded. The periodic temperature cycling may lead to a few large cracks originating from the mechanically weak spots in the glass, or a network of microcracks that crisscross the glass. The scarring impedes light transmission and reduces the short-circuit current. Moreover, these microcracks allow moisture to diffuse into the module over the entire surface area.

In the following section, we will discuss the elementary physics of delamination and glass cracking as examples of mechanical reliability challenges of a solar module. A module exposed to localized hail impact vs. non-uniform wind pressure vs. distributed hotspots (discussed in Chapters 18 and 19) have very different (electro-thermal-mechanical) stress environments. The mechanical response of a module subject to such complex stress environments can only be analyzed through detailed numerical modeling. Therefore, our goal in this chapter will be to focus on the basic physics that qualitatively explains the trends observed in the qualification experiments and field tests.

24.3 Cracking of a single layer of glass, backsheet, or solar cell

24.3.1 Stress is uniform in a defect-free thin film

Consider a thin piece of glass under a uni-axial tensile stress, S_0 (see Fig. 24.2(a)). If a local stress $S(x)$ expands the differential section dx by an amount $du(x)$, then by definition, the local strain is du/dx. Hooke's law relates the local stress to the local strain by Young's modulus, E. Therefore,

$$S = E\frac{du}{dx}. \tag{24.1}$$

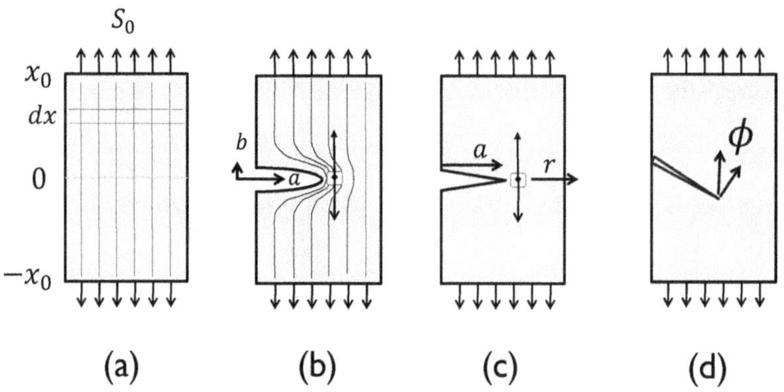

Figure 24.2: (a) A free-standing plate with a tensile stress, S_0. (b) A plate with an ellipsoidal opening. Also shown is the stress distribution around the opening. (c) A triangular crack of length a. The stress distribution reduces away from the crack point. (d) The angle of the crack with respect to the stress axis determines the stress magnitude.

The internal stress within a free-standing glass membrane is a position independent constant (i.e., every segment dx is held together by equal and opposite stress, S). Therefore,

$$\frac{dS}{dx} = E\frac{d^2u}{dx^2} = 0. \tag{24.2}$$

Since $S(x_0) = S_0$ and $S(-x_0) = -S_0$, therefore Eq. (24.2) can be solved to obtain

$$S(x > x_0) = S_0$$
$$S(x < x_0) = -S_0.$$

In this specific case, every point within the film experiences identical local stress. The films will fracture only when S exceeds S_B, the fracture strength of the material. Since S_B for most materials are very high and the typical stress in a solar cell is unlikely to exceed the value, one may incorrectly conclude that we have nothing to worry about.

Homework 24.1: Strain magnitude can be calculated by solving Eq. (24.2)

Show that the maximum displacement of a film with stress S_0 is given by

$$u_{max} = \frac{S_0}{E}x_0. \tag{24.3}$$

Solution. Since $S = E\,du/dx = S_0$, therefore, cross-multiplying and integrating, we find that

$$u(x) = \frac{S_0}{E}\,x. \tag{24.4}$$

Homework 24.2: Mathematical analogies can be used to solve the strain–strain equation

Show that Fick's law for particle diffusion, Fourier's law for heat conduction, Poisson's equation for electrostatics are mathematically equivalent to Hooke's law for stress–displacement relationship. Use these analogies to reinterpret Eq. (24.3) and Eq. (24.4).

Solution. The particle diffusion equation is given by Fick's law: $F = -D\, dn/dx$, where, F is the flux, D is the diffusion coefficient, and n is the local carrier concentration. Similarly, the heat diffusion equation is given by Fourier's law: $Q = -\kappa\, dT/dx$, where Q is the heat flux, κ is the thermal conductivity, and T is the local temperature. Similarly, electrical flux equation is given by Poisson's equation:, $Q = -\epsilon\, dV/dx$, where Q is the charge, ϵ is the relative dielectric constant, and V is the potential.

It is clear that Hooke's law (Eq. (24.1)) is analogous to Poisson's equation, Fourier equation, Fick's law, etc. In fact, Young's modulus E plays the same role as the diffusion coefficient D in Fick's law, or thermal conductivity κ in Fourier's law. Indeed, one can map the strain–strain equation into one of these equivalent flux equations with known solution, and then reinterpret the results in terms of original mechanical quantities. Based on the analogy, we see that Eq. (24.2) can be viewed as continuity of particle flux or continuity of heat flux in a source-free region.

24.3.2 Even a microcrack reduces material strength dramatically

The situation changes dramatically if the glass has a pre-existing crack (or hole). Such a crack, invisible to the naked eye, can occur during manufacture or shipment or cleaning. For simplicity, let us consider an ellipsoidal hole in the middle of the plate (see Fig. 24.2(b)). The edges of the ellipsoid are movable; therefore, they cannot support any force. The tension force applied to the plate edges must now flow around the hole and concentrate around the edges. The *extra* force is easily calculated as $F = S_0 \times a$. This is spread over the hole of width b, so that $\Delta S \approx S_0\, a/b$. Therefore, the total stress is

$$S = S_0 \left(1 + \frac{2a}{b}\right).$$

The factor of 2 reflects the symmetric stress across the film. Since $\rho \equiv b^2/a$, where ρ is the radius of curvature of the long edge of the hole, therefore,

$$S = S_0 \left(1 + 2\sqrt{\frac{a}{\rho}}\right). \qquad (24.5)$$

In other words, the local stress around the hole has increased by the stress concentration factor, K_t

$$K_t \equiv \frac{S_{\max}}{S_0} = 1 + \frac{2a}{b} = 1 + 2\sqrt{\frac{a}{\rho}}. \tag{24.6}$$

For $a \gg \rho$, the local stress concentration at the crack tip is so high that $S_{\max} \to S_B$ at orders of magnitude lower external stress level S_0. The new breakdown stress is obtained by rewriting Eq. (24.6) in the following form:

$$S_0 = \frac{S_{\max}}{K_t} = \frac{S_{\mathrm{B}}}{K_t} \equiv \frac{S_{\mathrm{B}}}{\left(1 + 2\sqrt{a/\rho}\right)}. \tag{24.7}$$

Homework 24.3: A numerical example illustrates the danger of crack tip stress localization

A 1 m^2 module has a small parabolic hole with $a = 1$ mm and $b = 1\,\mu m$. Calculate the stress concentration factor, K_t. If $S_{\mathrm{B}} = 170$ GPa, use Eq. (24.7) to calculate the failure stress, S_0, for the defective film.

24.3.3 An actual crack has even higher stress concentration

In practice, a crack is characterized by a sharper tip, as shown in Fig. 24.2(c). We can view the crack tip as a high aspect-ratio ellipsoid with $b \to 0$, or equivalently, $\rho \to 0$. Now the stress concentration factor increases dramatically at the edge of the "ellipsoid." A careful calculation shows that one may express the local stress as

$$S(r, \theta) = \frac{K_{\mathrm{I}}}{\sqrt{2\pi r}} \times f(\theta) \tag{24.8}$$

where r and θ are radial and angular distances away from the crack, see Fig. 24.2(c). For $\theta = 0,\ f(\theta) = 1$. The stress intensity factor, K_{I} is given by

$$K_{\mathrm{I}} = cY(\phi)\, S_0\sqrt{\pi a} \tag{24.9}$$

where c is a geometry-dependent factor of the order of 1, and $Y = \cos^2(\phi)$ is defined by the angle between the crack and the applied force. For the configuration in question, $\phi = \pi/2 \implies Y = 1$. Once $K_{\mathrm{I}} > K_{Ic}$, the critical stress intensity factor, the crack propagates uninhibited and the film breaks apart. Typically, $K_{Ic} \approx 1$ MPa.(meter)$^{0.5}$ for crystalline silicon.

To summarize, a glass is a strong material with a yield strength of $S_{\mathrm{B}} = 17$ GPa. Yet, local imperfections (e.g., scratches, bubbles, grain boundaries, microcracks, dislocations, etc.) increase the local stress and reduce the yield strength by orders of magnitude (e.g., just 12 MPa, as shown in Homework 24.4).

Homework 24.4: It is easy to calculate the stress around an ellipsoidal crack

The stress at the tip of the ellipsoidal crack is given by Eq. (24.5). The following expression generalizes the stress concentration away from the tip (and correctly reproduces the asymptotic limits at $r \to 0$ and $r \to \infty$):

$$\frac{S(r, \theta = 0)}{S_0} = 1 + 2\sqrt{\frac{a}{\rho + r}}.$$

By comparing with Eq. (24.8), show that the stress factor is given by $K_\mathrm{I} = 1.41\sqrt{\pi a}\, S_0$. Calculate the failure stress for a 1 mm pre-existing crack in a silicon wafer.

Solution. By comparison,

$$S = \frac{K_\mathrm{I}}{\sqrt{2\pi r}} \approx S_0 \left(1 + 2\sqrt{\frac{a}{\rho + r}}\right).$$

Assuming that the tip is sharp ($\rho \to 0$) and the intensity factor large ($K_\mathrm{I} \gg 1$), we get the desired expression.

For a 1 mm crack, $a = 10^{-3}$ m. Therefore, $K_\mathrm{I} = 1.41\sqrt{3.14 \times 10^{-3}}\, S_0 = 0.079\, S_0$. At the failure stress, $K_\mathrm{I} = K_{Ic}$; therefore $S_0 = K_{Ic}/0.079$ MPa. For silicon, $k_{Ic} = 1\mathrm{MPa} \cdot (\mathrm{meter})^{0.5}$; therefore, $S_0 = 12.65$ MPa.

24.3.4 Fatigue failure: Cracks begin to grow under repeated cycling

Typical pressures on a module by wind or snow (few kPa) is much smaller than the critical stress needed to fracture the glass or the backsheet. The stresses however occur many times over the module lifetime. The microcracks grow slowly, i.e., a becomes larger over time, so that $K_\mathrm{I}(t) \to K_{Ic}$ until the glass cracks suddenly.

The theory of fatigue predicts the number of cycles (N_f) needed for a pre-existing crack to become large enough to cause catastrophic failure. The problem was first analyzed by P.C. Paris in 1941, as follows. Assume that the crack size a increases with the number of cycles N as follows:

$$\frac{da}{dN} = A\, \Delta K_\mathrm{I}^m, \tag{24.10}$$

where $\Delta K_\mathrm{I} (\equiv K_{\mathrm{I,max}} - K_{\mathrm{I,min}} \approx \Delta S\, \sqrt{2\pi a})$ is the difference between maximum and minimum intensity factors associated with maximum and minimum cyclic stresses ($\Delta S = S_\mathrm{max} - S_\mathrm{min}$). The empirical constants A and m are obtained by stressing a sample with a periodic load and observing how the crack length a increases with cycle number, N (see Fig. 24.3(a)). Inserting Eq. (24.9) in Eq. (24.10), and integrating until the sample fails,

$$N_{\rm f}(\Delta S) = \frac{1}{A\sqrt{\pi^m}} \frac{(A_{\rm f}^{\Gamma} - a_i^{\Gamma})}{\Gamma} \frac{1}{\Delta S^m}. \tag{24.11}$$

Here, $a_{\rm i}$ is the initial crack size, while $a_{\rm f}$ is the largest crack length beyond which the sample fails catastrophically. The sign of $\Gamma \equiv (1-m)/2$ defines the two different regimes of failure: For $\Gamma > 0$, the growth rate is small and $N_{\rm F}$ determined by the critical size $a_{\rm f}$. On the other hand, for $\Gamma < 0$ implies $a_{\rm i}^{\Gamma} > a_{\rm f}^{\Gamma}$, and therefore the size of the initial crack defines the number of cycles that can be sustained by the sample.

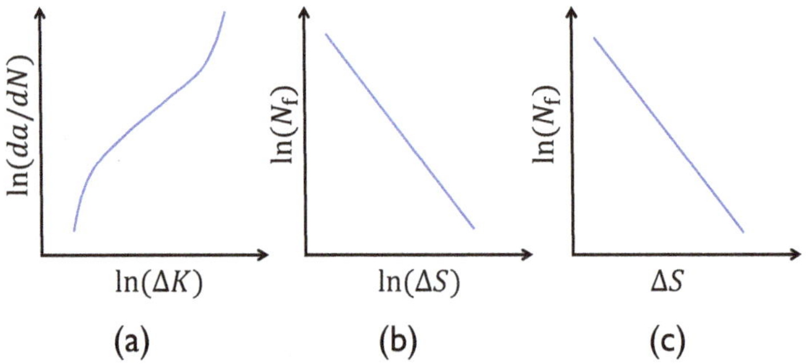

Figure 24.3: (a) Rate of crack growth as a function of number of loading cycles. (b) The $N_{\rm f}$ vs. S relationship predicted by the Paris model. (c) However, typical $N_{\rm f}$ vs. S for glass is described by a slightly different functional relationship.

24.3.5 Each material is defined by its $N_{\rm f} - S$ relationship

By taking log of both sides of Eq. (24.11), we see that $N_{\rm f}$ vs. S can be described as a power-law (for constant m), namely,

$$\ln(N_{\rm f}) = B - m\ln(\Delta S) \rightarrow B - m\ln(S) \tag{24.12}$$

where B is a constant. The final equation on the right-hand side assumes that the maximum and minimum loads are S and 0, respectively.

If the crack growth exponent $m(a)$ depends on the crack size a, as shown in Fig. 24.3(a), its exact integration produces a complicated result. For these systems, the following equation captures the functional dependence of the $N_{\rm f} - S$ relationship:

$$\ln(N_{\rm f}) = B\,(1 - (S/C))\,. \tag{24.13}$$

For example, glass fracture is defined by such a log–linear relationship.

Homework 24.5: The log–linear N_f vs. S characteristics predicts the number cycles before failure

Glass covers produced by a manufacturer is characterized by the following N_f vs. S relationship: At 90 MPa loading, it fails at 10 cycles and at 50 MPa loading, it fails at 1000 cycles. Use this fatigue-induced glass fracture data to determine the coefficients B and C in Eq. (24.13). If the daily wind load varies from 0 to 5 KPa, determine the time to failure of the glass cover.

24.3.6 Miner's law predicts lifetime under variable loading

Paris's law presumes a two-level (maximum and minimum) cyclic loading. In practice, a module sees a more complex multi-level loading. For example, wind pressure varies hourly; temperature varies throughout the day as well as over the seasons; accumulated snow stays put for days at a time. Miner's law applies to this more general situation.

Let us assume that there are M levels of variable loads: $S_1, S_2, \ldots, S_M$. During the module lifetime, we expect $n_1, n_2, \ldots, n_M$ cycles of loading with stress intensity factors of $\Delta K_1(S_1)$, $\Delta K_2(S_2), \ldots, \Delta K_M(S_M)$, etc. Miner's law predicts that the combined number of cycles must satisfy

$$\sum_{i=1}^{M} \frac{n_i(\Delta K_I)}{N_{f,i}(\Delta K_I)} \leq 1 \tag{24.14}$$

where $N_{f,i}(\Delta K_I)$ is the number of cycles to failure for the exclusive loading by the i-th stress level, easily read off from the N_f vs. S curve of the corresponding material.

Miner's law suggests that the loads "consume the lifetime" independently. One can use the relationship to check if the integrated stress would be low enough to ensure crack-free operation over the 25-year lifetime.

Homework 24.6: Miner's law holds for nonlinear N_f vs. S relationship

If N_f vs. S relationship is linear, Miner's law is easily proved. Show that Miner's law holds even for nonlinear N_f vs. S relationship.

24.3.7 Distribution of failure cycles: Weibull distribution

Since defect generation is a stochastic process, therefore, two samples with identical stress and nominally identical microcrack distribution will still fail at slightly different N_f. While Paris's law provides a macroscopic description regarding how the average crack size increases with loading cycles, the microscopic process is far more complicated. Bond breaking is a stochastic random process, the propagation of the crack depends on myriad local factors, including the crystal orientation. Empirically, one finds that the cumulative failure distribution after N cycles, $F(N)$, is given by a Weibull distribution:

$$1 - F(N) = e^{-(N/N_{\mathrm{f}})^{\xi}} \tag{24.15}$$

where $N_{\mathrm{f}}(S)$ is obtained from the N_{f} vs. S relationship and the empirical constant ξ defines the shape of the failure distribution.

The distribution has important implications for practical design. With $\beta > 1$, some samples will fail long before others do. We must design for these early failures, because unexpected cracking of even a few glass sheets will adversely affect the reputation of the manufacturer.

Given the importance of the failure distribution, it has become increasingly important to use photoluminescence and infrared imaging techniques to map the initial microcrack distribution as the materials are being manufactured. This is because the microcracks are so small that they are invisible to the eye. The damaged regions however scatter light (for glass) or contribute to recombination and self-heating (in c-Si cell). Therefore, the cracks become visible in infrared or photoluminescence imaging. Detailed analysis shows that the size distribution and failure stress are exactly correlated by the stress intensity factor, given in Eq. (24.9).

24.4 Delamination is a form of cracking

24.4.1 Delamination of two interfaces

A solar module consists of multiple layers of materials bonded to each other. Understanding delamination of such a system requires numerical simulation. However, we can gain valuable insights just by considering a pair of material bonded together by an interfacial region. If the coefficients of thermal expansion (CTEs) are different, the bilayer will bend in a way that the film with larger CTE will be in compression, while the film with smaller CTE will be under tension. This bending will create tremendous stress that would try to slide the high CTE material away the low CTE material along the interfacial plane. Such delamination could be catastrophic, as moisture will quickly diffuse through the delaminated plane and create air bubbles to reflect sunlight and corrode the contacts to suppress charge collection.

24.4.2 Calculating the lateral stress involves a few simple steps

Let us consider two films with the following properties: $\alpha_{1,2}$ are the thermal coefficients, $t_{1,2}$ are the thicknesses, $E_{1,2}$ are Young's moduli, and η and G are the thickness and the shear modulus of the interfacial region. Let us take apart the

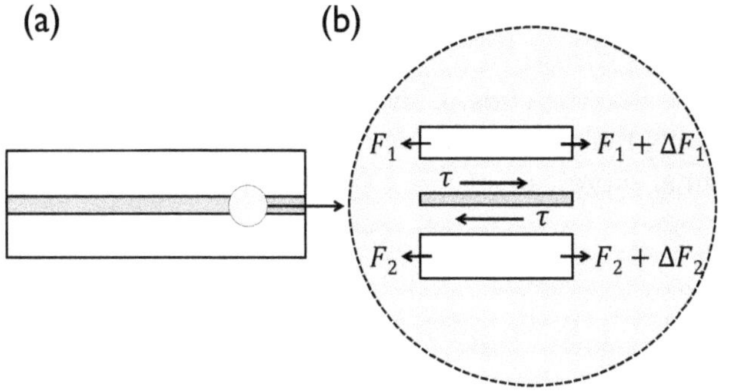

Figure 24.4: (a) A two-layer system bonded by an interfacial layer. (b) Forces that act on the individual layers.

two films and the interfacial layers, and consider the interfacial shear stress (τ) that develops across an element. For the top and bottom layers, the principle of force balance suggests:

$$\frac{dF_1}{dx} = \tau$$
$$\frac{dF_2}{dx} = -\tau \tag{24.16}$$

and

$$\frac{du_1}{dx} = \frac{F_1}{E_1 t_1} + \alpha_1 \, \Delta T$$
$$\frac{du_2}{dx} = \frac{F_2}{E_2 t_2} + \alpha_2 \, \Delta T. \tag{24.17}$$

Inserting the expressions for the force derivative into equations for displacements, we obtain second order differential equations for u_1 and u_2. Subtracting the two expressions, and realizing that the expression in the Hooke's law (applied to the interfacial layer) is given by

$$\frac{\tau}{G} = \frac{u_1 - u_2}{\eta}, \tag{24.18}$$

we find that

$$\frac{d^2\tau}{dx^2} = \beta^2 \tau, \tag{24.19}$$

where the constant β is given by

$$\beta \equiv \frac{G}{\eta}\left(\frac{1}{E_1 t_1} + \frac{1}{E_2 t_2}\right).$$

The solution of Eq. (24.19) is given by

$$\tau = A \sinh(\beta x) + B \cosh(\beta x). \tag{24.20}$$

The constants are determined by the boundary conditions. First, note that the middle of the film is stress-free (i.e., $\tau = 0$ for $x = 0$), which implies that $B = 0$. Second, we see that the ends of the films are free (i.e., $F_1(L) = F_2(L) = 0$). Therefore, by taking the derivative of Eq. (24.18) and inserting Eq. (24.17) and then evaluating the final expression at $x = L$ (where $F_1 = F_2 = 0$), we find that

$$\left(\frac{d\tau}{dx}\right)_{x=L} = \frac{G}{\eta}\left(\frac{du_1}{dx} - \frac{du_2}{dx}\right)_{x=L} = \frac{G}{\eta}(\alpha_1 - \alpha_2)\Delta T. \tag{24.21}$$

From Eq. (24.20), we know that $(d\tau/dx)_{x=L} = A\beta\cosh(\beta L)$. Combining with Eq. (24.21) and so that

$$A = \frac{G}{\eta}\frac{(\alpha_1 - \alpha_2)\Delta T}{\beta\cosh(\beta L)}.$$

Reinserting A back into Eq. (24.20), we find that

$$\frac{\tau(x)}{G} = \Delta T\left(\frac{\alpha_1 - \alpha_2}{\beta\eta}\right)\frac{\sinh(\beta x)}{\cosh(\beta L)}. \tag{24.22}$$

The maximum stress at the ends ($x = L$) is given by

$$\frac{\tau_{\max}}{G} = \Delta T\left(\frac{\alpha_1 - \alpha_2}{\beta\eta}\right)\tanh\beta L. \tag{24.23}$$

with $\tanh(\beta L) \to \beta L$ for typical systems.

Homework 24.7: It is possible to calculate the maximum stress intuitively

Show that for small βL, Eq. (24.23) reduces to

$$\frac{\tau_{\max}}{G} \to \Delta T\left(\frac{\alpha_1 - \alpha_2}{\eta}\right) L.$$

Is the actual stress larger or smaller than the asymptotic limit?

Solution. By definition, $\tau_{\max}/G = (u_1 - u_2)/\eta$. However, $u_1 = \Delta T\alpha_1 L$ and $u_2 = \Delta T\alpha_2 L$. Taken together, we obtain the asymptotic limit without solving any differential equation whatsoever!

24.4.3 Shear stress intensity factors and interfacial Paris law

Once the interfacial stress (τ) is known, the analysis for fatigue delamination mirrors that of the microcrack formation discussed in the last section. This is because Fig. 24.5 shows that the rate of interfacial delamination can be described by a power-law form analogous to Paris's equation (see Eq. (24.10)), namely,

$$\frac{da}{dN} = A(\Delta K)^m$$

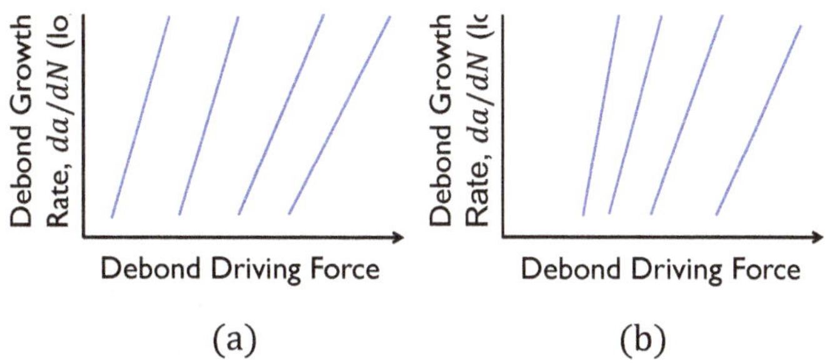

Figure 24.5: The rates of delamination plotted as a function of debonding energy, $G = \beta\tau^2$ where τ is the force of delamination. The delamination rate increases with (a) the temperature, and (b) the relative humidity.

where a is the delamination/debond length, N is the number of interfacial stress cycles, and ΔK is the stress intensity factor related to the interfacial stress (τ_{max}). The multiple lines in Fig. 24.5(a) are measured at different temperatures, while those in Fig. 24.5(b) are measured at different relative humidities. In other words, the power-law prefactor (A) and power-law exponent (m) depend on temperature and relative humidity.

One can either integrate the interfacial Paris equation, or measure the interfacial failure experimentally, to obtain a N_{f} vs. τ plot. Once again, the N_{f} vs. τ allows us to determine the number of cycles (N_{f}) the interface will survive before it delaminates. For example, analogous to Fig. 24.3(c) and Eq. (24.13), the fatigue equation may have the log–linear form :

$$\tau - \tau_{\mathrm{min}} = -b \ln \frac{N_{\mathrm{f}}}{N_{\mathrm{max}}} \tag{24.24}$$

where N_{max} is the maximum number of cycles with the minimum load, τ_{min}.

24.5 Stress-induced delamination of the metal grid lines

In Chapter 23, we discussed corrosion-induced delamination of metal grids placed on top of the silicon substrate. A related concern is stress-induced delamination of the metal grids and solder bonds. Three materials (EVA, metal, and silicon) with very different CTEs and stress–strain characteristics meet at the corner, see Fig. 24.6(c). Although the stress environment is complex, a number of groups have analyzed the problem numerically and fitted the results by analytical functions. One finds that the stress is given by a form similar to Eq. (24.8), namely,

$$\tau(r,\theta) = \frac{K_1}{r^{\lambda_1}} f_1(\theta) + \frac{K_2}{r^{\lambda_2}} f_2(\theta) \tag{24.25}$$

Homework 24.8: An example illustrates the key concepts of fatigue delamination

In Sahara, the temperature varies from 35°C during the day to 0°C during the night. Tests have shown that a particular interface within the solar module survives 600 stress cycles as the temperature cycles between 85°C and −40°C. Assume that a log–linear N_f vs. τ characteristic (Eq. (24.24)) determines the fatigue delamination, with $N_{\max} = 10^8$ for $\tau_{\min} \to 0$. Do you expect this module to survive the day-night temperature variation in Sahara for 25 years? Note that 25 years is equivalent to 8760 cycles (365 × 25 years).

Solution. To compare two temperature excursions, note that

$$\frac{(\tau_1 - \tau_{\min})}{(\tau_2 - \tau_{\min})} = \frac{-b\log(N_{f1}/N_{\max})}{-b\ln(N_{f2}/N_{\max})}.$$

Since, $N_{\max} = 10^8$ at $\tau_{\min} \to 0$, therefore the equation simplifies to

$$\frac{\tau_1}{\tau_2} = \frac{\ln(N_{f1}/10^8)}{\ln(N_{f2}/10^8)}.$$

Moreover, $\tau_1/\tau_2 = \Delta T_1/\Delta T_2$ by Eq. (24.23). Therefore,

$$\frac{85-(-40)}{(37-0)} = \frac{\ln(600/10^8)}{\ln(N_{f2}/10^8)} = \frac{\ln(600/10^8)}{\ln(N_{f2}/10^8)},$$

therefore

$$N_{f2} = 10^8 \times \exp\left((37/125) \times \ln(600/10^8)\right) = 2846641 \gg 8760 \text{ cycles}.$$

The module will safely survive the temperature cycling of 25 years without delamination.

where r and θ are distance from the triple-point junction in polar coordinates, K_1 and K_2 are the stress intensity factors, and $\lambda_1 \approx 0.1$–0.2 and $\lambda_2 = 0.003$–0.1 are called the order of singularities. Compared to $\lambda_1 = 0.5$ for microcrack propagation (see Eq. (24.8)), the stress decays much faster away from the triple point, as shown in the numerical simulation in Fig. 24.6. If the corner stress intensity factor exceeds a critical value, the metal will delaminate from the substrate and encapsulant. As seen in Chapter 23, the metal delamination reduces the short-circuit current and the power output of the solar cell.

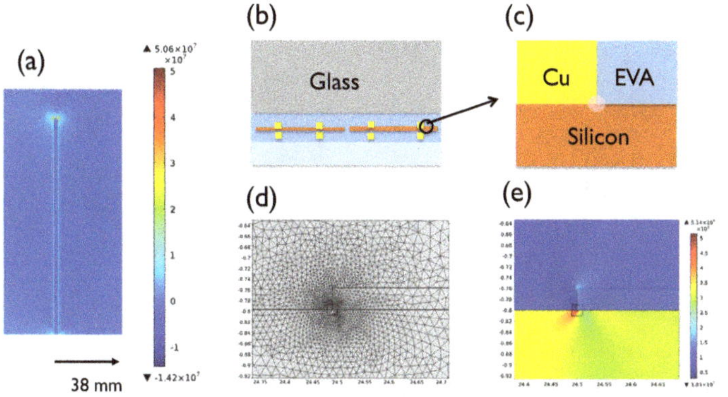

Figure 24.6: (a) Top view of a c-Si solar cell, with a metal line (busbar) on top of the silicon substrate. Numerical simulation shows very high stress concentration at the top edge of the busbar. (b) Side view of the same solar cell. Glass cover, encapsulant, silicon substrate, and backsheet are shown. (c) The junction of EVA, metal line, and substrate creates a complex stress environment. (d) The stress concentration is solved by discretizing the differential equation that describes the stress–strain relationship in three dimensions. (d) A plot of stress concentration at the triple junction.

24.6 Stress accumulation leads to solder bond failure

We also recall from Chapter 23 that solder bond failure caused by daily and seasonal temperature cycling increases the series resistance and reduces power output. A simple empirical model determines the number of thermal cycles needed for solder bond failure. A paper in *Microelectronics Reliability*, 62, pp. 124–129, 2016, uses a two-dimensional numerical modeling (similar to Fig. 24.7) and a Darveaux–Paris model to show that backside solder joints accumulate damage four times faster than the frontside solder joints, making the backside solder bonds vulnerable to fatigue failure. Their numerical simulation shows that the **yearly** accumulated damage (D) is a function of the maximum stress, τ_{max}, and can be expressed in terms of the Coffin–Manson empirical formula:

$$D(\tau_{\text{max}}) = C\,(\Delta T)^n\, N(T_{\text{r}})^b\, e^{-E_a/k_{\text{B}}T_{\text{max}}} \tag{24.26}$$

where $C = 405.6$, $n = 1.9$, $b = 0.33$, and the activation energy $E_a = 0.12$ eV are empirical constants. The factor $\Delta T \equiv T_{\text{max}} - T_{\text{min}}$ is obtained by calculating the daily difference in the module temperature and then averaging the difference over a year. Similarly, T_{max} is the average daily maximum temperature over a year. Finally, $N(T_{\text{r}})$ is the number of times the weather exceeds a critical temperature $T_{\text{r}} = 54.8°\text{C}$. We recall that the module temperature depends on the wind speed and relative humidity, and is typically 15–25°C higher than the ambient temperature.

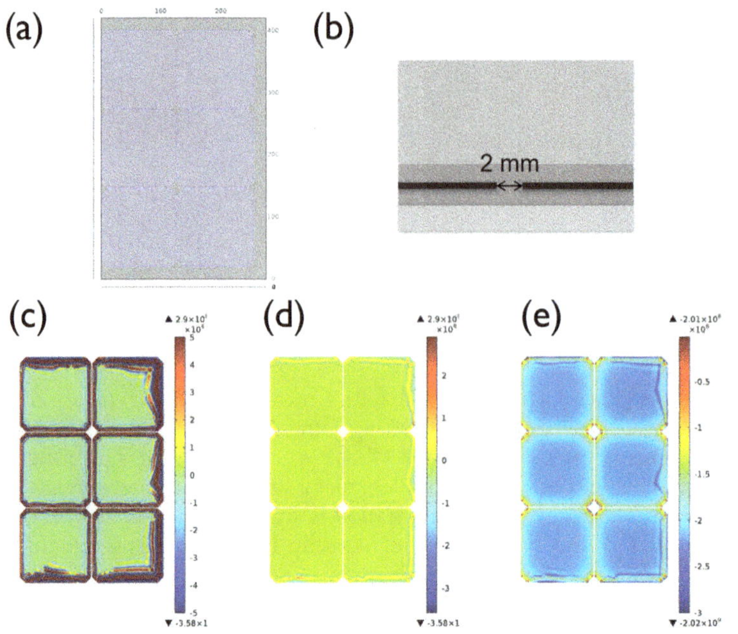

Figure 24.7: (a) Top view of a three layer stack consisting of a backsheet, EVA, and solar cell. (b) Side view of the same panel. The gap between two cells are also shown. (c), (d), and (e) show stress distribution in the respective layers.

Homework 24.9: Location-specific stress accumulation for solder bond failure

Based on the information above, show that for Chennai, India, ($T_{\text{max}} = 59°\text{C}$, $\Delta T = 35.5°\text{C}$, $N(T_{\text{r}} = 54.8) = 2060$) accumulates $D \approx 40$ kPa per year. Also show that the accumulated damage in Chennai over 25 years (i.e., 1 MPa) can be replicated in the laboratory by $\sim$ 630 cycles of temperature cycling from $-40°\text{C}$ to $85°\text{C}$.

Solution. Inserting the Chennai parameters in Eq. (24.26), we find that indeed $D \approx 40$ kPa. For the accelerated test, $\Delta T = 85 - (-40) = 125°\text{C}$, $T_{\text{max}} = 273 + 125 = 398$ K, and $C = 405.6$ per year. Therefore,

$$25 \times 40 \times 10^3 \text{ Pa} = 405.6 \times (125)^{1.9} \times N^{0.33} \times e^{-0.12/(0.0259 \times 398/300)}$$

can be solved to obtain $N \approx 630$ cycles. The same equation shows that increasing the maximum temperature just by 10 degrees to 135°C halves the number of cycles needed to accumulate the same damage.

24.7 Numerical modeling is essential for predictive modeling and quantitative insights

Engineers desire a simple analytical formula for cracking and delamination because they allow approximate (but efficient) predictions, and define the key functional variables and the importance of various material parameters. In practice, a module involves multiple layers and interfaces. The polymer encapsulants are viscoelastic, defined by stress and temperature-dependent modulus. Other material properties could also be temperature dependent. Only a numerical simulation, similar to Figs. 24.6 and 24.7, can capture the complex thermomechanical stress environment. These numerical simulations offer many insights. For example, Eitner *et al.* have shown that

1. The gap between the cells (Δv) depends on the module temperature, T and the location of the cell within the module. The spatially averaged change can be described by an empirical relationship:

$$\langle \Delta v \rangle = 0.001\, T^2 + 0.407\, T - 5.77 \tag{24.27}$$

   where Δv is expressed in μm, and T in °C. Obviously, Δv depends on the position of the cell within the module. For example, when the temperature changes from 150°C to -40°C, the gap changes by $\Delta v = 120\ \mu$m for cells at the center of the module, while $\Delta v = 170\ \mu$m for cells at the module edge.

2. At low temperature (e.g., -40°C), the solar cells experience compressive stress (up to 75 MPa), while the backsheet experiences tensile stress (up to 45 MPa). In contrast, the 4 mm thick glass is essentially stress-free.

3. The EVA accommodates strains of up to 23%, which proves that it acts as a compliant buffer layer. The mechanical properties are dominated by the thick glass layer.

24.8 Conclusions: Stress-induced delamination is an important PV reliability issue

In this chapter, we discussed the importance of glass/backsheet cracking, interface and grid delamination, and solder bond failure. The stress–strain environment is complex and can be defined quantitatively only by numerical simulation. Nevertheless, a simple understanding of the rate of delamination/cracking helps one interpret the qualification tests and redesign the module to ensure its mechanical integrity. In particular, cracking is known to (a) increase series resistance through finger delamination and solder-bond failure, (b) reduce the short circuit current due to enhanced reflection from delaminated interfaces, and (c) increase the junction recombination due to point defect formation.

Now that we understand the importance of reversible, metastable, and permanent reliability issues, we will explain how they can be integrated to predict the lifetime of a solar farm. We will also discuss how a carefully crafted qualification strategy can replicate the environmental stressors the module may be exposed to and ensure that the module survives its projected lifetime without any issue.

References

[1] G. R. Irwin. Analysis of stress and strains near the end of a crack traversing a plate, 1957.

[2] P.J.G. Schreurs. Fracture Mechanics Lecture notes course 4A780, Eindhoven University of Technology, 2012.

[3] Yaping Luo and Ganesh Subbarayan. A study of multiple singularities in multi-material wedges and their use in analysis of microelectronic interconnect structures. *Engineering Fracture Mechanics*, 74(3):416–430, February 2007.

[4] M. Demant, T. Welschehold, M. Oswald, S. Bartsch, T. Brox, S. Schoenfelder, and S. Rein. Microcracks in Silicon Wafers I: Inline Detection and Implications of Crack Morphology on Wafer Strength. *IEEE Journal of Photovoltaics*, 6(1):126–135, January 2016. Conference Name: IEEE Journal of Photovoltaics.

[5] Y. Murakami, editor. *Stress intensity factors handbook*. Pergamon, Oxford [Oxfordshire]; New York, 1st edition, 1987.

[6] G. C. Sih, P. C. Paris, and F. Erdogan. Crack-Tip, Stress-Intensity Factors for Plane Extension and Plate Bending Problems. *Journal of Applied Mechanics*, 29(2):306–312, June 1962.

[7] H. Altenbach, M. Köntges, S. Kajari-Schröder, M. Pander, and U. Eitner. Thermomechanics of PV Modules Including the Viscoelasticity of EVA. In *26th European Photovoltaic Solar Energy Conference and Exhibition*, pages 3267–3269, October 2011. ISBN: 9783936338270 Publisher: WIP.

[8] J. Käsewieter, F. Haase, and M. Köntges. Model of Cracked Solar Cell Metallization Leading to Permanent Module Power Loss. *IEEE Journal of Photovoltaics*, 6(1):28–33, January 2016. Conference Name: IEEE Journal of Photovoltaics.

[9] F. J. Higuera. Approximate solution for the natural convection flow around an array of obstacles on a vertical wall. *Physics of Fluids*, 16(2):486–489, January 2004. Publisher: American Institute of Physics.

CHAPTER 25

Qualification of Module Reliability

Chapter Summary

- Modules are characterized carefully and extensively before they are deployed in the field.
- The qualification tests are developed to mimic the anticipated field conditions, including thermal cycling, damp-heat, UV degradation, etc. The acceleration parameters are used in the lifetime models.
- Integrated stress sequence replicate the correlated degradations in actual environmental conditions and provide a stringent test of the module quality.
- The qualification datasheet contains a wealth of information for determining the viability of a module technology in a geographical location.

25.1 Extensive characterization is necessary to ensure module reliability

In the last seven chapters, we discussed the physics of various degradation modes, including partial shading, potential-induced degradation, corrosion, stress fracture, and so on. In practice, the degradation modes proceed in parallel. A solar module and/or panel are designed to survive for 25 to 50 years in a variety of weather conditions in various locations in the world. Temperature varies significantly between day and night, or seasonally between summer and winter. Some places are dry and cold, others hot and humid. Some places experience high UV radiation, others don't. Wind speed, snowfall, storms, etc. depend on geographical location. The industry has developed a set of qualification tests that accelerates individual and collective degradation modes to ensure that a well-made module/panel survives in variety of weather conditions.

First, recall that the reliability model predictions are based on intrinsic material properties of a module, e.g., corrosion rate of the front metal interconnect or moisture diffusivity through the EVA. The goal of qualification tests is to ensure that faulty manufacturing do not produce inferior metal contacts with significantly higher corrosion rate or low-quality EVA that allows faster moisture diffusion. A poorly manufactured module will not only fail quickly, but an electrical short would pose a grave safety threat to technicians who install/replace them.

Second, qualification tests are designed to isolate and accelerate specific degradation modes. For example, damp heat stress conditions are so chosen that it primarily affects and accelerates dark corrosion. Through these **accelerated tests**, one may be able to study the kinetics of corrosion in hours that would have taken years to occur under natural stress conditions. The tests allow one to determine the key rate constants of the degradation modes.

Finally, we know that a module is subjected to local weather conditions (e.g., temperature, relative humidity, etc.) that fluctuate periodically throughout the year. As a result, various modes of degradation may accelerate at different times of the year. An *integrated qualification* test mimics this natural variability and ensures that the module will survive this sequential, correlated stresses.

Jordan and Wohlgemuth in their review articles have noted that many of the tests were developed during late the 1970s and early 1980s when the Jet Propulsion Laboratory bought a large number of modules and insisted that they meet specific reliability goals. In this chapter, we will summarize the key tests and rationalize their use. You can find more details in the qualification document called *IEC 61646, Ed. 2, 2008 International Standard*. One other thing: you will recall seeing some of the homeworks below in the previous chapters. We reproduce them here to explain their implications for accelerated testing.

25.2 Insulation resistance ensures safe operation in dry/wet conditions

The dry and wet resistance tests shown in Fig. 25.1 ensure that the solar cells are well insulated from the frame. In an installation, the frame is grounded. Each cell produces less than a volt. In a series-connected system, however, the total voltage may reach 600 to 1000 V. You may recall from Chapter 22 that this voltage difference leads to potential-induced degradation. The goal of the insulation resistance test is to ensure that cells and the frame have not been shorted electrically.

The **dry insulation test** is done at the ambient temperature and with 75% relative humidity, RH. The output terminals of a solar cell are connected (shorted) together. This shorted terminal is then connected to the positive terminal of a voltage source, while the frame is connected to the negative terminal of the same source, see Fig. 25.1. The frame voltage is increased until it reaches the maximum voltage (1000 V plus twice the maximum rated voltage). During this test, the specific resistance must exceed $40\,\mathrm{M\Omega \cdot m^2}$.

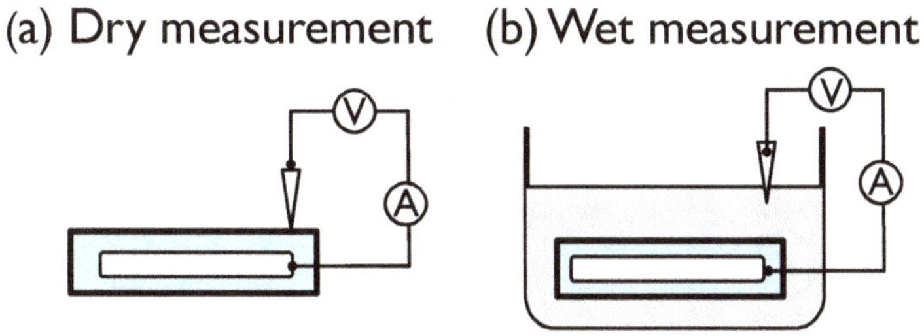

Figure 25.1: Measurement of module insulation between the metal frame and the solar cells: (a) when the module is dry, vs. (b) when the module is wet.

The **wet insulation test** ensures that the module will operate safely even when it is wet and its surface is covered with water (after a rain, for example). To mimic the situation, the module is put in a water solution (with added chemicals to improve electrical conductivity). This time, instead of applying negative voltage to the frame, a negative electrode is placed in the solution, see Fig. 25.1(b). The measured resistance again must exceed $40\,\mathrm{M\Omega \cdot m^2}$.

25.3 Thermal cycling ensures modules are resistant to cracking and delamination

When scattered clouds pass over a solar farm in the middle of an otherwise hot sunny day, a module experiences repeated dips in temperature over timescales of minutes. Even without clouds, midday and midnight temperatures may differ greatly, for example from 37°C to 0°C in the Sahara Desert. The seasonal temperature may range from −30 to 50°C. A thermal cycling test ensures that the modules will survive thermal mismatch, fatigue, and other stresses caused by the repeated change in temperature.

Figure 25.3 shows that a typical thermal cycling (TC) test consists of 200–600 cycles of −40 to +85°C with 10 minutes at extreme temperatures and maximum transition rate of 100°C/hour. The module is placed in a test chamber with precisely controlled stress parameters, as shown in Fig. 25.2. Periodically, modules are removed from the test chamber and their characteristics, such as light and dark IV, electroluminescence, etc., are recorded for analysis. Note that this is an accelerated test: the high temperature of 85°C causes extreme shear stress. If a module survives this extreme stress without delamination or cracking, then one can reasonably expect that the module will survive typical temperature ranges expected during operation.

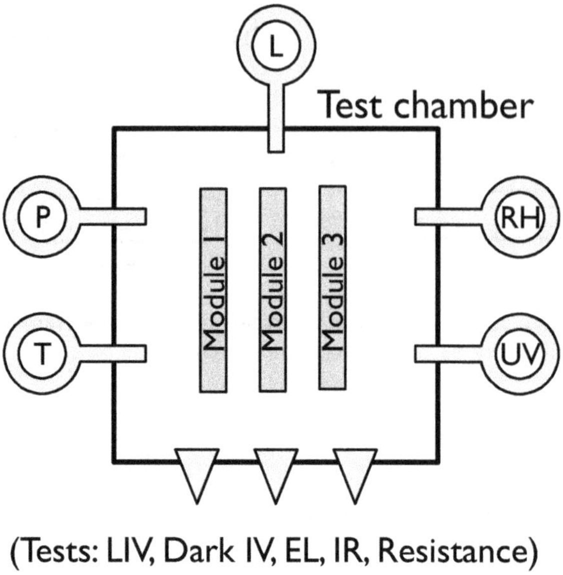

Figure 25.2: A qualification test chamber controls the environmental variables (e.g., pressure, temperature, humidity) precisely. The modules are periodically taken out for characterization tests, such as light IV (LIV), dark IV, electroluminescence, infrared imaging, insulation resistance tests.

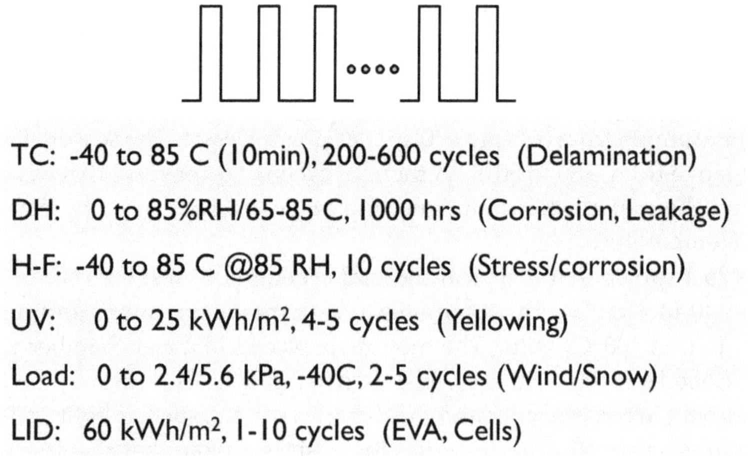
TC: -40 to 85 C (10min), 200-600 cycles (Delamination)
DH: 0 to 85%RH/65-85 C, 1000 hrs (Corrosion, Leakage)
H-F: -40 to 85 C @85 RH, 10 cycles (Stress/corrosion)
UV: 0 to 25 kWh/m2, 4-5 cycles (Yellowing)
Load: 0 to 2.4/5.6 kPa, -40C, 2-5 cycles (Wind/Snow)
LID: 60 kWh/m2, 1-10 cycles (EVA, Cells)

Figure 25.3: A summary of individual qualification tests, with the test parameters specified. TC: thermal cycling test; UV: ultraviolet test; DH: damp heat test; LID: light-induced degradation test; Load: pressure test; H-F: humidity-freeze test, etc. The tests are described in the text.

Homework 25.1: An example illustrates the key concepts of thermal cycling and fatigue delamination

In Sahara, the temperature varies from 35°C during the day to 0°C during the night. Tests have shown that a particular interface within the solar module survives 600 stress cycles as the temperature cycles between 85°C and −40°C. Assume that a log–linear N_f vs. τ characteristic (Eq. (24.24)) determines the fatigue delamination, with $N_{\max} = 10^8$ for $\tau_{\min} \to 0$. Show that this module will survive the day–night temperature variation in Sahara for 25 years (or, equivalently, 365 × 25 = 8760 cycles of stress).

Solution. To compare two temperature excursions, note that

$$\frac{(\tau_1 - \tau_{\min})}{(\tau_2 - \tau_{\min})} = \frac{-b\ln(N_{f1}/N_{\max})}{-b\ln(N_{f2}/N_{\max})}.$$

Since, $N_{\max} = 10^8$ at $\tau_{\min} \to 0$, therefore the equation simplifies to

$$\frac{\tau_1}{\tau_2} = \frac{\ln(N_{f1}/10^8)}{\ln(N_{f2}/10^8)}.$$

Moreover, $\tau_1/\tau_2 = \Delta T_1/\Delta T_2$ by Eq. (24.23). Therefore,

$$\frac{85-(-40)}{(37-0)} = \frac{\ln(600/10^8)}{\ln(N_{f2}/10^8)}.$$

Therefore,

$$N_{f2} = 10^8 \times \exp\left((37/125) \times \ln(600/10^8)\right) = 2846641 \gg 8760 \text{ cycles}.$$

The module should safely survive the temperature cycling of 25 years without delamination.

25.4 Damp heat test identifies if a module is susceptible to excessive corrosion

It rains much more in some area than others. In Miami, Florida, the relative humidity often approaches 100%. The excess moisture accumulates over a module and seeps in through the connector plug, any crack in the backsheet or the glass cover, or any gap in the frame sealant. Once inside, moisture diffuses through the polymer encapsulant and corrode the interconnects. The damp heat test mimics/accelerates this reliability issue.

In the damp heat (DH) test, the module is placed in a controlled chamber (see Fig. 25.2), where relative humidity can increase from 0 to 85% while the temperature is held relatively constant between 65 and 85°C (see Fig. 25.3). The test is done at relatively high temperature, so that the moisture diffusion is accelerated and its effect on the module can be observed at a shorter time. The modules are tested periodically until the 1000-hour (~6 weeks) test is completed.

Homework 25.2: Results of damp heat test can be used to predict module lifetime under specific weather conditions

During the JPL block buy program, it was observed that during damp heat testing, module degradation rate doubled for a 10°C increase in temperature and that the degradation increases by the same amount for a 1°C increase in temperature as it does for 1% increase in relative humidity. Develop a degradation model based on the information provided. If Miami, Florida has an average temperature of 29°C and an average relative humidity of 60%, show that if a module survives the damp heat test for $t_0 = 1000$ hours, it will survive Miami weather for 20 years.

Solution. It is easy to verify that the following expression for time to failure is consistent with the information provided.

$$\frac{t_\text{f}}{t_0} = \left(\frac{1}{2}\right)^{((\text{RH}-\text{RH}_0)+(T-T_0))/10}. \tag{25.1}$$

Therefore, at Florida with 29°C average temperature and 60% RH, 1000 hours damp heat test translates to a lifetime is

$$1000 \times \frac{2^{((85-29)+(85-65))/10}}{(24 \times 365)} = 22.14 \text{ years.}$$

A second way to calculate the damp heat lifetime is to use Eq. (23.3), which is a variant of the Peck's equation used in the microelectronics industry:

$$t_\text{f} = Ae^{E_\text{A}/k_\text{B}T} \times [\text{RH}]^{-B}. \tag{25.2}$$

Here, t_f is expressed in years. The prefactor $A \equiv b \ln P_{th} + c$, where b and c are constants, and P_{th} is maximum (threshold) power loss beyond which the module is presumed to have failed. For example, Kimball *et al.* report in "Global acceleration factors for damp heat tests for PV modules" (Proc. of PVSC, 2016), the following parameters for their damp heat tests. For $P_{th} = 20\%$, $A = 6.4 \times 10^{-10}$. Other constants in Eq. (25.2) are: activation energy, $E_\text{A} = 0.8$ eV; and humidity exponent, $B = 2.6$.

(continued on the next page)

Homework 25.2 (*continued from the previous page*)

Evaluating the Eq. (25.2) for field and accelerated conditions and taking their ratio, we find that

$$\frac{t_\mathrm{f}}{t_0} = e^{\frac{E_\mathrm{A}}{kT_\mathrm{f}} - \frac{E_\mathrm{A}}{k_\mathrm{B}T_0}} \times \left(\frac{\mathrm{RH_f}}{\mathrm{RH_0}}\right)^{-2.6}.$$

Therefore,

$$t_\mathrm{f} = \frac{(1000)}{(24 \times 365)} \times \left(\frac{85}{60}\right)^{2.6} \times e^{\frac{0.8}{8.61\times10^{-5}} \times \left(\frac{1}{273+29} - \frac{1}{273+85}\right)} = 34.74 \text{ years}.$$

The two approaches give comparable results, as expected.

Homework 25.3: Activation energies for onset delay and degradation rates

The data from a damp heat experiment, used to accelerate the corrosion in solar module, can be used to extract the four parameters of Eq. (23.2), namely, $\Delta P_\infty, T_\mathrm{i}, R_\mathrm{D}, \Delta P(T_\mathrm{i})$. Using damp heat tests at different temperature, one finds that $t_\mathrm{i}(T_\mathrm{D}) \propto e^{E_i/k_\mathrm{B}T_\mathrm{D}}$ and $R_\mathrm{D}(T_\mathrm{D}) \propto e^{E_R/k_\mathrm{B}T_\mathrm{D}}$. Explain why the activation energies are different? In the paper, "Is Damp Heat Degradation of c-Si Modules Essentially Universal?" published in *Proceedings of PVSC* 2019, Asadpour *et al.* explain the algorithm to calculate the activation energies by constructing a universal curve of existing data.

25.5 Humidity freeze test provides integrated testing for corrosion and cracking

Figure 25.3 shows that humidity-freeze (H-F) experiment is a sister test related to damp heat (DH) test. In the damp heat test, RH is cycled, while temperature is held fixed. In humidity-freeze experiment, RH is held fixed at 85%, while the temperature is cycled between −40 to 85°C. A small number of cycles (1–10) are used. Humidity-freeze experiments are motivated as follows: Assume that RH at a given place has reached very high value, making the cell susceptible to moisture diffusion and corrosion. Fortunately, the temperature will cycle from high to low values daily, which will in turn slow moisture diffusion. However, the temperature cycling may degrade the interfaces and crack the backsheet, allowing easier moisture transport.

Homework 25.4: Standardized tests have very specific requirements for test waveforms

Refer to IEC 61646 Ed. 2, 2008 International Standard. Find three characteristic differences between the stress waveform used for thermal cycling and humidity-freeze experiments.

25.6 UV measurements determines the rate of EVA yellowing

As discussed in Chapter 20, UV light breaks the bonds of the polymer encapsulant and the broken bonds render the module yellowish-brown. Light transmission is suppressed and the short-circuit current is reduced.

Many high-altitude regions of the world are showered with UV radiation (280 nm to 400 nm, or equivalently, 4.3 eV to 3.2 eV). It constitutes approximately 5.5% of the solar spectrum, or $55\,\mathrm{W/m^2}$ for a typical AM1.5G radiation. In the desert area, with an average of 6 hours/day of sunlight, the total exposure is approximately

$$\frac{55}{1000}\ \mathrm{kW/m^2} \times 6\ \mathrm{hrs/day} \times 365\ \mathrm{days/year} = 120\ \mathrm{kWh/m^2/year}.$$

Or, equivalently, $120 \times 25/1000 = 3\ \mathrm{MWh/m^2}$ over the 25 year lifetime.

Homework 25.5: Determining the UV test duration under accelerated test conditions

The typical accelerated test must not exceed 25 $\mathrm{kWh/m^2}$, because experiments show that the degradation saturates beyond the flux level. How many days of testing would you need to reach 3 $\mathrm{MWh/m^2}$ exposure, the integrated UV dose experienced by a module during its 25-year lifetime in a typical desert environment?

Solution. The equivalent exposure is reached in $3 \times 10^6/25 \times 10^3 = 120$ days or, approximately 4 months.

Homework 25.6: One can calculate the time-dependent loss of photocurrent due to UV degradation

Calculate the loss of short-circuit current, by using the linear approximation of Eq. (20.2), namely,

$$J_{\mathrm{sc}}(t) = J_{\mathrm{sc},0}\,(1 - a \times D'_{\mathrm{uv}}(t)) \tag{25.3}$$

where $a \equiv \gamma t_{\mathrm{p}} \sigma$ is the linear degradation coefficient to be obtained from experiments, and the integrated dose (D'_{uv}) is given by (cf. Eq. (20.1)):

$$D'_{\mathrm{uv}} = \int_0^t I(t) \times f_{\mathrm{uv}}\, dt$$

where $I(t)$ is the variable light intensity. For example, $I(t) = 1000\ W/m^2$ for AM1.5 spectrum, and $f_{\mathrm{uv}} = 0.055$ is the fraction of UV light in the solar spectrum.

Experiments show that short-circuit current decreases linearly until the dose reaches 100 $\mathrm{MJ/m^2}$ or 27.8 $\mathrm{kWh/m^2}$. Thereafter, UV degradation saturates to approximately 3%. Use this information to calculate the magnitude of a in the above equation.

25.7 Mechanical loading test ensures that the module will survive typical wind load

A module is stressed by variety of mechanical loads. Since wind speed changes frequently, it is best tested with a dynamic load of 0 to 2.5 kPa for 2–5 cycles. The load is relatively small, therefore unless the module fails because of poor quality, it will not fail under repeated wind load. Similarly, accumulated snow may remain in place for weeks. Its effect is tested with a uniform load of 5.6 kPa, but at $-40°$C, as expected.

Finally, in a hailstorm, a hail impacts the module locally and creates localized stress points. This localized stress buckles the laminate and shear stress so produced can delaminate the film. The glass can fracture under localized compressive stress as well. Hail tests are done with 25 mm ice balls impacting the modules with the terminal velocity of 23 m/s. A number of positions around the modules are specified and results collected. Hail as large as 75 mm (mass 203 g) moving at a velocity of 39.5 m/s is used for specialized tests.

Homework 25.7: A Hail test involves consideration of kinetic impact and localized stress

The hail test is conducted with 25 mm ice balls with a mass of 7.53 g impacting the module at 23 m/s. What is the impact pressure?

Solution. One can determine the momentum of the hail from the following equation:

$$F = 7.53 \times 10^{-3} \times 23 = 0.17\,\text{N} \cdot \text{s}.$$

If we assume that the impact radius is 10 mm, then

$$P = \frac{0.17}{(\pi/4)(10 \times 10^{-3})^2} = 2.165\,\text{kN} \cdot \text{s/m}^2.$$

This load is comparable to snow or wind load. The actual impact will be somewhat lower, because part of the incident energy $E = 7.53 \times 10^{-3} \times 23^2 = 3.98$ J will be lost to recoil and fragmenting the snowball.

For the largest hails mentioned above, $F = 0.203 \times 39.5 = 8$ N·s. The force is almost 50 times larger than that of a smaller hail.

25.8 Integrated stress sequence replicate actual environmental conditions

The individual tests described above ensure module integrity against temperature, humidity, and other stress. You may be surprised to see that none of the tests involve light exposure. Moreover, various degradation mechanisms are likely to proceed in parallel. For example, UV damage may be accelerated by moisture ingress.

A more refined approach (combined accelerated stress tests, C-AST) strings together multiple individual tests to establish the correlation effect. Figure 25.4 defines such a protocol. A set of 8–10 modules are initially characterized in terms of light I-V, dark I-V, electroluminescence, dry/wet insulation resistance, etc. The modules are soaked in light until its output is stabilized. Then, one module is set aside as a control or reference sample, and 1–2 modules are assigned for (composite) PID, UV, mechanical loading tests, the detail sequence of which are shown in Fig. 25.5. Finally, a few modules are not stressed, but operated normally to obtain the energy output. At the end of each test, the modules are recharacterized to produce the final report.

Figure 25.5 shows how the individual tests are strung together for a specific test. For example, the PID test shown in Fig. 25.4 (second branch in the middle stack) actually consists of 5 cycles of damp heat test under high bias. The modules are taken out periodically (50, 100, 200, 300, 400 hours) and are fully characterized.

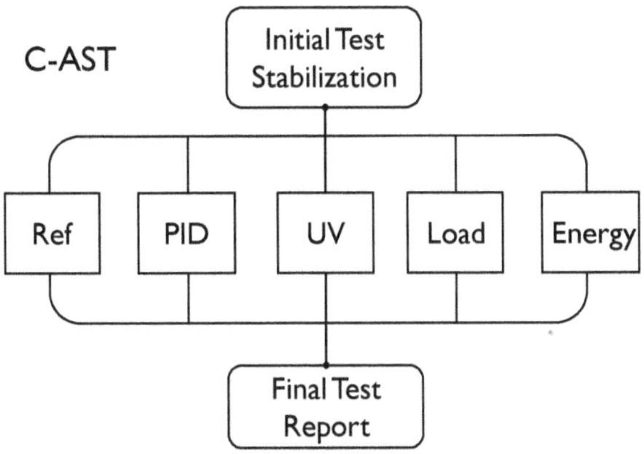

Figure 25.4: An illustrative example of combined acceleration stress testing protocol. For details of PID, UV, and Load tests, see Fig. 25.5.

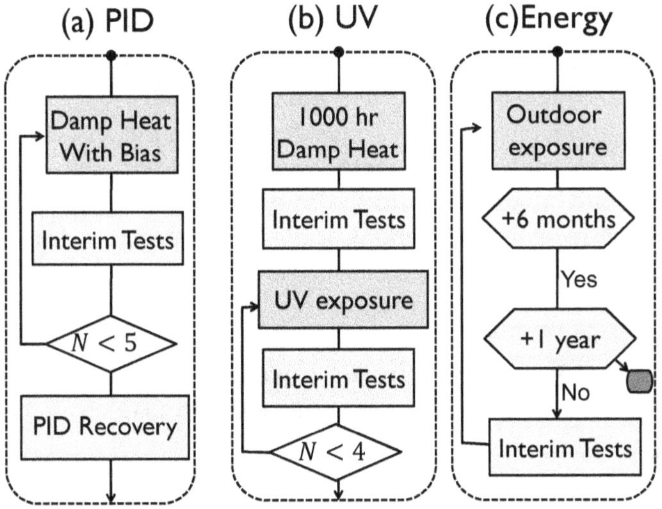

Figure 25.5: Details of various test sequence. Individual characterization tests (e.g., DH, UV exposure) have already been discussed in the text.

Afterward, in the PID recovery step, the module is operated at the maximum power point for a total exposure of 25 kWh/m^2 to see if PID effect is reduced under light exposure.

Similarly for the UV test sequence, a 100-hour damp heat test precedes the UV test. Subsequently, the module is exposed to UV dose of 25 kWh/m^2 for 4 cycles for a total exposure of 100 kWh/m^2.

Finally, the energy output of the module is monitored in a long-term field experiment. The modules are exposed in the outdoor conditions and its energy output is recorded continuously. Every six months, the modules are characterized by the full battery of tests. The experiments may continue for years until the energy produced falls below a specified minimum.

25.9 Conclusions: Rigorous module qualification is an essential prerequisite for reliable field operation

We discussed in Chapter 16 that the levelized cost of electricity (LCOE) of a solar cell technology depends both on the efficiency of the solar cell and the long-term reliability of the solar module. Once installed in a geographical location, a module may degrade in multiple ways, including corrosion, potential-induced degradation, UV degradation, and so on. A field failure is costly; therefore it is important to check the reliability of the solar cells through careful qualification tests.

In this chapter, we explained the importance and rationale of various qualification tests (e.g., damp heat, humidity freeze, etc.) and how to use the accelerated test results to extrapolate to field operating conditions. Since multiple degradation processes can co-exist, the community has gradually developed combined accelerated test protocols that better mimic the field conditions.

Since we now understand how various degradation modes depend on the depend the intensity of sunlight, temperature, and relative humidity of the local environment, we should be able to predict the time-dependent energy output of a solar farm installed anywhere in the world. Despite careful design and a rigorous qualification protocol, modules installed in solar farms may degrade differently than expected. One must continuously monitor the fielded modules to identify unexpected rates of degradation that may change the projected lifetime of the farm and to learn from the experience to improve the manufacture of the next generation of solar cells. We will discuss these issues in the next two chapters of the book.

References

[1] C. R. Osterwald and T. J. McMahon. History of accelerated and qualification testing of terrestrial photovoltaic modules: A literature review. *Progress in Photovoltaics: Research and Applications*, 17(1):11–33, 2009. _eprint: https://onlinelibrary.wiley.com/doi/pdf/10.1002/pip.861.

[2] Vikrant Sharma and S. S. Chandel. Performance and degradation analysis for long term reliability of solar photovoltaic systems: A review. *Renewable and Sustainable Energy Reviews*, 27:753–767, November 2013.

[3] Sarah Kurtz, Jennifer Granata, and Michael Quintana. Photovoltaic-reliability R&D toward a solar-powered world. In *Reliability of Photovoltaic Cells, Modules, Components, and Systems II*, volume 7412, page 74120Z. International Society for Optics and Photonics, August 2009.

[4] D. C. Jordan and S. R. Kurtz. Photovoltaic Degradation Rates — An Analytical Review. *Progress in Photovoltaics: Research and Applications*, 21(1):12–29, 2013. _eprint: https://onlinelibrary.wiley.com/doi/pdf/10.1002/pip.1182.

[5] J. H. Wohlgemuth, D. W. Cunningham, P. Monus, J. Miller, and A. Nguyen. Long Term Reliability of Photovoltaic Modules. In *2006 IEEE 4th World Conference on Photovoltaic Energy Conference*, volume 2, pages 2050–2053, May 2006. ISSN: 0160-8371.

CHAPTER 26

Predicting the Lifetime of Solar Farms

Chapter Summary

- The lifetime of the solar module can be obtained by location-specific integration of various degradation pathways.
- The local weather conditions (e.g. moisture, temperature, UV intensity, etc.) are easily obtained from multiple satellite-derived databases.
- The cell and module information are available from the research laboratories and product data-sheet.
- The degradation of photocurrent due to UV exposure, series-resistance increase due to Solder-bond failure, etc. can be included in the 5-parameter model so that the correlated degradation is correctly predicted.
- These lifetime model does not account for extrinsic failures, such as losses due to mishandling or incorrect installation of the fielded modules.

26.1 PV Lifetime depends on local climate and module technology

From Chapter 18 to Chapter 24, we discussed how various intrinsic degradation mechanisms reduce the power output of a solar module. In the previous chapter, we also saw how a fully assembled module is carefully tested to ensure that extrinsic factors, such as poor manufacturing or lower quality materials, do not compromise the integrity of the product.

Over the years, the output power output $P(t)$ of a fielded module will fluctuate periodically due to soiling, shadowing, etc. and reduce gradually due to corrosion, PID, yellowing, delamination, and so on. Since the cleaning and maintenance cost, $C_{\text{om}}(t)$, remains fixed (cf. Eq. (16.4)), the farm becomes too costly to maintain when the power output reduces below P_{crit}. The farm must be shut down and the modules recycled. In this chapter, we wish to calculate t_0, the intrinsic lifetime of the solar farm given the local environmental conditions. We will also determine the lifetime energy output of the solar farm ($E(Y)$), given by the area under the $P-t$ curve.

The lifetime t_0 depends on two factors: local climate determined by the geographical location of the farm and the module technology determined by the manufacturer. (As an aside, weather indicates short-term conditions of the atmosphere and is used to forecast short-term power output of a solar farm, while climate is the average daily weather for an extended period of time at a certain location that determines the slow degradation of the solar modules.) Once we know the local climate ($\mathbb{C}$: temperature, relative humidity, UV intensity, etc.) along with the module configuration ($\mathbb{G}$: cell type, arrangement of the cells, encapsulant type and thickness, etc.), we can calculate the degradation rates due to yellowing (D_1), PID (D_2), corrosion/solder bond failure (D_3), and so on; see Fig. 26.1. Each of these processes will reduce the short-circuit current or increase the series resistance in a predictable way. As the solar cell parameters degrade over the years, it is possible to estimate $P(t)$ for the module/farm due to the combined effects of various degradation modes. Once $P(t_0) \leq P_{\mathrm{crit}}$, we can determine the location-specific $t_0(\mathbb{C}, \mathbb{G})$ of a solar farm. In short, the correlation among the local climate, module configuration, and the time-dependent output power of a solar farm is complicated, but generally predictable. The goal of this chapter is to explain the components of such a predictive lifetime model.

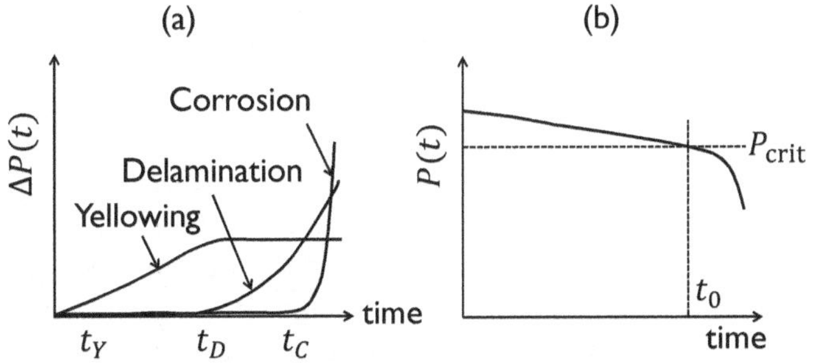

Figure 26.1: (a) Various degradation mechanisms, each with its characteristic time dependence, erode the power output of a solar module. (b) As a result, the total output power keeps decreasing until t_0 when the farm cannot be operated profitably any longer.

26.2 Local climate information ($\mathbb{C}$) can be obtained from public databases

There are several public databases that have recorded local climate information over many years; see Table 26.1. For example, the National Solar Radiation Database (NSRDB) is a collection of five databases containing information regarding a variety of geographic locations with different spatiotemporal resolution. The

Table 26.1: Multiple databases aggregate weather information from a variety of weather stations and satellites (as of June 2021). NSRDB database is located at `https://maps.nrel.gov/nsrdb-viewer` and NASA POWER database is located at `https://power.larc.nasa.gov/`.

Database (Satellite)	Year Coverage	Spatial Resolution	Temporal Resolution	Region Covered
NSRDB PSM (v3) (GOES)	1998–2016	4 km × 4 km	30,60 min	North/South America Latitude: 60 N to 21 S
NSRDB PSM (v3.2) (GOES)	2018–2019	4 km × 4 km	5,15,30,60 min	Continental USA
NSRDB PSM Full-Disc (GOES)	2019	2 km × 2 km	15,30,60 min	North/South America
NSRDB PSM v3 (METEOSAT)	2017–2019	4 km × 4 km	15,30,60 min	Europe, Africa, middle/Western Asia
NSRDB PSM v3 (HIMAWARI)	2016–2019	2 km × 2 km	15,30,60 min	Australia, Eastern Asia
NSRDB SUNY (METEOSAT 5,7)	2000–2014	10 km × 10 km	60 min	South Asia
NASA	1981–present	50 km × 50 km	3-hr[a], Daily, Monthly[a]	World

[a]The 3-hourly and monthly information based on 22-year (1983–2005) average data.

data are collected from a combination of satellite information and local weather stations. NASA also maintains a smaller database of satellite-derived information.

The data contained in these databases can be divided into two groups. Chapters 12–14 showed that information regarding insolation, clearness index, albedo, ambient temperature, wind speed, etc., are needed to calculate hourly or daily energy output of a solar module. In addition, information regarding relative humidity, temperature, soiling rate, snow fall, etc., are needed to calculate PV degradation.

It is clear from the list of parameters in Table 26.2 that a single database may not have all the parameters of interest. Sometimes the data may be incomplete or corrupted. The tables can be "repaired" in various ways: one can use information from complementary databases that tabulate the corresponding information. Analytical approximations for some of this information are also available. Nevertheless, since the degradation processes are slow, even imprecise data can be used to reliably predict the overall degradation. Indeed, the monthly averages smooths over local fluctuations and provides sufficient accuracy for long-term prediction.

Table 26.2: List of information contained in various databases as of June, 2021. NREL NSRDB database is located at `https://maps.nrel.gov/nsrdb-viewer` and NASA POWER database is located at `https://power.larc.nasa.gov/`.

Database (satellite)	Parameters
NSRDB PSM v3 (GOES)	(Clear sky: DHI, DNI, GHI) (W/m^2), Dewpoint (°C), Temperature (°C), Pressure (mbar), Relative Humidity (%) Albedo, Precipitation (cm), Wind direction (deg), Wind Speed (m/s), Solar Zenith (deg)
NSRDB PSM v3 (GOES)	(Clear sky: DHI, DNI, GHI) (W/m^2), Dewpoint (°C), Temperature (°C), Pressure (mbar), Relative Humidity (%), Albedo Precipitation (cm), Wind direction (deg), Wind Speed (m/s), Solar Zenith (deg)
NSRDB Full Disk (GOES)	(Clear sky: DHI, DNI, GHI) (W/m^2), Dewpoint (°C), Temperature (°C), Pressure (mbar), Relative Humidity (%), Albedo Ozone, Precipitation (cm), Wind direction (deg), Wind Speed (m/s), Solar Zenith (deg)
NSRDB PSM v3 (METEOSAT)	(Clear sky: DHI, DNI, GHI) (W/m^2), Dewpoint (°C), Temperature (°C), Pressure (mbar), Relative Humidity (%), Albedo Ozone, Precipitation (cm), Wind direction (deg), Wind Speed (m/s), Solar Zenith (deg)
NSRDB PSM v3 (HIMAWARI)	(Clear sky: DHI, DNI, GHI) (W/m^2), Dewpoint (°C), Temperature (°C), Pressure (mbar), Relative Humidity (%), Albedo Ozone, Precipitation (cm), Wind direction (deg), Wind Speed (m/s), Solar Zenith (deg)
NSRDB SUNY (METEOSAT)	(Clear sky: DHI, DNI, GHI) (W/m^2), Dewpoint (°C), Temperature (°C), Pressure (mbar), Relative Humidity (%), Albedo Snow (m), Precipitation (cm), Wind direction (deg), Wind Speed (m/s), Solar Zenith (deg)
NASA	(AM0, Clear sky: DHI, DNI, GHI) (kW-hr/(m^2 day)), clearness index, precipitation (mm/day), min. and max. daily Temperature (°C), Wind Speed (m/s), Solar noon, Solar zenith angle (deg), Solar Azimuth angle (deg), Surface albedo, Thermal radiative flux, Cloud amount

Homework 26.1: Exploring the NSRDB Data Viewer

Find the latitude and longitude of the city you live in. Next visit the NREL NSRDB website located at `https://maps.nrel.gov/nsrdb-viewer`, then select the "Download Data" tab at the top left of the page. Use the "NSRDB Data Download (point)" to select the city in the map and fill up the form. The data will be emailed to you in zipped excel files. Plot the results to see how the weather of your city has changed over the years.

26.3 Cell, module, and farm information ($\mathbb{G}$) are available from Research Labs, Solar Cell Manufacturers, and System Installers

Given the weather information, the module degradation depends on the module technology and the farm configuration. The module and farm information can be roughly divided into two groups: *physical* and *electrical*. During the design phase of a solar cell technology, a research laboratory collects a significant amount of electrical information about the cell and the module. This information includes intensity and temperature-dependent I-V characteristics, reverse-bias breakdown voltage, etc. The degradation rates of the cell/module under a variety of stress conditions are also available. If used appropriately, this information is sufficient to extract the physics-based compact models of a pristine and degraded solar module.

Physical information regarding the module technology is equally important for physics-based modeling of module reliability. We need to know the module architecture (e.g., glass-on-glass or glass backsheet, etc.) as well as its geometrical and material properties, such as the thickness of glass cover, the composition of the encapsulant, the type of module framing, etc. This information is available from the datasheet and various characterization experiments reported in the literature. Given the electrical and physical parameters, we can use the formula developed in Chapters 17–24 to calculate the magnitude of PID leakage current, onset time for corrosion, etc.

Finally, depending on the technology and geographical location, the modules must be optimally spaced and tilted. The tilt/spacing determines soiling and the electrical connection of the modules and the integration of the microinverters define the maximum voltage. Together, they dictate PID and light corrosion. The module/panel spacing also determines mutual shading, temperature-distribution over a module, and shade-induced degradation of a solar cell.

Homework 26.2: Learning to read the datasheet from a PV manufacturer

Find the datasheet for a solar module from a well-known manufacturer. Summarize the performance (e.g., efficiency, temperature coefficient, etc.) and reliability (e.g., PID leakage, shade tolerance, etc.) information contained therein. What other pieces of information would have been useful to calculate the lifetime of the solar module?

Homework 26.3: Pristine I-V-T data can be encapsulated in a physics-based compact model

For a c-Si solar cell obeying superposition principle, create a series of I-V curves at different temperatures T_{D} and intensities c by using the equation:

$$I = cI_{\mathrm{sc}} - I_0(e^{q(V+IR_{\mathrm{s}})/k_{\mathrm{B}}T_{\mathrm{D}}} - 1) - (V + IR_{\mathrm{s}})/R_{\mathrm{sh}}.$$

Assume that $I_{\mathrm{sc}} = 38$ mA/cm^2, $R_{\mathrm{sh}} = 1$ kΩ cm^2, $R_{\mathrm{s}} = 1\,\Omega\cdot$ cm^2.

Consider these synthetic data as if they came from I-V measurement conducted at different temperatures. Use **PV-PanelSim:** `https://nanohub.org/resources/pvpanelsim` to determine the parameters of the model. Are the parameters comparable to those used to synthesize the data?

26.4 There are different ways to predict PV lifetime

26.4.1 Module Lifetime can be predicted by empirical degradation models

Figure 26.2 shows that the weather and geometry information can be used to calculate the various degradation modes. The power lost due to each individual mode (D_{i}) can be added to calculate the total power loss in the system. In other words, the total power lost due to N degradation modes operating in parallel is given by

$$\Delta P(t) = \sum_{i=1}^{N} \Delta P_i(\mathbb{C}, \mathbb{G})(t) \qquad (26.1)$$

where $\Delta P_i(\mathbb{C}, \mathbb{G})$ is the degradation associated with the i-th mode. For these calculations, one obtains the parameterized degradation rates as a function of weather/climate variables ($\mathbb{C}$: temperature, RH, etc.) and technology specification ($\mathbb{G}$: cell type, EVA thickness, etc.). As we will explain in the following section, the superposition assumed in Eq. (26.1) is not accurate.

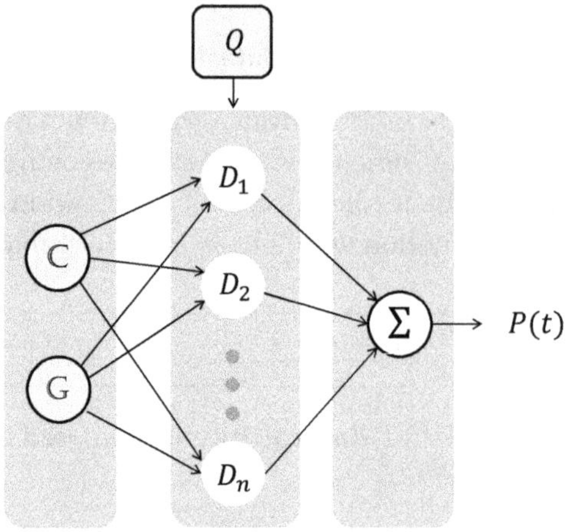

Figure 26.2: Given the climate ($\mathbb{C}$) and module ($\mathbb{G}$) information, we can calculate the module degradation rates, (D_i). The net power loss is obtained by summing over power lost to various degradation modes. The qualification tests Q discussed in Chapter 25 provide relevant parameters for the degradation rates.

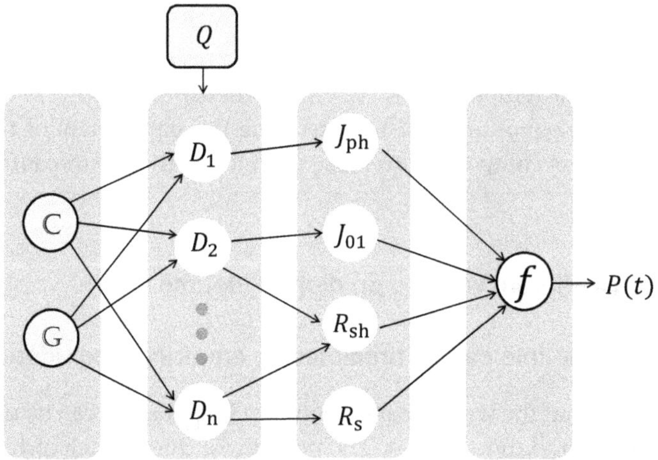

Figure 26.3: If the degradation rates ($D_1, D_2, \ldots$) are parameterized in terms of degradation of various components of a circuit model ($J_{ph}, \ldots, R_{sh}, R_s$), one can calculate their correlated effect on the output power of the solar farm.

26.4.2 Lifetime prediction is improved by physics-based compact models

An important limitation of the empirical approach described in Fig. 26.2 is that the approach does not account for the correlation among the degradation modes. While all the degradation modes eventually affect power output of the solar cell, they do so by independently affecting various parameters of the solar cell. For example, as shown in Fig. 26.4, some degradation modes affect a single device parameter, e.g., yellowing affects the short-circuit current of a solar cell, PID primarily affects the shunt resistance, solder bond failure affects the series resistance. Other degradation modes affect multiple parameters simultaneously. For example, in Chapter 23, we explained how corrosion/delamination simultaneously reduce photocurrent collection, decrease shunt resistance, and increase the series resistance. One can calculate the output power as the output of the time-evolving compact model, explicitly preserving the nonlinear dependencies among the variables. Note that the approach accounts for the time-dependent power losses associated with the nonlinear time dependence of the relevant parameters. A summary of the various degradation models are shown in Table 26.3.

A number of reliability software based on time-dependent compact models have been published in the literature. For a module installed in a given location, the databases provide historical weather information regarding environmental variables ($\mathrm{RH}(t)$, $T(t)$, $S_0(t)$). Reliability software, such as **PVLife** from SunPower, uses these inputs to calculate how the compact model parameters evolve as a function of time. Since the model is evaluated at each time step, one can calculate efficiency degradation and determine the lifetime t_0.

Table 26.3: Summary of degradation models implemented in PurduePVLife.

Degradation	Functional Relationship	Reference
Soiling	$J_{\mathrm{ph}}(t) = J_{\mathrm{ph}}(t=0)\left[1 - e^{-Q_i^* f_i(t)t}\right]$ $\simeq J_{\mathrm{ph}}(t=0)\left[1 - e^{-\alpha t}\right]$ **Notes.** α: Location-specific soiling constant. Q_i^*: Mie extinction efficiency. $f_i(t)$: Time-dependent surface coverage	Ch. 17 Eq. (17.8) HW. 17.7
Partial Shading	$R_{\mathrm{D,ps}} = A_{\mathrm{ps}} e^{-E_{\mathrm{A,ps}}/k_{\mathrm{B}} T_{\mathrm{D}}}$ $R_{\mathrm{sh}} = R_{\mathrm{sh}}(0)/\left[1 + \int_0^t R_{\mathrm{D,ps}} dt\right]$ (linearized form) $\frac{R_{\mathrm{sh}}(t)}{R_{\mathrm{sh}}(\infty)} = \left[1 + \left(\frac{R_{\mathrm{sh}}(t=t_{\mathrm{i}})}{R_{\mathrm{sh}}(\infty)} - 1\right) e^{-\int_{t_{\mathrm{i}}}^{t} R_{\mathrm{D}}(u)du}\right]^{-1}$ **Notes.** A_{ps}: Empirical prefactor, $E_{\mathrm{A,ps}}$: activation energy for defect formation, T_{D}: device temperature including self-heating	Ch. 18 Ch. 19 Eq. (18.6)

Table 26.3: Summary of degradation models implemented in PurduePVLife.

Degradation	Functional Relationship	Reference
UV EVA Yellowing	$D_{\mathrm{uv}}(t) = \int_0^t dt'\sigma I(t') \times f_{\mathrm{uv}}(t')$ $J_{\mathrm{ph}}(t) = J_{\mathrm{ph}}(t=0)\left[1 - e^{-\gamma\, t_{\mathrm{EVA}} D_{\mathrm{uv}}(t)}\right]$ $J_{\mathrm{ph}}(t) = J_{\mathrm{ph}}(t=0)$ $\left(1 - cD_{\mathrm{uv}}(t) - d\left(1 - e^{-\gamma\, t_{\mathrm{EVA}}\, D_{\mathrm{uv}}}\right)\right)$ **Notes.** J_{ph}: Photocurrent; D_{uv}: Integrated defects by UV radiation; γ: absorption rate per unit thickness; t_{EVA}: EVA thickness. $I(t)$: Time-dependent intensity; $f_{\mathrm{uv}} \approx 0.055$: Typical fraction of UV light. Some groups use the model up to $D_{\mathrm{uv}} = D_{\mathrm{uv}}^{crit}$ with no further degradation thereafter. The second form allows an onset delay before the exponential degradation, characterized by c and d parameters, begins.	Ch. 20 Eq. (20.2) Eq. (20.3)
UV Surface Damage	$J_s - J_{s,t}^x = (m\,t)^n$ J_s: Surface recombination; $n = 0.25 - 0.5$, $m \approx 2$: Empirical parameters.	Ch. 20 Eq. (20.4)
LID	$J_{02} \propto \tau^{-1} = A_l e^{-E_{A,l}/k_{\mathrm{B}}T} n_{\mathrm{ph}}^{2/3} t^{n_l}$ **Notes.** J_{02}: Diode recombination current, $E_{A,l}$: bond dissociation activation energy, n_{ph}: time-integrated photon number, n_l: time exponent.	Ch. 21 Eq. (21.8)

Table 26.3: Summary of degradation models implemented in PurduePVLife.

Degradation	Functional Relationship	Reference
PID	$R_{\text{sh}}(t) = R_{\text{sh}}(0)/(1+\int_0^t R_{D,p}(u)du)$ (linearized form) $\frac{R_{\text{sh}}(t)}{R_{\text{sh}}(\infty)} = \left[1+\left(\frac{R_{\text{sh}}(t=t_{\text{i}})}{R_{\text{sh}}(\infty)}-1\right)e^{-\int_{t_{\text{i}}}^{t} R_{\text{D}}(u)du}\right]^{-1}$ $\Delta P(t)/\Delta P_{\infty} = [1+((\Delta P_{\infty}/\Delta P(t_{\text{i}}))-1)\times$ $\exp(-R_{D,p}(t-t_{\text{i}}))]^{-1}$ $R_{D,p} = A_p V^{n_p} e^{-E_{A,p}/k_{\text{B}}T_{\text{D}}} \times f(\text{RH})$ $f(\text{RH}) = [\text{RH}]^{B_p}$ or $1/\left(1+Me^{-c\,\text{RH}}\right)$ or $\text{RH}/(1-\text{RH}+\epsilon)$ **Notes.** R_{sh}: Shunt resistance; $R_{D,p}$: Shunt degradation rate; n_p: PID voltage exponent; $E_{A,p}$: Activation energy for Na drift; B_p: Moisture exponent for PID. A_p is a empirical parameter specific module geometry. Na^+ diffusion enhances shunt formation. Also, T_{i}: onset delay, $\Delta P(t)$: Power degradation at time t. Both RH and T_{D} are time-dependent parameters. Typical values: $n_p = 1-2$, $E_{A,p} = 0.88\text{eV}$, $B_p = 7.12$.	Ch. 22 Eq. (22.1) Eq. (22.8) HW. 23.10
Corrosion Power loss	$\frac{\Delta I_{\text{sc}}(t)}{\Delta I_{sc,\infty}} = \left[1+\left(\frac{\Delta I_{sc,\infty}}{\Delta I_{sc,i}}-1\right)e^{-\int_{t_{\text{i}}}^{t} R_{D,c}(u)du}\right]^{-1}$ $\frac{\Delta P(t)}{\Delta P_{\infty}} = \left[1+\left(\frac{\Delta P_{\infty}}{\Delta P_{(t_i)}}-1\right)e^{-(t-t_i)R_{\text{D}}}\right]^{-1}$ $R_{D,c} = A_c e^{-E_{A,c}/k_{\text{B}}T_{\text{D}}} \cdot g(\text{RH})$ $g(\text{RH}) = [\text{RH}]^{n_c}$ or $1/\left(1+Me^{-c\text{RH}}\right)$ or $\text{RH}/(1-\text{RH}+\epsilon)$ **Notes.** R_{sh}: Shunt resistance; $R_{D,c}$: Shunt degradation rate; ΔP_{∞}: Saturated power loss; t_i: onset delay. P_i, A_c, n_c are empirical constants, $E_{A,c}$: Activation energy for corrosion. Typical values: $B_p = 2.2\pm0.8$, $E_{A,p} = 0.89\pm0.11\text{eV}$, $A_c(5\%) = 3.7\times10^{-10}$, $A_c(10\%) = 5.1\times10^{-10}$, $A_c(20\%) = 6.4\times10^{-10}$.	Ch. 23 Eq. (23.1) HW. 23.2

Table 26.3: Summary of degradation models implemented in PurduePVLife.

Degradation	Functional Relationship	Reference
Solder-bold Failure Series Resistance	$R_s(t) = R_s(0)\left[1 + \int_0^t R_{D,s}(u)du\right] \approx$ $R_s(0)\left[1 + A_s\, e^{-E_{A,s}/k_B T_D}[\mathrm{RH}]^{n_S} t^n\right]$ $\frac{R_s(t)}{R_s(\infty)} = \left[1 + \left(\frac{R_s(\infty)}{R_s(t=T_i)} - 1\right) e^{-\int_{T_i}^t R_D(u)du}\right]^{-1}$ $R_{D,s} = C_s \cdot (\Delta T)^n N(T_r)^{b_s} e^{-E_{A,s}/k_B T_D}$ **Notes.** The increase in series resistance, R_s, is attributed to solder bond failure. Here, A_s and n_S are empirical corrosion constants; $E_{A,s}$, the activation energy for solder bond failure; C_s, empirical constant; n_S, temperature exponent; $N(T_r)$, number of times the temperature crosses the critical temperature T_r; and b_s, corresponding exponent. Typical values: $n_S = 1.9, 0.33, T_r = 54.8°\mathrm{C}$.	Ch. 23 Eq. (23.14) HW. 23.5.6 Eq. (24.26)
Fracture	$\ln(N_F) = B(1 - S/C), 1 - F(N) = \exp\left(-(N/N_F)^{\xi}\right)$ **Notes.** Glass fracture under repeated tensile stressing. N_F: Average number of cycles to failure, S: Cyclic stress, B, C: Empirical material dependent constants. $F(N)$: Cumulative fraction failed after N-cycles, ξ: Weibull shape factor for failure distribution.	Ch. 24 Eq. (24.13) Eq. (24.15)
Delamination	$\frac{\Delta T_1}{\Delta T_2} = \frac{\tau_1 - \tau_{\min}}{\tau_2 - \tau_{\min}} = \frac{\ln(N_{f2}/N_{\max})}{\ln(N_{f1}/N_{\max})}$ **Notes.** ΔT: temperature range; τ: shear stress; N_F: number of cycles to delamination.	Ch. 24 Eq. (24.24) HW. 24.8

Table 26.3: Summary of degradation models implemented in PurduePVLife.

Degradation	Functional Relationship	Reference
Glass cracking	$R_s(t) = R_s(0)(1 + A_{g,s}\, t)$ Cracking of the metal grid leads to increase in series resistance. Parameter $A_{g,s}$ depends on the stress caused by temperature cycling. $J_{ph} = J_{ph}(0)\,[1 - A_{g,ph} t]$ Glass cracking reduces photocurrent by increasing light scattering and reflection. A maximum loss of 8–10% has been reported. $J_{02} = J_{02}(0)[1 + A_{g,r} t]$ The cracked surface increases junction recombination.	Ch. 24

26.4.3 Physics-based numerical modeling of PV reliability

So far, we have treated the degradation modes as independent (see Homework 26.5), but in practice this is not the case. For example, yellowing may enhance PID and corrosion. The broken bonds in a degraded polymer provide additional sites for moisture diffusion. Water in turn enhances ion transport, thereby increasing both PID and light corrosion. Similarly, solder bond failure enhances delamination by hotspot formation. Figure 26.4 does not capture these mutual interactions. The mutual interaction would have to be represented by arrows connecting D_1 to D_2 and D_3, for example. A complete mathematical model, formulated as a set of coupled multivariable differential equations, is needed to capture these interactions explicitly.

A variety of device simulators, such as **PC1D**, **ADEPT**, etc. can predict electron–hole transport as well as self-heating process within a solar cell. Similarly, a variety of module simulators, such as **Griddler** and **PVPanelSim**, can simulate the effect of various grid designs as well as the consequences of corrosion and partial shadowing. Theoretically, a three-dimensional model should be able to self-consistently account for the coupled moisture diffusion, stress distribution, ion transport due to potential difference, etc. to obtain the spatially and temporally resolved degradation. While several groups have developed three-dimensional finite element models for various individual processes, a self-consistent model that involves all the relevant processes is still missing.

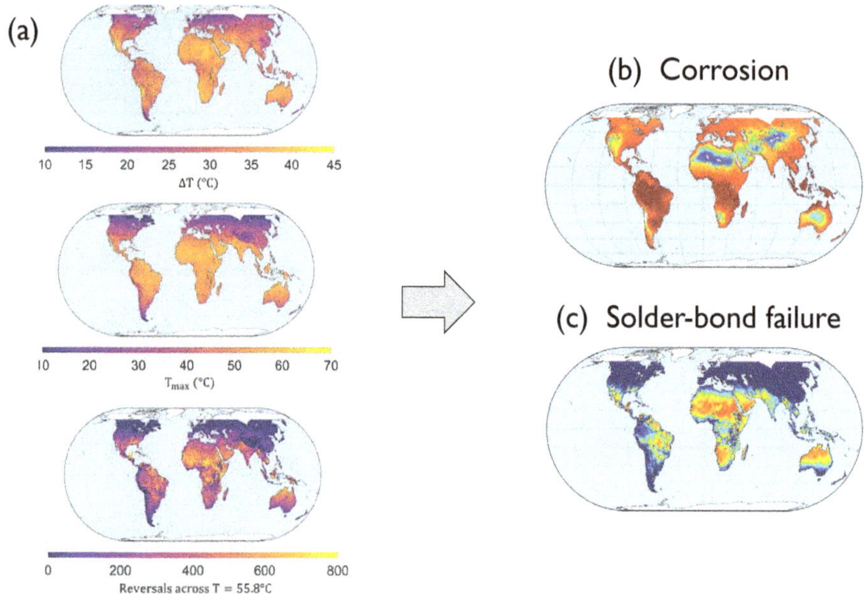

Figure 26.4: Two degradation modes (D_i) and location specific lifetime ($t_{i,j}$) of a solar module based on the equations summarized in Table 26.3: (a) The factors that determine the degradation modes, (b) corrosion limited lifetime, based on Eq. (23.1), and (c) solder-bond failure limited lifetime, based on Eq. (24.26). Blue indicates longer lifetime, while red indicates shorter lifetime. For a given location, the lifetimes of the two degradation modes are somewhat complementary: corrosion is accelerated by temperature and humidity, while solder-bond failure is driven by temperature fluctuation, but not by humidity. Taken from R. Asadpour *et al.* IEEE Journal of Photovoltaics, 2021.

Homework 26.4: A time-dependent compact model predicts module output

Use the compact model-based PV reliability software **PurduePVLife** to calculate the degradation of a solar module under yellowing, PID, and cracking. For a standard c-Si solar cell, assume the following parameters for cell degradation: $A_{g,s} = 1.6 \times 10^{-6}$, $A_{g,ph} = 4.56 \times 10^{8}$, and $A_{g,j} = 2.51 \times 10^{-6}$. How would the output evolve if the system is illuminated at 800 W/m^2 and RH = 50% and $T = 28°$C?

26.5 Conclusions: Predictive models do not account for extrinsic failures

In this chapter, we saw how the cell, module, and farm data can be integrated with weather information to predict location-specific power degradation. If the location-specific lifetime is satisfactory, one must reduce the design to practice. Qualifying the fully fabricated cells and tracking the performance of the modules that have been installed are necessary to ensure that the theoretical lifetime prediction did not overlook any important degradation pathway. Equally important: the theoretically predicted lifetime is achieved only if failures of other components (e.g., inverters, mechanical support) do not compromise the operation of the solar farm. In the next chapter, we will use the self-diagnostic information from installed modules to compare the model predictions with actual field degradation.

Homework 26.5: Lifetime calculation is simplified if the degradation rates are linear in time

The power loss due to a degradation mode can be described by $\Delta P_i(t) = f(t/T_\mathrm{i})$, where $t_i(T, \mathrm{RH}, \ldots)$ is the time needed to reach a critical power loss, c. Obviously, the time depends on climate variables: Given $f(t/T_\mathrm{i}) = (t/T_\mathrm{i})$, that is, if the degradation rate is linear in time and the climate variables are fixed over the duration of stress, show that the ultimate failure time is given by $1/t_\mathrm{f} = \sum_i (1/T_\mathrm{i})$. Generalize the results to account for variable climate conditions during the stress period.

Solution. Assuming that the $\Delta P(t = t_\mathrm{f}) = c$, where c is a constant, we find that

$$c = \sum_{i=1}^{N} ct_\mathrm{f}/T_\mathrm{i}.$$

Rearranging, we get the final answer.

If the climate variables change over time, we can divide the stress duration over P number of time intervals so that the weather variables are constants over each time period. We can calculate the total degradation by summing over each degradation period, namely,

$$1/t_\mathrm{f} = \sum_{i}^{N} \sum_{j}^{P} 1/t_{i,j}$$

where $t_{i,j}$ is related to the degradation mode i for the time duration j.

References

[1] M. A. Mikofski, D. F. J. Kavulak, D. Okawa, Y. Shen, A. Terao, M. Anderson, S. Caldwell, D. Kim, N. Boitnott, J. Castro, L. A. L. Smith, R. Lacerda, D. Benjamin, and E. F. Hasselbrink. PVLife: An integrated model for predicting PV performance degradation over 25+ years. In *2012 38th IEEE Photovoltaic Specialists Conference*, pages 001744–001749, June 2012. ISSN: 0160-8371.

[2] E. Hasselbrink, M. Anderson, Z. Defreitas, M. Mikofski, Y. Shen, S. Caldwell, A. Terao, D. Kavulak, Z. Campeau, and D. DeGraaff. Validation of the PVLife model using 3 million module-years of live site data. In *2013 IEEE 39th Photovoltaic Specialists Conference (PVSC)*, pages 0007–0012, June 2013. ISSN: 0160-8371.

[3] Benoît Braisaz, Chloé Duchayne, Mike Van Iseghem, and Khalid Radouane. PV aging model applied to several meteorological conditions. In *Proceedings of the 29th European Photovoltaic Solar Energy Conference (EU PVSEC), Amsterdam, the Netherlands*, pages 22–26, 2014.

[4] Ismail Kaaya, Julián Ascencio-Vásquez, Karl-Anders Weiss, and Marko Topič. Assessment of uncertainties and variations in PV modules degradation rates and lifetime predictions using physical models. *Solar Energy* 218, 354–367, 2021.

[5] Sascha Lindig, Ismail Kaaya, Karl-Anders Weiß, David Moser, and Marko Topic. Review of statistical and analytical degradation models for photovoltaic modules and systems as well as related improvements. *IEEE Journal of Photovoltaics* 8(6), 1773–1786, 2018.

CHAPTER 27

Inverse Modeling and Monitoring the Health of a Solar Farm

Chapter Summary

- Diagnostic inverse models can be used to interpret the power vs. time characteristics of a solar farm and determine the dominant degradation pathways.
- A combination of statistical and physics-based models are essential to determine the degradation mechanisms of a solar cell.
- The physics-based method is supported by two techniques: full I-V-T method and suns-Vmp based maximum power-point method.
- The statistical techniques can be used to obviate the needs for temperature/irradiance sensors.
- Physics-based Statistical Machine Learning is a powerful technique for diagnostic modeling of a solar farm.

27.1 Why forward/predictive modeling is insufficient: The need for inverse/diagnostic modeling

As discussed in Chapter 26, a fully predictive model will transform how modules are manufactured and deployed. A collection of well-calibrated, physics-inspired degradation models can estimate the upper bound for the farm lifetime (based on the subset of uncoupled degradation mechanisms) by summing over the power loss associated with each degradation mode. The predicted lifetime can be used to calculate location-specific LCOE. A forward predictive model not only specifies the lifetime and output energy of a solar farm based on a specific technology, but it also allows the manufacturer to adapt the module design for specific climates. For example, in a location with high humidity, the module is likely to fail by corrosion. Therefore, module lifetime will improve with improved framing, higher quality backsheet, and optimized processing of grid fingers. The increased module cost is offset by increased energy production from a more reliable module.

Four issues make the forward modeling difficult. First, despite significant progress, the current models are not fully predictive. Indeed, the science of various degradation modes (e.g., corrosion under light illumination, PID in thin films) are still evolving. Second, some of the degradation modes may be stochastic (e.g., glass cracking, partial shadowing, etc.) and thus unpredictable at shorter timescales. The longer-term seasonal averages, however, should be highly predictable. Third, the coupling among the degradation modes (e.g., how yellowing affects corrosion by allowing moisture to diffuse faster) is only qualitatively understood. Finally, the data needed for predictive modeling are often not available: The manufacturers may not have preserved all the qualification data, and even if they did, the information may be either incomplete or summarized as "pass/fail".

Therefore, while the forward predictive modeling remains the ultimate goal, we can obtain significant/complementary insights into the degradation processes by analyzing the data stream of an *existing* farm and treating each farm as a natural field experiment. If the power loss depends on the module design ($\mathbb{G}$) and environmental variables ($\mathbb{C}$), so that $\Delta P = f(\mathbb{C}, \mathbb{G})$, the goal is to determine the function f, given the field information regarding $\mathbb{G}$, $\mathbb{C}$ and ΔP. This data-driven diagnostic "inverse modeling" approach to be discussed in this chapter is complementary to the physics-based predictive "forward modeling" approach discussed in the previous chapter. In essence, forward modeling involves predicting ΔP given $f(.)$, while inverse modeling determines $f(.)$ given ΔP. In the following section, we will use a very simple example to illustrate the importance of data-driven diagnostic inverse models. Ultimately, these inverse models can be used to determine the location-specific degradation modes, predict its remaining lifetime, and guide the future choice of module technology.

27.2 Solar farm on Planet X defined by two weather variables, T and RH

Let us consider an idealized planet X where local climate variables T and RH vary from one location to the next, but they do not change with time (i.e., no seasons, a pretty boring weather!). For simplicity, let us assume that the solar cells in this world only degrade by corrosion which increases the (fake) shunt resistance, R_{sh}. The inhabitants have not read Chapter 22, and therefore they do not know that power loss due to corrosion is predicted by Eqs. (23.1)–(23.3), namely,

$$\frac{\Delta P(t)}{P_0} = A\,[\text{RH}]^B\, e^{-E_{\text{A}}/k_{\text{B}}T} t^{\beta} \tag{27.1}$$

where A, B, and β are technology-specific constants. In the absence of a theoretical model, the farm operators of planet X can adopt one of the two statistical approaches. (Their physics may not be good, but their statistics is!)

27.2.1 Statistical approach: Linear regression for a single farm

Realizing their solar farm is a giant experimental testbed, the farm operators carefully record how the power output drops over time. The black dots in Fig. 27.1(a) is a record of the power output, P_1, P_2, P_3, P_4, as a function of time. To find the pattern in the output power loss (ΔP), relative to the power output of the pristine module P_0, they assume a simple formula

$$\frac{\Delta P}{P_0} = f(T, \mathrm{RH}, t) \equiv at + b[\mathrm{RH}] + cT + d \tag{27.2}$$

where a, b, c, and d are unknown constants. The equation captures the idea that degradation increases with time, relative humidity, and temperature. The model is not quite right because $\Delta P(t \to \infty) = \infty$: a module output cannot go below zero, right? Actually, this is not a big concern because the module has to be replaced when it has lost 20% of its rated power. Perhaps a linear model can describe this relatively small degradation. To find a, b, c, and d, one can write the following matrix equation:

$$\begin{bmatrix} a & b & c & d \end{bmatrix} \times \begin{bmatrix} t_1 & t_2 & t_3 & t_4 \\ \mathrm{RH} & \mathrm{RH} & \mathrm{RH} & \mathrm{RH} \\ T & T & T & T \\ 1 & 1 & 1 & 1 \end{bmatrix}$$
$$= \begin{bmatrix} \Delta P_1 - e_1 & \Delta P_2 - e_2 & \Delta P_3 - e_3 & \Delta P_4 - e_4 \end{bmatrix}. \tag{27.3}$$

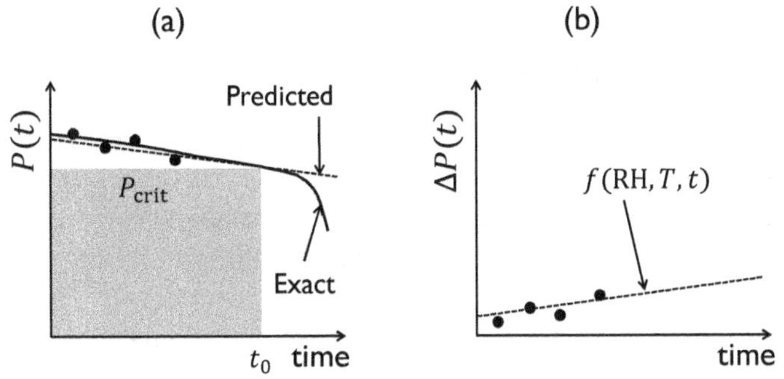

Figure 27.1: (a) Periodically recorded output power of a solar farm. (b) The loss of power can be interpreted by various statistical models, e.g., linear regression.

Here, ΔP_i is the measured degradation at time t_i; $e_1 \dots e_4$ denote the difference (error) between the model (Eq. (27.2)) and the actual data. The engineers in planet X can use the method of linear regression to find a set of a, b, c, d that minimizes the difference between model prediction and actual data. Specifically, if the square of the total error is

$$e_T^2 = e_1^2 + e_2^2 + e_3^2 + e_4^2,$$

then $de_T^2/da = 0$, $de_T^2/db = 0$, $de_T^2/dc = 0$, and $de_T^2/d(d) = 0$ will produce four independent equations, the solution of which will produce the optimum value for a, b, c, d. With these coefficients at hand, the installers could predict the future $\Delta P(t)$.

27.2.2 Statistical Approach: Log–log regression for a single farm

One could argue that Eq. (27.2) predicts $\Delta P(t)$ could not possibly keep increasing linearly with time. This may prompt us to look for other different functional forms that would make the long-term degradation more reasonable, e.g.,

$$\Delta P(t)/P_0 = a\,[\mathrm{RH}]^b\,T^c\,t^d. \tag{27.4}$$

For, $0 < d < 1$, the degradation will appear to saturate over time, consistent with experimental observations. The results may not be as accurate as Peck's equation (because Eq. (27.4) does not contain exponential temperature dependence), but the predictions may still be reasonable. By applying a log on both sides of Eq. (27.4), we can convert the nonlinear equation into a linear equation analogous to Eq. (27.2), and optimize the coefficients.

$$\ln(\Delta P/P_0) = a + b\ln[\mathrm{RH}] + c\ln(T) + d\ln(t). \tag{27.5}$$

The linear or log–log model described above applies to observations from a single farm. If historical data from multiple farms are available, one can use a more powerful approach based on statistical machine learning techniques.

27.2.3 Statistical approach to data-based farm modeling

If our planet X already has many solar farms, then it is possible to learn from the collective experience. Each datapoint in Fig. 27.2(a) is taken from a solar farm (operating at a location defined by its weather variables RH and T) at the end of its 10-year operation, for example. The gray points represent farms which have failed, i.e., its power level has reduced below a critical power, P_{crit}, while the white points represent farms with $P(t = 10 \text{ years}) > P_{\text{crit}}$. The question is the following: Should we install a farm at a location defined by the black triangle if we wish the farm to survive for 10 years? We have to answer the question exclusively based on statistics.

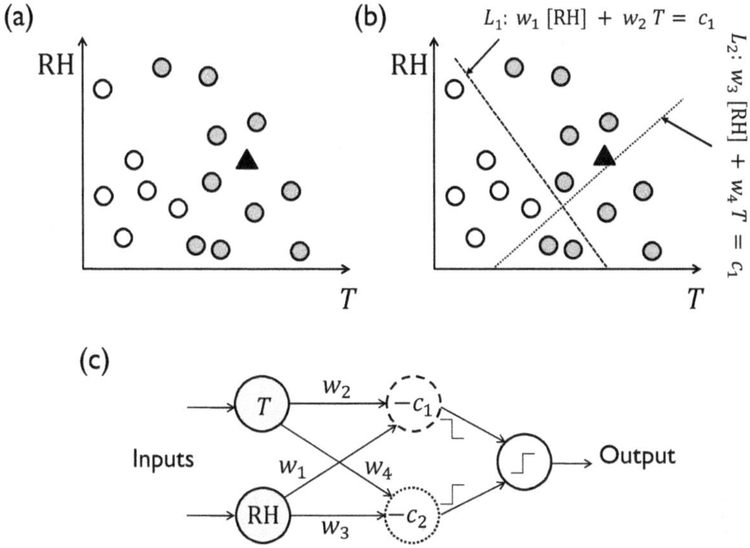

Figure 27.2: Machine learning approach to farm modeling. (a) Empirical 10-year reliability data from farms installed in various weather conditions. The information can be used to see if a new farm, indicated by the triangle, will produce power reliably for 10 years or not. (b) A region defined by the intersection of two lines differentiates between farms that survived for 10 years and those that did not. (c) A simple network model represents the regions described by the lines in Fig. 27.2(b).

Let us assume the fractional power loss depends linearly and independently on time, relative humidity, and temperature, i.e.,

$$\frac{\Delta P(t)}{P_0} = at + b[\text{RH}] + cT + d.$$

If the power output of the farm degrades by $\Delta P(t_\text{p})$ after a time t_p, then we can rewrite the equation as follows:

$$w_1[\text{RH}] + w_2 T = \frac{\Delta P(t_\text{p})}{P_0} - d - at_\text{p} \equiv c_1.$$

As a result, it may be possible to demarcate the passing vs. failing farms by a straight line, i.e., $w_1[\text{RH}] + w_2 T = c_1$. Then, decision making would be easy: For any new location defined by ([RH], T), the system will pass if $w_1[\text{RH}] + w_2 T < c_1$, whereas it will fail if $w_1[\text{RH}] + w_2 T > c_1$. We can determine w_1 and w_2 by adjusting the lines so that it puts the maximum number of pass-points on one side, while putting a maximum number of fail-points on the other side. Mathematically, one can define an error function that penalizes wrong categorization and minimize the error for the best fit line.

Unfortunately, the pass/fail boundary of the data in Fig. 27.2(a) is complicated and a single straight line does not separate them. We will need at least two straight-lines (i.e., $L_1 : w_1[\text{RH}] + w_2T = c_1$ and $L_2 : w_3[\text{RH}] + w_4T = c_2$) to differentiate passing vs. failing farms. Only those farms which are simultaneously below the first line and above the second line are acceptable.

Once the coefficients are known, they can be represented graphically with three sets of vertical bubbles (see Fig. 27.2(c)). The first set are input parameters (RH and T). Each bubble in the middle column represents one line used to divide the points. The two lines, L_1 and L_2 in Fig. 27.2(b) are represented by two bubbles of the middle column of Fig. 27.2(c). Given the weights (w_1 and w_2) of the connectors and the coefficient ($-c_1$) next to the top bubble, it is easy to see that the bubble implements the equation for L_1. The step symbol next to the bubble indicates that the output is binary, i.e., output of the bubble is 1 for all points below the line, or it is 0 otherwise. The third column contains the decision node, which aggregates the binary output of each bubble in the middle column to produce the final "pass/fail" decision. One may say that this network has "learned" the experience of building solar farms in various parts of the planet X. Clearly, this "artificial neural network" would not approve a proposed farm to be built at a location with the weather conditions defined by the black triangle.

The approach described above is one of the ways a computer (i.e., a machine) can learn from past experiences and make useful decisions, with no appreciation of the physics whatsoever. Now that is not necessarily a bad thing. When a stone is about to hit us, we do not worry about the Newton's law, but simply step away based on the painful past experiences. Similarly, if all the previous farms have failed in a comparable location, it may not be wise to build a new one. The general approach of machine learning or artificial intelligence uses a similar approach based on statistical inference. We used a variant of the statistical approach in Homework 14.6 to predict the energy output of a solar farm.

Homework 27.1: Using multiple vertical arrays of bubbles, also called neurons

Explain the operation of the network shown in Fig. 27.3 and graph the final expression.

When a new technology is introduced and/or a significant modification is made to an existing technology, we may not have sufficient statistical data to use the current approach directly. However, the statistics from the previous data do hide a significant physical insight. For example, it is easy to see that the farms that survive for a long time operate at low temperature and low humidity. Therefore, regardless of the details, it may be possible to use a physics-justified generalization of the degradation data from the older technology to complement the statistical data available for the new technology.

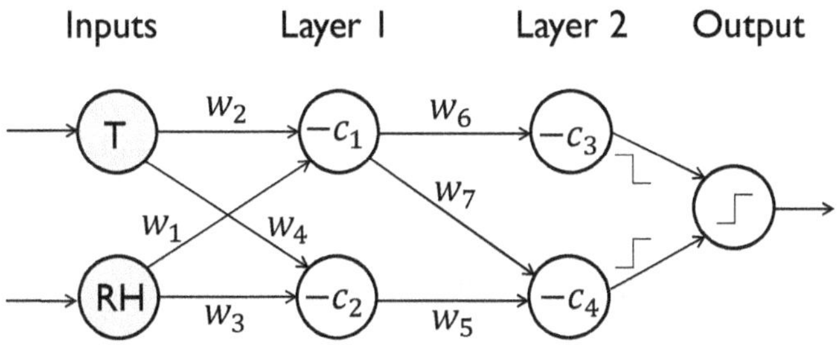

Figure 27.3: A two-layer network model can define more complex pass-fail regions.

27.3 Physics-based inverse modeling of solar farms

We have just seen that machine learning uses the historical data to create a "statistical model" of a system. When the physics is too complicated or the problem insufficiently specified, statistical modeling (with coefficients trained by historical data) is the best we can do. Fortunately, this is not the case for solar cells: we can not only predict how solar cells operate under a variety of weather conditions, but we also know how the cells degrade over time. Therefore, instead of using a general statistical network as in Fig. 27.3, we can use a physics-based network as shown in Fig. 27.4. Here we will use the historical data not to get the statistical weights $(w_1, w_2, \ldots)$, but the coefficients of time-dependent degradation of

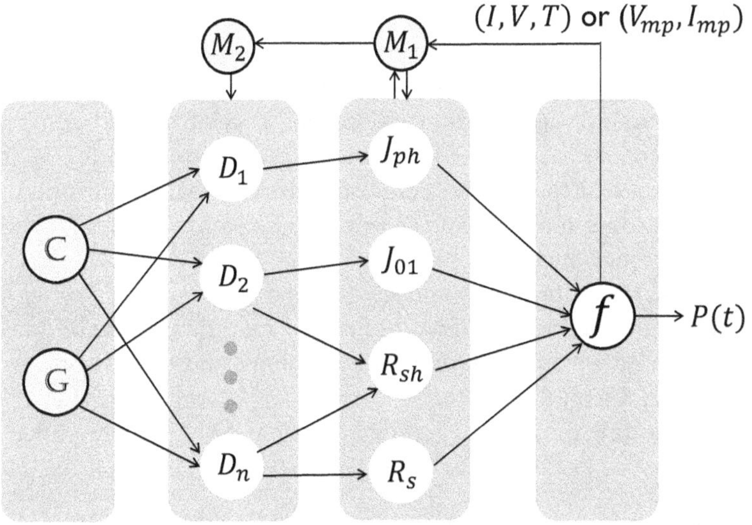

Figure 27.4: The parameters of the physics-based model can be calibrated or "trained" by using the historical $I - V - T$ data or $V_{\text{mp}} - I_{\text{mp}}$ data. The algorithms M_1 and M_2 allow us to calibrate the degradation parameters to reproduce the historical data. These "trained" parameters can be used to predict the future power output of the solar farm.

various compact model parameters (bubble M_1) and eventually coefficients of various degradation rates (bubble M_2). In other words, at any time t_c, we can adjust the coefficients of the degradation modes and the compact model parameters so that it reproduces the historical power output $P(t < t_c)$ given the module information and historical climate data (i.e., $\mathbb{C}(t < t_c)$). With these "trained" parameters, we can predict the power output $P(t > t_c)$ given the anticipated climate in the future $\mathbb{C}(t > t_c)$. In the forward modeling, the degradation coefficients are obtained by qualification tests (Q) before the module was installed. Unfortunately, a complete set of qualification data may not be available, especially for older farms. The calibration of the degradation coefficients by a farm's own historical data provides a powerful complementary dataset to physics-based predictive modeling. Note that the training and prediction continues throughout the lifetime of the module. At t_c, the availability of additional data allows continuous refinement of the predictions just made.

27.4 There are two ways to calibrate the model coefficients

If $P(t_i)$ is the only data available from a solar farm, a statistics-based approach may be the best one can do. Fortunately, a modern solar farm continuously records the maximum power point voltage, $V_{\mathrm{mp}}(t_i)$, and maximum power point current, $I_{\mathrm{mp}}(t_i)$, as well as $P(t_i)$ throughout the day. Indeed, databases containing decades-long history of $V_{\mathrm{mp}}(t_i)$ and $I_{\mathrm{mp}}(t_i)$ data of hundreds of solar farms are publicly available from multiple sources, including Sandia National Laboratory and National Renewable Energy Laboratory.

In addition, the widespread use of microinverters allows one to collect module-by-module degradation information. Moreover, specialized test equipment can temporarily disconnect a module from the grid, measure its full I-V characteristics, and reconnect it to the grid before the circuit breaker is tripped. Finally, information regarding monthly and yearly energy yields ($E(t)$) of various solar farms are also publicly available.

One can view the information collected (either full $I - V - T$ time series, or V_{mp}, I_{mp} time series) as built-in "EKG" of a solar module (see Fig. 27.6). Coupled with the weather data from the farm weather stations and the manufacturer information regarding the modules, the $I - V$ or $I_{\mathrm{mp}} - V_{\mathrm{mp}}$ traces provide a wealth of information regarding the time-dependent degradation of solar modules.

27.4.1 Physics-based forecasting: Full $I - V - T$ method

For simplicity, let us assume that a solar farm operator installs a few "canary" modules and records the full $I - V - T$ characteristics at fixed periodic intervals, t_i. These intervals are typically monthly, bi-yearly, or yearly. Given the $I - V - T$

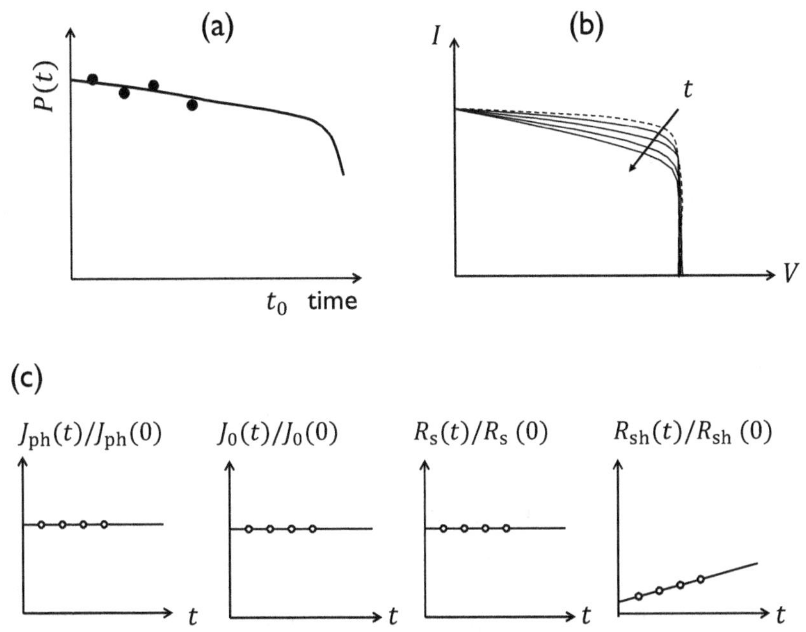

Figure 27.5: When the (a) output power and (b) full I-V characteristics are available, (c) the corresponding compact model parameters can be obtained by fitting the I-V curves.

characteristics, one can extract and monitor the time evolution of the parameters of the solar cell, namely, $J_{sc}(t_i)$, $J_0(t_i)$, $R_s(t_i)$, $R_{sh}(t_i)$, and so on.

For an illustrative example, consider a standard crystalline silicon solar module operated in the field. Assume that one has periodically recorded the output power of the module $P(t_i)$ as well as its $I(t_i) - V(t_i)$ characteristics. The I-V curves are normalized with respect to the standard intensity and temperature using the algorithm discussed in Sec. 27.5.1. The results are plotted in Figs. 27.5(a, b).

Now, one can use the five-parameter (i.e., $J_{sc} - J_0 - n - R_s - R_{sh}$) compact model discussed in Chapter 7 (see Figs. 7.1 and 23.8) to fit the $I - V$ data and the time-dependent parameters to explain how the module has degraded over time due to the local weather, see Fig. 27.5(c). In this specific case, we find that only the shunt resistance has increased over time, while other parameters have remained unchanged.

We have noted in Chapter 23 that the increase in the (fake) shunt resistance is related to finger corrosion. Also recall that the corrosion rate is correlated with the average temperature (T_D) and average relative humidity (RH) between the time-interval $\Delta t_i = t_i - t_{i-1}$. In other words,

$$R_{sh}(t_i) = A\,(\mathrm{RH})^B e^{-E_A/k_B T_D}\, t_i^d. \tag{27.6}$$

A linear regression determines the unknown constants: A, B, and d. We can now use these degradation parameters to forecast the (fake) shunt resistance increase due to finger corrosion and use a physics-based compact model to predict the performance degradation over time. Note that unlike the statistical model, the functional form is exact, and therefore the prediction will be more accurate than the statistical approach discussed earlier.

> **Homework 27.2: Extract the circuit parameters based on the given I-V characteristics**
>
> Download the I-V characteristics data and the data analysis tool PVAnalyzer `https://nanohub.org/resources/pvanalyzer`. Use the MATLAB code to extract the compact model parameters.

27.4.2 Physics-based forecasting: The $V_{\mathrm{mp}} - I_{\mathrm{mp}}$ method

In reality, farms are not typically equipped with specialized setup that can collect $I(t_i) - V(t_i) - T(t_i)$ curves. Instead, we can use two types of information to construct these curves. First, the complete $I - V - T$ information of the pristine modules are generally available from the manufacturer. This information is used to obtain the $J_{\mathrm{sc}} - J_0 - n - R_{\mathrm{s}} - R_{\mathrm{sh}}$ (five parameters) of the pristine module before

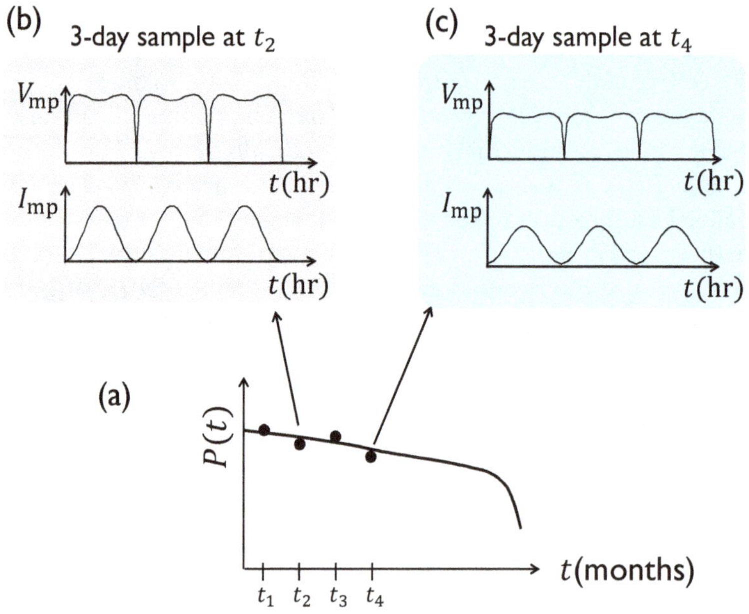

Figure 27.6: (a) Power loss is recorded at monthly intervals, t_i. To reconstruct the full I-V characteristics at those time points, 3 days of $V_{\mathrm{mp}} - I_{\mathrm{mp}}$ are recorded. The illustration shows the recorded $V_{\mathrm{mp}} - I_{\mathrm{mp}}$ data used to construct the full I-V data for the second (b) and the fourth month (c).

it was installed. Second, we use the natural morning-to-evening variation of the illumination and module temperature, as reflected in the continuously recorded $V_{\text{mp}}(t_i), I_{\text{mp}}(t_i)$ information to construct $I(t_i) - V(t_i) - T(t_i)$, as follows.

Recall that we need $I(t_i) - V(t_i) - T(t_i)$ information in a monthly or yearly interval, because the degradation proceeds slowly for a well-made commercial solar module. Let us choose three days of $V_{\text{mp}}, I_{\text{mp}}$ data around each t_i. If the data is taken every 15 minutes and there are 8 hours of sunlight in a day, there will be $4 \times 8 \times 3 = 96$ datapoints, see Figs. 27.6(b, c). In addition to the initial device parameters specified by manufacturer's datasheet, this 96-point dataset at each observation time t_i is sufficient to determine the five unknown parameters of the PV compact model at time t_i. In other words, given the intensity and temperature information, only a unique set of five parameters (i.e., $J_{\text{sc}} - J_0 - n - R_{\text{s}} - R_{\text{sh}}$) can produce the output that match the experimentally record $V_{\text{mp}} - I_{\text{mp}}$ dataset recorded over the 3-day period at time t_i. The parameters in turn allow us to determine the full I-V characteristics for the observation time t_i. With this I-V-T characteristic specified, one can follow the algorithm described in the preceding section to determine the time-dependent evolution of compact model parameters and predict the future loss of the module power output.

Homework 27.3: Extraction of the circuit parameters based on V_{mp}-I_{mp} characteristics

Read the paper by X. Sun, R. Chavali, and M. Alam, published in *Progress in Photovoltaics: Research and Applications*, 27(1), 55–66, 2019, to fully understand the use of Suns-Vmp method to reconstruct the I-V characteristics of a fielded module.

The paper referred to in Homework 27.3 reconstructs the I-V characteristics of a fielded module based exclusively on the V_{mp} and I_{mp} data stream and then finds the circuit parameters by varying the five parameters of a solar cell model. Modern characterization equipment also allows direct measurement of the full I-V characteristics on a minute-by-minute basis. With the directly measured data, the reconstruction is unnecessary and the key parameters can be estimated directly. Experimentally one finds that as sunlight intensity (c) increases from morning to noon everyday (see Fig. 27.7(a)), the cell efficiency traces a path from point A through point D as a function of c, as shown in Fig. 27.7(b). The path is traced from D to A during the second half of the day. The points A, B, C, and D can be used to determine R_{s} and R_{sh}, as follows.

We know that the intensity (c) dependent efficiency is given by $\eta(c) = V_{\text{mp}} I_{\text{mp}} / P_{\text{in}} = c I_{\text{sun}} V_{\text{oc}}(c) \text{FF} / c P_{\text{sun}} \propto \text{FF}(c)$. We recall from Homework 23.9 that

$$\text{FF} = \text{FF}_0 \left(1 - c\frac{R_s}{R_L}\right)\left(1 - \frac{R_L}{cR_{sh}}\right) \equiv \text{FF}_0 \,(1 - ac)\,(1 - b/c) \qquad (27.7)$$

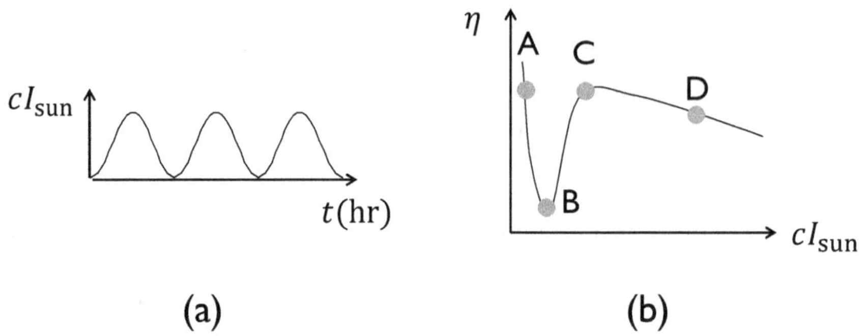

Figure 27.7: (a) Daily change in light intensity leads to (b) the characteristic features of the efficiency vs. intensity curve. The early morning rise in efficiency (point A) is related to absorption of sub-bandgap light through the longer than usual travel length through the atmosphere, i.e., AM $\to 1/\cos(\theta_Z \to 90)$. The valley is determined by the shunt resistance. The peak point C is determined by the balance between the series and shunt resistance. At higher intensity, point D is affected by both series resistance and the module temperature, as discussed in Sec. 5.2.

where $b = R_L/R_{sh}$ and $a = R_s/R_L$, where $R_L \equiv V_{mp}/I_{mp}$ evaluated at one-sun ($c = 1$) illumination. With $R_{sh} = 0$ and $R_s = 0$, R_L is easily calculated by the thermodynamic expressions for V_{mp} and I_{mp} given by Eqs. (4.1) and (4.2), respectively. The efficiency is maximized at point C with the corresponding maximization of the FF with c, namely, $c_{cric} = \sqrt{b/a} = \sqrt{R_L^2/(R_{sh}R_s)}$. To calculate R_s and R_{sh} separately, one corrects the efficiencies between points C and D due to self-heating by using the temperature coefficient β (see Chapter 5, Eq. (5.1)) and then estimate $r_s = R_s/R_L = d(\mathrm{FF})/dc$ for $c \to 1$ (point D). Finally, one uses point C to determine the shunt resistance, R_{sh}. In other words, fitting the temperature-corrected efficiency and FF from the field data, one can directly obtain R_{sh} and R_s by fitting the parameters in Eq. (27.7).

We conclude this section with two comments. First, the constants a and b in Eq. (27.7) are obtained from the general thermodynamic limit discussed in Chapter 4. For practical Silicon solar cells, the result can be improved by using the empirical formula proposed by Green *et al.* with $b \equiv (V_{oc} + 0.7)\mathrm{FF}(0)/(V_{oc} \cdot R_{sh})$ and $a \equiv 1.1R_s$. Second, the efficiency increases at very early morning (point A, Fig. 27.7(b)) because the light transmission through the thicker atmosphere (see Fig. 1.4 and Homework 1.7) disproportionately increases the absorption of the sub-bandgap ($E < E_g$) photons. Therefore, P_{in} is reduced without corresponding reduction in P_{out} associated with above-bandgap ($E > E_g$) photons, and the efficiency rises significantly. Therefore, let us not get overly excited if our fielded solar cell exceeds the S-Q limit very early in the morning — it will probably go away a little bit later!

Homework 27.4: Critical points of intensity-dependent efficiency and time evolution of circuit parameters

Given $\mathrm{FF} = \mathrm{FF}_0(1 - a^* c\, R_s)(1 - (b^*/c\, R_{sh})\mathrm{FF}_0)$, where $a = a^* R_s$ and $b = b^*/R_{sh}$. Derive the expressions of the critical points in Fig. 27.7.

27.5 There are several ways to determine the power degradation

In the preceding sections, we assumed that the normalized power degradation $\Delta P(t)/P_0$ are known and we only need to correlate the degradation to the underlying causes, such as ambient temperature and relative humidity. In practice, calculating the appropriately normalized $\Delta P(t)/P_0$ of fielded modules is nontrivial because the $\Delta P(t)$ fluctuates significantly depending on the instantaneous solar irradiance and ambient temperature. Two approaches have been proposed to separate the slow degradation of efficiency from the fluctuating instantaneous power output obtained from the field data.

27.5.1 Performance ratio method

The performance ratio (PR) is defined by the ratio of the power outputs at time t and time zero, both calculated under standard test conditions (STC), namely,

$$\mathrm{PR} \equiv \frac{P_{\mathrm{STC}}(t)}{P_{\mathrm{STC}}(t=0)} = \frac{P_{\mathrm{meas}}(T_{\mathrm{D}}, I_{\mathrm{POA}}; t)}{P_{\mathrm{STC}}(t=0)} \left[\frac{1}{1+\beta(T_{\mathrm{D}} - T_{\mathrm{STC}})}\right] \left[\frac{I_{\mathrm{ref}}}{I_{\mathrm{POA}}(t)}\right] \tag{27.8}$$

where the second and third factors on the right correct the module temperature, $T_{\mathrm{D}}(t)$ (with temperature coefficient β; see Secs. 5.2 and 5.3) and the plane-of-array (POA) intensity at the time of the measurement. It is easy to evaluate PR if I_{POA} and $T_{\mathrm{D}}(t)$ are obtained from the pyranometer and the module-mounted thermocouple.

It is possible to approximately calculate PR even if temperature and irradiance information are incomplete, corrupted, or unknown. The approximation relies on identifying the power output of the solar modules during comparable clear cloudless days over the years and calculating the change in power output as a measure of intrinsic power loss due to efficiency degradation. The clear-sky days are identified by first removing very low irradiance data ($< 200\ \mathrm{W/m^2}$) associated with poor light or clouds, and then calculating the clear-sky index as the ratio of the actual power output measured experimentally and the predicted location-specific module output from farm-level modeling, see Chapters 13–14. If the index is close to 1, the model-predicted $I_{\mathrm{POA}}(t)$ is used in lieu of actual pyranometer reading. Similarly, the hourly temperature is predicted by using the empirical relationship:

$$T_{\mathrm{A}}(t,h) = \frac{T_{day}(t) - T_{night}(t)}{2} \cdot \cos\left(\frac{h+8}{24} \cdot 2\pi\right) + \frac{T_{day}(t) + T_{night}(t)}{2}. \tag{27.9}$$

Here, T_{day} and T_{night} are monthly average of daytime and night-time temperatures in degree C, h is the hours from midnight, and 8 is an empirical factor that accounts for the difference between peak irradiance and the rise of the ambient temperature. Given the ambient temperature, the module temperature is obtained by the self-heating formula developed in Sec. 5.2 (Eq. (5.15)), i.e., $T_{\mathrm{D}} - T_{\mathrm{A}} = P_{in}(1-\eta) \times f(v_{\mathrm{w}})$ where the constant $f(v_{\mathrm{w}})$ is related to the convective and radiative heat transfer.

By calculating the rolling loss of power 365 days apart, one can get a statistical distribution of the year-on-year (YOY) degradation of the solar modules. A publicly available software called `RdTools` implements this methodology.

27.5.2 A statistical machine learning-based approach obviates the need for temperature/irradiance sensors

A complementary approach called statistical clear-sky fitting (SCSF) is based on generalized principal component analysis and relies on the fact that on average (notwithstanding global warming!), the monthly/seasonal irradiance and the temperature variation are roughly the same from one year to the next. These facts alone should allow one to extract the module power loss without the need for calculating the clear-sky index or determine the cell temperature. The dominant low-frequency components of the power-data (e.g., yearly, seasonal, etc.) are easily determined by the first few principal components of the data. The results show, remarkably, that the approach offers comparable results.

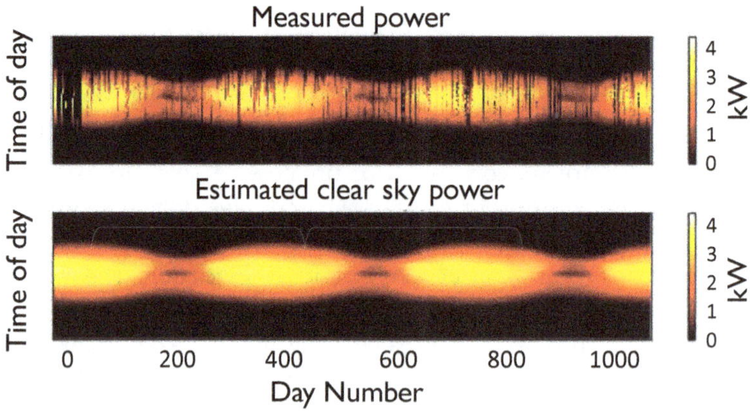

Figure 27.8: The periodicity and regularity of daily and yearly power output allow one to calculate YOY efficiency degradation.

We conclude this section by emphasizing that the approaches discussed above are not predictive, i.e., the information obtained from one solar farm cannot be used to predict its future degradation or the degradation of a farm located at a different geographical area. Physics-based degradation models are essential for predictive lifetime modeling of solar farms.

Homework 27.5: A simple example illustrates the key concept

Assume that

$$P(Y,t) = P_0[1 - d(Y + t/365)][1 + 0.2\cos(2\pi t/365)] + (P_0/20)[\text{rand}(0,1) - 0.5] \tag{27.10}$$

defines the degraded power output (with linear yearly degradation factor d and the intensity fluctuation defined by the random number generator `rand`) at any day t of the year Y since installation of the solar farm.

1. With $P_0 = 1000$ W/m^2, $d = 0.03$, and $Y_{max} = 8$ years, plot $P(Y,t)$ to see how the degradation evolves over the years.

2. Use various multi-day averaging of the signal to extract the periodic signal hidden underneath the fluctuating power output. Subsequently, plot $(P_{i+365} - P_i)/P_i$ for $i = 1, 2, \ldots 365$ to find the approximate degradation rate, d.

3. Create a 365 by 8 matrix with each column containing the normalized power output for a given year. Use the singular value decomposition (SVD) and principal component analysis (PCA) to show that one can obtain equivalent information as in the previous step.

27.6 Conclusions: The future of machine learning for PV technologies

The science of analyzing the historical data of a solar farm to predict the viability, output energy, and lifetime of a solar farm has just begun. Mining the PV field data is similar to reading the heartbeat of a person. Hidden in the historical $P(t)$ data and the $(V_{\text{mp}}, I_{\text{mp}})$ waveform are the current status and future prognosis regarding the technology. The current focus on statistical modeling and machine learning will help us identify the patterns of degradation.

It is important to realize, however, that only a physics-based model can be sufficiently predictive to calculate the output energy of a farm over many years in the future. In addition, if a physics-based model tells us exactly how the modules are failing, the manufacturer will be able to tailor the modules for specific geographical locations. Moreover, if the exact cause of failure is known, one can adopt a science-informed recycling program that can reuse a module by fixing only the components that had failed. Such recycling will reduce the carbon footprint of a solar cell technology and make them more economically and environmentally viable.

In the preceding three chapters, we discussed accelerated testing, forward reliability modeling, and inverse reliability modeling of solar modules. Reliability issues make the development of solar cells challenging, but it also offers a competitive edge to those who can address the reliability issues effectively.

References

[1] L. S. Bruckman, N. R. Wheeler, J. Ma, E. Wang, C. K. Wang, I. Chou, J. Sun, and R. H. French. Statistical and Domain Analytics Applied to PV Module Lifetime and Degradation Science. *IEEE Access*, 1:384–403, 2013. Conference Name: IEEE Access.

[2] Xingshu Sun, Raghu Vamsi Krishna Chavali, and Muhammad Ashraful Alam. Real-time monitoring and diagnosis of photovoltaic system degradation only using maximum power point — the Suns-Vmp method. *Progress in Photovoltaics: Research and Applications*, 27(1):55–66, 2019. _eprint: https://onlinelibrary.wiley.com/doi/pdf/10.1002/pip.3043.

[3] M. G. Deceglie, T. J. Silverman, B. Marion, and S. R. Kurtz. Real-time series resistance monitoring in PV systems without the need for IV curves. In *2015 IEEE 42nd Photovoltaic Specialist Conference (PVSC)*, pages 1–4, June 2015.

[4] C. M. Singal. Analytical expression for the series-resistance-dependent maximum power point and curve factor for solar cells. *Solar Cells*, 3(2):163–177, March 1981.

[5] D. C. Jordan, C. Deline, S. R. Kurtz, G. M. Kimball, and M. Anderson. Robust PV Degradation Methodology and Application. *IEEE Journal of Photovoltaics*, 8(2):525–531, March 2018. Conference Name: IEEE Journal of Photovoltaics.

[6] Michael G. Deceglie, Dirk Jordan, Ambarish Nag, Christopher A. (ORCID:0000000298678930) Deline, and Adam Shinn. RdTools: An Open Source Python Library for PV Degradation Analysis. Technical Report NREL/PR-5K00-71468, National Renewable Energy Lab. (NREL), Golden, CO (United States), May 2018.

[7] B. Meyers, M. Deceglie, C. Deline, and D. Jordan. Signal Processing on PV Time-Series Data: Robust Degradation Analysis Without Physical Models. *IEEE Journal of Photovoltaics*, 10(2):546–553, March 2020. Conference Name: IEEE Journal of Photovoltaics.

[8] Martin A. Green. *Solar cells: operating principles, technology, and system applications*. Prentice-Hall, Englewood Cliffs, NJ, 1982.

CHAPTER 28

The Road Ahead ...

In this book, we discussed the atom-to-farm physics of solar cells. Sunlight is free, but a solar cell is not. Therefore, there is an incentive to make the modules efficient and long-lived to reduce the cost of solar energy. We focused on large land-based solar farms, but smaller-scale solar systems (called micro- or pico-solar) are also important. Typically, they consist of a 100–200 W solar module, a battery, and power outlets for cell phones, TV, computers, and LED bulbs. The power consumption of modern electronics has decreased dramatically, so that a small module can power multiple components. Financially supported by micro-credit loans, these pico-solar systems are supporting access to education, healthcare, food storage, etc., for energy-poor populations across Asia, Africa, and Latin America.

A different kind of solar farm is required for densely populated cities fighting environmental pollution. In a city, land prices are so high that a traditional solar farm may not be feasible. Two solutions have emerged: floating solar farms and building-integrated PV. Floating solar farms installed in rivers or shallow beaches are now a common sight in many parts of the world. Building an integrated PV, where the windows are decorated by thin-film solar cells, offers another solution. The innovative "footprint" of these systems may provide economically and environmentally sustainable solutions for densely populated cities.

Balancing the needs for food, energy, and water is a challenge not only for large cities, but also for many countries that depend on agriculture and cannot afford to trade farmland for solar farms. New solar cell technologies must be optimized to use frequencies not used by plants, so that crops can grow unhindered, while solar cells still produce sufficient energy to sustain crucial activities, including irrigation. This multi-objective optimization of food, energy, and water (FEW) is an important topic for current research.

By 2050, the world population may reach 10 billion people. No matter how efficient or long-lived, solar cells alone cannot supply the energy needs of this large population. For a sustainable world, free energy harvested from other sources, such as wind, tide, nuclear, etc., must complement the energy produced by direct sunlight.

Now that you know the atom-to-farm physics of a solar cell, you should be able to make wiser decisions required to achieve specific goals. Now that you know the enormous economic potential of solar energy for humanity, the next time you take a walk outside on a sunny day, I hope that you will realize that the droplets of light filling the sky and illuminating the ground all around you are indeed golden ...

References

[1] Emre Gençer, Caleb Miskin, Xingshu Sun, M. Ryyan Khan, Peter Bermel, M. Ashraf Alam, and Rakesh Agrawal. Directing solar photons to sustainably meet food, energy, and water needs. *Scientific Reports*, 7(1):3133, June 2017.

[2] Kim Trapani and Miguel Redón Santafé. A review of floating photovoltaic installations: 2007–2013. *Progress in Photovoltaics: Research and Applications*, 23(4):524–532, April 2015.

[3] Amol A. Phadke, Arne Jacobson, Won Young Park, Ga Rick Lee, Peter Alstone, and Amit Khare. Powering a Home with Just 25 Watts of Solar PV: Super-Efficient Appliances Can Enable Expanded Off-Grid Energy Service Using Small Solar Power Systems. December 2017.

[4] S. Kumaravel and S. Ashok. An Optimal Stand-Alone Biomass/Solar-PV/Pico-Hydel Hybrid Energy System for Remote Rural Area Electrification of Isolated Village in Western-Ghats Region of India. *International Journal of Green Energy*, 9(5):398–408, July 2012. Publisher: Taylor & Francis _eprint: https://doi.org/10.1080/15435075.2011.621487.

Symbols and Units

Symbol	Quantity	Units
Chapter 1	Overview: Sun, earth, and solar cells	
T_S	Temperature of the sun	Kelvin
I_s	Sunlight intensity	W/m^2
ε	Emissivity	unitless
σ	Stephen–Boltzmann constant	$W\ m^{-2}\ K^{-4}$
r_s	Solar radius	m
r_p	Planet radius	m
d	Sun-to-planet distance	m
D_n	Day of the year	From Jan. 1
D	Number of days in a year	365 for Earth
R	Sun–planet distance	m
d_{max}	Max. sun–planet distance	m
d_{min}	Min. sun–planet distance	m
$\Delta = (d_{max} - d_{min})/2R$	Eccentricity	
$\theta = r_s/d$	Solar angle	rad
$I_0(d)$	Intensity on planet	W/m^2
I_{DNI}	Direct normal incident	W/m^2
I_{DHI}	Diffused horizontal irradiance	W/m^2
k_t	Clearness index	unitless
$U_{BB}(E)$	Blackbody energy density	J/m^3
$N_{BB}(E)$	Blackbody number density	$number/m^3$
$F_{BB}(E)$	Blackbody flux density	$number/(m^2 \cdot s)$
$I_{BB}(E)$	Blackbody power density	$J/(m^2.s) = W/m^2$
E_{peak}	Blackbody peak energy	$2.82\ k_B T$ (eV)
E_{avg}	Blackbody average energy	$2.7\ k_B T$ (eV)
N_{avg}	Blackbody delta-function Density	

Symbols and Units (*continued*)

Symbol	Quantity	Units
Chapter 2	2-level model of a solar cell	
$U(E_i, E_j)$ $D(E_i, E_j)$	Up- and down-transitions	number/s
n_{ph}	Photon occupation probability	unitless
μ_1, μ_2	Chemical potentials	eV
V_{oc}	Open-circuit voltage	V
R	Isotropic photon flux	photons/sond/atom
η_{c}	Carnot efficiency	unitless
T_{LED}	LED temperature	K
η_{b}	Bilayer efficiency	unitless
η_{s}	Series-connected efficiency	unitless
Chapter 3	3D solar cells	
θ_{S}	Solid angle of the sun	steradians
θ_{D}	Solid angle of a device	steradians
$n_{\text{S}}(E)$	Particle flux from the sun	photons/m^2/eV/s
$n_{\text{amb}}(E)$	Particle flux from ambient	photons/m^2/eV/s
$n_{\text{D}}(E)$	Particle flux from device	photons/m^2/eV/s
$E \times n_{\text{S}}(E)$	Energy flux from the sun	W/m^2/eV
$E \times n_{\text{amb}}(E)$	Energy flux from ambient	W/m^2/eV
$E \times n_{\text{D}}(E)$	Energy flux from device	W/m^2/eV
$\eta(E_{\text{g}}, V)$	Voltage-dependent efficiency	unitless
$\eta_{max}(E_{\text{g}})$	Maximum efficiency for a bandgap	unitless
η_{SQ}	Shockley–Quiesser efficiency limit	unitless
P_{out}	Output power of a solar cell	W/m^2
$P_{E<E_{\text{g}}}$	Sub-bandgap power	W/m^2

Symbols and Units (*continued*)

Symbol	Quantity	Units
Chapter 4	Shockley-Quiesser Triangle	
V_{oc}	Open-circuit voltage	V
V_{mp}	Voltage at maximum power-point	V
I_{sc}	Short-circuit current	A
I_{mp}	Current at maximum power-point	A
I_{sun}	Maximum current constant	A
β_{sun}	Bandgap coefficient for current	$(\mathrm{eV})^{-1}$
I_0, V_0	Normalized voltage and current	A, V
η_{N}	Efficiency limit of N-junction tandem cell	unitless
$\eta_{\mathrm{N}=\infty}$	Efficiency of infinite-junction tandem cell	unitless
R	Albedo coefficient	unitless
η_Q	External quantum efficiency	unitless
η_R	Absorbtance efficiency	unitless
LW	Lambert W function	unitless
$n_{\mathrm{S}}(E)$	Blackbody photon density at (E, T)	number/m^2/eV
Chapter 5	Self-heating of solar cells	
T_{A}	Ambient temperature	Kelvin
$\eta(T_{\mathrm{D}})$	Temperature-dependent efficiency	unitless
$\eta(T_{\mathrm{A}})$	STC efficiency	unitless
$R(T_{\mathrm{D}})$	Degradation rate	number/s
E_{A}	Activation energy	eV
P_{abs}	Absorbed power	$\mathrm{W/m^2}$
v_{w}	Wind velocity	m/s
δ	Interfacial thickness	m
k_{air}	Air thermal conductivity	W/(m K)

Symbols and Units (*continued*)

Symbol	Quantity	Units
ξ_{air}	Air kinematic viscosity	m^2/s
$1/f(v_{\text{w}})$	Sandia self-heating function	$W/(m^2 \cdot K)$
h	Convection coefficient	$W/(m^2 \cdot K)$
P_{carnot}	Carnot power	W/m^2
P_{angle}	Angle entropy power	W/m^2
N_{out}	Scale factor for APV	unitless
Chapter 6	Limits of light absorption	
P	Absorption probability	unitless
R	Reflectively	unitless
n_{r}	Real part of the index	unitless
n_{i}	Imaginary part of the index	unitless
α	Absorption coefficient	m^{-1}
f_{A}	Absorption enhancement factor	unitless
θ_{c}	Angle for internal reflection	radian
θ_{esc}	Escape angle	radian
β	Number of bounces before escape	unitless
θ_{D}	Angle of refracted light	radian
Chapter 7	p-n, p-i-n solar cells	
$J_{\text{total}}, J_{\text{ph}}, J_{\text{dark}}, J_{\text{n}}, J_{\text{p}}$	Various currents	A
$R_{\text{s}}, R_{\text{sh}}$	Shunt and series resistances	ohm
β	Efficiency temperature coefficient	number/°C
$N_{\text{D}}, N_{\text{A}}$	Doping density	$/cm^3$
$\mu_{\text{n}}, \mu_{\text{p}}$	Carrier mobility	$cm^2/(V \cdot s)$
$D_{\text{n}}, D_{\text{p}}$	Carrier diffusion coefficient	cm^2/s
$L_{\text{n}}, L_{\text{D}}$	Diffusion lengths	m
W	Cell thickness	m
$\mathcal{E}$	Electrical field	V/m
V_{bi}	Built-in voltage	V

Symbols and Units (*continued*)

Symbol	Quantity	Units
$n_{\mathrm{L},0}, n_{\mathrm{R},0}$	Carrier densities	number/cm^3
v_0	Carrier thermal velocity	m/s
$E_{\mathrm{B,R}}, E_{\mathrm{B,L}}$	Potential barriers	eV
$J_0 = J_{00} \exp(-Eg/kT)$	Dark current	A
k_{B}	Boltzmann constant	J/kelvin
T_{D}	Device temperature	Kelvin
η	Efficiency	unitless
$E_{\mathrm{ph}} = \hbar\omega$	Single-energy photon	eV or Joule
J_{total}	Total current	A/cm^2
$G_{x,V}$	Generation rate	number/cm^3/s
$R_{x,V}$	Recombination rate	number/cm^3/s
J_{dark}	Dark current	A/cm^2
J_{ph}	Photocurrent current	A/cm^2
R_{sh}	Shunt resistance	A/cm^2
R_{s}	series resistance	A/cm^2
v_0	Thermal velocity	cm/s
μ_{n} and μ_{p}	Chemical potential	eV
γ_{RL} and γ_{LR}	Flux injection probability over the barrier	unitless
$L_1 = \mu E \tau$	drift–recombination distance	cm
$L_2 = k_T q/E$	Bimolecular recombination distance	cm
$\Delta J_{\mathrm{ph}}, \Delta J_{\mathrm{dark}}$	Excess recombination and generation	A/cm^2
V_{emi}	Thermionic emission velocity	cm/s
$V_{\mathrm{diff}} = D/L_{\mathrm{n}}$	Diffusion velocity	cm/s
v_{eff}	Effective (diffusion parallel emission) velocity	cm/s
η_{s}	Series-connected efficiency	unitless

Symbols and Units (*continued*)

Symbol	Quantity	Units
Chapter 8	Organic solar cells	
$U(E_i, E_j)$, $D(E_i, E_j)$	Up- and down-transitions	number/s
n_{ph}	Photon occupation probability	unitless
μ_1 and μ_2	Chemical potentials	eV
T_{S}	Temperature of the sun	Kelvin
V_{oc}	Open-circuit voltage	V
$\mathcal{S}$	Size of VHJ fingers	cm
N_{F}	Finger density	number per cm^2
S_{F}	Surface area per finger	cm^2
l_{ex}	Exciton diffusion length	cm
J_{03}, J_{04}	Saturation current PHJ	A/cm^2
B	Bimolecular recombination constant	$1/cm^3.s$
$V_{\mathrm{n}}, V_{\mathrm{p}}$	Voltage drop across n and p in PHJ	V
R	Isotropic photon flux	photons/s/atom
η_{c}	Carnot efficiency	unitless
η_{s}	Efficiency (series-connection)	unitless
R	Isotropic photon flux	photons/s/atom
η_{c}	Carnot efficiency	unitless
T_{LED}	LED temperature	Kelvin
η_{b}	Bilayer efficiency	unitless
Chapter 9	Shunt resistance	
$I_{\mathrm{dark}}^{\mathrm{theory}}$	Dark current theory	A/cm^2
I_{scl}	space charge current	A/cm^2
I_{sh}	Shunt current	A/cm^2
α_{T}	Temperature ratio	unitless
α	Voltage asymmetry ratio	unitless
γ	Defect exponent	unitless

Symbols and Units (*continued*)

Symbol	Quantity	Units
Chapter 10	Series resistance	
L	Panel width	m
W	Panel length	m
δ	Gap between cells in thin-film module	m
P_ρ	Resistive power loss	W
$P_{\rho,\text{cell}}$	Resistive power loss in a cell	W
$P_{\rho,T}$	Resistive power loss in a module	W
n	Order (power-law) of the cell shape	unitless
N	Number of cells in a module	unitless
P_k	Resistive power loss along level k metal (c-Si module)	W
N_k	Number of metal lines in level k (c-Si module)	unitless
δ_M	Gap between cells in c-Si module	m
Chapter 13	Solar farm – Standalone module	
θ_e	Elevation angle	degree
θ_Z	Zenith angle	degree
D_n	Day of the year	
δ	Declination angle	degree
β_*	Panel tilt-angle	degree
R_s	Ground-projected row-spacing	m
h_y	Vertical height to the module top	W
h	Module height	m
ξ	Distance along the module height	m
p	Panel array pitch	m
$I_\text{dir}, I_\text{diff}, I_\text{alb},$	Direct, diffusion, and albedo illumination	W/m^2
$\text{VF}, F_{i\to j}$	View factor	unitless

Symbols and Units (*continued*)

Symbol	Quantity	Units
Chapter 14	Solar farm – bifacial	
R_{A}	Ground albedo	unitless
γ_{s}	Azimuth angle	degree
I_{2-axis}	Illumination on 2-axis tracking module	$\mathrm{W/m^2}$
I_{1-axis}	Illumination on 1-axis tracking module	$\mathrm{W/m^2}$
$T_{\mathrm{r}}, T_{\mathrm{s}}$	Sunrise, sunset times	in hours, from midnight
LER	Land equivalent ratio	unitless
LPF	Light productivity factor	unitless
AY	Agricultural crop yield	$\mathrm{kg/m^2}$
YY_{APV}	Energy yield with crop present	$\mathrm{kW \cdot h/m^2}$
PAR	photosynthetically active radiation	$\mathrm{kW \cdot h/m^2}$
Chapter 15	Energy storage	
Q	Stored energy	J
L	Latent heat of fusion	J/kg
c_p	Specific heat	$\mathrm{J/(kg \cdot K)}$
$J_{ox,A}$	Anode oxidation current	$\mathrm{A/cm^2}$
$J_{red,A}$	Anode reduction current	$\mathrm{A/cm^2}$
b_{ox}, b_{red}	Tafel slopes	V
$\mu_A, \mu_C, \mu_{\mathrm{th}}$	Solution potential	V
$J_{o,A}$	Exchange current density	$\mathrm{A/m^2}$
η_{ec}	Battery efficiency	unitless
Chapter 16	Levelized cost of electricity	
$C(Y)$	Cost of a farm	$
d, r	Degradation and discount rates	per year
$C_{\mathrm{om}}, C_{\mathrm{rv}}$	Operation-maintenance, residual value	$

Symbols and Units (*continued*)

Symbol	Quantity	Units
$E(Y)$	Integrated energy output	J
Y	Lifetime	years
COE	Cost of energy	\$/kW.h
$C_{BOS,V}$	Fixed and variable BOS	\$
C_{mod}, C_L	Module and land cost	\$/watt
$\mathbf{C}_\mathrm{M}, \mathbf{C}_\mathrm{L}$	Module and land cost	\$/m
$W_\mathrm{M}, W_\mathrm{L}$	Module and land scale factors	power/m
$M_\mathrm{L} = \mathbf{C}_\mathrm{M}/\mathbf{C}_\mathrm{L}$	Module and land cost ratio	unitless
C^*_{mod}	Module cost per watt	\$/W
X	Years in future	
q_M	Learning coefficient	unitless
g	Fractional cost reduction	unitless
$X_0, c, \eta_\mathrm{max}$	Goetzberger constants	year, year, and unitless
$n_\mathrm{r} = N_\mathrm{mod}(X)/N_\mathrm{mod}(0)$	Module ratio	unitless
g	Yearly differential installation	unitless
$LCOE$	Levelized cost	\$/kW.h
$LCOE^*$	Levelized cost	m/kW.h
$c(Y)$	Lifetime revenue	\$
R	Present cost	\$/watt
Chapter 17	Soiling and cleaning	
$L_\mathrm{s}(t)$	Fractional remaining power	unitless
R	Revenue earned per watt	\$
V_S V_L	Value earned and lost	\$
$H(t_\mathrm{c})$	Marginal cost	unitless
a	Soiling constant	per day
ρ_c	Critical soiling density	gm/m^2
N_V	Soiling particle density	number/cm^3
n_S	Soling area density	number/m^2
s	Particle size	m

Symbols and Units (*continued*)

Symbol	Quantity	Units
$\alpha_{\mathrm{s,i}}$	Extinction constant	/m
f_{i}	Fraction of soiled area	unitless
m, v	Single-particle mass and volume	kg, m^3
μ, σ	Log–normal params	
$N(r_i)$	Dust particle areal density	number/m^2
Chapter 18	Partial shadowing	
V_{bi}	Built-in voltage	V
V_{BD}	Breakdown voltage	V
N	Number of cells in a module	unitless
p	Number of cells shadowed	unitless
N_D	Donor doping density	number/m^3
N_A	Acceptor doping density	number/m^3
n_i	Intrinsic carrier density	number/m^3
h	Convection heat transfer coefficient	W/K
t_{s}	Time under shadow	s
$W_{\mathrm{sh}}, L_{\mathrm{sh}}$	Shadow width, length	m
Chapter 19	Hotspot degradation by shunts	
l^2	Small area	cm^2
$V_{\mathrm{oc,w}}$	Weak-diode v_{oc}	V
L	Radius of influence	cm
R_{s}	Series resistance	ohm-m^2
$R_{\mathrm{D}} = k_{\mathrm{B}}T_{\mathrm{D}}/J_{\mathrm{ph}}$	Diode resistance	ohm-m^2
Chapter 20	UV degradation	
γ	UV Beer–Lambert extinction efficiency	cm^2
I_{uv}	UV photocurrent	A/cm^2
N_{uv}	UV photon density	m^{-2}
σ	Bond dissociation efficiency of a photon	$1/(\mathrm{J} \cdot \mathrm{m})$

Symbols and Units (*continued*)

Symbol	Quantity	Units
f_{uv}	UV photon fraction	unitless
N_{uv}	Areal flux of UV photons	$\text{m}^{-2}\text{s}^{-1}$
D_{uv}	UV-created Defect density	number/m^2
$\sigma(E)$	Bond-dissociation efficiency	number/(J.m)
t_{EVA}	EVA thickness	m
k_p	Time constant for UV degradation	s^{-1}
ΔJ_{02}	UV-induced recombination current	A/m^2
E_*	Critical energy	eV
m	Si interface degradation per factor	unitless
$k = 0.25 - 0.5$	Si interface degradation exponent	unitless
R_{uv}	UV degradation rate	s^{-1}
k_D	Thermal self-dissociation probability	unitless
n_{ph}	Number of photon above a critical energy	number/m^3
R_{D}	Rate of thermal bond dissociation	number/s
D_{d}	Displacement dose damage, DDD	MeV/g
NIEL	Non-ionizing energy loss	$\text{MeV} \cdot \text{cm}^{-2}\text{g}^{-1}$

Chapter 21	Light induced degradation	
N_{BB}	Number of broken bonds	number/m^3
N_{H}	Number of mobile protons	number/m^3
$k_{\text{F}}, k_{\text{R}}$	Forward and reverse reaction rates	$\text{m}^3\text{/s}$
k_c	Dimerization rate constant	$\text{m}^3\text{/s}$

Symbols and Units (*continued*)

Symbol	Quantity	Units
Chapter 22	Potential induced degradation	
$\Delta P(t)$	Time-dependent power output	W/m^2
ΔP_∞	Saturated power output	W/m^2
R_D	Degradation rate	number/s
t_i	Incubation time	s
B	Humidity exponent	unitless
E_A	Activation energy	eV
t_{MTF}	Mean time to failure	s
K	Module connected in series	unitless
M	Number of cells in a module	unitless
R_L	Load that maximizes output	ohm
R_{sh}, R_s	Shunt and series resistance	ohm
L_p	Polymer thickness	m
V_p	Voltage across the polymer	V
J_{Na}	Sodium ion flux	A/m^2
Chapter 23	Corrosion	
B	RH exponent	unitless
t_f	Mean failure time	hours
D_w	Moisture diffusion coefficient	m^2/s
W	Internal RH	unitless
V_m	Maximum voltage at the finger tip	V
Chapter 24	Fracture and delamination	
CTE	Coefficient of thermal expansion	K^{-1}
E	Young's modulus	GPa
S	Stress	N/m^2
S_B	Yield Stress	N/m^2
ρ	Radius of curvature	m
K_t	Stress concentration factor	unitless
K_I	Stress intensity factor	$MPa \cdot \sqrt{m}$

Symbols and Units (*continued*)

Symbol	Quantity	Units
$a_{\rm i}$	Initial crack size	m
$a_{\rm f}$	Critical crack size	m
$S_{\rm max}, S_{\rm min}$	Maximum, minimum stress during cycling	N/m^2
$N_{\rm f}$	Cycles to failure	unitless
a	Fracture size	m
N	Number of cycles	unitless
ξ	Stress failure distribution	unitless
$N_{\rm F}$	Cycles to failure	unitless
E	Young's modulus	GPa, kg m$^{-1}s^{-2}$
η	Interfacial thickness	m
$\tau, \tau_{\rm max}$	Interfacial stress	N/m^2
G	Shear stress	N/m^2
CTE	Thermal expansion coefficient	K^{-1}
$S(r, \theta)$	Stress distribution function	N/m^2
λ	Order singularities	unitless
$T_{\rm r}$	Critical temperature	K
Chapter 26	Physics-based forward modeling	
$\mathbb{C}$	Climate/weather	unitless
$\mathbb{W}$	Module geometry	unitless
D_i	Degradation modes	unitless
Q	Qualification test	unitless
N	Number of degradation	unitless
Chapter 27	Monitoring the health of solar farms	
PR	Performance ratio	unitless
$P_{\rm STC}$	Power output under standard test condition	W/m^2
FF$_0$	Initial fill factor	unitless
$G_{\rm POA}$	Plane-of-array intensity	W/m^2

Symbols and Units (*continued*)

Symbol	Quantity	Units
T_{A}	Instantaneous ambient temperature	K
T_{day}	Average monthly daytime temperature	K
T_{night}	Average monthly nighttime temperature	K
h	Hours since midnight	hours
μ, σ	Log–normal distribution parameters	

Index

www.ingramcontent.com/pod-product-compliance
Lightning Source LLC
Chambersburg PA
CBHW081217220326
41598CB00037B/6808